Breeding of Perennial Horticultural Crops

The Editors

Dr S.K. Singh is presently the Head, Division of Fruits and Horticultural Technology, IARI, New Delhi. He received M.Sc. and Ph.D. degrees in Horticulture from Banaras Hindu University, Varanasi and joined Agricultural Research Service in 1994. He is involved in P.G. teaching and improvement of mango and grape. He has developed three extra-early grape varieties and production technologies of mango and grape. Dr Singh is working on biotechnological application including Genomics and marker-assisted breeding in fruit crops.

Dr V.B. Patel is presently Senior Scientist, Division of Fruits and Horticultural Technology, ICAR-IARI, New Delhi. He received M.Sc. (Ag) Hort from GBPUA&T, Pantnagar and Ph.D. from IARI, New Delhi. He joined Agricultural Research Service of ICAR in 2000. He also served as University Professor and Chairman of Hort. Dept. as well as Associate Director (Research), Bihar Agricultural University, Sabour (Bihar). He is associated with development of five varieties and several production technologies in fruit crops.

Dr A.K. Goswami is presently working as a Scientist at the Division of Fruits and Horticultural Technology, ICAR-IARI, New Delhi. He received his M.Sc. (Hort.) from CSAUA&T, Kanpur and Ph.D. degree from GBPUA&T, Pantnagar and joined Agricultural Research Service of ICAR in 2010. He is serving ICAR for the last 9 years and is engaged in research, teaching and extension and has involved in development of various lines of guava and papaya as well as standardization of canopy architecture in guava.

Dr Jai Prakash is presently Principal Scientist at the Division of Fruits and Horticultural Technology, IARI, New Delhi. He received M.Sc. (Hort.) from CSAUA&T, Kanpur and Ph.D. degree in Horticulture from Banaras Hindu University, Varanasi. He joined Agricultural Research Service of ICAR in 2003. He is serving ICAR for the last 16 years. He is involved in research and teaching and has developed improved papaya and pineapple varieties.

Dr Chavlesh Kumar is presently working as a Scientist at the Division of Fruits and Horticultural Technology, ICAR-IARI, New Delhi. He received his M.Sc. (Hort.) from CCSHAU, Hisar (Haryana) and Ph.D. from ICAR- IARI, New Delhi, and joined Agricultural Research Service of ICAR in 2016. He is serving ICAR for the last 3 years and actively engaged in varietal improvement program of grape and guava for desirable horticultural traits.

Breeding of Perennial Horticultural Crops

– Editors –

Dr S.K. Singh

Dr V.B. Patel

Dr. A.K.Goswami

Dr. Jai Prakash

Dr. Chavlesh Kumar

BIOTECH

2019

BIOTECH BOOKS®

ISBN 978-81-7622-454-3

Published by: **BIOTECH BOOKS®**
4762-63/23, Ansari Road, Darya Ganj,
New Delhi - 110 002
Phone: 09899295259
E-mail: biotechbooks@yahoo.co.in

Typeset at: **Classic Computer Services**
Delhi - 110 035

Printed at: **Replika Press Pvt. Ltd.**

PRINTED IN INDIA

Preface

Plant Breeding is the purposeful genetic improvement of crops through various techniques including selection, hybridization, mutation, polyploidization and molecular tools. The inherent problems associated with the breeding of perennial horticultural crops result in slow progress in developing new genotypes. In general, breeding of perennial crops is tedious owing to their complex cytogenetics and flower structures, highly heterozygous nature, long juvenile phase and poor understanding of genetics of important horticultural traits. These perennial horticultural crop varieties which are dominating the commercial cultivation are due to the result of traditional breeding methods. Still the traditional breeding methods are key to success of varietal development of many perennial horticultural crops. However, with the advent of the new biotechnological tools and their integration in the genetic improvement of perennial horticultural crops there is immense scope in increasing the breeding efficiency.

The biotechnological tools like molecular markers, help the breeders to effectively characterize genetic resources and assist the selection of potential parents for use in the improvement programme. The efforts on utilizing molecular markers facilitate the construction of high density linkage maps thus helping in identification of QTLs associated with the desired traits. The tightly linked molecular markers with traits of interest would facilitate the effective use of MAS in perennial horticultural crops for early screening of potential progenies. Furthermore, genetic engineering enables to tailor the novel genotype, while incorporating the foreign gene(s) without causing the much genetic load. The advanced techniques like genomic editing allow us to edit the crop genome thereby assisting the precise development of new genotypes. Furthermore, *in vitro* techniques, parasexual hybridization, polyploidy manipulation and mutation breeding are another arena, which can also be successfully explored for the genetic improvement of perennial horticultural crops.

In view, the present book is designed to give the glimpses of all basic and advance breeding methods of improvement in perennial horticultural crops, which may help the breeders, researchers and students who are associated with the perennial horticultural crop breeding. The aim of this book is to deliberate on recent advances and current knowledge in key areas of genetic improvement of perennial horticultural crops. The content has been designed to highlight the advances made in breeding approaches and varietal development in different perennial horticultural crops.

The Editorial Team acknowledges with thanks the Indian Council of Agricultural Research, New Delhi for providing the financial support for the conduct of the Winter School at the Division of Fruits and Horticultural Technology, IARI, New Delhi, which formed the basis of this book. The Editors also acknowledge the help and encouragements provided by the Director, Joint Director Research and Joint Director (Education) & Dean, Post Graduate School, ICAR-IARI, New Delhi at various stages for their support in bringing out this publication in the present shape. We are thankful to all scientists who have contributed the chapters to this book. Critical review and suggestions from the readers would help us in updating and revising the book in future.

Editors

Contents

	Preface	*v*
	List of Contributors	*xi*
1.	**Perennial Horticultural Crops Improvement: Current Scenario** *K.L. Chadha*	**1**
2.	**Improvement of Perennial Vegetable Crops** *Pritam Kalia and Shrawan Singh*	**23**
3.	**Genetic Resources of Tropical Fruits** *S.K. Malik and Rekha Chaudhury*	**35**
4.	**Genetic Resources of Arid and Semi-Arid Fruits** *O.P. Awasthi and Reena Prusty*	**49**
5.	**Genetic Resource Conservation** *Sandhya Gupta*	**61**
6.	**Biotechnology for Fruit Crop Improvement** *S.K. Singh, A.K. Goswami, Satyabrata Pradhan and Chavlesh Kumar*	**71**
7.	**Embryo Rescue** *S.K. Singh, Manish Srivastav and Chavlesh Kumar*	**85**
8.	**Genomics Assisted Breeding** *C. Bharadwaj, Supriya Sachdeva, A.K. Goswami, M.K. Verma and S.K. Singh*	**99**

9. **Genome Editing** 107
Shelly Praveen, T. Vinutha and Suneha Goswami

10. **Molecular Mapping** 113
Rakesh Singh

11. **Linkage Mapping** 121
Vignesh Muthusamy, Rajkumar Zunjare and Firoz Hossain

12. **Mapping Populations for Gene Tagging** 131
Gyan P. Mishra, H.K. Dikshit and Muraleedhar Aski

13. **Functional Genomics** 143
Jasdeep Chatrath Padaria

14. **Cisgenesis: Its Potential in Fruit Crop Improvement** 149
S.K. Singh, Megha R., Satyabrata Pradhan, Chavlesh Kumar, Jai Prakash, V.B. Patel and A.K. Goswami

15. **Phenomics: The Next Generation Phenotyping (NGP)** 159
Viswanathan Chinnusamy, Rabi Narayan Sahoo, S. Naresh Kumar, S. Gopala Krishnan, S. Lekshmy, R. Dhandapani, Sudhir Kumar, Harikrishna, Ranjith Kumar Ellur, Swati Goel, Krishna K. Jangid, Rahul Tiwari, Anil Rai, A.R. Rao, Sudeep Marwaha, Sanjeev Kumar, Brejesh Lall, Santanu Chaudhury, Archana Chugh, Swati Bhugra, Padmini Swain, Sushanta K. Dash and Jitendriya Meher

16. **Mutation Breeding** 183
Awtar Singh and Kanhaiya Singh

17. **Nucleo-cytoplasmic Interactions in Hybrid Breeding** 197
Naveen Singh, Rajendra Singh, Yashpal, Navinder Saini, A.K. Goswami and Awtar Singh

18. **Somaclonal Variations** 203
Zakir Hussain and Pawan Yadav

19. **Breeding Priorities and Strategies** 209
M.R. Dinesh

20. **Parent Selection Criteria and Shortening the Breeding Cycle** 227
Awtar Singh

21. **Banana Improvement** 241
Vikramaditya Pandey and S.K. Singh

22. Citrus Improvement 251

A.K. Dubey, Nimisha Sharma and R.M. Sharma

23. Cucurbits Improvement 269

A.D. Munshi, T.K. Behera, A.K. Sureja, Jogendra Singh and B.S. Tomar

24. Grape Improvement 283

M.K. Verma, S.K. Singh, V.B. Patel and Chavlesh Kumar

25. Guava Improvement 315

A.K. Goswami, A. Nagaraja, Madhubala Thakre and Chavlesh Kumar

26. Litchi Improvement 333

Vishal Nath, Alok Kumar Gupta, Manvendra Singh, Awtar Singh, Swapnil Pandey, Evening Stone Marboh, Devesh Pandey and Ankit Kumar Pandey

27. Mango Improvement 357

Manish Srivastav, S.K. Singh and A. Nagaraja

28. Papaya Improvement 365

Jai Prakash, Kanhaiya Singh, A.K. Goswami and Preeti Singh

29. Apple and Walnut Improvement 377

R.M. Sharma and A.K. Dubey

30. Low Chill Temperature Fruits Improvement 399

M.K. Verma, V.B. Patel, Chavlesh Kumar and Tanushree Sahoo

31. Rootstock Improvement 413

V.B. Patel, Chavlesh Kumar, M.K. Verma and S.K. Singh

32. Breeding Vegetables for Protected Cultivation 429

Harshawardhan Choudhary, T.K. Behera and A.D. Munshi

33. Fruit and Vegetable Varieties for Processing 441

Ram Asrey

34. Variety/Hybrid Maintenance 449

B.S. Tomar

35. Protection of Perennial Horticultural Crops through Intellectual Property Rights 457

Chandan Roy

Index 471

List of Contributors

Aski, Muraleedhar
Division of Genetics, ICAR-Indian Agricultural Research Institute, Pusa, New Delhi – 110 012

Asrey, Ram
Food Science and Postharvest Technology Division, ICAR-Indian Institute of Agricultural Research, Pusa, New Delhi – 110 012

Awasthi, O.P.
Division of Fruits and Horticultural Technology, ICAR-Indian Agricultural Research Institute, Pusa, New Delhi – 110 012

Behera, T.K.
Division of Vegetable Science, ICAR-Indian Institute of Agricultural Research, Pusa, New Delhi – 110 012

Bharadwaj, C.
Divison of Genetics, ICAR-Indian Agricultural Research Institute, Pusa, New Delhi – 110 012

Bhugra, Swati
Indian Institute of Technology, New Delhi

Chadha, K.L.
Former Deputy Director General (Hort.), Indian Council of Agricultural Research, New Delhi

Chaudhury, Rekha
ICAR-National Bureau of Plant Genetic Resources, Pusa Campus, New Delhi – 110 012

Chaudhury, Santanu
Indian Institute of Technology, New Delhi

Chinnusamy, Viswanathan
ICAR-Indian Agricultural Research Institute, Pusa, New Delhi – 110 012

Choudhary, Harshawardhan
Division of Vegetable Science, ICAR-Indian Institute of Agricultural Research, Pusa, New Delhi – 110 012

Chugh, Archana
Indian Institute of Technology, New Delhi

Dash, Sushanta K.
ICAR-National Rice Research Institute, Cuttack

Dhandapani, R.
ICAR-Indian Agricultural Research Institute, Pusa, New Delhi – 110 012

Dikshit, H.K.
Division of Genetics, ICAR-Indian Agricultural Research Institute, Pusa, New Delhi – 110 012

Dinesh, M.R.
ICAR-Indian Institute of Horticultural Research Hessarghatta, Bengaluru – 89, Karnataka

Dubey, A.K.
Division of Fruits and Horticultural Technology, ICAR-Indian Institute of Agricultural Research, Pusa, New Delhi – 110 012

Ellur, Ranjith Kumar
ICAR-Indian Agricultural Research Institute, Pusa, New Delhi – 110 012

Goel, Swati
ICAR-Indian Agricultural Research Institute, Pusa, New Delhi – 110 012

Goswami, A.K.
Division of Fruits and Horticultural Technology, ICAR-Indian Institute of Agricultural Research, Pusa, New Delhi – 110 012

Goswami, Suneha
Division of Biochemistry, ICAR-Indian Agricultural Research Institute, Pusa, New Delhi – 110 012

Gupta, Alok Kumar
ICAR-National Research Centre on Litchi, Muzaffarpur – 842 002, Bihar

Gupta, Sandhya
Tissue Culture and Cryopreservation Unit, ICAR-National Bureau of Plant Genetic Resources, New Delhi

Harikrishna
ICAR-Indian Agricultural Research Institute, Pusa, New Delhi – 110 012

Hossain, Firoz
Division of Genetics, ICAR-Indian Agricultural Research Institute, Pusa, New Delhi – 110 012

Hussain, Zakir
Division of Vegetable Science, ICAR-Indian Agricultural Research Institute, Pusa, New Delhi – 110 012

Jangid, Krishna K.
ICAR-Indian Agricultural Research Institute, Pusa, New Delhi – 110 012

Kalia, Pritam
Division of Vegetable Science, ICAR-Indian Agricultural Research Institute, Pusa, New Delhi – 110 012

Krishnan, S. Gopala
ICAR-Indian Agricultural Research Institute, Pusa, New Delhi – 110 012

Kumar, Chavlesh
Division of Fruits and Horticultural Technology, ICAR-Indian Agricultural Research Institute, Pusa, New Delhi – 110 012

Kumar, S. Naresh
ICAR-Indian Agricultural Research Institute, Pusa, New Delhi – 110 012

Kumar, Sanjeev
ICAR-Indian Agricultural Statistics Research Institute, New Delhi

Kumar, Sudhir
ICAR-Indian Agricultural Research Institute, Pusa, New Delhi – 110 012

Lall, Brejesh
Indian Institute of Technology, New Delhi

Lekshmy, S.
ICAR-Indian Agricultural Research Institute, Pusa, New Delhi – 110 012

Malik, S.K.
ICAR-National Bureau of Plant Genetic Resources, Pusa Campus, New Delhi – 110 012

Marboh, Evening Stone
ICAR-National Research Centre on Litchi, Muzaffarpur – 842 002, Bihar

Marwaha, Sudeep
ICAR-Indian Agricultural Statistics Research Institute, New Delhi

Megha R.
Division of Fruits and Horticultural Technology, ICAR-Indian Agricultural Research Institute, Pusa, New Delhi – 110 012

Meher, Jitendriya
ICAR-National Rice Research Institute, Cuttack

Mishra, Gyan P.
Division of Genetics, ICAR-Indian Agricultural Research Institute, Pusa, New Delhi – 110 012

Munshi, A.D.
Division of Vegetable Science, ICAR-Indian Institute of Agricultural Research, Pusa, New Delhi – 110 012

Muthusamy, Vignesh
Division of Genetics, ICAR-Indian Agricultural Research Institute, Pusa, New Delhi – 110 012

Nagaraja, A.
Division of Fruits and Horticultural Technology, ICAR-Indian Institute of Agricultural Research, Pusa, New Delhi – 110 012

Nath, Vishal
ICAR-National Research Centre on Litchi, Muzaffarpur – 842 002, Bihar

Padaria, Jasdeep Chatrath
ICAR-National Research Centre on Plant Biotechnology, New Delhi – 110 012

Pandey, Ankit Kumar
ICAR-National Research Centre on Litchi, Muzaffarpur – 842 002, Bihar

Pandey, Devesh
ICAR-National Research Centre on Litchi, Muzaffarpur – 842 002, Bihar

Pandey, Swapnil
ICAR-National Research Centre on Litchi, Muzaffarpur – 842 002, Bihar

Pandey, Vikramaditya
Division of Horticultural Sciences, Krishi Anisandhan Bhavan-II, Indian Council of Agricultural Research HQ, Pusa, New Delhi – 110 012

Patel, V.B.
Division of Fruits and Horticultural Technology, ICAR-Indian Agricultural Research Institute, Pusa, New Delhi – 110 012

Pradhan, Satyabrata
Division of Fruits and Horticultural Technology, ICAR-Indian Agricultural Research Institute, Pusa, New Delhi – 110 012

Prakash, Jai
Division of Fruits and Horticultural Technology, ICAR-Indian Agricultural Research Institute, Pusa, New Delhi – 110 012

Praveen, Shelly
Division of Biochemistry, ICAR-Indian Agricultural Research Institute, Pusa, New Delhi – 110 012

Prusty, Reena
Division of Fruits and Horticultural Technology, ICAR-Indian Agricultural Research Institute, Pusa, New Delhi – 110 012

Rai, Anil
ICAR-Indian Agricultural Statistics Research Institute, New Delhi

Rao, A.R.
ICAR-Indian Agricultural Statistics Research Institute, New Delhi

Roy, Chandan
Department of Plant Breeding and Genetics, Bihar Agricultural University, Sabour, Bhagalpur – 813 210, Bihar

Sachdeva, Supriya
Divison of Genetics, ICAR-Indian Agricultural Research Institute, Pusa, New Delhi – 110 012

Sahoo, Rabi Narayan
ICAR-Indian Agricultural Research Institute, Pusa, New Delhi – 110 012

Sahoo, Tanushree
Division of Fruits and Horticultural Technology, ICAR-Indian Agricultural Research Institute, Pusa, New Delhi – 110 012

Saini, Navinder
Division of Genetics, ICAR-Indian Agricultural Research Institute, Pusa, New Delhi – 110 012

Sharma, Nimisha
Division of Fruits and Horticultural Technology, ICAR-Indian Institute of Agricultural Research, Pusa, New Delhi – 110 012

Sharma, R.M.
Division of Fruits and Horticultural Technology, ICAR-Indian Institute of Agricultural Research, Pusa, New Delhi – 110 012

Singh, Awtar
Division of Fruits and Horticultural Technology, ICAR-Indian Agricultural Research Institute, Pusa, New Delhi – 110 012

Singh, Jogendra
Division of Vegetable Science, ICAR-Indian Institute of Agricultural Research, Pusa, New Delhi – 110 012

Singh, Kanhaiya
Division of Fruits and Horticultural Technology, ICAR-Indian Agricultural Research Institute, Pusa, New Delhi – 110 012

Singh, Manvendra
ICAR-National Research Centre on Litchi, Muzaffarpur – 842 002, Bihar

Singh, Naveen
Division of Genetics, ICAR-Indian Agricultural Research Institute, Pusa, New Delhi – 110 012

Singh, Preeti
Division of Fruits and Horticultural Technology, ICAR-Indian Institute of Agricultural Research, Pusa, New Delhi – 110 012

Singh, Rajendra
Division of Genetics, ICAR-Indian Agricultural Research Institute, Pusa, New Delhi – 110 012

Singh, Rakesh
Principal Scientist, ICAR-National Bureau of Plant Genetic Resources, New Delhi

Singh, S.K.
Division of Fruits and Horticultural Technology, ICAR-Indian Institute of Agricultural Research, Pusa, New Delhi – 110 012

Singh, Shrawan
Division of Vegetable Science, ICAR-Indian Agricultural Research Institute, Pusa, New Delhi – 110 012

Srivastav, Manish
Division of Fruits and Horticultural Technology, ICAR-Indian Agricultural Research Institute, Pusa, New Delhi – 110 012

Sureja, A.K.
Division of Vegetable Science, ICAR-Indian Institute of Agricultural Research, Pusa, New Delhi – 110 012

Swain, Padmini
ICAR-National Rice Research Institute, Cuttack

Thakre, Madhubala
Division of Fruits and Horticultural Technology, ICAR-Indian Institute of Agricultural Research, Pusa, New Delhi – 110 012

Tiwari, Rahul
ICAR-Indian Agricultural Research Institute, Pusa, New Delhi – 110 012

Tomar, B.S.
Division of Vegetable Science, ICAR-Indian Institute of Agricultural Research, Pusa, New Delhi – 110 012

Verma, M.K.
Division of Fruits and Horticultural Technology, ICAR-Indian Agricultural Research Institute, Pusa, New Delhi – 110 012

Vinutha, T.
Division of Biochemistry, ICAR-Indian Agricultural Research Institute, Pusa, New Delhi – 110 012

Yadav, Pawan
Division of Vegetable Science, ICAR-Indian Agricultural Research Institute, Pusa, New Delhi – 110 012

Yashpal
Division of Genetics, ICAR-Indian Agricultural Research Institute, Pusa, New Delhi – 110 012

Zunjare, Rajkumar
Division of Genetics, ICAR-Indian Agricultural Research Institute, Pusa, New Delhi – 110 012

Breeding of Perennial Horticultural Crops *Pages* **1–22**
Editors: S.K. Singh, V.B. Patel, A.K.Goswami, Jai Prakash and Chavlesh Kumar
Published by: **BIOTECH BOOKS, NEW DELHI**

Chapter **1**

Perennial Horticultural Crops Improvement: Current Scenario

K.L. Chadha

Former Deputy Director General (Hort.),
Indian Council of Agricultural Research, New Delhi

INTRODUCTION

Fruit crop improvement is more than one hundred years old today. It was in 1905 that varietal improvement work was initiated in guava by late Dr. G.S. Cheema and Deshmukh and followed by Burns and Prayag on mango in 1921 at Ganesh Khind Fruit Experiment Station, Pune. As a result L-49 guava was developed through a selection from cv. Allahabad Safeda in 1927. It may be of interest for you to know that this variety was renamed after Dr. Cheema after my written request to Mr. Nayak, the then Chief Minister of Maharashtra (after my visit to Maharashtra during 1967, when I was guest of Dr. Cheema) who honoured Dr. Cheema for his monumental initiative and contribution and named the variety as *Sardar*. These initiatives set the tone for emphasis on fruit crop improvement in India. Since then fruit crop improvement has been carried out in a number of crops through introduction, selection, hybridisation and mutation breeding. In earlier times improvement in fruit crops was brought about largely by making selections from the indigenous variability. Simultaneously exotic germplasm was also introduced. These activities performed by local growers, often patronized by the then rulers and some enthusiastic horticulturist continued over the years and resulted in the development of several popular *cvs* of different fruits.

After independence of India fruit crop improvement got a fillip with the setting up of a network of Institutes/Project Directorates/National Research Centres and Regional Station of Institutes in different regions of the country. Today we have institutes devoted to Tropical Fruits (IIHR, Bengaluru), Sub-tropical Fruits (CISH, Lucknow), Temperate Fruits (CITH, Srinagar), Arid Fruits (CIAH, Bikaner) and Citrus (CCRI, Nagpur). Besides, there are Project Directorates/NRCs for Banana (NRCB, Trichi), Grape (NRCG, Pune), Litchi (NRCL, Muzaffarpur), Pomegranate (NRCP, Solapur). Several Department of Horticulture/Fruit crops in state Agri/ Horti Universities also took up crop improvement work on a number of crops, *e.g.* mango, guava, grape, aonla, pomegranate, *etc.* which has resulted in a number of superior varieties. Some of the stalwarts, who have contributed to crop improvement significantly and require special mention are Drs. S.K. Mukherjee, R.N. Singh, P.K. Majumdar, and D.K. Sharma at IARI, New Delhi; Mansha Ram at IARI, Pusa; C.P.A. Iyer, Rajender Singh and M.R. Dinesh (IIHR); R.K. Pathak (Faizabad), T.R. Chadha, D.K. Uppal and Nazeer Ahmad (HP and J&K) besides a number of workers like K.C. Naik, N.Kumar from Coimbatore and Periyakulum and MPKV, Rahuri.

Problems in Fruit Breeding

Fruit crop breeding is a long drawn and difficult process due to distinct floral biology, out crossing, high heterozygosity, sterility, parthenocarpy, apomixes, polyembryony, poor fruit set, excessive fruit drop, low rate of sexual propagation with mostly one seed from one fruit, lack of selection procedures for isolation of desired hybrids and long pre-bearing age. Although the breeding methods are the same as for other group of crops, resulting offsprings in fruit breeding require huge space for evaluation. There is also need to refine the selection procedure in different crops coming in the way of improvements as understanding of the genetics and inheritance pattern of the quantitative traits in fruit crops has been extremely difficult. Hence, the selection of parents in a breeding programme based on the phenotype becomes difficult and progeny performance or manifestation of hybrid vigour becomes unpredictable. The progenies are also influenced to a great extent by environment. One of the advantages of the heterozygosity is that in the F_1 generation itself segregation is maximum and the desirable traits in the progenies can be fixed by vegetative propagation. However, a time has come to shift our emphasis from conventional breeding to modern methods of breeding based on biotechnological interventions.

Crop Improvement throught Conventional Methods

Plant genetic resources from basic raw material to meet current and future needs of crop improvement. Diversity comprises of native land races, selections, elite cultivars and wild relatives of plants. There is also lot of diversity due to seedling population and spontaneous mutations. In fact, unlike other crops where there is need to create variability in fruit crops it is the management of diversity which is more challenging. Significant achievement made through the above method is discussed below:

Introduction, Evaluation and Conservation of Genepool

India is an important centre of diversity and has rich variability of several fruits. There is a large diversity of several native but underutilized fruits namely aonla, bael, jackfruit, jamun, karonda and phalsa. Simultaneously, exotic introduction of new fruit species and varieties have been made in several fruits adding to the existing variability. As a result, a rich reservoir of species and cultivars of several fruit crops has been built up at various research centres listed below:

- ✰ Apple and other Temperate fruits: CITH, Srinagar; YSPUHF (Shimla, Solan)
- ✰ Banana: NRC Banana, Trichi
- ✰ Ber: CIAH, Bikaner; HAU, Hissar
- ✰ Citrus: CCRI, Nagpur
- ✰ Grape: IARI, New Delhi; IIHR, Bengaluru; NRC Grape, Pune
- ✰ Guava: CISH, Lucknow and IIHR, Bengaluru
- ✰ Mango: IARI, New Delhi; CISH, Lucknow; IIHR, Bengaluru; APAU Sangareddy
- ✰ Pomegranate: MPKV, Rahuri

Conservation, evaluation and utilisation of biodiversity, which is an important genetic wealth is very important. Many species and varieties in different fruits are rapidly being lost. There are several methods of conservation, which include pollen storage, conservation at haploid stage (accomplished through pollen) and *in situ* conservation, while evaluation of these collections have resulted in identification of several cvs suited to different agro-climatic regions of the country much remains to be done in respect of evaluation and utilization of large varietal wealth in India. Some of the prominent exotic introductions are listed below:

Apple

One of the most significant contributions in the temperate fruits was made by Mr. Samuel Evan Stokes, a resident of Philadelphia, USA who settled in Kotgarh and introduced Delicious group of apples in 1916. It is from these first few saplings that the sweet Delicious group of apple *e.g.* the Golden Delicious, Red Delicious, Richard and Royal Delicious were introduced in the area. A large number of new apple cultivars namely Fuji, Oregon Spur, Red Chief, Red Gold, Red Spur, Royal Gala, Scarlett Gala, Starking Delicious, Top Red and Vance Delicious were introduced from Denmark, Japan, Yugoslavia, West Germany, Europe and USA keeping in view our preference for red coloured, spur bearing and apple scab tolerant varieties.

Banana

Some of the prominent introduction in banana include varieties like Lady Finger having resistance to bunchy top virus from Australia; Grand Naine from France and Valery from West Indies. Of these varieties, Grand Naine is now extremely popular and is grown commercially all over the country.

Citrus

Several varieties and rootstocks of citrus have been introduced from USA and Japan. These include hybrids like Kinnow, Blood Red and Washington, Naval oranges, grapefruits Marsh Seedless, Foster, Duncan, Starking Ruby, Red Blush and lemon, Lisbon and Eureka.

Grape

A large number of varieties have been introduced in grape from different countries. Some of the notable introductions are Thompson Seedless, Perlette, Beauty Seedless, Delight, Himrod cultivars from USA and Kismish Chorni and Kismish Beli cultivars from erstwhile USSR. Other commercial grape cultivars introduced are Red Globe, Crimsom Seedless, Ruby Seedless, Riesling, Centennial Seedless, Flame Seedless, Fantasy Seedless, Rose, Pinot Noir, Shiraz, Chenin Blanc, Cabernet Sauvignon, Chardonnay, Merlot, Sauvignon, Blanc, and Cabernet Blanc from USA and European countries. Many of these varieties are now under commercial cultivation for table, raisin and wine purposes.

Mango

Many exotic cultivars of mango such as Tommy Atkins, Haden, Sensation, Keitt, Zill, Carabao and Julie have been introduced from Florida (USA) and Peru. Sensation and Tommy Atkins have been used in hybridisation programme for imparting red peel colour to indigenous varieties.

Papaya

In papaya, Solo, Sunrise, Sunset varieties have been introduced.

Pineapple

Almost all commercial varieties of pineapple grown in India, *e.g.* Kew, Giant Kew, Queen, Mauritius, and Spanish are exotic introduction.

Arid Zone Fruits

While a large number of arid zone furits are indigenous to India, several cultivars were introduced in date palm, from Sultanate of Oman, Saudi Arabia and Egypt, fig from USA and Iran and mulberry varieties, Black English, California Giant, Large Black, and Wellington *etc.* have been introduced.

Temperate Fruits

A number of varieties were also introduced in other temperate fruits. These include William's Bartlett, Conference, Keiffer, Baggugosha, Chinese sand pear in **pear**, New Castle, Royal, Kaisha, Charamgz in **apricot**; Early Rivers, July Elberta, JH Hale in **peach** and Allison, Bruno, Abbott, Hayward, Monty *etc.* in **kiwifruit**. Besides, a number of low chilling **peach** varieties namely Flordasun, Floradred, Sun Red and 16-33 were introduced from USA in subtropical regions by PAU Ludhiana. In olive, varieties like Coratina, Apression, Pendotino, and Lessino introduced in Jammu and Kashmir and UP hills produced good quality oil yield.

Selection of Seedling and Elite Clones

Clonal selection is an important tool in fruit crop improvement. While the exact basis of clonal variants is still not known, seedlings resulting from spontaneous mutations may be the reason for such variants.

India is an important centre of diversity of crop plants and has rich variability in several fruits. This genepool has provided cvs of several fruits grown commercially. Lot of emphasis has been laid on variety improvement in fruit crops through selection of superior chance seedlings and elite clones and their use through vegetative propagation. Clonal selections from commercial varieties are of great value as they differ from the parental material only in few important characteristics. Superior clones have been selected in a number of almost two dozen fruit crops and many of them are now being grown commercially. Some of the seedling and clonal selections which have become commercial in different crops are listed below:

One of the drawback in clonal selection is that if the mother clone is susceptible, the selection is also susceptible. However some clonal selections can be used as pre-breeding lines for traits different than their mother clones.

Important Seedling Selection and Elite Clones

Fruit	*Superior Clone(s)*
Aonla	Krishna Kanchan (Faizabad); Thar Aishwarya (Godhra)
Bael	Banarasi and Kagzi (UP); Pant Shivani (Pant Nagar)
Ber	Goma Kirti (Godhra)
Citrus	
Acid lime	Vikram, Pramalini (Parbhani), Sai Sarbati (Rahuri), Jai Devi (Tamil Nadu), Tenali (AP), Pusa Abhinav (IARI, New Delhi)
Lemon	Pant Lemon-1 (Pantnagar)
Mandarin	Mudkhed Seedless (Maharashtra)
Sweet orange	Yuvraj Blood Red (Ludhiana)
Custard Apple	NKM-1 and 2 (Maharashtra)
Grape	Dilkhush (Andhra Pradesh), Manjari Naveen (Pune), Sharad Seedless, Cheema Sahebi (Maharashtra); Pusa Seedless (New Delhi), Tas-e-Ganesh, Manik Chaman, Sonaka (Maharashtra)
Jackfruit	Palur-1 Jack
Jamun	Konkan Bhadoli (Vengurle), CISH-J42 (Lucknow), Narendra Jamun (Faizabad) and Goma Priyanka (Gujarat)
Karonda	Pant Manohar (Pantnagar)
Mango	Niranjan, off season bearing (Parbhani); Sunderraja, a red blushed selection (Rewa), Dashehari-51, regular bearing and high yielding (Lucknow); Cardoz Mankhurd, (Goa); Paiyur-1, a clonal selection (Paiyur); Maneka (Sabour)
Pomegranate	G-137, Ganesh (Pune) and Jyothi (Bengaluru)
Sapota	PKM-1 (Periyakulum); CO-1,2 and 3 (Bengaluru)
Walnut	Series of clones selected CITH

Fruit Breeding

Developments of fruit varieties through hybridisation has also been underway in some fruit crops at different centres under National Agriculture Research system in the country and have resulted in several varieties of which many are grown commercially. Some of these are as follows:

Apple

Breeding work has been in progress at Regional Horticultural Research Station, Mashobra, Shimla; Fruit Research Station, Shalimar, Kashmir, and Horticultural Experiment and Training Centre, Chaubatia, Uttarakhand with the objectives of developing varieties with better shelf-life, early maturity, high dessert quality and resistance to scab. Hybridization work initiated at Kashmir during 1956 resulted in the release of two hybrids namely Lal Ambri and Sunehari. Similar work started at Mashobra, Himachal Pradesh in 1960 resulted in four hybrids, namely, Ambstarking, Ambroyal, Ambred and Ambrich. To evolve low chilling varieties having coloured and sweet fruits with good keeping quality, work was initiated at Chaubatia in 1970. As a result Chaubatia Princess, Chaubatia Agrim, Chaubatia Swarnima and Chaubatia Anupam were released for cultivation. Further, systematic breeding aimed at development of scab resistant apple varieties at SKUAST, Srinagar resulted in the release of Firdaus and Shireen varieties. Almond hybridization programme aimed at developing low chilling varieties resulted in development of Hybrid-15, Hybrid A-258 and H-98 with high productivity and quality at PAU, Ludhiana.

Apricot

The major objective in apricot improvement has been climatic adaptation to help it escape spring frost. As a result, work at RFRS, Chaubatia (Uttarakhand) taken up in 1969 resulted in the release of three varieties namely, Chaubatia Madhu (early ripening and highly productive), Chaubatia Kesri (Mid-season variety), and Chaubatia Alankar (low chilling and early ripening).

Custard Apple

A variety of custard apple, Arka Sahan (Gem x Mammoth) has been released from IIHR, Bengaluru and has gained popularity throughout the country and now in the production chain.

Grape

Hybridization work in grape was taken up by IIHR, Bengaluru and IARI, New Delhi. Four hybrids namely Arkavati, Arka Kanchan, Arka Shyam and Arka Hans were released from IIHR, which have fruitful basal buds and could be trained on head system. Subsequently, seven more grape hybrids namely Arka Neelmani, Arka Sweta, Arka Majestic, Arka Chitra, Arka Soma, Arka Trishna and Arka Krishna have been developed for table, juice and wine purposes. At IARI, New Delhi, two promising hybrids namely Pusa Urvashi and Pusa Navrang (tenturier) have been released during mid nineties while, two more hybrids (Pusa Aditi and Pusa Trissar) have also been identified for release in 2014. Some of these hybrids namely Arkavati from IIHR, Bengaluru and Pusa Navrang from IARI, New Delhi have given good performance and are under adoption by grape growers in certain states.

Guava

Improvement through hybridization has been undertaken at various centres has resulted in the development and release of soft seeded and large fruited varieties Safed Jam and Kohir Safed (Sangareddy, A.P.), soft seeded white fleshed Arka

Amulya and soft seeded red blushed Hybrid 16-1 (IIHR, Bengaluru); Hisar Safeda and red dotted peel and pink fleshed Hisar Surkha (CCS HAU, Hisar), besides a dwarfing rootstock Pusa Srijan from IARI, New Delhi.

Papaya

Systematic work on breeding of papaya varieties with high yield and good quality for wider adaptability was taken up at IARI, Regional Station, Pusa, Bihar in 1966. Among the collections Ranchi variety was found to be most promising. As a result of inbreeding and selection for eight generations during 1966-82, uniform gyno-dioceious lines Pusa Delicious, Pusa Dwarf, Pusa Giant and Pusa Majesty with desirable attributes were developed. Development of papaya cultivars with good quality and high papain content was also taken up at TNAU, Coimbatore resulting in release of CO-1 to CO-7 series through inbred selection and hybridization. Papaya improvement work carried out at IIHR, Bengaluru has resulted in the development of two hybrids IIHR-39 named as Surya and IIHR-54 which have superior fruit quality. Another promising hybrid HPSC-3 has been developed at ICAR Research Complex, Tripura for high yield potential and resistance to mosaic virus.

Mango

Systematic hybridization work in India started at Sabour in Bihar and resulted in two hybrids Mahmood Bahar and Prabha Shankar by crossing Bombai and Kalapady varieites. Subsequently, work at Ananthrajupet, Andhra Pradesh resulted in the release of four mango hybrids namely Neeleshan, Neelgoa, Neeludin and Swarnjehangir. Subsequently mango hybridization was taken up at New Delhi (IARI), Vengurle (MPKV) and Bengaluru (IIHR). Major emphasis has been given to development of regular and precocious bearing, dwarf, and large fruited varieties with red blush, good keeping quality and freedom from spongy tissue disorder. More than 30 hybrid cultivars have been released by various institutes which include:

- ✰ Andhra Pradesh: AU Rumai (Sangareddy)
- ✰ Bihar: Sundar Langra, Alfazli, Sabori and Jawahar (BAU, Sabour);
- ✰ Gujarat: Neelphanso, Neeleshan and Neeleshwari from Paria;
- ✰ Karnataka: Arka Arun and Arka Puneet, (spongy tissue disorder free) Arka Anmol (having good quality) and Arka Neelkiran (attractive red blush) (IIHR, Bengaluru)
- ✰ Maharashtra: Ratna (free from spongy tissue disorder), Sindhu (seedless) and Konkan Ruchi from KKV, Dapoli; Sai Sugandh from Rahuri, Ratna released from Vengurle,
- ✰ New Delhi: Mallika, (regular bearing, semi-vigorous with high TSS); Amrapali, (regular bearing variety, suitable for high density planting); Pusa Arunima (red coloured peel and excellent shelf life), Pusa Shresth, Pusa Lalima, Pusa Pratibha and Pusa Peetamber (IARI, New Delhi)
- ✰ Tamil Nadu: PKM-1 and PKM-2 (Periyakulum);
- ✰ Uttar Pradesh: Ambika and Arunika from CISH, Lucknow

Passion Fruit

Kaveri, a hybrid passion fruit bred at IIHR, Chethalli by crossing Green and Purple varieties can endure rains and has better keeping quality. It is now extensively cultivated in the N.E. regions.

Peach

In peach, hybridization work at Saharanpur resulted in release of Saharanpur Prabhat, while that at CITH, Srinagar of variety CITH P-1 and Pant Peach at Pantnagar (Uttarakhand).

Pomegranate

Hybridization work in pomegranate in Karnataka and Maharashtra has resulted in the development of a number of new varieties. This includes selections No. 61 namely, Mridula from Rahuri, Maharashtra and Ruby-dark red non sticky arils with soft seeds from IIHR, Bengaluru from F_2 populations.

Ploidy Breeding

A fascinating phenomena in fruit plant has provided an important pathway for evolution and speciation. In nature, somatic mutations, union of unreduced gametes, irregularities of mitosis or meiosis produce polyploids. Attempts have been made to exploit polyploidy in fruit crop improvement through use of physical and chemical mutagenesis. Using colchicines or orizalam, fusion of protoplast upto different species or genera to form hybrids. Somatic hybridisation is an important tool or polyploidy manipulation resulting in allotetraploids and autotetraploids.

Polyploidy is important in conferring desirable characteristic in fruit crops *e.g.* seedlessness and increased fruit size in guava, pear, apple and has provided advantages ovular fertility in banana. Similarly tetraploid grapes produce distinctly large berries *e.g.* Mawal Seedless, Perle, Case, Early Niabel, *etc.* Difference in ploidy levels of parents of some species adversely affect desirable crosses. Such crosses lead to development of abnormal endosperm; poor pollen tube growth and embryo abortion at different development stages. Manipulating ploidy level of one of the parent may help to overcome these barriers in hybridisation. Treatments with colchici have helped to some extent in such situations.

In cases where interspecific or intergenric hybridisation is to be carried out studies dealing with somatic chromosome number determination (cytogenetic) and its ploidy level are of fundamental consideration. Success of such crosses depends on genomic relatedness particularly chromosome structural similarity, meiotic homology, genetic compatibility and recombination. Thus comparative genetic and Karyotopic characterisation of such taxa is an important step towards perspective in gene introgassion.

Scion vs. Rootstock Breeding

The important point which has been evading attention of fruit breeders in India is the recognition of the fact that in most fruit crops there are two parts requiring improvement *i.e.* scion and rootstock. The objectives in both cases are different.

The objectives in scion breeding include attractive colour and high yield; Dwarfing for amicability to high density planting; Precocity and regularity in bearing; Seedlessness, soft seeds and better pulp quality; richness in nutrients. Much of the work on fruit crop breeding carried out so far has been related to these objectives.

On the other hand rootstock in fruit crops are primarily used either because of their capability to impart dwarfness or their ability to utilise nutrients from soil and water efficiently or their resistance for abiotic and biotic stresses. They can be used either directly or in breeding programmes in their vital role in production and productivity. Selecting or breeding rootstocks is more difficult than selecting a scion variety. Evaluation of rootstocks for desirable characters and stionic effects on candidate fruit crops also requires much time.

Redefining Traint Priorities

Exploitation of Wild Species

Wild species can play an important role in developing genetically superior varieties particularly as donor for vigour and of disease pest resistance. Unfortunately most wild species have neither been systematically collected nor evaluated either by morphological or molecular means. Some wild species of mango are already in danger of extinction while many others grow wild even to-day *e.g. M.sylvatica* (NE) and *M.andaminica* (Andaman group of Islands).

Incidentally wild species have been extensively used for breeding scions and rootstocks in citrus, banana, grape, papaya, blueberries and raspberries in other countries. We have not even taken full advantage of information generated from this work by foreign institutions or introduced promising materials requiring priority in different crops.

Exploitation of wild species and incorporating their desirable genes in cultivated species however is different task as quantitative traits are harder to transfer. Also resultant hybrids are sterile with low yield and poor quality. Further mechanical isolation, chromosomal aberration and genetic and cytoplasmic incompatibilities hinder hybridization between wild and cultivated species.

Breeding of Scion Varieties

Yield and Quality Improvement

Some of the traits of commercial importance in scion varieties include fruit size, colour, yield and quality components. Efforts have been made to improve these in a number of crops *e.g.* mango, guava, papaya, grape, pomegranate through conventional breeding. However, polyploidy breeding is a potential tool for improving yield and quality, which are critical commercial traits. Even large fruit size is an important trait to fetch premium price in certain fruits like grape, kiwifruit, *etc.* besides size, novel colour and nutrient content adds value to the crop variety. While tetraploid varieties resulting in larger berries and early ripening have been developed in grape, certain defects like poor growth habit, irregular flower setting and reduced yield prevent their commercial acceptability.

Many fruit crops play an important role in nutritional security, being important sources of vitamins, mineral and antioxidants compounds like anthocyanims, flavonoids and procyanadins. Some wild species of *Vitis* (grape), strawberry and *Rubus* and cultivated varieties of apple, grape, kiwi, orange, peach, pear and *Rubus* are rich in antioxidants. These genes can be introduced through hybridisation. Source of some antioxidants reported elsewhere are listed below:

Anthocyanins	Grape	*V.vinifera x V. labrusca; V.labrusca x V. riperia*
	Raspberry	*R.innominates*
Anti-oxidants	Strawberry	*Wild* spp. *F. virginiana* spp. *glaca*
	Grape	*V. amurensis*
	Raspberry	Red and black raspberries of genus *Rubus*
Total Phenolics	Raspberry	*R. caucasicus*
Ascorbic Acid	Raspberry	*R. idaus*

Seedlessness

Seedlessness is an important trait in commercial fruit crops. It can be induced through mutation and polyploidy breeding by crossing tetraploids x diploids. Autopolyploids play a crucial role in development of triploids. Triploids are generally mere vigorous than diploids and sterile which features helps in producing seedless cultivars. Triploidy has been commercially exploited in citrus, banana, grape and guava. Some seedless varieties produced through Mutation/ Hybridisation are listed below:

Fruit	*Hybridisation*	*Mutation*
Grape	Pearlette, Beauty Seedless	
Grapefruit		Duncan, Star Ruby
Lemon		Eureka
Mandarin	Gold Nugget, Pixie, Winola	Shiyueju
Orange	Clara, Tacle	Hongju 418, 420, Pineapple
Pummelo x Grapefruit	Oroblanco, Melogold,	

Breeding of Rootstocks

Tree Architecture

Rootstocks have the ability to modify the vegetative growth and affect scion vigour resulting in dramatic change in planting systems of fruit crops by their ability in tree size control making it possible to go for high density planting system leading to meadow orchards. Emphasis therefore needs to be laid on selection of dwarf trees/rootstocks for close spacing. Spacing distance will however, vary based on scion vigour *i.e.* high, medium and dwarf; fertility status and water holding capacity of the soil. Some work in this regard has been carried out successfully in UK in apple, where a number of dwarfing, semi-dwarfing and standard rootstocks have been developed and are being successfully exploited commercially are listed below:

Dwarfing Rootstocks	*Semi-Dwarfing Rootstocks*	*Standard Size Rootstocks*
B.9, B.10, G.16, G.41, G.65, G.214, G.935, M.9, M.27, Mark, Ottawa 3, P.2, P.22, Supporter 1, Supporter 4, V.3	B.118, B.490, G.11, G.30, G.202, G.210, G.890, G.969, J-TE-H, M.26, M.27, M.4, M.7 EMLA, M.9 (interstem)/MM.106 rootstock, M.27, M.9(interstem)/MM.111 rootstock, MM.106 EMLA, V.1	Antonovka 313, MM.111 EMLA, P.18, Seedling

Breeding for Abiotic Stress Tolerance

As climate changes become more severe, several abiotic stresses *e.g.* high temperature, drought, cold and salinity limit plant growth, yield and fruit quality. Losses are often inter-related and hinder normal, physiological, biochemical and molecular processes in plants. There is also severe cellular damage impairing plant development and fruit set. Work done in breeding of varieties for abiotic stress tolerance is briefly given below:

Drought Tolerance

Water scarcity is becoming a key limiting factor in fruit crop production. It is being caused due to decline in rainfall pattern and increase in number of warmer days. Drought plays an important role in scion growth and yield during water scarcity period. With the drought incidence becoming frequent in the country, it has become important to identify drought tolerant varieties and species and developing drought tolerant hybrids has assumed significant importance. The mechanism involved in drought tolerance include, drought escape, dehydration, avoidance, drought tolerance, recovery and resistance besides increasing water up take and reducing water loss. Blueberry, a fruit not grown commercially in India is highly susceptible to drought.

A number of drought tolerance species/hybrids have been identified in different crops. These include *C. reshni, C. aurantium, C. limonia, C. jambhiri and P. trifoliata* of Citrus and *V. arboretum, V. dorrowi* of Raspberry. Some highly, moderately drought tolerant species/hybrids are listed below:

Reaction to Drought	*Crop*	*Rootstock*
Highly tolerant	Citrus	Flying Sunki
	Grape	110 R, Ramsey, 1103 Paulsen, 140 Ruggeri, Kober 5BB
Moderately tolerant	Apple	M 9, M 26, M 111
	Citrus	Carrizo Citrange, Swingle Citrumelo
	Grape	Teleki 5C, SO4
Moderately tolerant to drought	Peach	4247T1
Tolerant	Citrus	SSW 1

Thus breeding of rootstocks adapted to different drought conditions in important fruit crops is the best option available to manage drought.

Cold Tolerance

Cold stress is a major limiting factor that affects productivity of many fruit crops. While severe cold restricts growth, freezing temperatures injure plants and kill trees. Many commercial fruits namely apple, citrus, guava, grape, papaya, blueberry and pear are susceptible to cold and freezing temperature. Low temperatures often affect plant growth and crop productivity. Plants differ in tolerance to chilling (0-15°C) and freezing (0°C) temperatures. Continuous effort is required to breed *cvs*, which is cold hardy and escape severe cold climate. However, resistance to cold is complex, quantitatively inherited and controlled by polygene in fruit crops. Vast germplasm is available in different fruit crops possessing genes for cold hardness as below:

Apple

On the basis of evaluation, the following cold tolerant rootstocks have been identified:

Temperature	*Rootstocks*
-16°C to -19 °C	G 935
-12°C to -14 °C	G 11, G 30, G 41, B 9, P 2 and M 26, B 9, Ottawa 3, P.2
-08°C to -11 °C	M 26, MM 106, MM 111, and M 7, Jork 9, Alnarp 2, M 26
-06°C to -08°C	M.7, MM 106

Grape

American grape species such as *V. labrusca, V. aestivalis,* and *V. riparia* are potential sources of cold hardiness. *V. riparia* species have been reported to withstand up to -40°C. Rootstocks like *V. riparia, Glorie, St. George,* and 1103P are tolerant to cold injury. Even though resistant sources are available, selection of cold hardy progenies from hybrid populations by lengthy screening and evaluation makes breeding difficult. Further the study of cold hardiness and inheritance pattern is challenging. Such studies are also few. Inheritance pattern of cold tolerance has been studied in a few crops is given below:

Crop	*Trait*	*Heritability*
Blueberry	Bud freezing tolerance	High
Apple	Winter survival	Mid – high
Peach	Dormant flower buds	Mid – high
	Cambium, xylem and bud injury	0.65 – 0.96
Plum	Winter survival	>0.0

Salt Tolerance

Salinity is one of the most serious environmental factors that affect the crop productivity worldwide. Most crops are sensitive to high salt concentrations in the soil. Sodium and chloride ions in the soils are the dominating factors in adversely affecting yield of fruit crops. More than 800 million hectares of land are affected by soil salinity throughout the world. Fruit crops like citrus, grape, apple, pear, peach, cherry, apricot, strawberry, plum, *etc.* are susceptible to soil salinity. There is thus need to give priority to breeding salt tolerant rootstocks. The level of susceptiblity to salinity has been worked out in some fruit crops and the threshold levels are given below:

Crop	*Threshold Level (dS/m)*
Grape fruit	1.2
Sweet orange	1.3
Almond, Apple, Black berry, Grape, Lemon, Mandarin	1.5
Apricot, Avocado	1.6
Peach	1.7
Plum	2.6

There are three types of salinity tolerance mechanism in crops *e.g.*; tolerance to osmotic stress, salt exclusion, tissue tolerance. Citrus is one of the most sensitive crops for salinity. A large number of citrus orchards in India is getting affected by salinity every year. The ability of citrus trees to tolerate root zone salinity is mainly dependent on rootstocks. Citrus rootstocks vary in their degree of tolerance for salinity. The Cl^- excluding ability of Rangpur lime and Na^+ excluding ability of *P. trifoliata* can be transferred to their hybrid derivatives. Work done in Australia, Canada, Spain and USA has resulted in development of several hybrids, namely Swingle, Citrumelo, Troyer, Risk and Carrizo Citranges besides US-897 and X637. Evaluation of different Citrus species, relatives and hybrid rootstocks for salinity tolerance has indicated that Cleopatra mandarin is salt tolerant and Sour orange is susceptible. The rootstocks US-897 and X639 have also been shown to be highly salt tolerant than Cleopatra mandarin. Some of the salt tolerant citrus rootstocks are listed below:

Rootstock	*Remarks*
Swingle Citrumelo	Moderately tolerant to salinity
Troyer Citrange	Tolerant to salinity
Rusk Citrange	Tolerant to salinity
Us-897	Highly tolerant to salinity
X 639	Highly tolerant to salinity
Carrizo Citrange	Moderately tolerant to salinity

Salinity is a serious concern in grape production also because of its moderate sensitiveness in saline soils. *Vitis vinifera*, the commonly cultivated grape is a poor Cl^- excluder and rapidly absorbs and accumulates salts in the plant system leading to poor vegetative growth, burning of leaves and low yields in contrast to *Vitis* species like *V. rupestris, V. cinerea, V. champini* and *V. berlandieri* which are good chloride ion excluders. Results of different rootstocks have revealed that Salt Creek and Freedom have high level of tolerance to salinity. Some of the commonly used grape rootstocks tolerant to salinity are listed below:

Rootstock	*Parents*
Salt Creek	*V. champinii*
Freedom	Complex hybrid of *V. champinii, V. riparia, V. labrusca, V. vinifera, V. solonis*
140Ru	*V. berlandieri* x *V. rupestris*
1103P	*V. berlandieri* x *V. rupestris*
SO4	*V. berlandieri* x *V. riparia*
41B	*V. vinefera* x *V. berlandieri*

Salinity causes yield loss in citrus, mango, avocado, stone fruits *etc.* The mango rootstock Gomera-1 is reported to be was a Cl^- excluder. Similarly 13-1, Olour and Kurukan are also reported for salt tolerance. The interspecific *Prunus* rootstocks like HS-314 and GF-677 have proved to be good source of salinity resistance. The extent of salinity has been on the increase in India rendering several suitable areas unfit for fruit cultivation particularly for lack of salt tolerant rootstocks. It therefore calls for priority attention to this aspect.

Breeding for Biotic Stress Tolerance

Resistance to biotic stresses is a major objective in any crop breeding programme world over. Resistance is the ability of host plant to reduce the infestation or damage or both by an insect or disease or any of the external force. Resistance level may vary from only slight plant tolerance to stress or total immunity. It can be result of one or more mechanism. Both vertical and horizontal mechanisms have been used by breeders for development of new varieties. Building desirable resistance is the only permanent solution to control pest incidence. Further, with call for organic farming and making food safer a priority, resistance breeding has assumed great importance. In this regard, development of rootstocks for certain biotic and abiotic stress is extremely important in fruit crops.

Pest Resistance

A number of pests adversely affects the productivity and quality of different crops. Important pest which have attracted attention of breeders worldwide include the following:

Economically Important Pests in some Fruit Crops

Crop	*Pests*
Apple	Wooly Apple Aphids
Banana	Corm Borer, Pseudostem Weevil
Citrus	Aphids, Citrus Butterfly, Leaf Miner, Psyllids, Stem Borer, Thrips, Whitefly
Grape	Phylloxera
Guava	Fruit Fly
Mango	Fruit Fly, Hopper, Stone Weevil
Papaya	Papaya Mealy Bug, White Fly
Pomegranate	Anar Butterfly

Besides nematodes (in citrus, banana and grape) and borers in certain fruit crops cause huge economic losses. Since chemical control of these pests can be a major hazard to the environment, breeding of pest resistant rootstocks becomes important. Several wild species have been identified as sources of resistance as discussed below:

Apple

The woolly apple aphid (WAA), (*Eriosoma lanigerum*, Hausmann) is the primary pest of apple worldwide. It affects roots and aerial parts of the plant. It causes more damage by feeding the roots. The control of this pest when feeding of roots under soil is very difficult either chemically or culturally. As a result, rootstock breeding programmes were started globally and Northern Spy was used as a source of resistance along with Malling series. The Malling-Merton rootstock series *e.g.* MM 104 (M2 x Northern Spy), MM 106 (M1 x Northern Spy), and MM 111 (Merton 793 x Northern Spy) derived from Northern Spy were resistant to WAA. However, they were not immune. The immunity to WAA was identified in *M. robusta* accession Robusta 5 which has *Er*2 resistant gene.

A breeding programme aimed at Cornell University, Geneva campus for combining multiple resistance and tree size control using different Merton series, *M. domestica* and *M. robusta* has resulted in release of several rootstocks with multiple level of tolerance and varying degree of tree size control listed below

Sl.No.	*Rootstock*	*Tree Vigour*	*Resistance to*	*Tolerant*
1	G 41	Dwarf	WAA and fire blight	
2	G 214	Dwarfing	WAA, fire blight, and crown rot.	
3	G 202	Semi dwarf	WAA and fire blight, replant disease and crown rot	
4	G 222	Semi dwarf, precocious	WAA and fire blight, replant disease and crown rot	
5	G 210	Semi dwarf	WAA and fire blight,	cold hardy
6	G 890	Semi dwarf, precocious	WAA and fire blight,	cold hardy
7	G 969	Semi dwarf, precocious	WAA, fire blight and crown rot	cold hardy
8	G 4210	Semi dwarf	WAA and fire blight	

Grape

In grapes, Phylloxera, *Daktulosphaira vitifoliae* (Fitch) is the most devastating pest of grape and it invaded Europe during the late 1860s and caused severe economic losses in grape vine industries. Intensive grape breeding programme was initiated by several breeders resulting in development of hybrid rootstocks like SO4, 5BB and 5C derived from *Vitis berlandieri* Planch. X *V. riparia* Michx. resistant to phylloxera. However, this pest is fortunately not available in India.

Nematodes

Nematodes constitute one of the major limiting factors in fruit crop production. They cause extensive root necrosis resulting in serious economic consequences that drastically reduce the yield. The roots damaged by nematodes are easy prey to fungi which invade the roots and accelerate root decay. Important nematode species which threaten the global fruit industry and cause huge economic losses are listed below:

- ☆ Root-knot nematodes (*Meloidogyne incognita* and *M. arenaria*),
- ☆ Burrowing nematode (*Radopholus similis*),
- ☆ Citrus nematode (*Tylenchulus semipentrans*),
- ☆ Dagger nematode (*Xiphinema index*) and lesion nematodes (*Pratylenchus spp*) are the major nematode pests that infest commercial fruit crops in general

Work on breeding of nematode resistant rootstocks in citrus, grape and banana has resulted in the development of following rootstocks elsewhere:

Rootstocks Resistant to Nematodes

Grape	*Degree of Resistance to Root Knot Nematode and Dagger Nematode*
O39-16	Highly resistant to Dagger
Schwarzmann, Freedom, Boerner	Highly resistant to both
1616C	Highly resistant and Moderately resistant respectively
Ramsey	Highly resistant
101-14 Mgt	Moderately resistant to both
Fercal	Moderately resistant
Banana	
H 21, H 59, H 65, H 74, H 84, H 95, H 109 and H 201 (Triploids)	Burrowing nematode
NARITA-1 (Triploid)	***Helicotylenchus multicinctus***
H 212, H-02-19, H-02-21, H-02-22, H-02-23, H-02-36; FHIA 01 (Tetraploid)	***R. Similis***
H 61, H 74 and H 95, BRS 1 and BRS 2 (Triploids)	Other nematodes

The Citrus nematode (*Tylenchulus semipenetrans*) is a worldwide problem that leads to poor growth of young plants and is responsible for poor performance of mature trees. The important source of resistance is *Poncirus trifoliata* and its derivatives. Although *P. trifoliata* is resistant to citrus nematode, its performance is very poor in alkaline and calcareous soils. It has been reported that the hybrid rootstocks of *C. reshni* x *P. trifoliata* and *C. volkameriana* x *P. trifoliata* are highly resistant to citrus nematodes. Further, some hybrid rootstocks of *C. reshni* x *P. trifoliata,* King mandarin x *P. trifoliata,* and *C. volkameriana* x *P. trifoliata* also showed very little incidence of nematodes and are moderately resistant.

Disease Resistance

A number of bacteria, fungi, virus, and mycoplasma like organisms affect fruit crop production severely. The major diseases which affect production and quality of fruits drastically are listed below:

Fruit	*Diseases*
Apple	Scab, crowned rot, fireblight and powdery mildew
Banana	Fusarium wilt, sigatoka and leaf spot
Grape	Powdery and downy mildew
Guava	Wilt
Mango	Anthracnose, Powdery Mildew
Papaya	Mosaic, ring spot, phytopthora
Strawberry	Crinkle virus, mottle virus and red steel virus

Identification and utilisation of source of resistance is a basic requisite for incorporating disease resistance through breeding programmes. It requires screening and evaluation of germplasm to determine the magnitude of genetic variation of resistance. Several sources of resistance to major diseases have been identified in different fruit crops. These are as under:

Resistant Wild Species Used as Rootstocks in different Fruit Crops

Crop	*Diseases*	*Resistant Wild Species Rootstocks*
Apple	Fire blight	*Malus robusta*
	Scab	*M. floribunda* (vf resistant gene)
Banana	Black sigatoka	M. acumita
	Fusarium Wilt (Race 1,2,4)	M. acumita, spp. malaccensis
Citrus	Phytophthora	*Trifoliate orange, swingle citrumello, Bento cintrange (hybrid)*
	Tristeza virus	Ranpur lime, rough lemon, Cleopatra mandarin, Trifoliate orange
Grape	Downy mildew	*V. Muscadinia, V. rotendifobia, V. amorensis*
Guava	Wilt	*Psidium friedricsthalianum*
Papaya	Ring spot	*Vasconcille cauliflora, Quercifolia pubscence*
Pear	Fire blight	*Pyrus communis*

A number of efforts have been made in different fruit crops to develop disease resistant/tolerant hybrids as listed below:

Fruit	*Hybrids*	*Resistance*
Apple	Borovinka, Hagloe Crab, Tita Zetei, Braburn, Britegold, Delicious, Empire Red, Enterprise, Fuji, Goldstar, Jonafree, Nittany and wealthy red	Powdery Mildew Scab
	Prima, Primula, Sir Prize, Jonafree, Dayton	Scab
Banana	Diploids- SH2095, H109 and Triploid- NARITA-1 Tetraploids - FHIA 01, 07 and 23	Fusarium wilt Resistant to Sigatoka Sigatoka Race 4
	FHIA 01, 17 and 18 and PITA 17	Sigatoka
Guava	Peipa, TS-G1, TS-G2	Wilt
	TS-G3	Wilt

Fruit	Hybrids	Resistance
Mango	Hybrid progenies, CISH-2035 and 1734	
Papaya	Cinta L 41, L 90	Moderately

Strategies for Fruit Breeding

Haploids and Double Haploids

In fruit crops, high heterozygosity, long generation cycle with long juvenile period and often self-incompatibility do not allow the haploid development easily through the conventional methods. The haploids and double haploids can make a significant contribution in shortening the time taken for fruit breeding. Using DH technology, completely homozygous plants can be established in one generation thus saving several generations of selfing required in conventional methods. Development of homozygous lines through gametic embryogenesis results in single step improvement of the crop.

In a conventional breeding programme, a pure line is developed after several generations of selfing. However, in fruit crops, there is no way of obtaining homozygous lines through conventional methods. Hence, fixation by doubling the chromosome complement presents a possibility of using haploids. In many of the important fruit crops mango, citrus, banana, grape, papaya, sapota, pomegranate, custard apple, apple, peach, pear, plum, apricot, kiwi fruit *etc.* it can be a very handy tool in the improvement.

Role of Biotechnology in Fruit Breeding

Marker Assisted Breeding (MAB)

Fruit crops being perennial and heterozygous remain juvenile for a long period, when propagated through seeds. Since seed production is indispensable in a breeding programme, shortening of juvenile period or finding alternate ways to induce early fruiting in seedlings is very important to achieve early assessment at seedling stage and avoiding evaluation of unwanted seedlings. A number of methods like topworking, induction of flowering by chemicals, shortening of juvenile phase by grafting have been employed. Using a combination of above techniques a number of superior varieties have been developed. These include:

- Blood red guava
- Pink pulped papaya
- Early maturing seedless grape
- Seedless mango
- Vitamin rich peach and plum
- Regular bearing mango hybrids

Assessment of genetic diversity of fruit crops is an important step in breeding. A diverse range of plant material including land races, modern cultivars, wild relatives and other wild species are available as the starting material. Conservation of germplasm is of little importance without characterisation and further utilization of concerned plant material. To intensify these activities DNA fingerprinting has been implemented in the characterisation of germplasm. The molecular markers are being used for maintenance of core collection for the assessment of genetic diversity and fingerprinting. However, phenotyping is still advantageous to validate performance. Development of molecular markers for different traits should continue to receive priority attention. A large number of molecular markers have been developed and applied for analysis of genetic diversity and relatedness and all markers have their strengths and weaknesses. In general, two main groups of molecular markers have been utilised for PGR characterisation:

- ✰ Markers for identification of accessious detection of structure of genetic reserves collection and relatedness of available plant material.
- ✰ Gene specific markers, applicable for Marked assisted selections (MAS).

The last two decades have witnessed initiatives to a transition from traditional phenotype to genotype assisted breeding. This has become possible due to availability of wide array of molecular markers, transcriptome and whole genome sequence data in many woody fruit plants.

A marker is a tag of identification of any trait of an organism which can be identified with confidence and relative ease and can be followed in a mapping population. DNA marker is a small region of DNA that shows sequence polymorphism among genotypes. Such DNA/molecular markers have contributed numerously in breeding of many crops due to their reliability and tight linkage to the trait of interest. Mapping and tagging of genes controlling important agronomic traits have been facilitated by an array of molecular markers. Marker assisted breeding in fruit crops is given considerable importance as it would improve the breeding efficiency in fruit crops.

A few genes and markers identified in fruit crops linked to various traits are given below:

Crop	*Trait*	*Gene*	*Marker*
Apple	Fruit colour	Rf	BC226
	Fruit acidity	Ma	OPT16-1000
	Mildew resistance	Pl-1, Pl-2, Pl-w, Pl-d, Pl-m	LAP – 2
	Fire blast resistance		SCAR/SSR
Citrus	Fruit acidity	Ac	RFZ20
	Seedless fruit	Fs	OPMO6r
	Citrus leprosis virus resistance		AFLP and RAPD
Peach	Flesh adhesion	F/f	OPB5a

Crop	Trait	Gene	Marker
Pear	Incompatibility		AFLP/SSR
Banana	Sugar content		RFLP
Grapes	Seedlessness, berry size, and ripening date		AFLP, SSR RAPDs, ISSR and SCAR
Strawberry	Day neutrality		AFLP
Apricot	Plum Pox virus		SSR

Genomics in Fruit Crops

Genomics are the next generation sequencing technologies being used to generate whole genome sequences for a wide range of crop species. It helps to understand genome structures, to make use of quantum of data produced and to describe their function and interactions. With the significant advancement in sequencing technologies in recent years, efforts have been made to release whole genome sequences of many fruit crops. Many more whole genome sequencing projects of fruit crops are also in progress. The biggest challenge in genomics of perennial fruit trees is the development of bio-informatic tools to manage huge volume of data generated by next generation sequencing technologies. There is a need to create genome resource platforms for the benefit of academician and researchers. A few crops whose genomes have been sequenced are discussed below:

Apple

Genes have been identified for scab, fire blight fungal disease resistance, and induction of anthocyanim accumulation.

Grape

Genome sequence (size 457 Mbp) has been studied in variety Pinot Noir which was bred close to full homozygosity.

Mango

A draft genome based on transcriptome and whole genome sequence data of Dashehari, Neelum and Amrapalli cultivars has been reported.

Musa

A global Musa genomic consortium, an international network with 40 institutions in 24 countries has worked for sequencing the banana genome. All available markers have been used to study phylogenetic relationships, diversity, evaluation, parentage analysis differentiation of A and B genomes and to develop trait specific markers.

Papaya

Papaya with a small genome size (372Mbp), is having only a few disease resistant genes besides fruit ripening skin colour change and sugar accumulation genes. The transgenic female *Sun Up* variety was chosen for sequencing genes.

The genome size of different fruit species is given below:

Fruit Crop	*Scientific Name*	*Genome Size (Mbp)*
Apple	*Malus domestica*	742.3
Banana	*Musa acuminata*	523.0
Grape	*Vitis vinifera*	487.0
Kiwifruit	*Actinidia chinensis*	616.1
Mango	*Mangifera indica*	439.0
Papaya	*Carica papaya*	372.0
Peach	*Prunus persica*	265.0
Pear	*Pyrus bretschneideri* Rehd.	600.0
Sweet Orange	*Citrus sinensis*	452.0
Wild Banana	*Musa balbisiana*	438.0
Woodland Strawberry	*Fragaria vesca*	240.0

Recommendations

Breeding of perennial fruit trees is quite different than breeding of annual crops. Besides, there are several constraints impeding fast testing and release of varieties. In addition, there are several inherent problems in some fruits crops such as structural differences in floral biology, apomixes, parthenocarpy, cross incompatibility, number of seeds per fruit and large acreage required for meaningful assessment. While a lot of genetic variability has been collected and maintained in a number of fruit crops, major emphasis in fruit breeding has so far been laid only on developing varieties with high yield and superior fruit quality. Time has come to have a fresh look on breeding priorities and strategies to harness the full potential of fruit breeding. Following recommendations are made to achieve the desired objectives:

- ☆ While a lot of genetic variability has been collected and maintained in a number of fruit crops, major emphasis in fruit breeding should be given for systematic evaluation of germplasm of fruit crops.
- ☆ Collection of wild germplasm, endangered species and other gene sources should be taken on priority and their descriptive evaluation is of utmost importance.
- ☆ There is urgent need to do basic research on inheritance of characters which will ease the choice of parents in hybridization programmes.
- ☆ There is need to clearly define objectives and priorities of breeding of fruit crops for different regions. Furthermore, the work should be assigned to different centres on the basis of their human resource capabilities and available facilities.
- ☆ Breeding of commercial fruits like mango, citrus, grape, pomegranate *etc.*; indigenous fruits like bael, jamun should be taken at priority.
- ☆ In view of climate change, breeding work should be taken at priority for developing climate resilient varieties and rootstocks.

- ☆ Rootstock breeding in tropical and subtropical fruits should be strengthen particularly for biotic stress tolerance.
- ☆ There should be integrated breeding approaches involving traditional and modern molecular methods for the improvement of fruit crops. Marker aided selection and enriching the genome resources of fruit crops should be taken on priority.
- ☆ There is dearth of manpower for fruit breeding programmes keeping in view the number of crops and aspects requiring attention. There is thus need for allocating more manpower for this work.
- ☆ There is need to re-orient courses for P.G. students desirous of taking up fruit breeding programmes covering basic courses related to genetics, plant breeding and molecular biology.
- ☆ Researchers involved in fruit breeding should be exposed to new modern technologies by deputing them to front ranking centres in the world.

It is hoped that the three day deliberations will result and refining the above recommendations further and chalking out a roadmap for future fruit breeding programmes in the country.

Breeding of Perennial Horticultural Crops *Pages* **23–33**
Editors: S.K. Singh, V.B. Patel, A.K.Goswami, Jai Prakash and Chavlesh Kumar
Published by: **BIOTECH BOOKS, NEW DELHI**

Chapter **2**

Improvement of Perennial Vegetable Crops

Pritam Kalia and Shrawan Singh

Division of Vegetable Science,
ICAR-Indian Agricultural Research Institute, Pusa, New Delhi – 110 012

INTRODUCTION

Vegetable crops are rich source of dietary microelements constituting significant part of the daily diet in every household. Although the seasonal or annual vegetables have largest share in vegetable basket, but there are few promising perennial vegetable crops which have great prospect in the context of nutritional security of the poor and vulnerable communities in climate change and economic fluctuations. Perennial vegetables, though, from the point of view of consumption are prepared and eaten like other vegetables, but they are perennial in growth habit. Some well-known perennial vegetables from the temperate world include asparagus, artichoke and rhubarb while list of such kind of vegetables is long in tropics such as drumstick, chakurmani, agathi, vegetable banana, cassava, taro *etc.* (Table 2.1). Being nutrient store house, these vegetables are an integral part of many cultural diets around the world, particularly in tropical agriculture.

Breeding Method for Perennial Vegetable Crops

Conventional breeding in conjunction with molecular biology has bright prospects of developing vegetable varieties high in nutraceuticals, edible colours and bioactive compounds suitable for fresh market as well as fusion/functional food industry. Increasing the health functionality of vegetable crops through breeding and/or genetic modification should create products that deliver greater health benefits than current varieties. Nevertheless, research efforts strive to improve our

Table 2.1: Some Perennial Vegetable Crops

Common Name	*Scientific Name*	*Family*	*Climate*	*Chromosome No.*	*Mode of Pollination*	*Plant Part Used*	*Plant Habit*
Drumstick	*Moringa oleifera* Lam.	*Moringaceae*	Tropical	2n=28	Cross	Pods, leaves	Tree
Curry leaf	*Murraya koenigi* Spreng	*Rutaceae*	Tropical	2n=18	Self	Leaves	Bush
Bread fruit	*Artocarpus altilis* Postperg.	*Moraceae*	Tropical	2n=56	Cross	Fruits	Tree
Jackfruit	*Artocarpus haterophyllus* Lamk.	*Moraceae*	Tropical	2n=56	Cross	Fruits	Tree
Agathi	*Sesbania grandiflora* (L.) Poiret	*Fabaceae*	Tropical	2n=14	Cross	Pods, flowers	Tree
Asparagus	*Asparagus officinalis* L.	Liliaceae	Temperate	2n=20	Spears	Bush	Herb
Jerusalem artichoke	*Helianthus tuberosus* L.	Asteraceae	Temperate	2n=102	Cross	Tubers	Bush
Globe artichoke	*Cynara scolymus* L.	Asteraceae	Temperate	2n=32	Cross	Flower bud	Herb
Khejri	*Prosopis cineraria* L.	Fabaceae	Sub-tropical	2n = 28	Cross	Immature pods	Tree
Kair	*Capparis decidua* (Forssk.) Edgew.	Capparaceae	Sub-tropical	2n=44	Cross	Immature fruits	Tree
Lasoda	*Cordia dichotoma* G.Forst.	Boraginaceae	Sub-tropical	2n=48	Cross	Immature fruit	Tree
Chive	*Alllium schoenoprasum L.*	Alliaceae	Temperate	2n=16	Cross	Bulbs	Herb
Shallot	Cepa L.	Alliaceae	Temperate	2n=16	Cross	Bulbs	Herb
Pointed gourd	*Trichosanthes dioica* Roxb.	Cucurbitaceae	Tropical	2n=22	Cross	Fruits	Vine
Ivy gourd	*Coccinia grandis* (L.) Voigt.	Cucurbitaceae	Tropical	2n=24	Cross	Fruit	Vine
Indian bean	*Dolichos lablab* L.	Fabaceae	Tropical	2n=22, 24	Self	Green pods	Vine
Sword bean	*Canavalia gladiata* (Jacq.) DC.	Fabaceae	Tropical	2n=22(44)	Self	Green pods	Vine
Winged bean	*Psophocarpus tetragonolobus* L. DC. 2n=18	Fabaceae	Tropical	2n=18	Self	Flower, pods, roots	Vine
Cassava	*Manihot esculenta* Crantz.	Euphorbeaceae	Tropical	2n=36	Cross	Tuber	Shrub
Yam	*Discorea* spp.	Dioscoreaceae	Tropical	2n=40 (male), 60 (female)	Cross	Tuber	Vine
Horse radish	*Armoracia rusticana* G.Gaertn., B. Mey. and Scherb.	Brassicaceae	Tropical	2n=32	Cross (sterile)	Roots	Herb
Chow-chow	(*Sechium edule* (Jacq)	Cucurbitaceae	Tropical	2n=26, 28	Cross	Fruits	Vine
Chekurmanis	*Sauropus androgynus* Meer	Phyllanthaceae	Tropical	2n=50	Cross	Leaves	Shrub

understanding of crop manipulations for nutrition and in many cases substantial progress have been made by using conventional and molecular tools.

A. Conventional Breeding

a. Introduction

Introduced materials may be valuable in selecting desirable plants. In case of non-uniform introductions the desirable plants can be selected, their progeny increased, purified and later tested against the standard or local varieties for selecting the most promising lines. It is equally important to collect and domesticate the wild relatives of the cultivated species having desirable genes for high nutraceuticals and edible colours. The chances of an introduction depends to a great extent upon the relationship between the agro-climates, particularly temperature and day length of the donor and receptor areas.

b. Mass Selection

Mass selection is used for development of varieties in seed propagated perennial vegetable crops. For this, the plants are selected on the basis of easily observable characters and the population obtained from the selected plants remains more uniform for these traits. It helps in improvement of local ecotypes or landraces and purification of the existing varieties. During this process, the rejection of plants having diseases and insect pest incidence helps in breeding varieties with resistance or low incidence of insect pests and diseases.

c. Clonal Selection

Perennial vegetables are mostly propagated asexually to avoid the ill effects of segregation and recombination, both being the inevitable consequences of sexual reproduction. The progeny of asexual reproduction is exactly identical to their parents in genotype, remains true to type and have more uniformity with a desirable trait. Asexual reproduction is common in almost all perennial vegetables and clonal selection is one of the common breeding methods. A clone is a group of plants produced from a single plant through asexual reproduction. The members of a clone have same genotype, lacks genetic variation, severe inbreeding depression, immortal and phenotypic variations are purely due to environmental effects. The clones may arise due to somatic mutations, mechanical mixture and occasional sexual reproduction. Clonal selection consists of two phases, first, utilize already existing variability and second, selection of best genotype from the variable population to produce a superior clone or variety. It takes around seven years to develop a new variety through clonal selection.

d. Hybridization

The improvement through hybridization involves the following three steps: a) selection of parents; b) production of F_1 progeny; and c) selection of superior clones. It is applicable to those crops which can produce sexually and for other crops mutagenesis or biotechnological approaches are useful. Selection of parents is generally based on their known performance both as varieties and as parents in hybridization programmes. It would be highly desirable to know the relative values of GCA and SCA in the crop to be improved. Use of partially inbred parents in hybridization is useful because it allows breeders to identify plants that would have a greater concentration of desirable alleles. After producing F_1 progeny through controlled pollination, the promising F_1s with outstanding individuals are then grown at a much larger scale for selection.

e. Backcross Method

This method is more common in annual crops to transfer the character(s) controlled by single gene or few genes but its applicability also extended to perennial vegetables propagated through seeds. However, special activities are required to cut-short the time required from seed to seed through use of phytotron, plant growth regulators or physiological stresses along with use of DNA markers. It is more convenient to execute this method in self-fertilizing crops. The backcross method is based on the principle of quick and efficient transfer of a gene through repeated backcross and then selection in F_2 and subsequent generation. With continuous backcrossing with the recurrent parent, the backcross population becomes more like recurrent parent with additional trait of donor parent.

Table 2.2: Improved Varieties of Perennial Vegetable Crops Bred through Conventional Breeding

Common Name	*Varieties*	*Yield (t/ha)*
Drumstick	Jaffna, Moolanoor Murangi, Chavakacheri Murungai, Chemmurungai, Palmurungai, KM1, PKM 1, PKM2, Rohit 1, Coimbatore 2,KKM1, Dhanraj, Konkan Ruchira	40-50
Curry leaf	Sen Kaampa, DWD1 (Suwasini), DWD2	5-10
Agathi	Local types	
Jackfruit	NJC1, NJC2, NJC3, NJC4, Varikka, Koozha, Navarikka	*40-50*
Asparagus	Martha Washingotn, Mary Washington, Reading Giant, Palmetto, Colossal, Argentenil, Berr's Mamoth, Mammoth, Mammoth White, Jersey Queen, Jersey Giant, New Jersey, Improved Perfection, UC 72, UC 66, SL-831, Limbrass 22, Limbrass 126, Lucullas 1813J	1.2-3.0
Globe artichoke	Green Globe, Purple Globe, Violet de Provence, Catanese, Spinosa Sarda, Bianco Tarantino, Precoce di Jesi, Bull, Tudella, E15, Brindisino, Romonesco	10-12
Rhubarb	Victoria, Linnaeus, Mc Donald, Ruby Valentine, Sunrise, Strawberry, Cherry Red	10-15
Chow-chow	Round White, Long White, Pointed Green, Broad Green, Creamy Green	25-35
Jerusalem artichoke	Columbia, Mammoth, Oregon, Mammoth	16-20
Chive	Wilau, Aliai, Zornaja Rostan, Sempra	
Pointed gourd	CHES Hybrid 1, CHES Hybrid 2, Faizabad Parwal 1, Faizabad Parwal 3, Faizabad Parwal 4, Rajendra Parwal 1, Rajendra Parwal 2, Swarna Rekha, Arka Leelachal Kirti	15-25
Ivy gourd	Sulabha, Arka Neelachal Kunkhi, Arka Neelachal Sabuja	10-12

Common Name	*Varieties*	*Yield (t/ha)*
Indian bean	Pusa Early Prolific, Rajni, Deepali, KDB 403, KDb 405, T1, Co series, JDL series, HD18	3-5
Sword bean	SBS-1	7-10
Winged bean	UPS-122, UPS31, Ravathy, WBC2, IIHR Selections 21, 60, 71, JCV44	10-15
Chekurmanis	Local types	10-15
Cassava	H97, H165, H226, Sree Vishakam, Sree Sahya, Sree Prakash, Sree Jaya, Sree Vijaya, Nidhi, KMC1, CO1, CO2, CO3, Sree Apporva, Sree Athulya, CTNAU Tapioca Yethapur 1	25-35
Yam	Sree Keerthi, Sree Roopa, Sree Silpa, Sree Latha, Sree Kala, Sree Subha, Sree Priya, Sree Dhanya, Da-25	25-40

B. Non-Conventional Breeding

a. Tissue Culture Techniques

Anther culture technique has been employed for production of haploids from elite plants in natural population or F_1 plants. Colchicine treatment is given to these plants to develop doubled haploid (DH) plants for use in breeding programme. Asparagus is a dioecious crop, the breeding objective is to develop male hybrids because they yield more and live longer than female plants and are comparatively more disease resistant. Tissue culture plays a key role in asparagus breeding and most of the new improved cultivars from United States and Europe actually are clonal hybrids, which are F_1 progenies from single crosses. Anther culture has been employed for production of haploids from male plants. The first anther-derived, all-male hybrid cultivar, 'Andreas' was released by Corriols *et al.* (1990). The homozygous female parent of 'Andreas' is a parthenogenetic doubled haploid, whereas the homozygous super male parent is an anther derived doubled haploid. Yield trials conducted at several locations in different years showed yield advantage for 'Andreas' compared to leading cultivars.

b. Mutation Breeding

Mutation is a sudden heritable change in a characteristic of an organism. In clonal propagated crops, the bud mutation or somatic mutation still has relevance because it allows use of useful mutations occurring in buds and somatic tissues.

c. Polyploidy Breeding

Polyploidy can be induced due to aberration in cell division. This may occur both in the mitosis as well as in meiosis. This method can be used successfully in vegetable breeding as a means of enhancing of growth and yield, fertility issues (restoring fertility or introducing sterility), changes in plant morphology or other characteristics like improved stress resistance. Two practical examples of breeding programmes involving polyploids in perennial crops are *Lolium perenne* (perennial ryegrass) and Buddleja (Leus *et al.*, 2012).

C. Marker Assisted Breeding

Marker-assisted selection (MAS) for traits controlled by major genes or quantitative trait loci is now commonly employed in out-breeding woody perennial fruit crops. The molecular markers such as random amplified polymorphic DNA (RAPD), inter-simple sequence repeats (ISSR), microsatellites or simple sequence repeats (SSR), sequence characterized regions (SCAR), cleaved amplified polymorphic sequences (CAPS), sequence tag sites (STSs), expressed sequence tags (ESTs), single nucleotide polymorphisms (SNPs) and diversity arrays technology (DArT) are used to study linkage with genes responsible for important horticultural traits. In MAS, a marker (morphological, biochemical or one based on DNA/RNA variation) is used for indirect selection of a trait of interest. The mapping populations such as Near Isogenic Lines (NILs) and Recombinant Inbreed Lines (RILs) are used to identify the molecular markers linked to genes of interest. However, MAS is now coming into its own as the key tool that renders GWS cost-effective, by enabling elimination of undesired genotypes from the original breeding population (selection population) following a relatively cheap MAS pre-screen with a few (less than 10) markers. Only those seedlings passing this filter are subjected to the much more expensive screen with the several thousand genetic markers required to enable application of genome wide selection (GWS).

a. Diversity Analysis and Association Mapping

Association mapping (AM) or linkage disequilibrium (LD) mapping uses natural genome-wide distribution of various genes together with other detectable loci (markers) in predicting the marker-trait associations. The AM can be deployed in crops that are perennial and have complex breeding systems (Singh and Singh, 2015). It is being used in different perennial fruit trees and also annual crops but their use in vegetable crops is still negligible. Understanding on extent of genetic diversity is essential to develop new and improved cultivars with desirable characteristics. Saini *et al.* (2013), Ganesan *et al.* (2014) and Singh *et al.* (2017) investigated genetic diversity in drumstick germplasm from different regions in India using molecular markers.

b. Sex-Determination in Perennial Vegetables

In perennial vegetables such as asparagus and pointed gourd, knowledge of sex forms before planting is essential for higher yield. In perennial vegetables like pointed gourd and asparagus, it is difficult to distinguish male and female plants at seedling stage, hence, use of molecular markers are promising. In pointed gourd, Kumar *et al.* (2014) reported two RAPD markers, male-specific marker $OPC05_{1000}$ and female-specific marker $OPC14_{400}$, together can reliably differentiate between male and female plants of *T. dioica*. In Asparagus, Gebler *et al.* (2007) performed Bulk Segregant Analysis (BSA) to identify the bands differentiating the male (XY, YY) from the female (XX) genotypes. A Marker $OPB\text{-}20_{700}$ was found in all female plants, and in 4 phenotypically male specimens but the intensity of the bands in male plants was much lower suggesting that this fragment can be linked to the X chromosome in the investigated specimens. In the female specimens with XX karyotype, template duplication occurs and hence the band intensity is twice as high as in the XY karyotype. Three male plants did not generate OPB-20_{700} fragment so they could potentially have the super male (YY) karyotype.

D. RNA Interference in Perennial Vegetables

The discovery of RNA interference (RNAi) and its regulatory potentials has opened new vista for crop improvement (Jagtap *et al.*, 2011). RNAi technology is precise, efficient, stable and better than antisense technology and even the chances of acceptance of products is more in public. The RNA silencing is a novel gene regulatory mechanism that limits the transcript level by either suppressing transcription (TGS) or by activating a sequence-specific RNA degradation process [PTGS/RNA interference (RNAi)] (Agrawal *et al.*, 2003).This technology has been employed successfully to alter the gene expressions for improved quality traits by increasing antioxidants in tomatoes (Niggeweg *et al.*, 2004) or suppressing over expression of negative traits such as sinapate esters in canola (Husken *et al.*, 2005) and alpha-linolenic acid in soybean (Flores *et al.*, 2008). The role of RNAi technology in extending the shelf-life of tomato by blocking the expression of ACC oxidase gene (Xiong *et al.*, 2005) and suppression of two ripening specific N-glycoprotein modifying enzymes, α-mannosidase and β-D-*N*-acytlhexosaminidase (β-Hex) was established by Meli *et al.* (2010). The role of RNAi in regulation of reproductive behaviour (male sterility and fertility) is under investigations because in changing climate the traditional sterility systems are getting affected which is affecting hybrid seed production and ultimately crop productivity. Negligible attempts were made to utilize this novel technology for improvement of perennial vegetable crops; however, it can be explored for improvement of perennial vegetables for biotic and abiotic stress tolerance and elimination of anti-nutrients.

E. Transgenic Approach

Three genes, encoding phytoene synthase (*CrtB*), phytoenedesaturase (*CrtI*) and lycopene beta-cyclase (*CrtY*) from *Erwinia* have been introduced in potato to produce beta carotene. Romeret al. (2000) developed transgenic tomato to enhance the carotenoid content and profile of tomato fruit, we have produced transgenic lines containing a bacterial carotenoid gene (*crtI*) encoding the enzyme phytoene desaturase which converts phytoene into lycopene. Gerjets and Sandmann (2006) developed genetically engineered potato for the production of commercially important keto-carotenoids including astaxanthin (3, 32-dihydroxy 4, 42-diketo-β-carotene). Lu *et al.* (2006) suggested transgenic cauliflower with *Or* transgenesis associated with a cellular process that triggers the differentiation of proplastids or other non coloured plastids into chromoplasts for carotenoids accumulation and also reported that *Or* can be used as a novel genetic tool to induce carotenoid accumulation in a major staple food crops as is being done potato. The post-transcriptional gene silencing is also used for change in nutrient biosynthesis pathways which improve the nutrient contents.

F. Gene Editing Techniques

i) Zinc Finger Nuclease (ZFN) Technology (ZFN-1, ZFN-2 and ZFN-3)

ZFNs are proteins which have been custom designed to cut at specific deoxyribonucleic acid (DNA) sequences. They consist of a "zinc finger" domain (recognising specific DNA sequences in the genome of the plant) and a nuclease that cuts double-stranded DNA. The rationale for the development of ZFN technology for plant breeding is the creation of a tool that allows the introduction of site-specific mutations in the plant genome or the site-specific integration of genes. As ZFNs act as heterodimers, two genes have to be delivered to the target cells, usually in an expression plasmid, with or without a short template sequence or a stretch of DNA to be inserted. Many methods are available for delivering ZFNs into plant cells, *e.g.* transfection, electroporation, viral vectors and *Agrobacterium* mediated transfer. ZFNs can be expressed transiently from a plasmid vector. Once expressed, the ZFNs generate the targeted mutation that will be stably inherited, even after the degradation of the plasmid containing the ZFNs. Alternatively, ZFN genes can be integrated into the plant genome as transgenes. In this case the offspring of the transformed plant includes plants that still carry the transgenes for the ZFNs and so have to be selected out, in order to obtain only nontransgenic plants with the desired mutation.

ii) TALENs in Gene Modification

A new EEN, *i.e.*, transcription activator-like effector nucleases (TALENs), has rapidly emerged as an alternative to ZFNs for genome editing recently. The broad applications of TALENs were based on the recognition of the functional principles of the type III transcription activator-like (TAL) effectors that are secreted by the plant pathogenic bacteria *Xanthomonas*. After having been pumped into host cells, the TAL effectors enter the nucleus and bind to effector-specific sequences in the host gene promoters and activate transcription. Due to easier manipulation, the genes modified by TALENs have been successfully used in both animal and plant species within three years of deciphering their function. These species include zebra fish, rat, human cells, rice, wheat, Arabidopsis, and horticultural crops such as potato and tomato.

iii) CRISPR/Cas9 Technology

More recently, a new class of genome-editing technology, *i.e.*, the CRISPR (clustered regularly interspaced short palindromic repeats)/Cas9 (CRISPR-associated) system, has been developed. The principle of the CRISPR/Cas9 system was derived from a type II prokaryotic organism adaptive immune system. Since the first report in early 2013, this technology has been widely applied in gene modification in both animals and plants, such as zebra fish, mice, human cells, Arabidopsis, tobacco, rice, wheat, and sweet orange. In plants, the modified traits include changing plant architecture, *e.g.*, Miao *et al.*, 2013 changed the tiller angle by modifying the LAZY1 gene in rice, and Shan *et al.*, 2013 changed the plant colour to white by modifying the OsSPD gene in rice. The first and foremost advantage of CRISPR/Cas system is based on simple RNA/DNA hybrids that confer sequence specificity. Secondly, it can simultaneously introduce multiple gene disruptions, which allows researchers to edit multiple genes in one plant line through a single transformation without the time consuming post-transgenic hybridization and screening processes. Therefore, CRISPR/Cas9 method is considered to be the most efficient, least expensive, and most user-friendly among genome-editing technologies. Compared to conventional and GM crops, it is advantageous in four domains *i.e.* precision, speed, cost and regulation.

Problems in Breeding of Perennials

Most perennial plants are out breeders and are highly heterozygous and they do not breed true from seed. Accordingly, the subtle combinations of genes which constitute a good cultivar must be perpetuated by vegetative propagation. Most of the traits which make up a good fruit are polygenic in their inheritance and few are simple Mendelian characters. Seedlings of woody perennial plants have a protracted juvenile phase during which they do not normally form flowers. With conventional techniques, the minimum generation times are from 4-15 years depending on the species. To this must be added the time required for orchard or vineyard evaluation of advanced selections, perhaps another 5 years. Other biological constraints in perennial crop breeding include self-incompatibility, polyploidy and parthenocarpy. Having, at last, raised a progeny the perennial crop breeder is faced with the problem of selection criteria. Perennial crops tend to be bred for quality rather than yield and there are few simple objective tests of fruit quality. Finally, there are conservative forces which tend to resist the introduction of new cultivars irrespective of excellence. Certain cultivars have become firmly entrenched in the market-place and consumers have little interest in unfamiliar products. Wholesalers tend to dislike new cultivars because marketing is made more complicated.

REFERENCES

Agrawal, N., Dasaradhi, P.V.N., Mohmmed, A., Malhotra, P., Bhatnagar, R.K. and Mukherjee, S.K. (2003). RNA interference: biology, mechanism, and applications. *Microbiol. Mol. Biol. Rev.*, 67: 657-685.

Flores, T., Karpova, O., Su, X., Zheng, P., Bilyeu, K., Sleper, D. A., Nguyen, H. T., Zhang, Z. J. (2008). Silencing of the Gm*FAD3* gene by siRNA leads to low a-linolenic acids (18: 3) of *fad3*-mutant phenotype in soybean *Glycine max* (Merr.). *Transgenic Research*, 17: 839–850.

Ganesan SK, Singh R, Choudhary DR, Bharadwaj J, Gupta V, Singode A (2014) Genetic diversity and population structure study of drumstick (*Moringa oleifera* Lam.) using morphological and SSR markers. *Ind Crops Prod.*, 60: 316–325.

Gebler, P., Wolko, L. and Knaflewski M. (2007). Identification of molecular markers for selection of supermale (YY) asparagus plants. *J Appl Genet.*, 48(2): 129-31.

Gerjets, T. and Sandmann, G. (2006). Keto-carotenoid formation in transgenic potato. *J. Exp. Bot.*, 57(14): 3639-3645.

Hüsken, A., Baumert, A., Strack, D., Becker, H. C., Möllers, C. and Milkowski, C. (2005). Reduction of sinapate ester content in transgenic oilseed rape (*Brassica napus* L.) by dsRNAi-based suppression of BnSGT1 gene expression. *Molecular Breeding*, 16: 127–138

Jagtap, U.B. Gurav, R. G. and Bapat, V.A. (2011). Role of RNA interference in plant improvement. *Naturwissenschaften*, 98: 473–492

Kumar, S., Singh, B.D., Sinha, D.P. and Rai, M. (2014). Sex expression-associated RAPD markers in pointed gourd (*Trichosanthes dioica*). 1 Cucurbitaceae 2008, Proceedings of the IXth EUCARPIA meeting on genetics and breeding of Cucurbitaceae (Pitrat M, ed), INRA, Avignon (France), May 21-24th, 2008

Leus, L., Baert, J, Van Laere, K. and Van Huylenbroeck, J. 2012. 'Polyploidy in breeding: examples in horticultural and agricultural crops', Pruhonice, Czech Republic, 7/05/12 - 10/05/12.

Lu, S., Eck, J.V., Zhou, X., Lopez, A.B., Halloran, D. M., Cosman, K. M. *et al.* (2006). The cauliflower *Or* gene encodes a DnaJ cysteine-rich domain-containing protein that mediates high levels of ß-carotene accumulation. *Plant Cell.*, 18: 3594-3605.

Meli, V., Ghosh, S. Prabha, T., Chakraborty, N. Chakraborty, S. And Datta, A. (2010). Enhancement of fruit shelf life by suppressing N-glycan processing enzymes. Proceedings of National Academy of Sciences of United States of America. 107(6): 2413–2418.

Niggeweg, R., Michael, A. J. and Martin, C. (2004). Engineering plants with increased levels of the antioxidant chlorogenic acid. *Nature Biotechnology*, 22: 746–754.

Miao J, Guo D, Zhang J, Huang Q, Qin G, Zhang X, Wan J, Gu H, Qu LJ.(2013). Targeted mutagenesis in rice using CRISPR-Cas system. *Cell Res.*, 23(10): 1233-6. doi: 10.1038/cr.2013.123.

Peters, S., Imani, J., Mahler, V., Foetisch, K., Kaul, S., Paulus, K.E., Scheurer, S., Vieths, S. and Kogel, K.H. (2011) Dau c 1.01 and Dau c 1.02-silenced transgenic carrot plants show reduced allergenicity to patients with carrot allergy. *Transgenic Research*, 20, 547–556.

Romer, S., Fraser, P. D., Kiano, J. W., Shipton, C. A., Misawa, N., Schuch, W. and Bramley, P.M. (2000). Elevation of the pro-vitamin A content of transgenic tomato plants. *Nature Biotechnology*, **18**: 666–669.

Saini RK, Saad KR, Ravishankar GA, Giridhar P, Shetty NP (2013) Genetic diversity of commercially grown *Moringa oleifera* L. cultivars from India by RAPD, ISSR and cytochrome P450-based markers. *Plant Syst. Evol.*, 299: 1205–1213.

Shan, Q., Wang, Y., Li, J., Zhang, Y., Chen, K., Liang, Z., Zhang, K., Liu, J., Xi, J.J., Qiu, J. L., Gao, C.(2013). Targeted genome modification of crop plants using a CRISPR-Cas system. *Nature Biotechnol.*, 31(8): 686-8.doi: 10.1038/nbt.2650.

Singh B.D., Singh A.K. (2015). Association Mapping. In: Marker-Assisted Plant Breeding: Principles and Practices. Springer, New Delhi.

Singh, S., Kartikeyan, K., Singh, D. R., Sihmachalam, P., Bianslal, N. K. and Jaisankar, I. (2017). Genetic Diversity in Drumstick of Andaman Islands and Their Relatedness with Probable Introduction Sites from Mainland India. Proc. Natl. Acad. Sci., India, Sect. B Biol. Sci. https: / /doi.org/10.1007/s40011-017-0950-0.

Xiong, A., Yao, Q., Peng, R., Li, X., Han, P. and Fan, H. (2005). Different effects on ACC oxidase gene silencing triggered by RNA interference in transgenic tomato. *Plant Cell Rep.*, 23: 639–646.

Breeding of Perennial Horticultural Crops *Pages* **35–47**
Editors: S.K. Singh, V.B. Patel, A.K.Goswami, Jai Prakash and Chavlesh Kumar
Published by: **BIOTECH BOOKS, NEW DELHI**

Chapter **3**

Genetic Resources of Tropical Fruits

S.K. Malik and Rekha Chaudhury

ICAR-National Bureau of Plant Genetic Resources, Pusa Campus, New Delhi – 110 012

INTRODUCTION

Globally about 600 tropical and subtropical fruit species are known to occur in their areas of genetic diversity (Nagy and Shaw, 1980). Southeast Asia is represented by more than 500 species of fruits (Arora and Ramanatha Rao, 1995), while the Hindustani region of diversity represents 344 species of fruits having vast potential for new crops (Arora, 1995). Indian sub-continent hold vast genetic diversity of several tropical and sub-tropical underutilized and underexploited fruits. Many of these species are being used by local people as minor fruits or vegetables as well as for various medicinal purposes. Several of these fruits still remain neglected as these still grow in natural wild or semi-wild conditions providing livelihood and nutritional support to small and marginal farmers, and tribals women and children. Most of these fruits are directly harvested from trees or shrubs growing wild or semi-wild in forests, marginal forest lands and homestead gardens by local people. Bulk of this produce is used by the family/community itself and in case of surplus, sold in the local markets with small commercial gains to families. However, changing dietary patterns amongst the new generations, large scale urbanization, easy access to so called elite fruits and expansion of irrigation facilities has further affected use and natural sustenance of these fruit species. Tropical fruit species like Khirni (*Manilkara hexanda*) fulfill the Vitamin A requirement of tribal women and children in the tribal dominated districts of Gujarat, Madhya Pradesh, Rajasthan and Maharashtra. Amla (*Emblica officinalis*), bael (*Aegle marmelos*), chironji (*Buchanania lanzan*), *Citrus* species, karonda (*Carissa congesta*), ker (*Capparis decidua*), lasora (*Cordia myxa*), phalsa (*Grewia asiatica*), pilu (*Salvadora oleoides*), jamun (*Syzygium cuminii*), kokam (*Garcina indica*), Malabar tamarind (*G. cambogia*) and tamarind (*Tamarindus *indica*) have tremendous nutritious value and form the small but special part of diet of local inhabitants. These potential crops of the future are awaiting their popularization and full utilization as they remain only as plants of local importance

Genetic resources management of these tropical and sub-tropical fruits has not been given desired attention due to their less commercial importance and limited research on genetic improvement of cultivars. There is an urgent need to realize socio-economic value of these fruits in view of their nutritional, cultural, religious and economic values for the local communities and adaptive capabilities of plants to changing climate. Most of these tropical species are providing ecosystem sustainability being fully adapted to harsh and stressed arid and semi-arid fragile habitats, the most vulnerable to the climate change scenario. It is, therefore, advocated to encourage participation of farmers and tribal communities in the conservation effort of these fruits by supporting and providing them good planting material, training them in better cultivation practices, motivating communities to protect and use their own plant species, facilitating marketing options and empowering them with available public support systems. Such community based participation would encourage farmers to protect, grow and promote indigenous genetic diversity of these potential species in their own farmlands, backyards, community lands *etc.* for harvesting fruits and other products. Encouragement to such farmers recently defined as the "Custodian farmers" (Sthapit *et al.*, 2013), would be ensuring maintenance, adoption and protection of the valuable indigenous agrobiodiversity serving dynamic in situ conservation and supporting livelihood to farmers.

Western Ghats and North eastern region of India are rich in diversity of several underutilized fruits species. Wild relatives of several major and minor fruits are also available as wild and semi-wild in these areas and used as minor fruits. Several wild species of citrus, mango and bananas occur in the humid and evergreen forests of north eastern India, Western Ghats, eastern Ghats, and Andaman and Nicobar Islands. With the prevalence of 23 species, 1 subspecies and 68 varieties of citrus, this region has got a special status as a treasure house of Citrus germplasm (Sharma *et al.*, 2004). Citrus gene sanctuary set up at Garo hills of Meghalaya to preserve the diversity of wild endangered species (*Citrus indica*), locally known as 'ghost orange' and other Citrus species (Singh, 1981). Some native *Mangifera* species are found in Tripura, Manipur, Mizoram, South Assam and other parts of India. Wild form of *M. indica* and its allied species *M. sylvetica* occur in Arunachal Pradesh, *M. khasiana* and *M. pentandra* in Assam. A dwarf and late maturing polyembryonic mango cultivar Moresh bears sweet fruit with high pulp content within 2 years from planting and is free from stone weevil. A large number of other tropical and sub-tropical fruits belonging to the genera *Garcinia, Artocarpus, Embelica, Annona, Grewia, Averrhoa, Aegle, Passiflora*, Ziziphus *etc.*, are found growing wild in the different regions of India (Arora, 1991). In Andaman and Nicobar island, Nicobari mangosteen (*Garcinia hombroniana*) consumed by Nicobaris is considered to be a progenitor of cultivated mangosteen. A wild relative of sapota, *Manilkara hexandra* used as salt tolerant rootstock with highly nutritious, small, elongated fruits with taste of sapota is distributed throughout the plains of India including Andaman and Nicobar Island.

Genetic Resources of Perennial Horticulture

India is endowed with a rich genetic diversity of fruits. Tropical fruits constitute a major proportion of the spectrum of fruit diversity available in India. Important tropical fruits mango, banana, citrus, papaya and guava alone are grown in 4.29 m ha out of the total area of 6.32 m ha under fruit crops and account for 81 per cent of the total annual fruit production (Annonymous 2010a). These and several other tropical fruit species not only meet the needs of local and export markets for fresh fruit but also contribute substantially to the fruit processing industry. Mango, citrus, banana and guava due to the presence of vast diversity and liking flavour and taste were supported and improved by local fruit growers and horticulturists for wide adoption. The less important, so-called underutilized fruits remained confined mainly to natural wild, semi-wild and semi-domesticated conditions albeit with large ever increasing variability. Besides their importance as potential horticulture species these plants are incidentally store houses of genes for adaptation to hot and hardy climates, salt tolerance, diseases tolerance and several essential nutritional values. However, ensuring continued growth in fruit production and to have vast diversity of fruits would require continuous selection and development of desired genotypes adapted to diverse agroclimatic conditions of the country. In this regard the role of genetic resources as the base material for crop improvement is invaluable and, hence, their collection, evaluation, documentation, conservation and utilisation is of prime importance. Systematic exploration and collection of germplasm in major fruit crops from indigenous and exotic sources have resulted in widening of the genetic base and selection of suitable genotypes. Efforts for conservation of existing diversity are being made using various *in situ* and *ex situ* methods. Selection of suitable conservation strategies depends upon reproductive and breeding mechanism and physiology of suitable plant propagules. These factors determine the sample size of the propagules to be stored, and the appropriate conservation technologies to be applied. Accordingly, different conservation strategies have been suggested and utilized by conservation biologists for achieving successful conservation of targeted species.

Conservation of horticulture genetic resources (HGR) and specifically the underutilized fruit species and wild species of major fruits which are still grown as natural wild and in semi-wild conditions would require adoption of complementary conservation strategies where, suitable *in situ* and *ex situ* conservation methods are to be employed to achieve the successful conservation goals. Within fruit crops, specific conservation strategy is to be developed and adopted based on extent of genetic diversity available, mechanism of propagation, reproductive biology and present biological status of the species. *In situ* conservation involves promoting growth of plant species in their natural habitats where evolutionary processes continue to operate, making it a dynamic system. Majority of the lesser-known tropical fruit species grow in the diverse climatic and edaphic conditions and are adapted to arid and semi-arid climates. Indian National PGR System advocates for the complementary conservation strategy using one than more methods to achieve comprehensive conservation of species. Ministry of Environment and Forest is mandated with *in situ* conservation while *ex situ* conservation activities are looked after by NBPGR. Horticulture genetic resources of fruits comprise candidates for conservation based on their biological status and propagation method (Figure 3.1).

Mango

Mango (*Mangifera indica* L.) is the most important fruit crop of India. It occupies the largest area (2.3 m ha) among fruit crops and production-wise, mango ranks second (15.0 mt) with nearly 21 per cent contribution to the total fruit production of the country (Indian Horticulture Data Base 2010). The nearly 1000 varieties of mango cultivated in India provide an unusual diversity of flavours and tastes, managing them is a challenge for farmers and PGR managers.

The origin of mango has been suggested to be Eastern India to Myanmar and possibly further in the Malay region (Popenoe 1920, Vavilov 1951). *M. indica* having originated in Assam-Myanmar (Assam-Burma) region Mukherjee (1951). To date, available taxonomic and molecular evidence suggests that mango probably evolved within a large area comprising north-western Myanmar, Bangladesh and north-eastern India (Mukherjee 1997).

Indian peninsula is the most important centre of diversity for *M. indica* having a large number of cultivated varieties as well as wild types. Truly wild mango trees have been found in north-eastern India, sub-Himalayan tracts, deep gorges of the Bahraich and Gonda hills in Uttar Pradesh and in outer hills in Kumaon and Garhwal in Uttaranchal (Brandis 1874, Kanjilal and Kanjilal 1937). Semi-wild trees are present in the forests of almost all parts of the Indian subcontinent. Other *Mangifera* species, namely *M. andamanica* King, *M. khasiana* Pierre, *M. sylvatica* Roxb. and *M. camptosperma* Pierre have also been reported from India (Mukherjee, 1985; Yadav and Rajan 1993). *M. sylvatica* occurs up to 1300 masl in the eastern Himalayas in Sikkim, Darjeeling district in west Bengal, Khasi Hills in Meghalaya, Upper Assam and Surma valley (Assam) and in the Andaman Islands (Mukherjee 1949, Agharkar and Roy 1951). *M. khasiana* is an endemic but insufficiently known species from Khasi hills and is distinguished from *M. sylvatica* in having smaller and narrower leaves, smaller flowers, inflorescence with fasciculate divergent branching, petals ovate lanceolate and disc glandular or with distinct lobes (Mukherjee 1985). *M. andamanica* is widespread in tropical, wet evergreen forests of Andaman Islands.

In India there are nearly 1000 varieties of mango are grown, however, only 20 varieties are commercially grown due to their specific ecogeographical requirements for optimum performance (www.mangifera.org). Mango being a cross-pollinated tree is highly heterozygous and naturally established seedlings have been the most important source of its diversity in India. Almost all the commercial cultivars of mango have arisen as a result of selection from these seedlings

Banana

Banana (*Musa* spp.), including plantain, is the third most important fruit crop of India in terms of the area under cultivation (0.77 m ha). However, with respect to production, it tops the list with an annual production of 26.47 mt (Anonymous, 2010a). The term banana is given to varieties that are used for fruit and eaten raw whereas plantain is used for those that are edible only when cooked. Besides its use as fruit and vegetable, banana has several other uses because of which it is often termed as 'Kalpatharu', plant with all imaginable uses. The fruit is rich in phosphorus, calcium and potassium and does not contain any fat. The stem is taken as a medicine for eliminating stones in kidney and bladder. Further, banana plant has an important role in many social and religious ceremonies (Amalraj *et al.*, 1993). Reference to plantain in ancient Indian literature like Valmiki's Ramayana (2029 BC), Kautilya's Arthasastra (250 to 300 BC) and the Tamil Classic Silappadikaram (500 to 600 AD) suggests very early domestication of the plant in India (Krishnamurthy and Seshadri 1958).The wild bananas are distributed in South-east Asia and the Pacific (Simmonds 1962). The cultivated bananas originated from *M. acuminata* Colla (2n = 22, *AA* genome) and *M. balbisiana* Colla (2n = 22, *BB* genome). The wild types are seed bearing with poor edibility; *Musa acuminata* was the first species to develop edible traits. It is probable that human selection favoured parthenocarpy, structural heterozygosity and seed sterility, leading to the production of edible seedless fruits (Simmonds 1994). Most cultivated bananas are triploid comprising three major groups, *AAA, AAB* and *ABB*. Diploids *AA, AB* and tetraploids *AAAB, AABB, ABBB* are less abundant.

India possesses rich diversity in wild *Musa* spp. in the north-eastern region while variability of cultivars, especially belonging to the genomic groups *AA, AB, AAB* and *ABB* exists in the southern region. Maximum genetic variability of *M. acuminata* and *M. balbisiana* occur in north-east India. *Musa flaviflora* Simmonds is localised to Manipur and Meghalaya. Several wild species have also been recorded from southern and western regions of India. These include *M. acuminata* subspecies *burmannica, M. balbisiana, M. nagensium* Prain, *M. sikkimensis* Kurz, *M. cheemanii* Simmonds, *M. ornata, M. laterita, M. velutina* Wendl. Ex Drude and *M. sanguinea*. Banana cultivation is predominantly practiced in Tamil Nadu, Karnataka, Kerala, Andhra Pradesh, Maharasthra, Bihar, West Bengal and north-eastern states. The variability in cultivars decreases as one proceeds from extreme south. It is also interesting to note that all the Indian *AA* and *AB* cultivars occur only in the extreme south. The most common commercial varieties grown in different states are; Poovan (*AAB*) in Bihar; Pachabale (*AA*) in Karnataka; Nendran (*AAB*) in Kerala; Kapuravalli (*ABB*), Monthan (*AAB*), Morris (*AAA*), Mysore Poovan (*AAB*), Nendran (*AAB*), Pachanadan (*AAB*), Rasthali (*AAB*) and Robusta (*AAA*) in Tamil Nadu; and Dwarf Cavendish (*AAA*), Poovan (*AAB*), and Rasthali/Morthaman (*AAB*) in West Bengal.

Citrus

Citrus is the third most important fruit crop of India with an estimated production of 9.63 mt from an area of 0.98 m ha (Anonymous 2010a). Mandrin (*Citrus reticulata* Blanco), sweet orange (*C. sinensis* Osbeck), acid lime (*C. aurantifolia* Swing) and lime (*C. limon* Burn. f.) are the major cultivated species of the country. Other species that are cultivated to a lesser extent include, seedless lime (*C. latifolia* Tan.), pummelo (*C. grandis* Osbeck), grapefruit (*C. paradisi* Macf.) and belladikithuli (*C. maderaspatana* Tan.). Citrus fruits have been put to numerous uses though primarily they are eaten fresh or prepared into juice concentrate. Pulp and seed are used for cattle feed and molasses as also for flavouring and for the production of pharmaceuticals, soaps and perfumes. Fermented orange juice produces vinegar and alcohol. Some species are known to have properties of curing fever and colic (Solley 1997). Quite a few species are grown for ornamental fruits and flowers.

South-east Asia, Australia, the intervening islands between Asia and Australia and central Africa are recognised as important centres of origin of *Citrus* and related genera (Swingle and Reece 1967). *Citrus reticulata, C. sinensis* and *C. aurantifolia*, the commercially important species grown in India, have originated in India and China (McPhee 1967). In India, north-eastern Himalayan region and foothills of the central and western Himalayan tracts are rich sources of *Citrus* diversity. As many as 17 *Citrus* species, their 52 cultivars and 7 probable natural hybrids are reported to have originated in the north-eastern region (Bhattacharya and Dutta 1956). A number of commercial varieties have also been developed from native germplasm through selection and crop improvement.

Jackfruit

Jackfruit (*Artocarpus heterophyllus* Lamk.) is a tropical tree indigenous to South and South-east Asia. Wild forms of jackfruit are commonly found in India, Thailand, the Philippines, Myanmar, Indonesia, Malaysia and to a considerable extent in Brazil. The prominent and economically important species of genus *Artocarpus* include *A. heterophyllus* and *A. altilis* (Parkinsons) Forsberg (breadfruit). Jackfruit is a tetraploid with 2n = 56.

A. heterophyllus is reported to have originated in Western Ghats of India and later got distributed to South-east Asia and other tropics; and has high degree of variability in South-east Asia. Being protandrous and mainly cross-pollinated, the plants show great variability in many characters including texture, odour and flavour of the pulp.

Garciania Species

Genus *Garcinia* L. belonging to family Clusiaceae has approximately 400 species having dioecious, evergreen trees growing in tropical parts of the world (Maheshwari, 1964; Whitmore, 1973, Richards, 1990a). Presence of agamospermy (seed apomixis) in the genus *Garcinia* is known or suspected in at least ten species which have been further classified as facultative and obligate agamosperms (Richards, 1990a). Approximately 30 species of *Garcinia* are cultivated and produce edible fruits (Arora, 1998). In India 36 species of *Garcinia* are reported. Malaysia and Africa with large number of endemic species appear to be the two main centres of development of the genus *Garcinia.* In India eleven species occur in the southern Western Ghats, out of which 6 species are endemic while sixteen species occur in the north-eastern India, out of which 2 are endemic and fifteen species are found in Andaman and Nicobar Islands out of which 7 are endemic. Overall 11 species of *Garcinia* namely *G. cadelliana, G. calycina, G. brevirostris, G. cambogia var. conicarpa, G. cambogia var. gummi-gutta, G. cambogia var. papilla, G. imbertii, G. indica, G. keenainia, G. kingii, G. talbotii, G. spicata* and *G. wightii* are facing survival threat due to overexploitation for wood, medicine, gum, resin, edible fruits and seed (Kundu, 2006). The increase in the level of endemism from 50 per cent to 65 per cent is an important indication of the shrinking population of these species posing challenge for conservation biologists. Therefore, the management of genetic resources of Indian species of *Garcinia* is urgently required.

Underutilized Fruits

Tropical underutilized fruits have vast genetic wealth in the Indian sub-continent, however, desired attention for their genetic resource management is lacking due to their comparatively less commercial importance and limited research on genetic improvement of cultivars. In India several fruits species have local importance and considered as important underutilized fruits namely *Aegle marmelos* (Bael), *Buchanania lanzan* (Chironji), *Capparis decidua* (Ker), *Carissa* species (Karonda), *Cordia* species (Lasora), *Diospyros melanoxylon* (Tendu), *Emblica officinalis* (Aonla), *Feronia limonia* (Wood apple), *Ficus palmata* (Wild Fig), *Garcinia* species (Kokam, Malabar Tamarind and Mysore Gamboge), *Grewia* species (Phalsa), *Madhuca indica* (Mahua), *Manilkara hexandra* (Khirni), *Phoenix sylvestris* (Wild date), *Pithecellobium dulce* (Manila Tamarind), *Salvadora* species (Pilu and Meswak), *Syzygium cumini* (Jamun), *Tamarindus indica* (Tamarind) and *Ziziphus* species (Ber). Most of these fruits have been originated in India and still found as wild or semi wild in limited area having local socio-economic importance.

Exploration and Collection

Indian is rich is diversity of underutilized fruits, large number of these fruit species are distributed in various phyto-geographical regions of India. Exploration and collection of underutilized fruits crops has not received much attention when compared with the other agriculture crops. ICAR-National Bureau of Plant Genetic Resources (NBPGR), the nodal national organization responsible for plant genetic resources management in India, has been undertaking explorations and collections in consultation and collaboration with scientists of relevant horticultural institutes of ICAR and SAUs for a holistic approach for PGR collection and utilization. The aim is to ultimately facilitate the PGR utilization by the collaborators and other stakeholders. Various horticultural institutions have collected the germplasm of underutilized fruits species under the All India Coordinated Project on Tropical Fruits with 22 centres covering 11 SAUs and 2 ICAR Institutes and also under the All India Coordinated Project on Arid Fruits with 13 centres of which 10 are located in SAU (s) and 3 are located in ICAR institutes as per their mandated crops. Overall more than 3000 accessions of underutilized fruits have been collected by various institutions in India and national identity (IC Numbers) have been assigned by the NBPGR since 1976. Important fruit crops where substantial diversity is collected are *Aegle marmelos*, *Anona* species, *Buchanania lanzan*, *Capparis decidua*, *Carissa* species, *Cordia* species, *Citrus* species *Diospyros melanoxylon*, *Emblica officinalis*, *Garcinia* species, *Grewia* species, *Madhuca indica*, *Manilkara hexandra*, *Pithecellobium dulce*, *Salvadora* species, *Syzygium cumini*, *Tamarindus indica* and *Ziziphus* species. The field genebanks at various ICAR institutions and SAU's are maintaining the collected germplasm and national identity is being assigned by these to have fairly good estimate of overall National Collection. However, there is an urgent need to have assessment of existing germplasm diversity in Indian germplasm of these underutilized fruits species for that collective efforts of various stake holders are required.

At NBPGR systematic exploration programmes to collect the germplasm of important underutilized fruits has been undertaken since last fifteen years. Specific exploration and collection programmes have been undertaken in the north, west, central and south Indian states of Rajasthan, Haryana, Gujarat, Uttar Pradesh, Madhya Pradesh, Himachal Pradesh (foot hills of Himalayas), Kerala, Karnataka and Maharashtra for the collection of several underutilized tropical and sub-tropical fruits (Figure 3.1).

The targeted fruit species were collected from wild and semi-wild areas of forest, marginal forest lands, homestead gardens, orchards and from farmer's fields. Details of fruiting period, importance and uses of these fruits species has been provided under the individual fruits. During the explorations general field

Figure 3.1: Distribution of Fruits in India (Modified from Arora, 1991).

observations, status of natural populations, probable threats to species, breeding behavior and associated indigenous technical knowledge on use was gathered. Fruits from selected plants were collected and after initial characterization of fruits the elite trees were identified based on physico-chemical characters. Vegetative cuttings or bud wood was also collected in some of these fruits for establishment in the field genebanks of concerned institute. Area of collection of these fruit species is depicted in the map (Figure 3.1).

Characterization and Evaluation

Characterization and evaluation of tropical underutilized fruits has not been undertaken in a systematic way till now except for some important fruits namely aonla, bael, ber and jamun. Hence, selection of elite genotypes or cultivars in most of these fruits is still lacking. In comparison to high value major fruit species less attention has been given to characterization and evaluation of these fruits due to the inaccessibility of these fruits species germplasm only being confined to the natural wild and semi-wild conditions. Recently, efforts have been made to undertake systematic collection, initial characterization and establishment of superior genotypes in the field genebanks for detailed evaluation. Due to the perennial nature and specific climatic and edaphic needs of these fruits it takes several years for flowering and fruiting in field genebanks causing delay in the characterization and evaluation of established germplasm. In some important underutilized fruit species vegetative propagation methods has been developed and field establishment of identified genotypes by using the scion from the original sources on local rootstocks has been undertaken. Characterization and evaluation of some collected germplasm and identified cultivars or released varieties of important underutilized fruits namely aonla, ber, bael, jamun, phalsa, khirni, karonda, mahua, Chironji and tamarind has been undertaken at various locations (Pathak and Pathak, 1993; Singh *et al.,* 1999; Singh and Singh, 2012; Malik *et al.,* 2010; Malik *et al.,* 2012a,b). Field performance and multi-location trials of identified cultivars or selections of several arid fruits is extensively being undertaken at various centers of AICRP on Arid Fruits and performance is being reported regularly for recommendation for suitable cultivars of particular region or state (Pareek and Nath, 1996; CISH website; CIAH website; Malik *et al.,* 2010). The concerted efforts in characterization and evaluation of germplasm led to release of some varieties in these fruits by CIAH and its CHES, Godhra (Aonla- Goma Aishwariya; Bael-Goma Yashi; Ber- Thar Sevika, Thar Bhubharaj, Goma Kirti; Jamun- Goma Priyanka; Tamarin-Goma Prateek), CISH (Jamun-CISH J-42, CISH-J-37), Dr. BS Konkan Krishi Vidyapeeth, Dapoli (jamun-Konkan Bahadoli; Karonda Konkan Bold; Kokum- Kankan Amruta, Konkan Hatis), Dr. PD Krishi Vidhyapeeth, Akola (Tamarind Akola Smruti-10) and MP Krishi Vidyapeeth, Rahuri (Aonla –Krishna, Kanchana, NA-7 and 10), NDAUT, Faizabad (Aonla-N-5, N-6, N-7, N-10; Bael- N-Bael-5, N-Bael-16, N-Bael-17; Ber selection-1, Ber selection-2, Karonda – N Karonda-1) and by several other institutions.

Molecular characterization studies in these fruit species have been rarely attempted, which may be of use for diversity analysis and identification of molecular markers for germplasm utilization. Few studies using the Random Amplified Polymorphic DNA (RAPD), the ISSR (Inter Simple Sequence Repeats) and Simple

Sequence Repeats (SSR) molecular markers have been undertaken in Jamun (Anonymous 2010; Shakya *et al.*, 2010; Khan *et al.*, 2011) and in *Manilkara hexandra* using RAPD markers (Malik *et al.*, 2013c) have been undertaken. There is an urgent need to take up morphological and molecular characterization of underutilized fruits with vast diversity for further utilization of germplasm. These underutilized fruits are traditionally known as highly nutritious and valued for medicinal purposes, which needs further authentication through phytochemical evaluation using modern tools and techniques to assess the real worth of available genetic diversity which can be exploited for selection and genetic improvement.

Challenges and Future Plans

Genetic resources of tropical fruits have not been given desired attention due to their less commercial importance and limited research on genetic improvement of cultivars. Several other factors such as change in climatic conditions, introduction of new irrigation methods, large scale developmental activities taking place in developing economies like India have led to further loss of genetic diversity of these species which are well adapted to the stressed arid and semi-arid ecosystems. Opening of the world markets and development of new biotechnological methods of genetic modification in commercial crops have further kept the attention away from these traditional underutilized crops. Farmer's preference for high valued crops, difficulties in management of trees, harvest of fruits and non availability of quality planting material are other compelling reasons for loss of genetic diversity of these local fruits. Therefore, management of genetic resources of these crops is facing serious challenge.

There is a need to have national plan on survey, exploration and collection of germplasm of these fruit species from diversity-rich areas mentioned above and Protected Areas in collaboration with Ministry of Environment, Forest and Climate Change. In view of the inevitable climate change scenario genetic diversity and habitats of these fruits species are threatened and this phenomenon is highlighted in several publications on climate change emphasizing high vulnerability of stressed ecosystem especially arid and semi-arid regions (Bates *et al.*, 2008). It has been further highlighted that rare and narrow ranged species such as wild and semi-wild species of tropical underutilized fruits would be highly vulnerable to such climate change and habitat fragmentation due to anthropogenic pressure. It is, therefore, suggested to envisage crop specific germplasm management plans starting with survey to demarcate species richness, collection of available genetic diversity its maintenance, characterization and evaluation for selection of elite genotypes. Conservation of germplasm using complimentary conservation methods of in situ, in situ-on-farm and ex situ conservation is desirable for achieving inclusive results. Ex situ conservation through regional germplasm repositories, field genebanks, in vitro conservation and cryogenebanking, based on species, priority and utility of the accessions may be taken up. Species based protected areas for *in situ* conservation are to be finalized based on diversity maps and policy of respective state government for on-farm conservation of local natural selections/cultivars/ farmer's varieties. Complementarities of conservation strategies using both *in situ* and *ex situ* approaches may provide a better/effective substitute for conservation of these genetic resources. For benefiting the farmers involved in genetic resources conservation, provisions under the Protection of Plant Varieties and Farmer's Right Act, 2001 of India and other policy interventions need optimum realization.

REFERENCES

Anonymous 2001. The Protection of Plant Varieties and Farmers' Rights Act, 2001 (No. 53 of 2001). Universal Law Publishing Co. Pvt. Ltd., New Delhi-India.

Anonymous 2004. The Biological Diversity Act, 2002 and Biological Diversity Rules, 2004, National Biodiversity Authority (2004). Frontline Offset Printers, Triplicane, Chennai-India, p.57.

Anonymous 2010. National Network Project on Underutilized Fruits: Final Report. Central Institute for Subtropical Horticulture, Lucknow. p.217.

Arora, R.K. 1991. Plant diversity in the Indian gene centre. In: Plant Genetic Resources-Conservation and Management: Concepts and Approaches (Paroda, R.S. and Arora, R.K. eds.), International Board of Plant Genetic Resources, New Delhi. p.25-54.

Arora,R.K. 1995. Promoting conservation and use of tropical fruit species in Asia. In Proc. *Expert Consultation on Tropical Fruit Species of Asia*, MARDI, Kuala Lumpur, Malaysia, 17-19 May 1994 (Arora, R.K. and Ramanatha Rao, V. eds.) IPGRI Office for South Asia, New Delhi, India. p.19-30.

Arora, R.K. 1998. Genetic resources of native tropical fruits in Asia: diversity, distribution and IPGRI's emphasis on their conservation and use. In: *Tropical Fruits in Asia: Diversity, Maintenance, Conservation and Use* (Arora, R.K. and Ramanatha Rao, V. eds.). Proceedings of the IPGRI-ICAR-UTFANET Regional Training Course on the Conservation and Use of Germplasm of Tropical Fruits in Asia held at Indian Institute of Horticultural Research, 18-31 May 1997, Bangalore, India. p.42-53.

Arora, R.K. and Ramanatha Rao, V., editors 1995. Proceedings of *Expert Consultation on Tropical Fruit Species of Asia*, held at Malaysian Agricultural Research and Development Institute, Serdang, Kuala Lumpur, Malaysia, 17-19 May, 1994.

Bates, B.C., Kundzewicz, W.Z., Wu, S. and Palutikof, eds. Climate change and water. Technical paper of the Intergovernmental Panel on Climate Change, IPCC Secretariat, Geneva, 2008, p.210.

Chaudhury, R., Malik, S.K. and Rajan, S. 2010. An improved pollen collection and cryopreservation method for highly recalcitrant tropical fruits species of mango (*Mangifera indica* L.) and litchi (*Litchi chinensis* Sonn.). *CryoLetters* 31: 268-278.

Chaudhury, R. and Malik, S.K. 2014. Implementing cryotechniques for plant germplasm storing seeds, embryonic axes, pollen and dormant buds. *Acta Hort.* (ISHS) 1039: 273-280.

Khan S., Vaishali and Sharma, V. 2011. Inter and intra-population variability of *syzygium cumini* using RAPD markers. *Progressive Agriculture* 11: 17-22.

Mahajan, R.K., Gangopadhyay, K.K., Kumar, G., Dobhal,V.K., Srivastav, Gupta, P.N. and Pareek, S.K. 2002. In: Minimal descriptors of Agri-horticultural crops, part III: Fruit Crops. National Bureau of Plant Genetic Resources, Pusa Campus, New Delhi.

Malik, S.K., Chaudhury, R. and Kalia, R.K. 2003.Seed storage behaviour and cryopreservation of tropical fruit species. p.175-190. In: Chaudhury, R., Pandey, R., Malik, S.K. and Bhag, Mal (eds.), In Vitro Conservation and Cryopreservation of Tropical Fruit Species. IPGRI Office for South Asia, New Delhi, India/ NBPGR, New Delhi, India.

Malik, S.K., Chaudhury, R., Dhariwal, O.P. and Kalia, R.K. 2006a. Collection and characterization of *Citrus indica* Tanaka and *C. macroptera* Montr.: wild endangered species of northeastern India. Genet. Resour. Crop Evol. 53: 1485-1493.

Malik, S.K. and Chaudhury, R. 2006b. The cryopreservation of embryonic axes of two wild and endangered Citrus species. Plant Genetic Resources 4: 1-7.

Malik, S.K., Chaudhury, R., Dhariwal, O.P., Bhandari, D.C.2010. Genetic resources of tropical underutilized fruits in India. NBPGR, New Delhi, 2010, p.178.

Malik, S.K., Chaudhury, R., Kalia, R.K. and Dullo, E. 2011.Seed storage characteristics and cryopreservation of Genetic Resources of Tropical Underutilized Fruits in India. *Acta Horticulturae*, 918: 189-198.

Malik, S.K., Choudhary, R., Kumar, S., Dhariwal, O.P., Deswal, R.P.S. and Chaudhury, R. 2012a. Socio-Economic and Horticultural Potential of Khirni [*Manilkara hexandra* (Roxb.) Dubard]: A Promising Under-Utilized Fruit Species of India. *Genetic Resources and Crop Evolution* (DOI 10.1007/s10722-012-9863-1).

Malik, S.K., Chaudhury, R., Panwar, N.S., Dhariwal, O.P., Choudhary, R. and Kumar, S. 2012b. Genetic resources of Chironji (*Buchanania lanzan* Spreng.): a socio-economically important tree species of central India. *Genetic Resources and Crop Evolution*. DOI 10.1007/s10722-012-9801-2

Malik, S.K., Kumar, S., Singh, I.P., Dhariwal, O.P. and Chaudhury, R. 2013a. Socio-economic importance, domestication trends and *in situ* conservation of wild *Citrus* species of Northeast India. Genetic Resources and Crop Evolution: DOI 10.1007/s10722-012-9948-x.

Malik, S.K., Bhandari, D.C., Kumar, S. and Dhariwal, O.P. 2013b. Conservation of multipurpose tree species to ensure ecosystem sustainability and farmers livelihood in Indian Arid Zone. In: S Nautiyal *et al*. (eds.) *Knowledge Systems of Societies for Adaptation and Mitigation of Impacts of Climate Change, Environmental Science and Engineering*, DOI: 10.1007/978-3-642-36143-2_16, © Springer-Verlag Berlin Heidelberg, p. 257-270.

Malik, S.K., Kumar, S., Choudhary, R., Kole, P.R., Chaudhury,R. and Bhat, K.V. 2013c. Genetic diversity in Khirni [*Manilkara hexandra* (Roxb.) Dubard], an important underutilized fruit species of India using Random Amplified Polymorphic DNA (RAPD) markers. *Indian Journal of Horticulture* : 70 (1), p.18-25.

Nagy, S. and Shaw, P.E. 1980. Tropical and Subtropical Fruits: Composition, Properties and Uses. AVI Publication Inc., Westport, Connecticut, USA.

Normah, M.N., Malik, S.K., Chaudhury, R., Salma, I. and Makeen, M.A. 2013. Conservation of Tropical Fruit Genetic Resources. In: Normah, M.N., Chin, H.F. and Reed, B.M. (eds.), Conservation of Tropical Plant Species. Springer, New York. p.137-170.

Pareek, O.P. and Nath, V. 1996. Coordinated fruit research in Indian Arid Zone-A two Decade profile. NRCAH, Bikaner: 1-104.

Pathak, R.K. and Pathak, R.A. 1993. Improvement of Minor Fruits. In: Advances in Horticulture Vol. I – Fruit Crops Part I 1993: (Chadha, K.L. and Pareekh, O.P. eds.) MPH, New Delhi. p.407-421

Rao, V.R. 1998. Complementary Conservation Strategy. In: *Tropical Fruits in Asia: Diversity, Maintenance, Conservation and Use* (Arora,R.K. and Ramanatha Rao, V. eds.). Proceedings of the IPGRI-ICAR-UTFANET Regional Training Course on the Conservation and Use of Germplasm of Tropical Fruits in Asia held at Indian Institute of Horticultural Research, 18-31 May 1997, Bangalore, India. p.142-151

Rao, R. and Sthapit, B. 2013.Conservation of tropical plant genetic resources: In situ approach. In: Normah, M.N., Chin, H.F. and Reed, B.M. (eds.), Conservation of Tropical Plant Species. Springer, New York. p.3-26.

Shakya, R., Siddiqui, S.A., Srivatawa, N. and Bajpai, A. 2010. Molecular characterization of Jamun (*Syzygium cumini* L. Skeels) genentic resources. *Int. J. Fruit Sci.*, 10: 29-39.

Sharma, B.D., Hore, D.K., Gupta, S.G. 2004. Genetic resources of *Citrus* of north-eastern India and their potential use. Genet Resour Crop Evol 51: 411–418

Singh, B. 1981. Establishment of First Gene Sanctuary for Citrus in Garo Hills. Concept Publishing Company, New Delhi, India.

Singh, I.S., Srivastava, A.K. and Singh, V. 1999. Improvement of some under-utilized fruits through selection. *Journal of Applied Horticulture.* **1(1)**: 34-37.

Singh, S. and Singh, A.K. 2005. Genetic variability in jamun (Syzygium cuminii Skeels) in Gujarat. *Progressive Horticulture.* 37: 44-48

Singh, S. and Singh, A.K. 2012. Studies on variability in jamun (*Syzygium cuminii* Skeels) from Gujarat. *Asian J. Horti.* 7: 186-189.

Sthapit, B., Lamers, H. and Rao, R.2013. Custodian Farmers of Agricultural Biodiversity: Selected Profiles from South and South East Asia. Proc. Workshop on Custodian Farmers of Agricultural Biodiversity, 11-12 Feb 2013. Bioversity International, New Delhi, India.

Breeding of Perennial Horticultural Crops *Pages* **49–59**
Editors: S.K. Singh, V.B. Patel, A.K.Goswami, Jai Prakash and Chavlesh Kumar
Published by: BIOTECH BOOKS, NEW DELHI

Chapter **4**

Genetic Resources of Arid and Semi-Arid Fruits

O.P. Awasthi and Reena Prusty

Division of Fruits and Horticultural Technology,
ICAR-Indian Agricultural Research Institute, Pusa, New Delhi – 110 012

INTRODUCTION

Arid lands the *"dry domains"* of India occupy nearly 12 per cent of the total area of the country located in the states of Rajasthan (62 per cent), Gujarat (20 per cent), Andhra Pradesh (7 per cent), Punjab (5 per cent), Haryana (4 per cent), Karnatak (3 per cent) and Maharashtra (0.4 per cent). Production system in these region faces several biophysical constraints (high solar radiation, low and erratic rainfall, high wind velocity, poor soil conditions *etc.*). Availability of irrigation water and irrigation potential in these regions are meager. Despite the several production constraints, the nutritive horticultural crop, particularly the underutilised perennial fruit trees such as jujube or ber (*Ziziphus mauritiana*), wild jujube or jharber (*Ziziphus numularia*), boradi (*Ziziphus rotundifolia*), Date palm (*Phoenix dactylifera*), Bael (*Aegle marmelos*), ker (*Capparis deciduas*), lasora or goonda (*Cordia dichotoma*), goondi (*Cordia gharaf*), khejri (*Prosopis cineraria*), kumut (*Acacia Senegal*), pilu (*Salvadora* spp.), Karonda (*Carissa congesta*), Phalsa (*Grewia subinaequalis*) and Cactus pear (*Opuntia ficus indica*) form an important component of the flora of arid regions and posses fairly good diversity in several pockets of the country. These fruit tress posses xeric characters, high 'bound water' in the tissues, reduced leaf area, deep tap root system and Complete maximal vegetative growth and reproductive phase during the period of maximum water availability. The genetic resources of fruit trees, their diversity distribution, status of the germplasm collection and their utilization in improvement are discussed here.

Annonas (*Annona* sp.)

The annonas primarily used as fruit are *Annona squamosa* L. (sugar apple or sitaphala), *A. cherimola* Mill. (cherimoya or lakshmanphala), *A. muricata* L. (soursop or mamphala or munolla) and *A. reticulata* L. (custard apple or ramphala) besides the hybrid Atemoyas (hanumanphala). The other related species are *A. glabra* L. (pond apple), *A. purpurea* Moc.*and Sessé* (soncoya), *A. scleroderma* Saff (posh té), *A. crassifolia* Mart. (araticum), *A. senegalensis* Pers. (wild soursop), *A. montana* Macfad. (mountain sop), *A. diversifolia* Safford (ilama) and *Rollinia* spp. (such as *R. mucosa* Baill. or biriba or wild sugar apple) and *Asimina* spp. (such as *A. triloba* Dun. or northern pawpaw).

Various species can be used as rootstocks to which desirable species can be grafted. Seedling rootstocks of annonas are generally derived from heterogeneous open-pollinated plants. *A. glabra* has dwarf characteristics and good adaptation to damp areas. *A. reticulata* can withstand diverse ecological conditions and survive long dry periods. It is very useful as a vigorous rootstock. It is well adapted to unfavourable soil conditions and can grow in soils having pH 5.0 to 8.0. Because of its high tolerance to variable soil types, it is considered to be a good rootstock for cherimoya and soursop *A. diversifolia* also has similar ecological amplitude, but has been less widely used.

The ideal characteristics of annona ideotype are:

- ☆ Vigorous and prolific plants, regular bearing, resistant to cold and dry conditions, as well as to pests and diseases;
- ☆ Architecture of the canopy with acceptable form, which does not need pruning, and which is easy to harvest;
- ☆ Abundant flowers with fertile pollen, and which attract insect pollinators;
- ☆ Out-of-season fruit harvest for specific locations;
- ☆ Fruit with symmetrical form, high natural fruit set, hard skin resistant to pests and diseases, as well as with long post-harvest life;
- ☆ Excellent fruit quality with regard to flavour, with fine, fibreless and firm pulp texture and a low number of free seeds in the pulp.

Each species will have a slightly different ideal type due to their inherent biological differences. For cherimoyas, early maturity, better fruit appearance and long post-harvest life for the tropics and greater cold tolerance in the subtropics are the main objectives (Nakasone and Paull, 1998).

Aonla (*Emblica officinalis* Gaertn)

Aonla or emblic or Indian gooseberry (*Phyllanthus emblica* L. syn. *Emblica officinalis* Gaertn., *Averrhoa acida L.*) is indigenous to India and therefore its variability is widespread (Pareek and Sharma, 2009). Aonla germplasm collections have been made at Faizabad (118 accessions), Sardarkrushinagar (20 acces-sions), and Aruppukottai (14 accessions). A germplasm reserve of 279 cultivars has been conserved at various centres in India. Seedling selections have been done from the natural variability in different regions of the country, mainly in Uttar Pradesh and Gujarat some which are commercially cultivated.

The types that were commonly grown earlier are Banarsi, having vigorous and spreading growth habit, Chakaiya, having more upright growth habit and Hathijhool (Francis) trees having drooping growth habit but suffer from the malady of fruit necrosis. Cultivar Chakaiya is observed to be free from this malady.

The recent promising seedling selections made in Uttar Pradesh are Krishna, Kanchan, NA 6, NA 7 (Neelum), NA 8, NA 9, Balwant (Agra Bold), *etc.* Selections Kanchan and NA 6 are believed to be variants of Banarsi. NA 7 is a precocious and high yielding selection made from the seedling population of Francis but is free from necrosis. Selections made in Gujarat are Anand 1, Anand 2 and Anand 3. The selection Goma Aishwariya is early and drought tolerant.

A very close relative of *P. emblica* found in Deccan Plateau and certain drier forest pockets of southern India is *P. indofischeri*, presumably different from *P. fischeri* Pax., which already existed in Africa. The fruits of *P. indofischeri* have been observed to be less vulnerable to fungal spots than those of *P emblica*. Thus this species could be usefully exploited. The other related species is *P. acidus* (L.) Skeels syn. *P. distichus* Muell. Arg., *Cicca acida* Merr., *C. disticha* (L. (Otaheite gooseberry or Malay gooseberry or star gooseberry) ḥaving smaller trees with rather spreading crown.

Bael (*Aegle marmelos* (L.) Correa)

The bael or Bengal quince (*Aegle marmelos* (L.) Correa) is indigenous to India and considerable variability exists in Uttar Pradesh, Bihar and West Bengal States where as a result of seedling selection, cultivars such as Mirzapuri, Darogaji, Ojha, Rampuri, Gorakhpur, Azamati, Khamaria, Varanasi, Kaghji Gonda, Kaghji Etawah and Deoria Large were developed. Genetic resources of bael have been maintained at NDUAT, Faizabad (23 accessions), CAZRI, Jodhpur (5 accessions).

Some superior selections such as NB4, NB5, NB7 and NB9, which produce good quality fruits of medium size, having low fibre and mucilage, have been made. Three promising lines, Basti Collection-1, Basti Collection-2 and Basti Collection-4, superior in physico-chemical characters, have also been identified. Few promising cultivars such as Pant Aparna, Pant Sujata, Pant Shivani and Pant Urvashi have been developed from Pantnagar. Recently new cultivar Goma Yashi have been developed by CIAH-CHES, Godhra.

A total 427 accessions of bael from its major growing areas in 19 states have been collected. Maximum accessions are collected from Jharkhand (180) followed by Uttar Pradesh (37), Odisha (33), West Bengal (25), Rajasthan (23), Gujarat (19), Bihar (18), Chattisgarh (16), Himachal Pradesh (12), Andhra Pradesh (11), Madhya Pradesh, Mahrashtra, Kerla and Andaman and Nicobar (9, 8 and 7). The potential areas that need to be explored are Bihar (Chhapra, Sivan), part of Chattisgarh, Madhya Pradesh (Chitrakoot), Uttar Pradesh 9Agra, Allahabad, Baharaich, banda, Basti, Bulandshahar, Deoria, Etawah, Gonda, Gorakhpur, Mirzapur), West Bengal (Murshidabad) and Rajasthan (Mount Abu and Jodhpur, Vashistha *et al.*, 2005).

Ber (*Ziziphus mauritiana* Lam.)

Based on different records, about 170 *Ziziphus* species are reported to exist. The species found wild in India are *Ziziphus apetala* Hook. f., *Z. funiculosa* Buch-Ham, *Z. incurva* Roxb., *Z. glabra* Roxb., *Z. horsifieldii* Miq., *Z. mauritiana* Lam., *Z. nummularia* Wight and Arn., *Z. oxyphylla* Edgew., *Z. oenoplia* Mill., *Z. rugosa*, *Z. trinervia* Roxb., *Z. truncata* Blatt and Hall. and *Z. xylopyra* Willd. Maheshwari and Singh (1965) recogonised six economically important species in India. These are *Z. mauritiana* (ber), *Z. nummularia* (*jharber*), *Z. oenoplia* (*makoh*), *Z. rugosa* (*suran ber*), *Z. sativa* Gaertn (*Z. vulgaris* Lam. *Kandiari*) and *Z. xylopyrus* Hochst. Ex A. Rich. Tent. (*kat ber*). The exploitable attributes of some of the species are given in Tables 4.1(a) and (b).

Table 4.1(a): Diversity for Improvement

Wild Relative	*Exploitable Attributes*
Z. nummularia	Dwarfness, deep tap root system
Z. lotus	Drought tolerance
	Early fruit Maturity
Z. jujuba, Z. mistol	Resistance to low temperature damage
Z. mauritiana	Vigrous tree frame
	Extended period of fruit availability
Z. rotundifolia	Tolerant to drought and low temperature
	Fruits of medicinal value

Table 4.1(b): Response of Ber Cultivars to Frost in Arid Region

Tolerant	Tikkadi
Susceptible	Sanaur-1,Jogia,Kathaphal, Kaithali
Moderately affected	Gola, Umran, Seb, Mundia, Banarsi Kadaka, Illiachi, Goma Kiriti

The ber or Indian jujube (*Z. mauritiana* Lam.) is indigenous to India and thus a wide variability of its genotypes exists throughout the Indian subcontinent. Germplasm holdings are maintained at Central Institute for Arid Horticulture (338), National Bureau of Plant Genetic Resources, Jodhpur, Rajasthan (146), Central Arid Zone Research Institute, Jodhpur, Rajasthan (68), Central Horticultural Experiment Station, Godhara, Gujarat (52), Indian Agricultural Research Institute, New Delhi (41) and at the Centres of All India Co-ordinated Research Project on Arid Zone Fruits at Jobner, Rajasthan (40), Hisar, Haryana (79), Bawal, Haryana (47), Sardar Krushinagar, Gujarat (64), Rahuri, Maharashtra (93), Bahadurgarh, Patiala, Punjab (41), Ludhiana, Punjab (34), Faizabad, Uttar Pradesh (32) and Arrupukotai, Tamil Nadu (26).

Date Palm (*Phoenix dactylifera* L.)

Date palm a potential fruit crop for dry arid region is considered to be the oldest cultivated tree fruit. It belongs to family Palmacae. There are three major species of date palm grown in India *viz.*, *Phoenix dectylefera* L., *P. sylvestris* L.Roxb, and *P.*

canariensis Chabaud. Phoenix dactylifera is very common and grown commercially because of its superior quality edible fruits. There are several species growing wild in different parts of the country. Date palm is high potential of its development in arid regions where suitable irrigation facilities exist. Proverbially said, date palm requires its head to be in fire and foot in water. Hot arid region of the country meets primary requirement of climate *viz.*, prolonged hot dry summers and moderate winter temperatures. Explorations for date palm have been conducted in and around *Kutch* region of Gujarat, where its grooves of seedling originated plants are very common. Sixty one accessions are being maintained at CIAH, 35 at Swami Kehwanand Rajasthan Agriculture University, Bikaner; 31 at Punjab Agriculture (PAU) Research Station, Abohar and 24 at Gujarat Agricultural University (GAU) Research Station, Mundra and few at CCSHAU, Hissar. However, IC nos have been issued to only 36 accessions of seven species. Future collection should target areas of Gujarat for P. dactylifera; sub-Himalayan mountains for P. acaulis Roxb. Ex Buch-Ham.; Assam, Manipur and Khasi hills for *P. humilis* var. Loureirii; western Ghats from Konkan southwards for *P. humilis* Royle var. pedunculata and *P robusta* (Bec.) BeCC.; Kumaon hills to Myanmar for *P humilis* to Myanmar for *P. humilis*, eastern India, for *P. reclinata* Jacq and *P. zeylanica* Trimen; Andaman and Nicobar Islands and parts of West Bengal for P. paludosa Roxb.; Jharkhand (Chhota Nagpur), Bihar and Western Ghats for *P. robusta*; Sikkim, Assam and Mishmi Hills for P. rupicola T Anders; and Gujarat, Maharashtra, Andhra Pradesh, Karnataka and Madhya Pradesh for *P. sylvestris* (Vashistha *et al.*, 2005). Important traits to be targeted while collecting germplasm includes resistance to Graphiola disease, early maturity and high TSS. Further introductions are needed from Pakistan, Tunisia, Oman, Egypt, Iraq and Algeria particularly, for early maturity type for dry date and Grphiola disease resistance.

Jamun (*Syzygium cuminii*)

Jamun belongs to the family Myrtaceae and is a native of India, Burma (Myanmar) and Sri Lanka. The tall tree with evergreen foliage is an excellent roadside tree and very often used as windbreak. It is widely grown both in north and south India. Distribution of some economic species of *Syzygium* has been shown in Table 4.2.

Table 4.2: *Syzygium* Species of Horticultural Interest found in India

Species	*Distribution*
S. cuminii	Indo-Gangetic plains of north India, Tamil Nadu in south, widely distributed
S. arnottianum	Western Ghats, the Nilgiris, Palni and Anamalai Hills
S. bracteatum	Western Ghats, Eastern/north-eastern India
S. operculatum	Grows wild in Nilgiri Hills of Tamil Nadu, Western Ghats
S. aqueum	Mainly in Assam, Sikkim and Meghalaya, Eastern/north-eastern India
S. fruiticosum	Grows as an avenue tree, widely distributed

The work on germplasm collection in jamun is very less. Being highly recalcitrant, it conservation protocol is yet to be standardised. Most of the germplasms is in field gene banks at institutes, like ICAR-CIAH, CHES Godhra, ICAR-CISH, Lucknow, ICAR-IIHR, Bangalore. The areas need to be explored are humid and sub humid parts of Uttar Pradesh, Bihar, Madhya Pradesh, Odisha, Maharashtra and West bangal. Potential areas of diversity are, Varanasi, Sultanpur, Jaunpur, Pratapgarh, Allahabad, faizabad, Gorakhpur, Deoria, barabanki and Lucknow in Uttar Pradesh; and Mahabaleeshwar, Sriampur, Satara, Pune and Aurangabad in Maharashtra. Cluster bearing, small stone, high TSS and longer storage life should be given priority while collecting germplasm.

Karonda (*Carissa carandus* L.)

On the basis of fruit colour, the cultivars of Karonda or Christ's thorn (*Carissa congesta* Wight. syn. *C. carandas* L.) have been classified as Green fruited, Pink fruited and White fruited. The shape and size of their fruits resemble very much. These also have variability in shape and size. There is, however, tremendous scope to select desirable types from the rich variability. Some promising clones such as K_1, K_2, K_3 (Joshi *et al.*, 1986), No. 3, 12, 13 and 16 (Karale *et al.*, 1989) and 3 clones named as Pant Manohar, Pant Sudarshan and Pant Suvarna have been selected.

The other related species is Natal plum or ciruela de Natal (*C. grandiflora* A. DC. *C. macrocarpa* (Eckl.) A. DC.). Germplasm collection consisting 3 *Carissa* species and 24 cultivars have been maintained at different centres in India (Anon., 1998). Thirty six accessions of C. congesta have been collected from parts of Rajasthan, Maharashtra and Karnataka. Areas for future explorations of Karonda are Uttar Pradesh (Varanasi, Mirzapur, Faizabad, Gorakhpur, Lucknow and Allahabad), Rajasthan (Jodhpur, Jaipur, Udaipur, Bikaner) and Maharashtra (Wardha, Jalgaon, Nagpur, manmad (Vashishtha *et al.*, 2005). Precocity in bearing, big fruit size, pink colour of skin and small seed should be looked for during germplasm collection. Accessions having high yield, large fruit size and high vitamin C content need to be introduced from South Africa. Genetic variability in plant growth, fruit size, shape and colour of Karonda has been reported (Bankar and Prasad, 1992; Bankar *et al.*, 1992).

Ker (*Capparis deciduas*)

Ker is an important fruit plant growing wildly on marginal land in arid region. Its raw fruits are consumed as vegetable and are used for pickling. The raw fruits are brined and sun dried to use it throughout the year. Dried ker fruits are sold at a very high price in domestic market and also have good export value. The fruits are dried either by blanching or by dipping in salt solution. The genus comprises of more than 300 species distributed throughout the warm regions of the world (Ushar, 1974). About 40 species occur in India, of which a few have economic importance and among them *Capparis deciduas* is most important. In India, it is found mainly in Rajasthan, Punjab, Haryana and Gujarat; Thar desert, arid regions of north-west plains and decan peninsula. Being an open pollinated plant, large variability occurs in seedlings for plant vigour, morphology, productivity and quality parameters. Thirty-five accessions of Capparis germplasm collected from parts of Rajasthan and Gujarat are conserved in cryobank at NBPGR. Recently, 33 accessions of C. deciduas were collected from Ahmadabad, banskantha, Kutch, Gandhinagar, Mehsana, Morbi, Patan, Surendernagar districts of Gujarat and Sirohi and Jalore districts of rajasthan. Variability in fruit colour and size has been recorded. Two distinct, tall and dwarf types have been identified at CCSHAU, Hissar. The traits, big fruit size, small seed and long storage life need to be targeted during the collection.

Kumat (*Acacia senegal* L.)

This much-branched thorny tree with pale smooth bark is known as *Kumat* or *Kumata*. It is commonly found on hillsides and stabilized sand dunes. It yields the true gum Arabic, which is an important commercial product and used in pharmaceuticals. Apart from gum production, seeds are used as food; leave and pods as fodder and wood for fuel wood and charcoal. It is hardy species surviving under most adverse conditions and an ideal for agro forestry systems. Its gum remains a valuable article of trade and is one of the main cash crops of the desert region. Central Arid Zone Research Institute (CAZRI), Jodhpur has developed a technology of inducing gum exudation by injecting ethephon solution into the main stem through a small size hole.

Lasoda (*Cordia* spp.)

Lasoda (*Cordia myxa* Roxb. syn. *C. dichotoma* Forsk. f) is also known as Gonda or Lehsua or Indian cherry. Lasoda leaves have sunken stomata and other characters associated with drought tolerance. Plants are deciduous. Great variation exists in natural population with respect to morphological characters particularly plant height, spread, leaf size, fruit size and quality parameters like fruit colour, pulp content, seed/pulp ratio and pickling quality.

Besides *Cordia myxa*, the other important species are *C. gharaf* Forsk. (goondi), *C. rothii, C. macleodii, C. vestita* and *C. wallichii.* Goondi trees are of medium size having dense foliage with crooked trunk. It is a small-fruited type of lasoda bearing very small (about 1 cm) and orange to light pink fruits that are reported to have medicinal value.

At CIAH, Bikaner 65 genotypes of lasoda have been collected. A collection of 45 genotypes has been reported by Saini *el al.* (2002) at CCS Agricultural University, Hisar. Two species of *Cordia* and 45 cultivars have been maintained in the repositories at different centres in India such as Kovilangulam, Kumarganj, Sardarkrushinagar, Jodhpur and Bichpuri (Anon., 1998). NBPGR Regional Station, Jodhpur and Central Institute for Arid Horticulture, Bikaner have identified some big fruited types of Lasoda having high productivity. A large fruited type Paras is known in Gujarat. Lasoda growing areas in states of Gujarat, Madhya Pradesh, Jharkhand, Chattisgarh and foothills of Uttarakhand can be targeted for further collections.

Phalsa (*Grewia subinaequalis* DC syn. *G. asiatica* L.)

Phalsa is a drought tolerant and suitable crop for cultivation in wasteland and salt affected soil. Despite the various traits, the popularity of this fruit is restricted owing to its perishable nature, small size of fruit and non-synchronous maturity, which necessitates repeated harvesting. Grewia has 140 species, out of which 40 are reported in India. Among them, *G. subinaequallis* is the only species of commercial importance. Other edible species are *G. damine* Graerth., a small tree, found in Rajasthan, Bihar and drier parts of South India; *G. elastic* Royle, found in the sub Himalayan tract, Central India and wetern Ghats; *G. flavescens* Juss., in upper Gangetic plans, Bihar and central and southern India; *G. hirsuta* Vahl., in warm parts of India ascending to sub Himalayan tract; G. rothi DC., in peninsular region; *G sapida* Roxb., in North western India eastwards to Assam hills and also in Eastern Ghats; *G sclerophylla* roxb., in tropical Himalaya Central India and western Ghats; G. flavescens Juss. in upper Gangetic plans, Bihar and central and Southern India; G hirsute Vahl., in warm parts of India ascending to sub-Himalayan tract; G rothi DC., in peninsular region; G sapida Roxb., in North-western India eastwards to Asam hills and also in Eastern Ghats; G sclerophylla roxb., in tropical Himalayas from Kumaon to eastern hills of Assam; G tenax Forsk., an evergreen shrub, in Rajasthan and arid regions of central, eastern and western India; G. tilii-folia Vahl., throughout India, and G. villosa wild, found in hot parts of north-western, central and western India. About 187 accessions of phalsa have been assembled through explorations. Some local selections have been named as Local, Sharbati, Tall, Dwarf, *etc.* Dwarf type is more productive and juicy.

Pilu (*Salvadora oleoides* Decne)

The genus *Salvadora* com-prises five species of which *Salva-dora oleoides* Decne (pilu or toothbrush tree) and *S. persica* L. (Indian mustard tree or saltbush) are important. Considerable variability is found in nature. Salvadora oleoides is distributed in arid regions of Punjab, rajasthan and Gujarat and the other species *S. persica* L. is more common in other drier tracts of the country (Chundwat, 1990). Fruits of S. persica is also eaten though it is less pulpy than *S. oleoides*

Prickly Pear (*Opuntia ficus indica* Mill)

The species of *Opuntia* that yield edible fruits are *O. ficusindica* Mill. (prickly pear or tuna or Indian fig), *O. robusta* Wendl., *O. lind-heimeri* Engl. (camusea), *O. streptacantha* Lema. (cardona), *O. leucotri-cha* DC, *O. tuna* Mill., *O. inermis* DC., *O. caribaea* Britt. and Rose, *O. macrocentra* Engelm., *O. clavata* Engelm., *O. phaeacantha* Engelm., *O. vulpina* Web., *O. amyclaea* Tan. (fafayuca), *O. megacantha* SalmDyck. (amarilla) and *Opuntia* spp. (Blanca I) (Pareek and Sharma, 2009).

From the variablity of prickly pear, several promising clones have been identified. The important ones are 1270 (high protein forage, fairly sweet pale green fruits), 1271 (fast growing, fairly sweet fruits), 1280 (thornless, pretty good yellow fruits), 1287 (thorny, best fruit variety) and 1308 (very fast growing vegetable variety). The germplasm needs to be enriched with accessions having thorniness, large fruit and better keeping quality and of vegetable type from Argentina, the USA, Mexico, Brazil, Israel and Chile.

Seabuckthorn (*Hippophae rhamnoides*)

Seabuckthorn a naturally growing thorny, deciduous bush, is an indigenous fruit crop of cold arid region and grows naturally in Leh and Ladakh. This wonder plant of the Himalaya belongs to the family Elaeagnaceae and is a dwarf (3-15 ft tall), much branched, thorny, dioecious shrub. The plant is able to tolerate moisture as well as temperature stress (-30°C). The fruits are small, round to oval in shape with a diameter of 6-8 mm, yellow to scarlet in colour and sour to highly acidic in taste. Fruits of seabuckthorn are a rich source of vitamins having vitamin C (800-1600 mg/g), vitamin E (162-255 mg/100g), vitamin A (11 mg/100g), vitamin B_1 (0.04 mg), vitamin B_2 (0.56 mg/100g) and organic acid (2-4 per cent). They also contain lysine and microelements such as iron, cobalt, selenium and molybdenum. Presence of antioxidants and flavonoids make it ideal for health. Oleaster (*Elaeagnusspp*), belonging to the Elaeagnaceae has been identified as a potential rootstock for sea buck thorn. Naturally growing areas of seabuck thorn have been surveyed and 200 morphotypes Collected from natural population. Nineteen genotypyes have been identified for various traits.

Approaches to Exploit Genetic Resources

1. Interspecific Hybridization

Crossing or mating between two different species of the same genus is referred to as interspecific hybridization. Because interspecific hybridization involves two species of the same genus, it is also termed as intrageneric hybridization. Main features of interspecific hybridization are given below:

- It is used when the desirable character is not found within the species of a crop.
- It is an effective method of transferring desirable genes into cultivated plants from their related wild species.
- Interspecific hybridization is more successful in vegetatively propagated species.
- Interspecific hybridization leads to introgression which refers to transfer of some genes from one species into the genome of another species.

2. Intergeneric Hybridization

Intergeneric hybridization refers to Crossing or mating between two different genera of the same family. Such crosses are rarely used in crop improvement because of various problems associated with them. The main features of inter-generic crosses are given below:

- Intergeneric hybridization is used when the desirable character is not found in different species of the same genus.
- This method is rarely used in crop improvement programme and that too transfer of some specific characters into cultivated plants from allied genera.
- Intergeneric hybridization has been generally used in asexually propagated species is more successful in vegetatively propagated species.

3. Embryo Rescue

Embryo can abort after fertilization due to incompatibility with the endosperm and supporting maternal tissues of the ovule and or ovary. This may be overcome by removing young embryos from developing ovules and growing them aseptically *in vitro* on a culture medium into intact plants. This technique is called embryo rescue.

4. Somatic Hybridization

Somatic hybridization refers to crossing of crop plants through fusion of somatic cells. In this method somatic cells are used for hybridization and the sexual process is bypassed. The fusion of somatic cells takes place through protoplasts. Somatic hybridization permits hybridization between any two plant species irrespective of their taxonomic relationship.

5. Transgenic

Genetically engineered organisms are referred as transgenic. In other words, a genotype developed by the process of genetic engineering is called transgenic. It may be a plant, an animal or microbes such as fungi, bacteria and viruses. Transgenic plants are obtained involving tissue culture and genetic engineering techniques.

6. Cisgenics

Cisgenics is the genetic modification of a recipient plant with a natural gene from a crossable -sexully compatible plant. Such a gene includes its introns and is flanked by its native promoter and terminator in the normal sense orientation. Cisgenic plants can harbour one or more cisgenes, but they do not contain any transgenes.

Future Thrusts

There is need to:

- ☆ To understand the genepool to identify taxa and a series of specific genotypes which can be of value as rootstocks.
- ☆ Evaluation of a large number of indigenous species so as to explore their possibility as rootstocks under different agro climatic conditions.
- ☆ To develop rootstocks which can control the overall tree morphology and its vigour and to maximise its adaptation to different edaphic conditions.
- ☆ Biotechnological tools may be applied to incorporate resistant genes in wanting commercial cultivars from the wild species relatives

REFERENCES

Anonymous. 1998. *Project Report.* Ninth Group Workers Meeting, AICRP on Arid Zone Fruits. CCS HAU, Hisar,16-18 October, 1998. NRC for Arid Horticulture, Bikaner, pp111.

Bankar, G. J. and Prasad, R. N. 1992. Sushka Kshetron mein Karonda ki Bagwani, *Udyan Lok* 4: 13-15.

Bankar, G. J., Verma, S. K. and Prasad, R. N. 1992. Vegetative and Fruit characterization of some Carissa species in Thar desert. *Progressive Horticulture*, 25: 7-9.

Bhattacharya, S. C. and S. Dutta. 1956. Classification of Citrus Fruits of Assam. ICAR Monogr. No. 25.

Chundawat, B. S. 1990. Arid Fruit Culture. Oxford and IBH Co. Ltd., New Delhi, India.

Mukherjee, S. K. 1949. A monograph of the genus Mangifera L. Lloydia 12: 73-136.

Mukherjee, S. K. 1985. Systematic and ecogeographic studies in crop genepools: 1. Mangifera L. IBPGR, Rome, Italy. 86 p.

Nakasone, H. Y. and R. E. Paull. 1998. "Annonas." In: *Tropical Fruits.* Edited by H. Y. Nakasone and Paull R. E. CAB International, London,UK.: pp. 45-75.

Pareek, O. P. and Suneel Sharma. 2009. *Underutilized Fruits and Nuts.* Aavishkar Publishers, Distributors, Jaipur, India. Vol. 1, pp. 366 and Vol 2, pp. 342.

Vashishtha, B. B., Saroj, P. L., Gunjeet Kumar and Awasthi, O. P. 2005. Arid Fruits: In: Plant Genetic resources: Horticultural Crops (Eds. B. S. Dhilon, R. K. Tyagi, S. Saxena and G. J. Randhawa), Noorsa Publishing House Pvt. Ltd., New Delhi, India. Pp168-189.

Zeven A. C. and Wet J. M.J. de. 1982. Dictionary of Cultivated Plant and their Regions of Diversity. Wageningen, 259.

Breeding of Perennial Horticultural Crops *Pages* **61–70**
Editors: S.K. Singh, V.B. Patel, A.K.Goswami, Jai Prakash and Chavlesh Kumar
Published by: **BIOTECH BOOKS, NEW DELHI**

Chapter **5**

Genetic Resource Conservation

Sandhya Gupta

Tissue Culture and Cryopreservation Unit,
ICAR-National Bureau of Plant Genetic Resources, New Delhi

INTRODUCTION

Plant genetic resources (PGR), as a vital segment of biodiversity in general and agrobiodiversity in particular, constitute the genetic material of plants having value as a resource for present and future generation of human being (Dhillon and Saxena, 2003). As genetic resource, the PGR may be of reproductive or vegetative propagules such as seeds, shoots, tissues, cells, pollen, DNA molecule *etc.*, containing the functional unit of heredity in addition to corresponding information and knowledge about their use that can be applied in crop improvement programme and other product development. The categories of PGR range from landraces and farmers' varieties, absolute cultivars, modern cultivars, breeding lines and genetic stocks, wild relative, weedy races and potential domesticate species, exotic and indigenous species (FAO, 1996; Sharma, 2007).

The agricultural paradigm has undergone a shift in focus from food security towards nutritional quality. Horticultural crops, besides improving biological productivity and nutritional standards also have enormous export potentials. Horticultural genetic resources (HGR) are a subset of agrobiodiversity that is related to horticulture plant species or their wild genepool, having genetic material of actual or potential value. Horticultural plants comprise groups of important crops that include fruits, vegetables, spices and condiments, ornamental plants, aromatic and medicinal plants (Rathore *et al.*, 2005). India's diversity in agro-ecologies provides greater scope for going a wide variety of horticultural crops than most other countries in the world. Presently, horticultural crops cover more than 15.7 million ha, which is roughly 7 per cent of the gross cropped area and contributes 18-20 per cent of the gross value of India's agricultural output. India is the second largest producer of fruits and vegetables. Diversification and value addition have been the major thrusts in Indian Horticulture in the 21st century; and plant genetic resources play a major role to achieve these objectives. Depletion of crop genetic resources of horticultural crops at a rapid pace, is a matter of global concern.

Conservation Strategies

Various conservations strategies are summarised in the Figure 5.1. Basically, germplasm of horticultural crops can be conserved using in situ and ex situ conservation approaches. Both the approaches are complementary which can be judiciously integrated so that the conservation efforts are effectively realized for sustainable use of variability. Therefore, complementary conservation strategies employing in situ and ex situ approaches have been applied to conserve horticultural crops diversity.

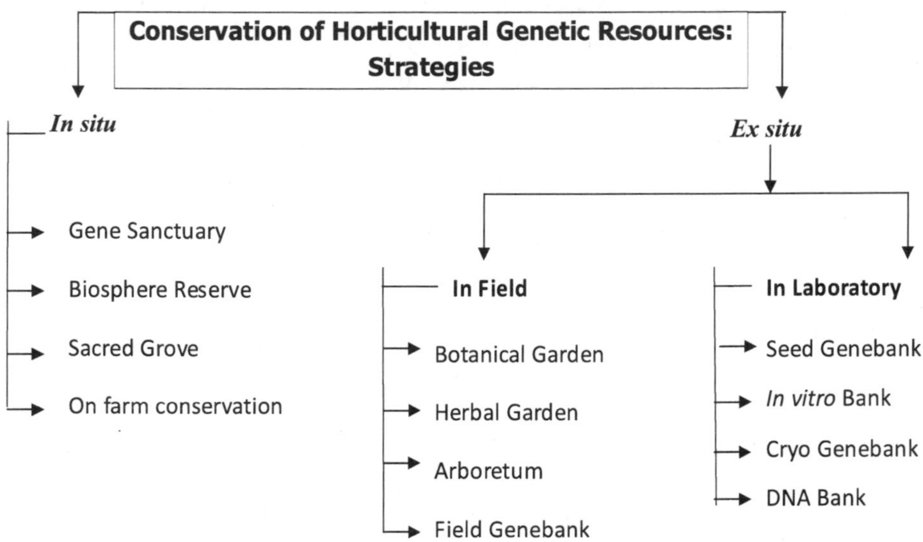

Figure 5.1: Conservation Strategies of Horticultural Genetic Resources.

In situ Approaches of Conservation

Plant species are promoted to grow in their natural habitats where evolutionary processes continue to operate in situ thus making it a dynamic system. Genetic variability is generated through mutation, pollen and seed dispersal, and recombination within and among population. Selection operates on this variability leading to the development of new plant types with improved adaptability. This has been achieved with the establishment of biosphere reserves, habitat parks, wild life parks and sanctuaries, protecting the diversity under natural habitat following ecosystem or habitat approach. Outside protected areas and national reserves, in situ conservation is carried on at the farm level, where landraces and locally improved material are cultivated, utilized and conserved as parts of traditional farming systems. This requires active farmers' participation to conserve landraces and traditional farmers' varieties. The novel genetic resources may be conserved even in the home gardens (Rathore *et al.*, 2005; Tao, 2003). On farm and home garden conservation is of particular importance, to conserve indigenous HGR and to provide diverse food and other products for household needs and local markets.

There are about 460 biosphere reserves in more than 90 countries. Some wild species of mango and their relatives occur in biosphere reserves, national parks or other reserves in India, Indonesia, Singapore, the Philippines, Thailand and Sri Lanka, but little targeted in situ genetic conservation is being carried out (Heywood and Dulloo, 2005). There are also sanctuaries and conservation areas named after the target plants they afford protection to. For example, the in situ gene sanctuary for Citrus was established in Tura hills range of Garo hills in Meghalaya, India, for conservation of genetic diversity. Northeast hill region is considered as one of the homes of *Citrus*. Areas rich in orchid species have been given protection as 'orchid sanctuaries' in various states of India such as Arunachal Pradesh, West Bengal, Sikkim, and Mizoram, and more are planned. Some of them are sacred forests and other are associated with orchid research centres and nurseries. Over 20 species of wild orchids are recorded in the Sessa Orchid Sanctuary, which extends over 100 km^2 in the Dafla Hills of Arunachal Pradesh. Sanctuaries have been established in Sikkim for rhododendrons and orchids. The term 'Medicinal Plant Conservation Area' has been applied to the network of 54 in situ reserves, each about 200 ha, which captures inter- and intra-specific medicinal plant diversity and which has been set up across different forest types and altitude zones in five states (Andhra Pradesh, Karnataka, Kerala, Maharashtra and Tamil Nadu) of peninsular India. Five such areas have also been established in Sri Lanka. In the high valleys of the Himalayas, members of local communities are being encouraged to protect medicinal and aromatic plants in their private/community lands known as *dhangs*, which basically serve as the areas for grazing and for collecting fodder. Women's groups in different villages are also being motivated to adopt neighbouring natural sites as in situ conservation areas for protecting them from excessive grazing and unscientific harvesting. Other such examples are community participation in the management of *Prunus africana* in the Mount Cameroon region (Gabriel, 2003), in situ conservation and use of medicinal plants by Afro-Colombian communities in Colombia (IDRC, 2001), and community-based conservation of medicinal plants in Kenya. In India, however, the current protected area network is unevenly distributed over states and biogeographic regions. Many biodiversity rich areas such as Ladakh, South Deccan, Gangetic Plains, Assam Hills, and Nicobar Island are not adequately covered.

With in situ approach however, there is always a threat of a species becoming extinct or its population declining due to demographic and environmental variations, habitat loss, over-exploitation, competition from invasive alien species, pests, genetic drift, inbreeding and human disturbances. In such situations, the alternative way to conserve diversity is to maintain it ex situ.

Ex situ Approaches for Conservation

Ex situ conservation aimed at maintaining plant genetic material outside its natural habitat. The genetic material is conserved ex situ either in field, seed, in vitro and cryo genebank. Other methods used for ex situ conservation include in pollen and DNA conservation. At present about 6 million accessions are stored in ex situ collections worldwide including some duplication among collections. The percentage of horticultural germplasm in seed genebanks is not very high. Mostly, these groups of species are conserved as living collection in the 1,800 botanical gardens around the world. Although the botanical gardens, private orchards and gardens play very significant role in the conservation of horticultural germplasm, however, in the majority of those gardens, only few selected species are maintained which are not adequate to represent whole variability existing within a species.

The CGIAR genebanks represent the most important international effort to conserve genetic resources of staple crops, forages and agroforestry species. In 1994, the international community recognized the need to protect the large and important crop diversity collections held by these genebanks and in response, the CGIAR Centres placed their collections in trust for the world community under the intergovernmental authority of the Food and Agriculture Organization of the United Nations (FAO). In 2006, that arrangement has been succeeded by the Centres' signing agreements with the Governing Body of the International Treaty on Plant Genetic Resources for Food and Agriculture (ITPGRFA), which bring the in-trust collections into the new Multilateral System. The Centres have the responsibility to ensure that the in-trust collections be maintained to the highest international standards and available to all according to the terms of the Treaty's provisions on access and benefit-sharing.

Seed Conservation

In the crops that produce seed, which are amenable to desiccation and can tolerate low temperature, germplasm conservation through seeds, is the most common approach. The desiccated seeds having 3-5 per cent moisture content are stored at -20°C for long-term conservation. Though it is among the most cost-effective and easy approach, development of more permafrost storage facilities like Svalbard may prove very useful. On the basis of temperature regime the storage is classified into medium and long-term storage: Medium-term storage at 4°C: active collection; Long-term storage at -20°C, and at -196°C (cryopreservation): base collection. However, in horticultural crops the orthodox seeds are being conserved for genepool conservation of crops such as apple *etc.*

In the early 1970s there were fewer than 10 genebanks with perhaps no more than half a million accessions (samples). A total of 1,308 genebanks are now recorded in the World Information and Early Warning System database. It is believed there are now more than 6 million accessions stored worldwide in ex situ germplasm collections (although many may be duplicates). Forty per cent of all accessions in genebanks are cereals, and 15 per cent are food legumes. Vegetables, roots and tubers, fruits and forages each account for less than 10 per cent of global collections. Medicinal, spice, aromatic and ornamental species are rarely found in long-term public collections. Aquatic plants of relevance for food and agriculture are likewise not found in such collections. In India, more than 400,000 accessions have been conserved in the seed genebank at ICAR-NBPGR, New Delhi including samples of fruits, vegetables, spices, and medicinal and aromatic plants.

In vitro Conservation

Efficient conservation of genetic resources in case of vegetatively propagated plants and recalcitrant seed species has been hampered due to problems faced during application of their conventional method of ex situ conservation in field genebank. To tackle these challenges, in vitro techniques have been increasingly used for conservation and its related activities like collecting and exchange of germplasm of these problem species. In vitro meristem culture technique offers the possibility of eliminating viruses and thus, exchange of virus-free germplasm. In vitro slow/ normal growth techniques offer up to medium-term storage option, avoiding risk of losses of germplasm on field genebank due to insects, nematodes, disease attacks and natural disasters. It is commonly use for vegetatively propagated species, nonorthodox seeded species and wild species which produce little or no seeds. While cryopreservation at ultra-low temperature, usually that of liquid nitrogen (196°C), is the only option currently available for the long-term conservation of these PGR avoiding exogenous contamination, requiring small space and minimum maintenance. At this very low temperature, all metabolic activities of cell cease, and theoretically the cell or tissue can be stored for an indefinitely period. Both in vitro conservation and cryopreservation techniques use tissue culture principles for conservation (Mandal, 2003; Gupta, 2011).

The art and science of plant tissue culture is based on devising media for each genotype that would elicit the optimal response in terms of growth rate of the explants. However, when tissue techniques are employed for conservation, the aim is to devise a medium that would decrease the growth rate of explants to the minimum, thereby increasing the subculture intervals. Slow growth techniques have been developed for medium-term conservation of crop species. The various methods used to achieve this include the following: Use of growth retardants, use of minimal growth media, use of osmotic regulators, reduction in oxygen concentration, size and type of culture vessels, type of enclosures, maintenance under reduced temperature and for reduced light intensity and combination of more than one treatment. Explants used for in vitro conservation must be of right type as well as physiological stage. The apical and axiliary meristems of very small size are the preferred explants for in vitro storage. In fact, organized explants have proved better than unorganized tissues, in terms genetic stability of the germplasm (Reed *et al.*, 2004).

Methods of *In vitro* Conservation

Maintenance of germplasm on more or less defined nutrient medium under controlled environmental conditions is referred as in vitro conservation. There are two major in vitro conservation strategies: (i) normal growth (ii) slow growth and (iii) cryopreservation. Slow growth techniques for short- to medium term-conservation storage allow clonal plant material to be stored under in vitro conditions with extended subculture intervals. Cryopreservation techniques may be applied for long-term conservation of plant material. Slow growth methods aim for reducing the growth rate of plant material so as to prolong the subculture interval without causing any adverse effect on the plant tissue.

Normal Growth

The objective of this method is to maintain germplasm cultures under normal growing conditions. Though it involves frequent subculturing which is laborious. However, the method is advantageous as the cultures are available for immediate multiplication and distribution and it avoids requirement of low temperature facility or the application of stresses.

Slow Growth

The main aim of this method is to maintain cultures under growth limitation so as to reduce requirement of frequent subculturing. Some of the various approaches to achieve slow growth are discussed below:

Low Temperature Incubation

This method is very simple, easy to use and may be applicable to wide range of genotype. Here, the in vitro cultures are maintained at low temperature that affects the metabolic activities which in turn restrict the growth of the plant. The storage temperature, generally, is crop specific, for example, cold tolerant species like *Fragaria* and *Prunus* can be stored between 0-4°C (Wilkins *et al.*, 1988; Reed, 1992).

Use of Growth Regulator

Some of the commonly used growth regulators are abscisic acid, dimethylamino succinamide, phosphon D, maleic hydrazide. These are used to reduce the overall growth rate of the *in vitro* plantlets thereby enhance the subculture intervals. However, the use of some of the growth retardants may cause mutation, owing to their mutagenic properties. They may also pose physiological problem if used for longer time.

Use of Minimal Growth Media and Restrictive Growth Conditions

Carbon source has marked effect on growth rate. Alteration of optimum dose could reduce the growth rate of cultures in many species. Inclusion of sugar alcohol like mannitol or sorbitol in culture media is quite effective in restricting the growth of many plant species *in vitro*. Use of minimal media is more effective at low temperature.

However, *in vitro* maintenance of germplasm under normal growing conditions is the best if the subculture interval may be extended up to a year or more (Gupta and Mandal, 2003). Besides being economical it can avoid all kinds of risks imposed by other slow growth methods. About 1,800 accessions of various horticultural crops are being conserved in vitro under normal growing conditions at NBPGR, India.

Cryopreservation

Cryopreservation refers to the non-lethal storage of biological tissues at ultra-low temperature, usually that of liquid nitrogen (LN) which is -196°C. Currently, it is the only option available for the long-term conservation of germplasm of vegetatively propagated and recalcitrant seed species. Due to storage at the temperature of the vapor phase (-150 to -180°C) or liquid phase (-196°C) of LN, cell divisions and metabolic activities are arrested and thus, plant material can be stored for unlimited periods of time. Conservation of germplasm using cryogenic approach required very limited space; the plant material stored is protected from exogenous contamination and needs very limited maintenance. It causes no change in viability, vigour and genetic makeup of the conserved materials. It also eliminates the need to test stored materials frequently, thus making storage cost-effective. There have been several reviews on the use of cryopreservation for storage of plant materials (Mandal, 2003; Reed *et al.*, 2004). The choice of material for cryogenic storage will depend on the plant species as well as the objectives of storage. For conservation of PGR, the explants can include shoot apices, auxiliary buds, dormant buds, somatic embryos, seeds, zygotic embryos, embryonic axes or pollens and DNA. Cryopreserved explants (but pollen and DNA) should eventually regenerate whole plants to be used and therefore, regeneration protocols need to be clearly defined prior to embarking on cryo-preservation. Regenerated plants should also maintain genetic integrity of the starting material.

Cryopreservation Techniques

The various techniques currently under investigation or in use for cryopreservation of various horticultural crops include: i) the classical freezing method, and (ii) vitification based techniques such encapsulation-dehydration, vitrification, encapsulation–vitrification, desiccation, pre-growth droplet freezing, and pre-growth desiccation, D-cryoplate and V-cryoplate method (Gupta, 2011).

Vitrification-based procedures such as PVS2-vitrification and encapsulation-dehydration, encapsulation-vitrification, droplet freezing *etc.* have greater potential for broad applicability. Vitrification involves the application of cryoprotectant solutions that increase cell viscosity to a critical point at which water forms a meta-stable glass on exposure to ultra low temperatures. Encapsulation-dehydration involves removing cell water through osmotic and evaporative dehydration to achieve the same state. The encapsulation-dehydration protocol comprises the successive steps. Conditions for each step require optimization to achieve maximal recovery of explants after cryopreservation. The basic steps of preconditioning, preculture, encapsulation, osmoprotection and dehydration before exposure to liquid nitrogen apply in all cases. The technique of encapsulation-dehydration has been applied to shoot apices of several genotypes of pear, apple, mulberry, blackberry and raspberry (Niino and Sakai, 1992; Gupta and Reed, 2006; Reed *et al.*, 2008; Gupta, 2014). The vitrification techniques have been developed for apple, pear (Niino and Sakai, 1992) and mulberry (Niino *et al.*, 1992) *etc.*

DNA Conservation

DNA storage is a relatively new technique that is rapidly gaining recognition. Several DNA libraries are being established which provide an easy access for scientists. DNA from the nucleus, mitochondria and chloroplasts are now routinely extracted and immobilized into nitrocellulose sheets where the DNA can be probed with numerous cloned genes. With the development of PCR (polymerase chain reaction), one can now routinely amplify specific genes or oligonucleolides from the entire mixture of genomic DNA. This approach, according to Engelmann *et al.* (2003), is relatively easy and of low cost. It is particularly useful for the conservation of specific genes and it allows easy access to specific material. The exchange of germplasm through DNA sequences is safe since infestations with pathogens can be simply avoided. Note however, that entire plant can not be regenerated from conserved DNA (Reed *et al.*, 2004; Guimaraes *et al.*, 2007).

Future Thrust

The use of wild relatives is of special concern as these are also rich in genetic diversity and carry genes particularly for resistance or tolerance to abiotic and biotic stresses. Their great genetic variability makes them important genetic resources for producing the desired traits in crops. Until recently, wild relatives of cultivated crops have rarely been used for breeding purposes because they frequently have many undesirable characteristics aside from the desired ones. However, success in breeding better adapted varieties depends partly on collection, conservation, dissemination and use of the wild relatives. The importance of wild HGR is also underlined by the reports that several wild forms of vegetables have been found to have higher contents of minerals, proteins and vitamins compared to cultivated forms. Thus, the conservation of HGR, whether cultivated forms, exotic species, underutilized indigenous plants, or wild relatives, is extremely important to meet the present and future needs of various crop improvement programmes. For a complementary approach a continued dialogue between environmentalists, conservationists, breeders and biotechnologists is required to identify best approach and its adaptation.

REFERENCES

Dhillon BS, Saxena S (2003). Conservation and Access to Plant Genetic Resources. In: BB Mandal, R Chaudhury, F Engelmann, B. Mal, KL Tao, BS Dhillon (eds.). Conservation Biotechnology of Plant Germplasm. NBPGR, New Delhi/IPGRI, Rome/FAO, Rome, pp. 318.

Engelmann F (2011). Use of biotechnologies for the conservation of plant biodiversity. *In vitro Cell Dev. Biol. Plant*, 47: 5–16.

Engels J, Visser B (2006). Genebank Management: Effective management of germplasm collection Training manual on "Conservation, Management and use of Plant Genetic resources in food and Agriculture'. Wageningen University and Research, Wageningen, the Netherlands.

FAO (1996). State of World's Plant Genetic Resources for Food and Agriculture. FAO, Rome Italy.

Gabriel L (2003). Community involvement in the management of Prunus africana in the Mount Cameroon region. Paper submitted to the XII World Forestry Congress 2003, Québec City, Canada (http: //www.fao.org/docrep/article/wfc/xii/0456-b1.htm).

Guimaraes EP, Ruane J, Scherf BD, Sonnino A, Dargie JD (eds.) (2007). Marker-assisted selection: current status and future perspectives in crops, livestock, forestry and fish. FAO, Rome, pp. 330.

Gupta S and Reed BM (2006). Cryopreservation of Shoot Tips of Blackberry and Raspberry by Encapsulation-Dehydration and Vitrification. *CryoLetters*, 27 (1): 29-42.

Gupta S (2011) Management of Temperate Fruits Genetic Resources in India. *Acta Hort.*, 918: 71-80.

Gupta S (2014) Cryobanking of germplasm through encapsulation-dehydration technique. *Acta Hort.*, 1039: 147-153.

Gupta S and Mandal BB (2003). *In vitro* methods for PGR conservation. In: R. Chaudhury *et al.* (eds) In Vitro Conservation and Cryopreservation of Tropical Fruit Species. IPGRI/NBPGR, New Delhi, pp. 71-80.

Heywood VH and Dulloo ME (2005) *In situ* conservation of wild plant species: a critical global review of good practices. IPGRI Technical Bulletin No. 11, IPGRI, Rome, Italy.

IDRC (2001). IDRC Support for Medicinal Plant Research. Regional Focus: Latin America. Factsheet Series 3 of 5. IDRC, Ottawa, Canada (http: //www.idrc.ca/biodiversity).

Mandal BB (2003). Cryopreservation Techniques for plant germplasm conservation. In: BB Mandal, R Chasdhury, F Engelmann, B Mal, KL Tao, BS Dhillen. (eds.) Conservation biotechnology of Plant Germplasm.

Niino T and Sakai A (1992). Cryopreservation of alginate-coated *in vitro* grown shoot tips of apple, pear and mulberry. *Plant Sci.*, 87: 199-206.

Rathore DS, Srivastava U and Dhillon BS (2005). Management of Genetic Resources of Horticultural Crops: Issues and Strategies. In: BS Dhillon, RK Tyagi, S Saxena and GJ Randhawa (eds.). *Plant Genetic Resources: Horticultural Crops*. Narosa Publishing House, New Delhi, pp. 1-18.

Reed BM (1992) Cold storage of strawberries in vitro: a comparison of three storage systems. *Fruit Var. J.*, 46: 98-102.

Reed BM (2008). *Plant Cryopreservation: A Practical Guide.* Springer, New York, 511p.

Reed BM, Engelmann F, Dulloo ME and Engels JMM (2004). Technical guidelines for the management of field and in vitro germptasm collections. IPGRI Handbooks for Genebanks No.7 International Plant Genetic Resources Institute, Rome.

Sharma SK (2007). Indian Plant Genetic Resources (PGR) System: Role of NBPGR. Training manual on *In vitro* and Cryopreservation Techniques for conservation of PGR NBPGR and Bioversity International, New Delhi India.

Tao KL (2003). Complementary Conservation Strategy for Plant Genetic resources. In: BB Mandal, R. Chaudhury, F. Engelmann, B Mal, KL Tao and BS Dhillon (eds.) *Conservation Biotechnology of Plant Germplasm*. NBPGR, New Delhi/IPGRI, Rome/FAO, Rome, p. 51.

Wilkins CP, Newbury HJ and Dodds JH (1988). Tissue culture conservation of fruit trees. *FAO/IBPGR Plant Genet. Resour. Newsl.* 73/74: 9-20.

Breeding of Perennial Horticultural Crops *Pages* **71–84**
Editors: S.K. Singh, V.B. Patel, A.K.Goswami, Jai Prakash and Chavlesh Kumar
Published by: **BIOTECH BOOKS, NEW DELHI**

Chapter 6

Biotechnology for Fruit Crop Improvement

S.K. Singh, A.K. Goswami, Satyabrata Pradhan and Chavlesh Kumar

Division of Fruits and Horticultural Technology,
ICAR-Indian Institute of Agricultural Research, Pusa, New Delhi – 110 012

INTRODUCTION

Until about 100 years ago, the systematic fruit orcharding concept came into existence. With the industrialization and later with the development of dwarfing rootstocks in various fruit crops, single cultivar orchards started to be planted. This led to a dramatic increase in pathogen and pest problems. Especially pathogens were favoured by the presence of a genetically homogeneous host population with close tree proximity. For instance, apple scab (*Venturia inaequalis*), a fungus that grows on the epidermis under the cuticle, damaging the fruits and leaves, increased from a tolerable rather cosmetic disease, to a factor which was able to destroy the whole apple orchard (Gessler, 2015). Contemporarily, farmers and industries developed control measures based on chemicals toxic to the fungus (fungicides), first copper in various formulation later synthetic and continually more specific chemicals (MacHardy *et al.*, 2001). Today to control this disease in conventional apple production, copper fungicides are used up to 6 kg ha^{-1} $year^{-1}$, in integrated approaches too had 10 to 12 treatments with specific (synthetic) fungicides. In central Europe and America, sporadic, however highly damaging epidemics of fire blight caused by the bacterium *Erwinia amylovora* demanded chemical control. As third, depending on the meteorological conditions of a geographical area and cultivar, powdery mildew caused by *Podosphaera leucotricha* also requires chemical protection.

Essentially the current apple production system relies heavily on chemical protection using synthetic molecules in integrated, on copper, sulfur and natural substances in conventional apple production.

Keeping in view of the exponential rise in world population along with the changing climate, which leads to evolution of new races of insect pests, transgenic fruit crops can be used as a strapping tool to complement and substitute the traditional fruit breeding to meet the worldwide demand for high productive of quality fruits. Fruit crop genotypes developed through transgenic can not only be used for yield enhancement and better nutritional quality but also to combat the climate change through development of various biotic and abiotic stress tolerance to genotypes. As most fruits are consumed fresh, the use of various pesticides for control by different diseases, insects and weeds constitute a serious threat to both consumers' health and environmental safety too. As the traditional fruit breeding have several inherent bottlenecks, advancement in biotechnology, especially use of transgenic research, will help the breeders in achieving success by vanquishing these limitations. This approach involves transfer, integration and successful expression of specific gene(s) of interest to an otherwise elite genotype, which results in a resistant or tolerance genotype without affecting other pre-exiting traits. However, introduction of a transgene into a breeding programme can be complicated and cost prohibitive, especially in crops with difficulty using backcrossing. Furthermore, regulation of a transgenic trait is event specific in many countries (Dias and Ortiz, 2014). From the biosafety and consumer acceptance point of view, the existence of selectable marker genes in the transgenic plants, which are required for the initial screening, is undesirable. Hence, the production of 'clean' marker-free transgenic fruit plants is an essential pre-requisite for their commercial acceptance.

Crop losses from insect pests and diseases are of big concern due to devastating financial loss for farmers and starvation in developing countries. Application of chemical pesticides is increasing annually. This is a serious cause for potential health hazards, and even run-off of agricultural wastes from excessive use of pesticides and fertilizers can poison the water supply and cause harm to the environment. Fruits usually we consume in raw form so it is more harmful if we eat them as such. Therefore, growing insect-pest and disease resistance GM fruits such as papaya, grapes, and apple *etc.* not only reduces the economic losses but also ensure to provide chemical-free fruits (Sharma *et al.*, 2016). A list of various fruit crops in which transgenic approach have been explored for insect pest resistance is given in Table 6.1.

Insect Pest Resistance

Defoliating insects are the most important insects causing economic losses in various fruit tree species. The damage often results into reduction of tree growth and survival, as well as in alterations in tree shape or fruit quality. The bacterium *Bacillus thuringiensis* (*Bt*) has been used successfully in commercial crop production for controlling insect pests. Nevertheless, its use in fruit crops is more limiting because of the large areas to be covered (Pena and Seguin, 2001). Effective *Bt* toxin gene expression was observed in apple, poplar, spruce and larch, and resulted in

Table 4.1: Application of Transgenic Approach for Disease and Insect Pest Resistance in Various Fruit Crops

Crop	*Gene*	*Target Trait*	*Reference*
Apple	HcrVf2	Scab resistance	Belfanti *et al.* (2004)
	MpNPR1	Fire blight resistance	Malnoy *et al.* (2007)
	PinB	Scab resistance	Faize *et al.* (2004)
	dpo	Fire blight resistance	Flachowsky *et al.* (2008)
	Lc	Fire blight resistance	Flachowsky *et al.* (2010)
	MhNPR1	Powdery mildew resistance	Chen *et al.* (2012)
Grapevine	*GFLV movement protein (MPc) gene*	Grapevine fan leaf virus (GFLV) resistance	Jardak-Jamoussi *et al.* (2009)
	Chitinase and b-1,3-glucanase	Downy mildew resistance	Nookaraju and Agrawal (2012)
	rpfF	Pierce's disease	Lindow *et al.* (2014)
Banana	*rcc2 or rcg3 class-I rice chitinase*	Black leaf streak resistance	Kovacs *et al.* (2013)
	PhDef1 and PhDef2 Petunia floral defensins gene	Fusarium wilt resistance	Ghag *et al.* (2012)
	MusaBAG1	Fusarium wilt resistance	Ghag *et al.* (2014)
	Cystatin	Nematode resistance	Roderick *et al.*, 2012)
Papaya	*PRSV coat protein (CP)*	Papaya ringspot virus resistance	Gonsalves *et al.* (2000)
	DmAMP1	*Phytophthora palmivora*	Zhu *et al.* (2007)
Strawberry	*Chitinase*	Powdery mildew resistance	Asao *et al.* (1997)
	Chit-42, or b-1,3-glucanase	Anthracnose resistance	Mercado *et al.* (2007)
	Thaumatin *II*	Gray mold resistance	Schestibratov and Dolgov (2005)
	β-1,3-glucanase gene *bgn13.1*	Crown rot disease resistance	Mercado *et al.* (2015)
Citrus	*hrpN*	Citrus canker resistance	Barbosa-Mendes *et al.* (2009)
	scFv	Citrus tristeza virus (CTV) resistance	Cervera *et al.* (2010)
	Attacin E	Citrus scab resistance	Mondal *et al.* (2012)
	NPR1	Citrus greening resistance	Dutt *et al.* (2015)
	ihpCP, siRNA	Citrus psorosis virus resistance	Reyes *et al.* (2011)

reduced larval feeding on transgenic material (Tzfira *et al.*, 1998; Schuler *et al.*, 1998). However, it is only recently that a pronounced increase in larval mortality has been observed in *Bt*-engineered walnut (Dandekar *et al.*, 1998). Losses in production of walnut (*Juglans regia* L.) due to insect pests are significant in the US and other walnut producing areas of world, despite the extensive use of chemical pesticides (McGranahan and Leslie 1991). Codling moth (*Cydia pomonella* L.) is a major pest of walnut, apple, pear and other crops. The major damage is caused by the larval stage of *C. pomonella* feeding on the fruit or walnut meat. Newly hatched *C. pomonella* tunnels their way to the core of the developing embryo (seed or nutmeat). The application of chemicals to control *C. pomonella* infestations often causes outbreaks of secondary pests, such as navel orange worm (*Amyelois transitella*) or walnut aphid (*Chromoaphis juglandicola*) and mites. Thus, controlling codling moth enhances the suppression of the entire spectrum of pests of walnut. A full-length synthetic version of the *cry*IA(c) gene, expressing a *Bt* insecticidal crystal protein (ICP), was transferred into walnut somatic embryos. Sixty one transgenic embryo lines or clones were obtained and bioassayed with first instar codling moth larvae. In 34 per cent of these lines, designated as 'class A', expression was high enough to obtain 80-100 per cent mortality. A total of 20 per cent were designated 'class B', which produced a mortality between 25 and 70 per cent and also caused a decreased rate of larval development. About 62 per cent of the class A embryo lines showed correspondingly high activity for the GUS gene. However, 38 per cent of the class A would have been missed if the level of GUS activity was used as the sole indicator. Detectable levels of protein corresponding to *cry*IA(c) could be found only in class A, but not in class B or C clones. Southern analysis of border regions revealed single insert for class A clones and multiple inserts for classes B and C (Dandekar *et al.*, 1998).

The manipulation of production levels of proteinase inhibitors, a group of small proteins having inhibitory effects on insect digestive enzymes, is complementary to the use of *Bt* toxin genes. Overexpression of proteinase inhibitor genes in transgenic poplar, resulted in reduced insect feeding (Delledonne *et al.*, 2001).

Given that recent molecular studies have improved our knowledge of stress- and defence-related specific metabolic pathways in fruit crops, the ability to modify and improve these pest resistance mechanisms is possible in the foreseeable future.

Disease Resistance

Apparent successes in the genetic engineering of fruit plants for virus resistance have come through the application of the concept of pathogen-derived resistance. This model proposes that, introduction and expression of viral sequences in fruit plants could interfere with the life cycle of the same or a closely related challenging virus, thus providing resistance to infection. A list of various fruit crops in which transgenic approach have been explored for disease resistance is given in Table 6.1.

Papaya is the first fruit crop in which transgenic varieties with PRSV resistance was allowed for commercial cultivation in Hawaii (Gonsalves *et al.*, 2000). Transgenic papaya lines with pathogen derived resistance carrying the coat-protein (*cp*) gene provides effective protection against PRSV strains over a significant period of the cycle of this perennial fruit crop (Lius *et al.*, 1997). Seeds of the transgenic varieties 'SunUp' and 'Rainbow' became available to Hawaii farmers in 1998, and growing of cultivar 'Rainbow' led to a significant increase in papaya production (Suzuki *et al.*, 2007). Prior to the introduction of PRSV-resistant transgenic papaya, growing papaya in Hawaii was no longer viable despite area-wide efforts to eradicate the PRSV infected trees in order to limit the spread of the virus. The cultivation of PRSV-resistant transgenic papaya dramatically reduced the incidence of PRSV in many areas, which also allowed some cultivators to return to growing non-transgenic papaya, which is important for export to Japanese market in which transgenic papaya is not allowed.

Sharka, caused by Plum pox virus (PPV) is one of the most destructive diseases of plum and other stone fruits throughout the world. Over the past 2-3 decades, an intensive international research project has focused on the development of transgenic plum having resistance to PPV (Scorza *et al.*, 1998). The PPV *coat protein (CP)* gene was isolated, sequenced, cloned and transformed into plum genome using *Agrobacterium*-mediated method (Ravelonandro *et al.*, 1992). This lead to the development of a cultivar named as 'Honey Sweet', which is highly PPV-resistant as successfully demonstrated during 15 years of study and field testing in the US and Europe. Through crossbreeding, its resistance was further transferred to other plum seedlings, to develop additional PPV-resistant cultivars (Ravelonandro *et al.*, 1998; Scorza *et al.*, 1998). For 12 years, the safety and efficacy of 'Honey Sweet' was observed in various laboratory, greenhouse and field conditions, which led to its regulatory approval in the USA during 2011. To transfer the PPV resistance transgene into new cultivars, breeding with "Honey Sweet" is underway in the US. This breeding is relatively easier and rapid due to availability of molecular marker and also the resistance is governed by a single dominant gene. The transgenic technology and post-transcriptional gene silencing tested in 'Honey Sweet' can also be used to confer resistant in various stone fruits which are potentially vulnerable to PPV. Reports showed that 'Honey Sweet' plum have excellent fruit quality with higher productivity (Scorza *et al.*, 1998). A list of fruit crops in which RNA interference is exploited for disease resistance is given in Table 6.3.

Xanthomonas wilt caused by the bacterium *Xanthomonas campestris* pv. *musacearum* is one of the most important diseases and threat to banana production in the Great Lakes Region of Africa. A transgenic banana exhibiting resistance to *Xanthomonas* wilt was developed by the International Institute of Tropical Agriculture (IITA), along with Uganda's National Agricultural Research Organisation (NARO) and other institutions. These transgenic bananas constitutively express the hypersensitivity response-assisting protein (*Hrap*) and the plant ferredoxin-like protein (*Pflp*) genes from sweet pepper (*Capsicum annuum*) (Tripathi *et al.*, 2010). Transgenic bananas showing *Xanthomonas* wilt resistance will undergo multi-location testing in Uganda for environmental and food safety according to the country's biosafety regulations (Tripathi, 2012).

Different strategies have been proposed to engineer plants for resistance to bacterial and fungal diseases, including production of antibacterial and antifungal proteins of non-plant origin. Genes encoding lytic peptides, such as cecropins and attacins of insect origin, have been used in transgenic apple (Norelli *et al.*, 1994) and pear (Reynoird *et al.*, 1999) for protection against the bacteria *Erwinia amylovora,* which is the causal agent of fire blight.

Recently, genes encoding a strongly antifungal endochitinase from the mycoparasitic fungus *Trichoderma harzianum* have been transferred to apple with the aim of controlling apple scab (caused by the fungus *Venturia inaequalis*). Although some level of resistance was achieved, constitutive expression of the endochitinase in apple cells was detrimental to plant growth (Bolar *et al.*, 2000).

Apple production in temperate climate with often wet springs needs a relevant input of fungicides to control apple scab and powdery mildew (Laurens, 1998; MacHardy, 1996). Such large chemical input is under critical scrutiny due to its potential environmental impact and ability to induce resistance in the pathogen. Additionally, if weather conditions are particularly warm and wet during the flowering, fire blight (caused by *Erwinia amylovora*) may cause severe damage to the apple orchard. Classical breeding has produced many scab resistant cultivars overwhelmingly based on the resistance from *M. floribunda* 821 (*Vf*) and some mildew resistant cultivars and efforts to breed also fire blight resistant cultivars are currently undertaken (Lespinasse, 1989; MacHardy, 1996; Gessler, 2015). Today after over 100 years of breeding efforts a range of scab resistant cultivars is available, however in most of them base their resistance is on a single gene named *Vf* originating from the *Malus floribunda* tree number 821 (Gessler and Pertot, 2012). Marker-assisted selection added a relevant quality by shortening time and allowing pyramiding of resistance genes. If the development of markers for MAS was the primary goal of genetic analysis in the 90's, identification and cloning of resistance genes is now the goal. Three resistance genes have been isolated and cloned into a susceptible apple cultivar: *HcrVf2* and *Vr2_C* responsible for the *Vf* and *Rvi15* scab resistance, respectively, and the fire blight resistance gene *FB_RM5* from *Malus* × *robusta* 5. All these cloned resistance genes could now be pyramided into a single cultivar through genetic engineering using biotechnology.

The research findings demonstrated that introduction of apple own genes with their own regulatory sequences without the presence of foreign genes in the end product (cisgenic) is possible and a popular cultivar can be ameliorated through addition of genes by genetic engineering. Cisgenesis represents a step toward a new generation of GM crops. The lack of selectable genes (*e.g.* antibiotic or herbicide resistance) in the final product and the fact that the inserted gene(s) derive from organisms sexually compatible with the target crop should rise less environmental concerns and increase consumer's acceptance. The used methodology allows the sequential addition of genes; *e.g.*, a second or more transformations. Vanblaere *et al.* (2011) reported the generation of a cisgenic apple plant by inserting the endogenous apple scab resistance gene *HcrVf2* under the control of its own regulatory sequences into the scab susceptible apple cultivar Gala. A previously developed method based on *Agrobacterium*-mediated transformation combined with a positive and negative selection system and a chemically inducible recombination machinery allowed the generation of apple cv. Gala carrying the scab resistance gene *HcrVf2* under its native regulatory sequences and no foreign genes. Three cisgenic lines were chosen for detailed investigation and were shown to carry a single T-DNA insertion and express the target gene *HcrVf2*. Recently, Kost *et al.* (2015) developed a cisgenic apple line C44.4.146 from a fire blight susceptible cultivar 'Gala Galaxy' using the cisgene *FB_MR5* from wild apple *Malus* × *robusta* 5 (*Mr5*), which confers resistance to fire blight. Apart from apple, cisgenesis was also explored in grape and other fruit crops (Table 6.2).

Table 6.2: Application of Cisgenesis for Disease Resistance in Various Fruit Crops

Fruit Crop	*Gene*	*Source*	*Trait*	*Reference*
Malus domestica Borkh.	HcrVf2	*Malus domestica*	Apple scab resistance	Joshi *et al.* (2011)
Vitis vinifera L.	*VVTL*-1	Vitis vinifera	Powdery mildew resistance	Dhekney *et al.* (2011)
Malus domestica Borkh.	HcrVf2	Apple cv. Gala	Apple scab resistance	Gessler *et al.* (2014)
Malus domestica Borkh.	Rvi6	*Malus floribunda* 821	Apple scab resistance	Krens *et al.* (2015)
Malus domestica Borkh.	Rvi6	*Malus floribunda* 821	Apple scab (Race 1) resistance	Wurdig *et al.* (2015)
Malus domestica Borkh.	FB_MR5	Apple cv. Gala Galaxy	Fire blight resistance	Kost *et al.* (2015)

In the US, first field trials show that the *Vf* scab resistance is, also holding up in the field assuming absence of *Vf* overcoming races of the scab fungus. However, even if currently no indication of any negative effect can be demonstrated or even postulated, the general opposition against GMO, and the rather cumbersome legislation requests, hampers the development of a cisgenic disease resistant apple cultivar. Public opinion enhanced by particular active groups is based on misconception of the general danger of the technology and not on the true value (or added value) of the product. In the case of apple this added value would be a relevant reduction of pesticide input in the orchard (Table 6.3).

Table 6.3: Application of RNA Interference for Disease Resistance in Various Fruit Crops

Fruit Crop	*Target Gene*	*Source*	*Target Trait*	*Reference*
Prunus domestica	PPV-CP	*Plum pox virus* (PPV)	Sharka (PPV) resistance	Scorza *et al.* (2001; 2013)
Malus domestica	iaaM and ipt	*Agrobacterium tumefaciens*	Crown gall Resistance	Viss *et al.* (2003)
Citrus paradisi	CTV	*Citrus tristeza virus* (*CTV*)	CTV resistance	Febres *et al.* (2008)
Citrus sinensis	CPsV-CP	*Citrus psorosis virus* (*CPsV*)	*Citrus psorosis* resistance	Reyes *et al.* (2011)
Malus domestica	MdMLO19	*Malus domestica*	Powdery mildew resistance	Pessina *et al.* (2016)

Apart from the various biotechnological tools, *viz.* RNA interference (RNAi), trans-grafting and cisgenesis/intragenesis, the targeted genome editing (TGE), has introduced the possibility of more precise and faster genetic modifications of plants (Limera *et al.*, 2016). TGE using site-specific nucleases (SSNs) is of particular importance for the introduction or modification of specific traits in fruit crops while maintaining unchanged general characteristics of a selected cultivar. Moreover, genome editing tools give the possibility to obtain transgene-free modified fruit tree genomes, which should increase consumer's acceptance. Recently, the RNA-guided CRISPR/Cas9 system has been widely used in TGE. To date, the CRISPR/Cas9 system has been adopted for TGE successfully in limited fruit crops, including sweet orange, grape, apple, grape fruit, banana and kiwi (Table 6.4).

Table 6.4: Applications of Genome Editing for Disease Resistance in Various Fruit Crops

Fruit Crop	*Target Gene*	*Target Trait*	*Reference*
V. vinifera	*MLO-7*	Resistance to powdery mildew	Malnoy *et al.* (2016)
Malus domestica	*DIPM*-1, 2, and 4	Resistance to fire blight	Malnoy *et al.* (2016)
C. sinensis	*CsLOB1* promoter	Citrus canker resistance	Peng *et al.* (2017)
C. paradisi	*CsLOB1*	Citrus canker resistance	Jia *et al.* (2017)
V. Vinifera cv. Thompson Seedless	*VvWRKY52*	Resistance to *Botrytis cinerea*	Wang *et al.* (2018)
Theobroma cacao	*TcNPR3*	Resistance to *Phytophthora tropicalis*	Fister *et al.* (2018)

In a recent study, genome modification of the effector binding element (EBE) of one single allele of *CsLOB1* gene in grapefruit Duncan (*Citrus paradisi* Macf.) alleviated the canker symptoms due to a specific transcription activator-like (TAL) effector. But as grapefruit is a hybrid (*C. maxima* × *C. sinensis*), the modified grapefruit line is still susceptible to wild-type *Xanthomonas citri* ssp. *citri* (*Xcc*), as only one *CsLOB1* allele was altered (Jia *et al.*, 2016). Thus, Jia *et al.* (2017) used CRISPR/Cas9/sgRNA technology to disrupt the coding region of both alleles of *CsLOB1* by targeting the conserved region of both alleles. Six independent lines, $D_{LOB}2$, $D_{LOB}3$, $D_{LOB}9$, $D_{LOB}10$, $D_{LOB}11$ and $D_{LOB}12$ were generated, which showed the mutation rate of 31.58, 23.80, 89.36, 88.79, 46.91 and 51.12 per cent, respectively. Amongst the lines, $D_{LOB}11$ and $D_{LOB}12$ were noted as the most resistant to *Xanthomonas citri* ssp. *citri*. No side effects and off-target mutations were detected in the genome edited grapefruit lines.

Very recently, Wang *et al.* (2018) demonstrated an efficient genome editing of Thompson Seedless grape (*Vitis vinifera*) suspension cells using the type II CRISPR/Cas9 system. Four guide RNAs were designed to target the *VvWRKY52* transcription factor gene for using with the CRISPR/Cas9 system *via* Agrobacterium-mediated transformation, using somatic embryos. Knockout of *VvWRKY52* in Thompson Seedless grape enhances resistance to *Botrytis cinerea* with no off target modifications.

Herbicides Tolerance

For some fruit crops, it is not cost-effective to remove weeds by physical means such as tilling, so farmers will often spray large quantities of different herbicides to destroy weeds, a time-consuming and expensive process that requires care so that the herbicide doesn't harm the crop plant or the environment. Crop plants genetically engineered to be resistant to one very powerful herbicide could help prevent environmental damage by reducing the amount of herbicides needed (Sharma *et al.*, 2016). For example, cranberry is genetically modified using bar, *aphII* gene possess herbicide tolerance properties (Zeldin *et al.*, 2002).

Conclusion

Genetically-modified fruit crops with the help of biotechnology have the potential to solve many of the world's hunger and malnutrition problems, and to help protect and preserve the environment by increasing yield and reducing reliance upon chemical pesticides and herbicides.

REFERENCES

Asao, H., Nishizawa, Y., Arai, S., Sato, T., Hirai, M., Yoshida, K. and Hibi, T. (1997). Enhanced resistance against a fungal pathogen *Sphaerotheca humuli* in transgenic strawberry expressing a rice *chitinase* gene. *Plant Biotechnology*, **14**(3): 145-149.

Barbosa-Mendes, J. M., Mourao Filho, F. D. A. A., Bergamin Filho, A., Harakava, R., Beer, S. V. and Mendes, B. M. J. (2009). Genetic transformation of *Citrus sinensis* cv. Hamlin with *hrpN* gene from *Erwinia amylovora* and evaluation of the transgenic lines for resistance to citrus canker. *Scientia Horticulturae*, **122**(1): 109-115.

Belfanti, E., Silfverberg-Dilworth, E., Tartarini, S., Patocchi, A., Barbieri, M., Zhu, J. and Sansavini, S. (2004). The *HcrVf2* gene from a wild apple confers scab resistance to a transgenic cultivated variety. *Proceedings of the National Academy of Sciences*, **101**(3): 886-890.

Bolar, J. P., Norelli, J. L., Wong, K. W., Hayes, C. K., Harman, G. E. and Aldwinckle, H. S. (2000). Expression of *endochitinase* from *Trichoderma harzianum* in transgenic apple increases resistance to apple scab and reduces vigor. *Phytopathology*, **90**(1): 72-77.

Cervera, M., Esteban, O., Gil, M., Gorris, M. T., Martinez, M. C., Pena, L. and Cambra, M. (2010). Transgenic expression in citrus of single-chain antibody fragments specific to Citrus tristeza virus confers virus resistance. *Transgenic Research*, **19**(6), 1001-1015.

Chen, X. K., Zhang, J. Y., Zhang, Z., Du, X. L., Du, B. B. and Qu, S. C. (2012). Overexpressing *MhNPR1* in transgenic Fuji apples enhances resistance to apple powdery mildew. *Molecular Biology Reports*, **39**(8): 8083-8089.

Dandekar, A. M., McGranahan, G. H., Vail, P. V., Uratsu, S. L., Leslie, C. A. and Tebbets, J. S. (1998). High levels of expression of full-length *cry*IA (c) gene from *Bacillus thuringiensis* in transgenic somatic walnut embryos. *Plant Science*, **131**(2): 181-193.

Delledonne, M., Allegro, G., Belenghi, B., Balestrazzi, A., Picco, F., Levine, A. and Confalonieri, M. (2001). Transformation of white poplar (*Populus alba* L.) with a novel *Arabidopsis thaliana* cysteine proteinase inhibitor and analysis of insect pest resistance. *Molecular Breeding*, **7**(1): 35-42.

Dhekney, S. A., Li, Z. T. and Gray, D. J. (2011). Grapevines engineered to express cisgenic *Vitis vinifera* thaumatin-like protein exhibit fungal disease resistance. *In Vitro Cellular and Developmental Biology-Plant*, **47**(4): 458-466.

Dias, J. S. and Ortiz, R. (2014). Advances in transgenic vegetable and fruit breeding. *Agricultural Sciences*, **5**: 1448-1467.

Dutt, M., Barthe, G., Irey, M. and Grosser, J. (2015). Transgenic citrus expressing an *Arabidopsis NPR1* gene exhibit enhanced resistance against Huanglongbing (HLB; Citrus Greening). *PloS One*, **10**(9): e0137134.

Faize, M., Sourice, S., Dupuis, F., Parisi, L., Gautier, M. F. and Chevreau, E. (2004). Expression of wheat *puroindoline-b* reduces scab susceptibility in transgenic apple (*Malus* × *domestica* Borkh.). *Plant Science*, **167**(2): 347-354.

Febres, V. J., Lee, R. F. and Moore, G. A. (2008). Transgenic resistance to *Citrus* tristeza virus in grapefruit. *Plant Cell Reports*, **27**(1): 93-104.

Fister, A. S., Landherr, L., Maximova, S. N. and Guiltinan, M. J. (2018). Transient expression of CRISPR/Cas9 machinery targeting *TcNPR3* enhances defense response in *Theobroma cacao*. *Frontiers in Plant Science*, **9**: 268.

Flachowsky, H., Richter, K., Kim, W. S., Geider, K. and Hanke, M. V. (2008). Transgenic expression of a viral *EPS depolymerase* is potentially useful to induce fire blight resistance in apple. *Annals of Applied Biology*, **153**(3): 345-355.

Flachowsky, H., Szankowski, I., Fischer, T. C., Richter, K., Peil, A., Hofer, M. and Hanke, M. V. (2010). Transgenic apple plants overexpressing the *Lc* gene of maize show an altered growth habit and increased resistance to apple scab and fire blight. *Planta*, **231**(3): 623-635.

Gessler, C. (2015). Biotechnology for a pesticide free orchard: a promise or an utopia. ***In***: *XIV EUCARPIA Symposium on Fruit Breeding and Genetics*, **1172**: 1-8.

Gessler, C. and Pertot, I. (2012). *Vf* scab resistance of *Malus*. *Trees*, **26**(1): 95-108.

Gessler, C., Vanblaere, T., Parravicini, G. and Broggini, G. A. L. (2014). Cisgenic 'Gala' containing the scab resistance gene from *Malus floribunda* 821 and the fire blight resistance genes from M. 'Evereste'. *Acta Horticulturae*. **1048**: 43-50. doi: 10.17660/ActaHortic.2014.1048.4

Ghag, S. B., Shekhawat, U. K. S. and Ganapathi, T. R. (2012). Petunia floral *defensins* with unique prodomains as novel candidates for development of *Fusarium* wilt resistance in transgenic banana plants. *PLoS One*, **7**(6): e39557.

Ghag, S. B., Shekhawat, U. K. S. and Ganapathi, T. R. (2014). Native cell-death genes as candidates for developing wilt resistance in transgenic banana plants. *AoB Plants*, **6**. doi: 10.1093/aobpla/plu037

Gonsalves, D., Ferreira, S., Manshart, R., Fitch, M. and Slightom, J. (2000). Transgenic virus resistant papaya: New hope for controlling papaya ringspot virus in Hawaii. *Plant Health Progress*. **1**: 105-121. doi: 10.1094/PHP-2000-0621-01-RV.

Jardak-Jamoussi, R., Winterhagen, P., Bouamama, B., Dubois, C., Mliki, A., Wetzel, T. and Reustle, G. M. (2009). Development and evaluation of a *GFLV* inverted repeat construct for genetic transformation of grapevine. *Plant Cell, Tissue and Organ Culture*, **97**(2): 187-196.

Jia, H., Orbovic, V., Jones, J. B. and Wang, N. (2016). Modification of the *PthA4* effector binding elements in Type I *Cs LOB 1* promoter using Cas9/sg RNA to produce transgenic Duncan grapefruit alleviating XccΔpthA4: dCs LOB 1.3 infection. *Plant Biotechnology Journal*, **14**(5): 1291-1301.

Jia, H., Zhang, Y., Orbovic, V., Xu, J., White, F. F., Jones, J. B. and Wang, N. (2017). Genome editing of the disease susceptibility gene *CsLOB1* in citrus confers resistance to citrus canker. *Plant Biotechnology Journal*, **15**(7): 817-823.

Joshi, S. G., Schaart, J. G., Groenwold, R., Jacobsen, E., Schouten, H. J. and Krens, F. A. (2011). Functional analysis and expression profiling of *HcrVf1* and *HcrVf2* for development of scab resistant cisgenic and intragenic apples. *Plant Molecular Biology*, **75**(6): 579-591.

Kost, T. D., Gessler, C., Jansch, M., Flachowsky, H., Patocchi, A. and Broggini, G. A. (2015). Development of the first cisgenic apple with increased resistance to fire blight. *PloS One*, **10**(12): e0143980.

Kovacs, G., Sagi, L., Jacon, G., Arinaitwe, G., Busogoro, J. P., Thiry, E. and Remy, S. (2013). Expression of a rice *chitinase* gene in transgenic banana ('Gros Michel', AAA genome group) confers resistance to black leaf streak disease. *Transgenic Research*, **22**(1): 117-130.

Krens, F. A., Schaart, J. G., van der Burgh, A. M., Tinnenbroek-Capel, I. E., Groenwold, R., Kodde, L. P. and Schouten, H. J. (2015). Cisgenic apple trees; development, characterization, and performance. *Frontiers in Plant Science*, **6**: 286.

Laurens, F. (1996). Review of the current apple breeding programmes in the world: objectives for scion cultivar improvement. In: *Eucarpia Symposium on Fruit Breeding and Genetics*, **484**: 163-170.

Lespinasse, Y. (1989). Breeding pome fruits with stable resistance to diseases. 3. Genes, resistance mechanisms, present work and prospects. *Bulletin SROP (France)*, **2**: 100-115.

Limera, C., Sabbadini, S., Sweet, J. B. and Mezzetti, B. (2017). New biotechnological tools for the genetic improvement of major woody fruit species. *Frontiers in Plant Science*, **8**: 1418.

Lindow, S., Newman, K., Chatterjee, S., Baccari, C., Iavarone, A. T. and Ionescu, M. (2014). Production of *Xylella fastidiosa* diffusible signal factor in transgenic grape causes pathogen confusion and reduction in severity of Pierce's disease. *Molecular Plant-Microbe Interactions*, **27**(3), 244-254.

Lius, S., Manshardt, R. M., Fitch, M. M., Slightom, J. L., Sanford, J. C. and Gonsalves, D. (1997). Pathogen-derived resistance provides papaya with effective protection against papaya ringspot virus. *Molecular Breeding*, **3**(3): 161-168.

MacHardy, W. E. (1996). *Apple scab: biology, epidemiology, and management*. American Phytopathological Society. APS Press, St. Paul, Minnesota.

MacHardy, W. E., Gadoury, D. M. and Gessler, C. (2001). Parasitic and biological fitness of *Venturia inaequalis*: relationship to disease management strategies. *Plant Disease*, **85**(10): 1036-1051.

Malnoy, M., Viola, R., Jung, M. H., Koo, O. J., Kim, S., Kim, J. S. and Nagamangala Kanchiswamy, C. (2016). DNA-free genetically edited grapevine and apple protoplast using CRISPR/Cas9 ribonucleoproteins. *Frontiers in Plant Science*, **7**: 1904.

McGranahan, G. and Leslie, C. (1991). Walnuts (Juglans). *Genetic Resources of Temperate Fruit and Nut Crops*, **290**: 907-974.

Mercado, J. A., Barcelo, M., Pliego, C., Rey, M., Caballero, J. L., Munoz-Blanco, J. and Pliego-Alfaro, F. (2015). Expression of the β-*1,3-glucanase* gene (*bgn13. 1*) from *Trichoderma harzianum* in strawberry increases tolerance to crown rot diseases but interferes with plant growth. *Transgenic Research*, **24**(6), 979-989.

Mercado, J.A., Martin-Pizarro, C., Pascual, L., de los Santos, B., Romero, F., Quesada, M. A., Pliego-Alfaro, F., Galvez, J., Rey, M., de la Vina, G., Llobell, A., Yubero-Serrano, E. M., Munoz-Blanco, J. and Caballero, J. L. (2007). Evaluation of tolerance of *Colletotrichum acutatum* in strawberry plants transformed with *Trichoderma*-derived genes. *Acta Horticulturae*, **738**: 383-388.

Mondal, S. N., Dutt, M., Grosser, J. W. and Dewdney, M. M. (2012). Transgenic citrus expressing the antimicrobial gene *Attacin E* (*attE*) reduces the susceptibility of 'Duncan' grapefruit to the citrus scab caused by *Elsinoe fawcettii*. *European Journal of Plant Pathology*, **133**(2), 391-404.

Nookaraju, A. and Agrawal, D. C. (2012). Enhanced tolerance of transgenic grapevines expressing *chitinase* and β-*1, 3-glucanase* genes to downy mildew. *Plant Cell, Tissue and Organ Culture*, **111**(1): 15-28.

Norelli, J. L., Aldwinckle, H. S., Destefano-Beltran, L. and Jaynes, J. M. (1994). Transgenic 'Mailing 26'apple expressing the *attacin E* gene has increased resistance to *Erwinia amylovora*. *Euphytica*, **77**: 123-128.

Pena, L. and Seguin, A. (2001). Recent advances in the genetic transformation of trees. *Trends in Biotechnology*, **19**(12): 500-506.

Peng, A., Chen, S., Lei, T., Xu, L., He, Y., Wu, L. and Zou, X. (2017). Engineering cankerresistant plants through CRISPR/Cas9targeted editing of the susceptibility gene *Cs LOB 1* promoter in citrus. *Plant Biotechnology Journal*, **15**(12): 1509-1519.

Pessina, S., Angeli, D., Martens, S., Visser, R. G., Bai, Y., Salamini, F. and Malnoy, M. (2016). The knockdown of the expression of *MdMLO19* reduces susceptibility to powdery mildew (*Podosphaera leucotricha*) in apple (*Malus domestica*). *Plant Biotechnology Journal*, **14**(10): 2033-2044.

Ravelonandro, M., Monsion, M., Teycheney, P. Y., Delbos, R. and Dunez, J. (1992). Construction of a chimeric viral gene expressing plum pox virus *coat protein*. *Gene*, **120**(2): 167-173.

Ravelonandro, M., Scorza, R., Renaud, R. and Salesses, G. (1998). Transgenic plums resistant to plum pox virus infection and preliminary results of cross-hybridization. *Acta Horticulturae*, **478**: 67-71.

Reyes, C. A., De Francesco, A., Pena, E. J., Costa, N., Plata, M. I., Sendin, L. and Garcia, M. L. (2011). Resistance to *Citrus* psorosis virus in transgenic sweet orange plants is triggered by coat protein-RNA silencing. *Journal of Biotechnology*, **151**(1): 151-158.

Reyes, C. A., De Francesco, A., Pena, E. J., Costa, N., Plata, M. I., Sendin, L. and Garcia, M. L. (2011). Resistance to *Citrus* psorosis virus in transgenic sweet orange plants is triggered by coat protein-RNA silencing. *Journal of Biotechnology*, **151**(1): 151-158.

Reynoird, J. P., Mourgues, F., Norelli, J., Aldwinckle, H. S., Brisset, M. N. and Chevreau, E. (1999). First evidence for improved resistance to fire blight in transgenic pear expressing the *attacin E* gene from *Hyalophora cecropia*. *Plant Science*, **149**(1), 23-31.

Roderick, H., Tripathi, L., Babirye, A., Wang, D., Tripathi, J., Urwin, P. E. and Atkinson, H. J. (2012). Generation of transgenic plantain (*Musa* spp.) with resistance to plant pathogenic nematodes. *Molecular Plant Pathology*, **13**(8): 842-851.

Schestibratov, K. A. and Dolgov, S. V. (2005). Transgenic strawberry plants expressing a *thaumatin II* gene demonstrate enhanced resistance to *Botrytis cinerea*. *Scientia Horticulturae*, **106**(2): 177-189.

Schuler, T. H., Poppy, G. M., Kerry, B. R. and Denholm, I. (1998). Insect-resistant transgenic plants. *Trends in Biotechnology*, **16**(4): 168-175.

Scorza, R., Callahan, A., Dardick, C., Cambra, M., Polak, J., Ravelonandro, M., Zagrai, I. and Malinowski, T. (1998). "Honey Sweet"-A transgenic plum pox virus resistant plum-from laboratory and experimental field plots to regulatory approval. *Acta Horticulturae*, **974**: 57-63.

Scorza, R., Callahan, A., Dardick, C., Ravelonandro, M., Polak, J., Malinowski, T. and Kamenova, I. (2013). Genetic engineering of Plum pox virus resistance: 'Honey Sweet' plum from concept to product. *Plant Cell, Tissue and Organ Culture*, **115**(1): 1-12.

Scorza, R., Callahan, A., Levy, L., Damsteegt, V., Webb, K. and Ravelonandro, M. (2001). Post-transcriptional gene silencing in plum pox virus resistant transgenic European plum containing the plum pox potyvirus coat protein gene. *Transgenic Research*, **10**(3): 201-209.

Scorza, R., Callahan, A., Levy, L., Damsteegt, V., Webb, K. and Ravelonandro, M. (2001). Post-transcriptional gene silencing in plum pox virus resistant transgenic European plum containing the plum pox potyvirus coat protein gene. *Transgenic Research*, **10**(3): 201-209.

Sharma, N., Singh, S. K. and Lal, S. (2016) Transgenic Research in Fruit Crops: Current Status. *Advancements in Genetic Engineering*, **3**: 155. doi: 10.4172/2169-0111.1000155

Suzuki, J. Y., Tripathi, S. and Gonsalves, D. (2007). Virus-resistant transgenic papaya: commercial development and regulatory and environmental issues. *Biotechnology and Plant Disease Management. CAB International, Wallingford,* 436-461.

Tripathi, L. (2012). Transgenics in crop improvement research at IITA. *IITA Research for Development (R4D) Review*, **8**: 58-60.

Tripathi, L., Mwaka, H., Tripathi, J. N. and Tushemereirwe, W. K. (2010). Expression of sweet pepper *Hrap* gene in banana enhances resistance to *Xanthomonas campestris* pv. musacearum. *Molecular Plant Pathology*, **11**(6): 721-731.

Tzfira, T., Zuker, A. and Altman, A. (1998). Forest-tree biotechnology: genetic transformation and its application to future forests. *Trends in Biotechnology*, **16**(10): 439-446.

Vanblaere, T., Szankowski, I., Schaart, J., Schouten, H., Flachowsky, H., Broggini, G. A. and Gessler, C. (2011). The development of a cisgenic apple plant. *Journal of Biotechnology*, **154**(4): 304-311.

Viss, W. J., Pitrak, J., Humann, J., Cook, M., Driver, J. and Ream, W. (2003). Crown-gall-resistant transgenic apple trees that silence *Agrobacterium tumefaciens* oncogenes. *Molecular Breeding*, **12**(4), 283-295.

Wang, X., Tu, M., Wang, D., Liu, J., Li, Y., Li, Z. and Wang, X. (2018). CRISPR/Cas9-mediated efficient targeted mutagenesis in grape in the first generation. *Plant Biotechnology Journal*, **16**(4): 844-855.

Wang, X., Tu, M., Wang, D., Liu, J., Li, Y., Li, Z. and Wang, X. (2018). CRISPR/Cas9-mediated efficient targeted mutagenesis in grape in the first generation. *Plant Biotechnology Journal*, **16**(4): 844-855.

Wurdig, J., Flachowsky, H., Sab, A., Peil, A. and Hanke, M. V. (2015). Improving resistance of different apple cultivars using the *Rvi6* scab resistance gene in a cisgenic approach based on the *Flp/FRT* recombinase system. *Molecular Breeding*, **35**(3): 95.

Zeldin, E. L., Jury, T. P., Serres, R. A. and McCown, B. H. (2002). Tolerance to the herbicide glufosinate in transgenic cranberry (*Vaccinium macrocarpon* Ait.) and enhancement of tolerance in progeny. *Journal of the American Society for Horticultural Science*, **127**(4): 502-507.

Zhu, Y. J., Agbayani, R. and Moore, P. H. (2007). Ectopic expression of Dahlia *merckii defensin DmAMP1* improves papaya resistance to *Phytophthora palmivora* by reducing pathogen vigour. *Planta*, **226**(1): 87-97.

Breeding of Perennial Horticultural Crops *Pages 85–97*
Editors: S.K. Singh, V.B. Patel, A.K.Goswami, Jai Prakash and Chavlesh Kumar
Published by: **BIOTECH BOOKS, NEW DELHI**

Chapter 7

Embryo Rescue

S.K. Singh, Manish Srivastav and Chavlesh Kumar

Division of Fruits and Horticultural Technology,
ICAR-Indian Agricultural Research Institute, Pusa, New Delhi – 110 012

INTRODUCTION

In plants, the embryo is a multi-cellular structure which germinates into full plant. It represents the sporophytic generation. Further, it is known that the embryo may arise from a specialized cell, *i.e.*, zygote or from a somatic cell. The former is known as a zygotic embryo, whereas the latter is termed a somatic, adventive or asexual embryo. An embryo is essentially a bipolar structure with root and shoots primordial structures.

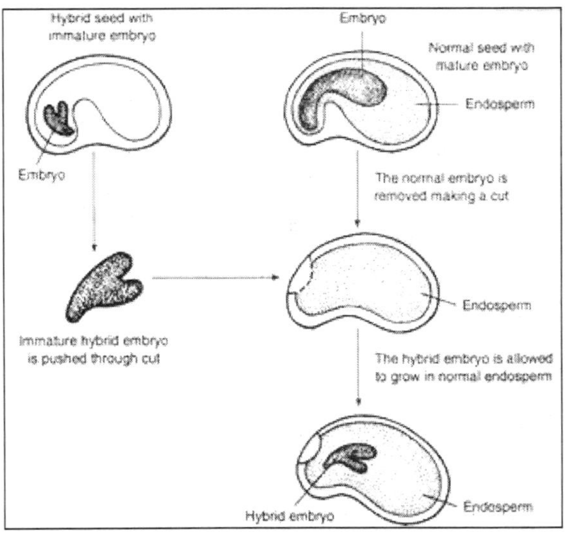

Normally, the embryos are very minute structures in grasses, while are very large in *Sequoia* sp. For breeders, F_1 hybrids, imbreds *etc.* are desired to achieve a certain objective. This becomes a point of consideration when the embryos are of desirable genotypes of cross combination. It is of concern, when a desired trait or traits are to be transferred from distant or wild species when a viable embryo cannot be formed. In such cases, the embryos (or pro-embryos as the case may be) are required to be rescued by culturing it *in vitro* from being aborted under controlled artificial medium.

The term 'embryo rescue'or '*in-ovulo* embryo rescue' refers to a number of *in vitro* techniques whose purpose is to promote the development of an immature or weak embryo into a viable plant. Embryo rescue has been widely used for producing plants from hybridizations in which failure of endosperm to properly develop causes embryo abortion. In embryo rescue procedures, the artificial nutrient medium serves as a substitute for the endosperm, thereby following the embryo to continue its development. Embryo rescue techniques are among the oldest and most successful *in vitro* procedures.

One of the primary uses of embryo rescue has been to produce interspecific and inter-generic hybrids. While interspecific incompatibility can occur for a wide variety of reasons, one common cause is embryo abortion. The production of small, shrunken seed following wide hybridization is indicative of a cross in which fertilization occurred but seed development was disrupted. Embryo rescue procedures have been very successful in overcoming this barrier to wide hybridization in a wide range of plant materials (Collins and Grosser, 1984). In addition, embryo rescue has been used to recover maternal haploids that have developed as a result of chromosome elimination following interspecific hybridization.

Historical Perspective

More systematic steps in embryo culture, however, started from the beginning of this century when Hannig (1904) cultured embryos of Cruciferae excised from a few taxa such as *Raphanus* spp. and *Chochlearia damica* in medium containing mineral salts and sugar under aseptic conditions and succeeded in obtaining plants. This study opened up a field for basic investigations in embryology. Inter-specific crosses may fail because of several reasons, but when the development of embryo is arrested owing to the degeneration of the endosperm, or when the embryo aborts at an early stage of development, embryo culture is the only technique to recover hybrid plants. Laibach (1925) for the first time, employed embryo culture for obtaining hybrids from a cross between *Linum perenne* and *L. austriacum*. Thereafter, this technique has been successfully adopted to raise hybrid embryos from interspecific hybrids that otherwise fail due to embryo abortion. In fruit crops, it was Tukey (1933a) successfully cultured cherry embryos and raised complete plants. His medium and procedure has been widely used and even applied to other crops. Using this procedure, Blake (1939) was the first to employ culture in a peach. Embryo culture is useful when there is poor embryo development or abortion. Embryo abortion occurs in early ripening genotypes of *Prunus* where the flesh matures before seed maturity (Tukey, 1933a), precluding their use as females in a breeding program. The poor germination of seeds from early ripening grapes is probably due to poor embryo development. Embryos have been rescued from early ripening *Prunus* (Tukey, 1934) and *Vitis* (Ramming and Emershad, 1984) cultivars, thus opening the scope for evolving seedless varieties in the crosses like seedless x seedless or seedless x seeded parents.

Embryo culture has helped in overcoming self-sterility of seeds, especially of crop plants propagated vegetatively, when the seeds do not germinate in nature. Many seedless grape cultivars are stenospermomic (Stout, 1936), *i.e.*, they are characterized by cessation of embryo development after fertilization and failure of seed development. In a wild relative of commercial banana, *Musa balbisiana*, the seeds do not germinate under natural conditions. However, if the embryos are excised and grown on a simple culture medium of mineral salts, seedlings can be obtained. Hence, any post zygotic barrier or natural sterility barrier could be overcome by resorting to culture of embryos.

Embryo Culture

Embryos have been rescued and grown into plants from stenospermocarpic seedless grapes (Emershad and Ramming, 1984; Spiegel-Roy *et al.*, 1985), allowing growth of progeny resulting from the hybridization of seedless with seedless grapes. The production of interspecific hybrids is useful for the transfer of desirable genes from wild to cultivated species. In many cases, wide crosses between species are difficult to produce because of many factors that act as barriers. Post-zygotic barriers such as endosperm abortion are a common occurrence, but they have been overcome through the use of embryo rescue and many interspecific hybrids have been produced (Collins and Grosser, 1984). Embryo rescue has also been used to overcome postzygotic barriers during self- and cross-incompatibility. Embryo culture has been used for rapid seed germination in peach; for production of virus-free citrus plants through culture of nucellar embryos (Bitters *et al.*, 1970, 1972); and for studying the developmental and growth requirements of embryos (Steward and Hsu, 1977). Embryo culture has also been used to shorten breeding cycles like apple (Nickell, 1951), where effective germination was reduced from years to months by overcoming dormancy. The possibility also exists for using embryo culture as a screening procedure.

Depending on the organ cultured, embryo rescue may be referred to as embryo, ovule, or ovary culture. The different uses of embryos culture could be as follows;

Attempting Incompatible Crosses

In inter-specific and inter-generic hybridization programmes, incompatibility barriers often prevent normal seed development and production of hybrids. Although there may be normal fertilization in some incompatible crosses, embryo abortion results in the formation of shriveled seeds. Poor and abnormal development of the endosperm caused embryo starvation and eventual abortion. Isolation of hybrid embryos before abortion and their in vitro culture may prevent these strong post-zygotic barriers. The most useful and popular application of embryo cultures is to raise rare hybrids by rescuing embryos of incompatible crosses.

Ovary and Ovule Culture

Embryos at very early stages development have not been amenable to culture. This is so since there is the problem in excision of the immature embryo and complex nutrient requirements. In such a situation it is always better to culture either the entire ovary or the ovule (*in ovulo*). This facilitates minimum damage to young embryos. The culture of pollinated ovaries and ovule *in vitro* has been also successful. Nitsch *et al.* (1951) developed the technique of ovary culture of different crops including *Fragaria chilensis* and *F. virginiana*. Ovaries excised after pollination showed satisfactory growth on a simple culture medium. However, the size of the fruit development *in vitro* was smaller than that of field grown fruits. The crosses between these species are often unsuccessful because of the abortion of hybrid endosperm and embryo at early stages of development. Hence, *in ovulo* technique offers an alternative to overcome such difficulties.

Nucellus Culture

Rangaswamy (1958) evolved a novel technique of studying nucellar polyembryony in citrus fruits. He excised the nucellus from pollinated ovules of *Citrus japonica* (*C. microcarpa*) and cultured on different media. The micropylar half of the nucellus proliferated on White's medium containing casein hydrolysate. The callus formed differentiated into pseudo-bulbils, which eventually developed into whole plantlets. Presently, it has been attempted in many fruit crops towards clonal propagation of selected materials, *e.g.*, *Aegle marmelos, Vitis vinifera, Mangifera* sp. *etc.* Nucellar culture also helps in the elimination of viruses, *e.g.* in *Citrus*.

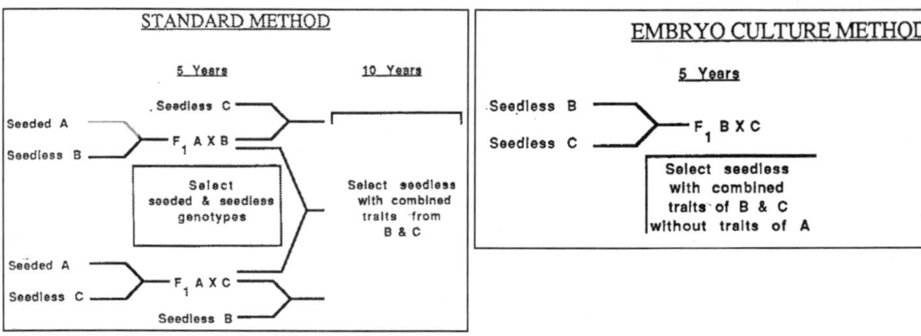

Controlled Pollination and Fertilization

It is one of the practical method to overcome crosses that are often confronted with several sexual barriers like (a) Non-synchronous flowering periods of the parent genotypess, (b) Short life span of pollen, (c) Failure or poor pollen germination, (d) Slow pollen tube growth, (e) Inhibition or bursting of pollen tube in style, and (f) Failure of male gametes to cause fertilization. Test-tube pollination and fertilization, first employed at the University of Delhi, have now been shown to have wider implications. With the success of intra-ovarian pollination, investigations were extended to demonstrate the feasibility of the technique in pollinating and fertilizing excised pistils and ovules *in vitro*. For successful pollination and fertilization *in vitro*, the following factors are important: (a) Culture of pollen grains and ovules at the right stage of development, (b) Ideal nutrient medium, and (c) Appropriate cultural conditions.

Endosperm Culture

Endosperm, usually a triploid tissue in angiosperms, can be suitable explant for raising triploid genotypes, which are seedless and self-sterile. Endosperm culture has been demonstrated in several fruit crops including grape, citrus, apple *etc.* The successful culture of endosperm tissue has opened up new avenues for plant breeders who can apply this technique to economically important crop plants, especially where conventional breeding methods prove futile.

Overcoming Dormancy

In vitro* culture of embryos helps in deciduous** fruit plants to overcome dormancy or chilling requirements or stratification before germination, like pome and stone fruits, like *Malus*, Quince, ***Prunus etc.

Shortening of Breeding

Seeds which takes 10-20 days for maturation places on *in vitro* culture to overcome maturation time. In seedless fruit breeding it can be used in just one generation. Inter-specific or inter-generic suffering incompatibilities can be minimized or even bridge species could be avoided so that desired genotypes could be salvaged.

Overcoming Seed Sterility

In early ripening fruit cultivars, seed do not germinate because their embryos are still immature. Using the embryo culture method it is possible to raise seedling from sterile seeds of early ripening stone fruits, peach, apricot and plum. 'Makapuno' coconuts are very expensive and most relished for their characteristics soft fatty endosperms in place of liquid endosperm in place of liquid endosperm. Under normal conditions the coconut seeds fail to germinate.

Production of Monoploids

An embryo culture has been used in production of monoploids of barley. With the cross *Hordeum vulgare*, fertilization occurs normally but thereafter chromosomes of H. bulbosum are eliminated, resulting in formation of monoploid *H. vulgare* embryo, which can be rescued by embryo cultures.

Factors Involved in Embryo Culture

Success of any embryo culture depends upon several basic factors which determine the recovery of desired hybrids/plants. It is essential that a worker who wants to undertake such studies need to generate basic flowering and pollination, *i.e.* pollen viability, pollen germination, flowering period, days to fertilization, data on incompatibility reactions, exogenous application of PGRs or micro-nutrients to enhance pollen germination or pollen tube growth, application of growth retardants like TDZ, ethrel, *etc.* to mature ovules making them fit for culture. Ovule of several temperate fruit crops requires low temperature/stratification treatment to complete their physiological needs for proper germination. All these basic information help in implementation of embryo rescue programme, since a lot of efforts are required during the initial period for initiating *in vitro* cultures of ovules post hand pollination.

Histological Observations on Embyro Abortion

These studies are basic in nature, which need to be undertaken in crops where no previous information exists in literature. Use of microscopic and fluorescence observations are required to ascertain the type of incompatibility reactions, pollen tube germination and extent of growth and fertilization;

Otimum Age of Embryo for Culture

It is the most basic information to be generated systematically for any crops where embryo abortion is common. It has to be standardized for all the parent genotypes and cross combination at least for two seasons. This can be ascertained after crossing/selfing and making culture after a week period at two day interval. This basic information will result in improving the efficiency with high degree of success and also handling the *in vitro* culture schedule in a breeding season. Workers must pinpoint the exact time of culture post pollination under a specific climatic conditions like tropical, sub-tropical and temperate. In grape, it has been shown that *in ovulo* embryo rescue can be done by three week period (18-24 days) compared to over 90-145 days under temperate conditions.

Culture Medium

Embryo is ontogenically a very young and immature tissue. Hence, basal medium needs to be selected as per the *in vivo* requirement of the developing embryos. Medium to high salt media are used. Murashige and Skoog (MS) (Murashige and Skoog, 1962) and Gamborg's B-5 (Gamborg *et al.*, 1968) media are the most commonly used basal media for embryo rescue studies. Besides, different modification are also used for macro/micro salts. Types and concentrations of media supplements required depend greatly on the stage of development of the embryo. There are two phases of embryo development. In the heterotropic phase, the young embryo, which is often referred to as a proembryo, is dependent on the endosperm. Embryos initiated at this stage require a complex medium. Amino acids, particularly glutamine, aspargine, are often added to the medium. Various vitamins may also be included. Natural extracts, such as coconut milk and casein hydrolysate, have sometimes been used instead of specific amino acids. Immature embryos require a culture medium having high osmotic potential. Sucrose often serves both as a carbon source and osmoticum. For heterotropic embryos, 8-12 per cent sucrose is commonly used. Other sugars, glucose, galactose, maltose *etc.* have been successfully used.

The second stage of embryo development is the Autotrophic phase, which usually begins in the late heart-shaped embryo stage (Raghavan, 1976). At this time, the embryo is capable of synthesizing substances required for its growth from salts and sugar. Embryo germination takes place on simple inorganic medium, supplemented with 2-3 per cent sucrose. Growth regulators have been extensively used in embryo rescue studies, especially for heterotropic embryos; however, their effects have been highly inconsistent. In general, low concentrations of auxins have promoted normal growth, gibberellic acid has caused embryo enlargement, and cytokinins have inhibited growth (Sharma *et al.*, 1996). Further, natural extracts like corn milk, tomato juice *etc.* are also added to facilitate embryo growth. More young is the embryos, more complex media formulations may be needed. The inter-specific hybrids are initially cultured on a high sucrose medium containing a moderate level of auxin and a low level of cytokinin. After 1 to 2 weeks on this medium, embryos stopped growing. Growth resumed after they were transferred on to a medium with a lower sucrose concentration, a low level of auxin, and a moderate level of cytokinin (Collins and Grosser, 1984). For inter-specific hybrids, it may be useful to develop media that can nurture embryos of one or both parental species. While, the nutritional needs of the hybrid may be different from the parents, the parental media formulations will serve as a good starting point for the hybrid.

Physical Factors (Temperature, Light, pH and Media matrix)

Temperature and light requirements vary among plant species. According to Sharma (1996), the growth requirements of embryos often mimic those of their parents, with embryos of cool-season crops requiring lower temperatures than those of warm-season crops. Cultures are often incubated at 25 to 30°C, although considerably lower temperatures are needed for some species. In species that normally exhibit seed dormancy, a cold treatment may be required. Cultures are initially cultured in dark to prevent precocious germination, but are moved to a lighted environment to allow chlorophyll development after 1 to 2 weeks in the dark.

Time of Culture

When attempting to rescue embryos in incompatible crosses, it is critical that the cultures be initiated prior to embryo abortion. It is more difficult to rear young embryos than those that have reached the autotrophic phase of development chances of success are maximized by allowing the embryo to develop *in vivo* as long as possible. Histological examinations can be used to determine the time of endosperm failure and embryo abortion; however, these evaluations can be very laborious. Cultures are often initiated at various intervals following pollination to maximize chances of recovering viable plants.

Culture Procedures

Embryo Culture

The most commonly used method, in which embryos are excised and placed directly onto culture medium. Fruit from controlled pollination of greenhouse- or field-grown plants is collected prior to the time at which embryo abortion is thought to occur. Since embryos are located in a sterile environment, disinfestation of the embryo itself is not required. In some cases, the entire ovary is surface-sterilized. In other cases, ovules are removed from the ovary under non-aseptic conditions and then disinfested. Careful excision of the embryo is critical to the success of embryo culture. The best point of incision into the ovule differs among species. In some cases, embryos can be extracted by cutting off the micropylar end of the ovule and then applying gentle pressure at the opposite end of the ovule. This results in the embryo being pushed out through the opening.

It is crucial that the embryo be placed directly into culture after its excision so that it does not become dry. For heart-shaped and younger embryos, the embryo should be excised with the suspensor intact (Hu and Wang, 1986). Because of the extreme importance and frequent difficulty of excising embryos without causing damage, it may be helpful to develop and practice an excision technique under non-aseptic conditions.

Embryo culture is sometimes preceded by ovule or ovary culture. One advantage of this technique (sometimes termed ovule-embryo or ovary-embryo culture) is that embryo excision is delayed until the embryo becomes large enough to remove without damage. Also, the presence of the integument during the ovule or ovary culture phase has been found to reduce the possibility of precocious germination (Ramming, 1990). Once excised, the embryo may benefit from being in direct contact with the medium. Also, for those species affected by dormancy, removing the embryo may overcome any inhibitory effects imposed by the surrounding ovular tissues. Nurse cultures have been used for rescuing embryos (Williams *et al.*, 1982). This technique involves inserting the embryo from an incompatible cross into endosperm removed from a related compatible cross. For example, the embryo of an interspecific hybrid may be inserted into endosperm from an intraspecific cross involving one of the parental species. The embryo and endosperm are then placed into culture together.

In-ovulo Culture

Embryos are difficult to excise when very young or from small-seeded species. To prevent damaging of embryos during excision, they are sometimes cultured, while still inside the ovule. This technique is referred to as ovule culture or *in-ovolo* embryo culture. As with embryo culture, ovaries are collected prior to the time at which embryo abortion is thought to occur. The ovary is surface-sterilized and the ovules removed and placed into culture. This step ranges from extremely easy to accomplish, for large-seeded species in which only a single ovule is present, to time-consuming and difficult, for small-seeded poly-ovulate species. Excision of the ovules may require the use of a stereomicroscope. Including placental tissue in ovule cultures has been found to be beneficial in some species (Rangan, 1984). Recent modifications of the standard ovule culture technique have been developed for use in peach (Pinto *et al.*, 1994). One technique, ovule perforation, requires making small holes in each ovule just prior to its placement on the culture medium. These perforations, which should be made with care not to damage embryos, increase water and nutrient uptake. Two types of ovule support systems have been developed. The filter paper support system involves culturing ovules on top of filter paper placed over liquid medium, whereas the vermiculite support technique entails placing ovules micropylar side down into a sterile vermiculite/liquid media mixture (vermiculite support). While the ovule perforation and vermiculite support systems may not be feasible in small-seeded species, ovule size should not pose a limitation for testing the filter paper support system in different species like peach.

Ovary Culture

In ovary or pod culture, the entire ovary is placed into culture. Fertilized ovaries are collected and any remaining flower parts removed. The ovary is placed onto culture so that the cut end of the pedicel is in the medium. After certain period, the seeds are removed from the fruit that develop into plant culture.

Applications of Embryo Culture

Peach and Apricot

The initial attempt by Tukey who could successfully culture cherry embryos (1933 b) *in vitro*, several others tried it in their breeding programmes. Blake (1939) used the procedure and Hough improved the culture medium and procedure (Smith *et al.*, 1969). Embryos with 75 per cent fill (PF_1) could be rescued. Peach embryo culture was started in 1975 using Smith, Bailey, and Hough (SBH) medium (Smith *et al.*, 1969) and procedures to develop progeny from early ripening genotypes, thus helped in developing good quality early ripening genotypes.

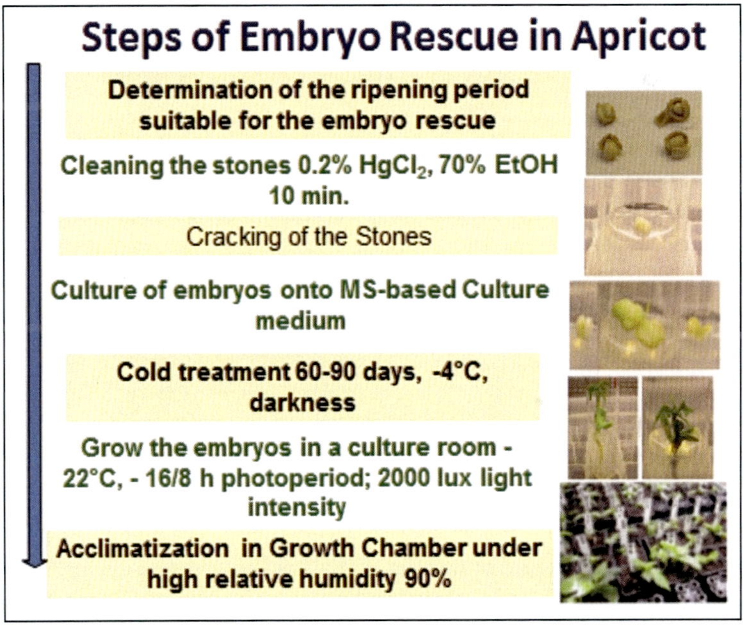

The procedure includes, surface sterilization of immature fruitlets with 70 per cent ethanol for 30 sec, followed with 0.525 per cent Na hypochlorite for 5 min. Thereafter, air-dried and excised using saw toothed flamed knife/secateurs under sterile conditions. The developing fruit is then cultured on the solid culture medium. Care must be taken to avoid damage of embryos while excision or taking damaged fruits. The low chill genotypes under California conditions the ideal stage for culture is 70- to 72-day after pollination. The procedure allowed the selection of genotypes ripening 10 to 30 days earlier than the parents. One can use Murashige and Skoog (MS) (1962) medium with the addition of 1,240 mg potassium succinate and 584.6 mg L-glutamine per liter and 3 per cent sucrose allowed embryos as small as 5 mm long to be grown successfully. Under north Indian conditions, the ideal culture age has been identified as 45-48 days. All over the world this technique is being followed. Germination of embryo ranges from 30-45 per cent. Two cultivars, Goldcrest peach and Mayfire nectarine were evolved from embryo culture and are among the earliest of the commercially grown cultivars in California (Ramming and Tanner, 1987).

Grape

Emershad and Ramming (1984) demonstrated this technique for rescuing early ripening grapes, which germinate very poor upon stratification. These genotypes are valuable as females and as males to increase the chances of developing early ripening grape cultivars. They used modified White's macroelements medium with Norstog's microelements and vitamins plus glycine and casein hydrolysate. They improved initial embryo germination from 0-16 per cent to 19-24 per cent and total plant production to 30-32 per cent depending upon the cross combination. Ovule after maturation were stratified for 3 months at 0.5°C, and then allowed to germinate. In some cases giving a small slit with knife give early emergence of seedling which is then hardened and plated out for evaluation. Every grape breeder's dream has been to interbreed stenospermic seedless grapes. This technique holds promise so that seedless parents could be used as seed parents. The proportion of seedless progeny could be enhanced considerably for making effective selection. Using ovule culture, many progenies could be developed from seedless × seedless crosses giving rise to 82 per cent seedlessness population instead of the 15 per cent average expected from seeded x seedless combinations (Loomis and Weinberger, 1979). This technique has special significance in sub-tropical grape breeding where earliness is the major trait. The ideal stage of culture is 18-24 days under sub-tropical conditions. For typical seedless genotypes, like Thompson Seedless, Centennial seedless, Flame Seedless *etc.* the stage in 10-14 days.

Conclusion

Embryo rescue procedures have been successfully used for producing interspecific and intergeneric hybrids and progeny of other incompatible crosses. It leads to successful application of this highly valuable *in vitro* procedure to additional crop species. The method can be useful to rescue seedless triploid embryos, produce haploids, overcome seed dormancy, determining seed viability, understanding embryo morphogenesis and precocious germination. While embryo culture is the most widely used embryo rescue procedure, ovule and ovary culture are more suitable for small seeded species or very young embryos. In addition to being able to culture embryos before abortion, techniques like *in vitro* pollination and fertilization can also be attempted. Crossing barriers could be eliminated and the potential to screen male and female gametes before hybridization might also be possible.

REFERENCES

Bitters, W.P., Murashige, T., Rangan, T.S., nauer, E. 1970. Investigations on established virus-free citrus plants through tissue culture. *Calif. Citrus Nurserymen's Soc.* **9**: 27-30.

Bitters, W.P., T. Murashige, T.S. Rangan, and E. Nauer. 1972. Investigations on established virus-free plants through tissue culture, p. 267-271. In: W.C. Price (Ed.). *Proc. 5th Conf. Intl. Organization Citrus Virologist*, Gainesville, Fla.

Bridgen, M.P. 1994. A review of plant embryo culture. *HortScience* **29**: 1243-1246.

Cain, D., Emershad, R.L. and Tarailo, R.E. 1983. *In-ovulo* embryo culture and seedling development of seeded and seedless grapes (*Vitis vinifera* L.). *Vitis* **22**: 9-14.

Collins, G.B. and Grosser, J.W. 1984. Culture of embryos. **In**: *Cell Culture and Somatic Cell Genetics of Plants, Vol. 1: Laboratory Procedures and Their Applications*. Ed. I.K. Vasil. Academic Press, New York, pp. 241-257.

Emershad, R.L. and Ramming, D.W. 1982. *In-ovulo* embryo culture of *Vitis vinifera* L. cv. 'Thompson Seedless'. *HortScience* 17: 576.

Emershad, R.L. and Ramming, D.W. 1984. In-ovulo embryo culture of *Vitis vinifera* L. cv. 'Thompson Seedless'. *Amer. J. Bot.* **71**: 873-877.

Emershad, R.L. and Ramming, D.W. 1994. Somatic embryogenesis and plant development from immature zygotic embryos of seedless grapes (*Vitis vinifera*). *Plant Cell Rep.* **14**: 6-12.

Emershad, R.L., Ramming, D.W. and Serpe, M.S. 1989. *In-ovulo* embryo development and plant formation from stenospermic genotypes of *Vitis vinifera*. *American J. Bot.* **76**: 397-402.

Gamborg, O.L., Miller, R.A. and Ojima, K. 1968. Nutrient requirements of suspension cultures of soybean root cells. *Exp. Cell Res.* **50**: 157-158.

Gray, D.J., Mortensen, J.A., Benton, C.M., Durham, R.E. and Moore, G.A. 1990. Ovule culture to obtain progeny from hybrid seedless bunch grapes. *J. Amer. Soc. Hort. Sci.***115**: 1019-1024.

Hannig, E. 1904. Zur Physiologie pflanzlicher Embryonen. I. Über die Kultur von Cruciferen-embryonen ausserhalb des Embryosacks. *Bot. Ztg*. 62: 45-80.

Hu, C. and Wang, P. 1986. Embryo culture: Technique and application. In: *Handbook of Plant Cell Culture* Vol. 4. Ed. Evans, D.A. Sharp, W.R. and Ammirato. P.V., Macmillan, New York, pp. 43-96.

Laibach, F. 1925. Das Taubwerden von Bastardsamen und die künstliche Aufzucht früh absterbender Bastardembryonen. *Z. Bot.* **17**: 417-459.

Murashige, T. and F. Skoog. 1962. A revised medium for rapid growth and bioassays with tobacco tissue cultures. *Physiol. Plant.* **15**: 473-497.

Murashige, T. and Tucker, D.P.H. 1969. Growth factor requirement of Citrus tissue culture. Proc. 1st Int. Citrus Symp. vol. 3, p. 1155–1161, H. D. Chapman, ed. Univ. Calif. Riverside: Publ. Dept. 1969.

Narayanaswamy, S.; and Norstog, K. 1964. Plant embryo culture. Bot. Rev. **30**: 587-628.

Nickell, L.G. 1951. Embryo culture of weeping crabapple. Proc. Amer. Soc. Hort. Sci. 57: 401-405.

Nitsch, J.P., Pratt, C., Nrrsch, C. and Shaulis, N.J. 1960: Natural growth substances in Concord and Concord Seedless grapes in relation to berry development. *Amer. J. Bot.* **47**: 566-576.

Pinto, A.C.Q, Rogers, S.M.D. and Byrne, D.H. 1994. Growth of immature peach embryos in response to media, ovule support method, and ovule perforation. *HortSci.* **29**: 1081-1083.

Raghavan, V. 1976. *Experimental Embryogenesis in Vascular Plants*. Academic Press, London.

Raghavan, V. 1980. Embryo culture, In: I.K. Vasil (ed.). *Perspectives in Plant Cell and Tissue Culture. Intl. Rev. Cytol.*, Suppl. 11B. Academic, New York, pp. 209-240.

Ramming, D.W. 1990. The use of embryo culture in fruit breeding. *HortSci.* **25**: 393-398.

Ramming, D.W. and R.L. Emershad. 1984. Embryo culture of early-ripening seeded grape genotypes. *HortSci.* **19**: 594. (Abstr.)

Rangan, T.S. 1984. Culture of ovules. In: *Cell Culture and Somatic Cell Genetics of Plants, Vol. 1: LaboratoryProcedures and Their Applications*. Ed. I.K. Vasil. Academic Press, New York, pp. 227-231.

Rangaswamy, N.S. 1958 Culture of nucellar tissue of *Citrus in vitro*. *Experientia* **14**: 111-112.

Sharma, D.R., Kaur, R. and Kumar, K. 1996. Embryo rescue in plants- a review. *Euphytica* **89**: 325-337.

Singh, Z. and Brar, S.J.S. 1992: *In vivo* development of ovule in seedless and seeded cultivars of grapes (*Vitis vinifera* L.). A particular reference to *in ovulo* embryo culture. *Vitis* **31**: 77-82.

Spiegel-Roy, P., Sahar, N., Baron, J. and Lavi, U. 1985. *In vitro* culture and plant formation from grape cultivars with abortive ovules and seeds. *J. Amer. Soc. Hort. Sci.* **110**: 109-112.

Stewart, J.M. and Hsu, C.L. 1977. *In-ovulo* embryo culture and seedling development of cotton (*Gossypium hirsutum* L.). *Planta* **137**: 113-117.

Stout, A.B. 1936. Seedlessness in grapes. *N Y State Agric. Exp. Stn. Tech. Bull.* 238: 1-68.

Tukey, H.B. 1933a. Embryo abortion in earlyripening varieties of *Prunus avium. Bot. Gaz.* **94**: 433-468.

Tukey, H.B. 1933b. Artificial culture of sweet cherry embryos. *J. Hered.* **24**: 7-12.

Tukey, H.B. 1934. Artificial culture methods for isolated embryos of deciduous fruits. *Proc. Amer. Soc. Hort. Sci.* **32**: 313-322.

Tukey, H.B. 1944. Excised-embryo method of testing the germinability of fruit seed with particular reference to peach seed. *Proc. Amer. Soc. Hort. Sci.* **45**: 211-219.

Tukey, H.B. and Lee, F.A. 1937. Embryo abortion in the peach in relation to chemical composition and season of fruit ripening. *Bot. Gaz.* **98**: 586-597.

White, P.R. 1954. *The Cultivation of Animal and Plant Cells*, Ronald Press, New York.

Williams, E.G., Verry, I.M. and Williams, W.M. 1982. Use of embryo culture in interspecific hybridization. **In**: *Plant Improvement and Somatic Cell Genetics*. Ed. I.K. Vasil, W.R. Scowcroft, and K.J. Frey. Academic Press, New York, pp. 119-128.

Breeding of Perennial Horticultural Crops *Pages* **99–106**
Editors: S.K. Singh, V.B. Patel, A.K.Goswami, Jai Prakash and Chavlesh Kumar
Published by: **BIOTECH BOOKS, NEW DELHI**

Chapter 8

Genomics Assisted Breeding

C. Bharadwaj[1], Supriya Sachdeva[1], A.K. Goswami[2], M.K. Verma[2] and S.K. Singh[2]

[1]Divison of Genetics; [2]Division of Fruits and Horticultural Technology, ICAR-Indian Agricultural Research Institute, Pusa, New Delhi – 110 012

INTRODUCTION

As precise phenotyping is essential and the cost of generating phenotyping data at every generation is very expensive, recent advances in genomics technologies and the availability of a wide range of genotyping platforms have made the cost of genotyping much less expensive compared with phenotyping. The recent developments in sequencing technologies have manifold increased the repertoire of various types of markers that are available in fruit crops including SSRs, SNPs, DArTs, hundreds of thousands transcript reads and BAC-end sequences saturated genetic maps, QTL maps as well as physical maps and the sequencing of genomes of fruit crops (partial or completed) has greatly helped in using marker assisted technologies to be applied in plant breeding. Germplasm resequencing for identification of genome wide SNPs and their subsequent utilization in genomic selection has the potential to break the yield barrier being experienced in these perennial crops. Genomics-assisted breeding for marker-assisted backcrossing (MABC) for introgressing QTL region, marker assisted recurrent selection, gene pyramiding, marker-assisted selection (MAS) and genomic selection can now be taken up in fruit crops too. The conventional crop improvement should take these tools to make greater genetic gains, increase selection potential and have faster breeding cycles so that the genetic improvement gains are increased in these crops.

One of the most pressing problems of the world today is adequate nutrition for exploding human population as 870 million people go hungry worldwide (http://www.fao.org). The growing world population together with the lack of expansion or even reduction of available arable lands needed to maintain agricultural sustainability, implies that the relative importance of plant breeding to raise crop yield potential and adaptiveness is now greater than in the past (Araus *et al.*, 2002). Plant adaptation is a key factor that will determine the future severity of the effects of climatic change on food grain production. Relatively inexpensive changes, such as shifting planting dates or switching to an existing crop variety, may moderate the negative impact of climatic change. However, improvements in crop productivity to meet the requirement of growing demand mentioned above will not be easy without further technological breakthroughs that allow yield ceilings to be shifted through the development of new fruit crop varieties (Rosenzweig and Parry, 1994).

Fruit science is one of the important sectors of horticulture which consists of cultivation, production and improvement of different types of fruit plants. The importance of fruit crops is widely acknowledged in many aspects of innovation, production, quality maintenance, for uplifting economic condition of farmers, entrepreneurs and in providing nutritional security to the people. With the growing population, demand for fruits as such and its products are gradually increasing. But there are several constraints to increased sustained fruit production and its improvement. This limitation is especially evident for tree fruits and nuts, products of breeding cycles that can extend to a dozen years or more. Like other sectors of horticulture, the fruit and nut industries face highly dynamic situations arising from such factors as decreasing labor availability, increasing environmental concerns, cost of energy, climate change and epidemics of new and invasive insects and diseases. In addition, the inability to re-program tree form and phenology quickly limits deployment of highly efficient production technologies. The generally reactive, rather than proactive, nature of response to these factors translates to the release of new cultivars only after such pressures have accumulated significant impact on production. In recent years significant technological advancement has been observed (concepts, approaches and applications) in fruit improvement which has been revolutionized the fruit production (Bliss 2009). In the last two decades, the availability of many genomic resources like genome sequences, high-throughput analysis of gene expression, sufficient numbers of molecular markers, express sequence tags (ESTs) and high-density genetic maps has paved the way to the genetic engineering and molecular breeding of fruit plants for crop improvement. The application of these modern biotechnological techniques to fruit plants can contribute efficiently to solve or reduce the problems faced by the fruit industries. In the last decade, the emphasis on crop improvement using novel genomic tools has shifted toward the identification and functional analysis of miRNAs, one of the hottest research fields in plant sciences. However, as compared to cereals and other crops, the progress of genomic studies in horticultural crops is relatively slow. Availability of next generation sequencing (NGS) technologies like FLX-454, Illumina, SOLiD and Helicose have brought hopes to generate genomic resources for many more horticultural crops in few years' time. Therefore, the horticulture breeders should equip themselves to make use of this extensive genome information in their varietal development programmes. The objectives of this review were to take stock of availability of genomic resources in horticultural crops, compile this information at one place, makes sequence information useful for breeders and to identify potential future challenges which one can face while making proper use of genomic resources. The term 'genomics' is coined by Dr.Tom Roderick, a geneticist at the Jackson Laboratory (Bar Harbor, ME) in 1986. Genomics is a discipline in genetics that applies recombinant DNA, DNA sequencing methods, and bioinformatics to sequence, assemble, and analyze the function and structure of genomes (the *complete* set of DNA within a single cell of an organism). The field includes efforts to determine the entire DNA sequence of organisms and fine-scale genetic mapping. The field also includes studies of intra genomic phenomena such as heterosis, epistasis, pleiotropy and other interactions between loci and alleles within the genome.

This paper discusses both the techniques that have been used for mapping important traits in fruit crops.

Types of Genomics

Structural genomics: is the study to characterize the structure of the genome.

Functional genomics: is a field of molecular biology that attempts to make use of the vast wealth of data produced by genomic projects (such as genome sequencing projects) to describe gene (and protein) functions and interactions.

Comparative genomics: is a field of biological research in which the genomic features of different organisms are compared. The genomic features may include the DNA sequence, genes, gene order, regulatory sequences, and other genomic structural landmarks.

Need of Genomic Studies

1. To know sequence of whole genome
2. To locate genes on chromosomes and transcribed regions
3. To locate promoters and function of motif
4. Genome manipulation and MAS
5. Evolutionary studies
6. To discover and use of genetic variation

Advanced Genomic Approaches for Fruit Breeding

1. NGS platforms
2. Advanced Molecular markers and Marker assisted breeding (MAB)
3. GWAS, QTL mapping and Association mapping
4. Comparative genomics
5. Genomics assisted selection

Approaches and Utilization of Genomics in Fruit Breeding

Genome Sequencing Projects in Fruit Crops

Since the publication of the first plant genome sequence, that of Arabidopsis thaliana (Arabidopsis Genome Initiative, 2000), technology based on plant genomics has advanced substantially and revolutionized our understanding of plant biology by unraveling basic mechanisms in plant development, tolerance to biotic and abiotic stresses and adaptation(Feuillet *et al.*, 2011; Hamilton and Buell, 2012). With the significant advancement and achievement in DNA sequencing technology in recent years, efforts have been made to release the draft sequences of many fruit plant species and many more sequencing projects for fruit plants are in progress. Among fruit plants, grapevine was the first crop genome to be sequenced and the first draft with 8 times high-quality sequence was released by the International Grape Genome Project (Jaillon *et al.*, 2007). Recently, Velasco *et al.* (2010) reported a high-quality draft genome sequence of the domesticated apple (*Malus domestica*). The putative gene content in apple (57,386) was higher compared to the model plant, *A. Thaliana* (Arabidopsis Genome Initiative, 2000), or other fruit plants of the family Rosaceae such as the pear (Wu *et al.*, 2013). Recently Azim *et al.*, 2014 has been characterized mango leaf transcriptome and chloroplast genome using next generation DNA sequencing. Blast searching against non redundant nucleotide databases and several Viridi plantae genomic datasets annotated 24,593 mango unigenes (80 per cent of total) and identified *Citrus sinensis* as closest neighbor of mango with 9,141 (37 per cent) matched sequences. The draft mango cp genome size is 151,173 bp with a pair of inverted repeats of 27,093 bp separated by small and large single copy regions, respectively. Out of 139 genes in mango cp genome, 91 found to be protein coding. Sequence analysis revealed cp genome of *C. sinensis* as closest neighbor of mango.

A representation of the location of genes or genetic markers with respect to each other, based on recombination frequencies. A genetic map is linear, and distances between loci are measured in recombination percentage = map units =centimorgans. The farther apart two loci are, the more likely that a crossover will occur between them. Conversely, if two loci are close together, a crossover is less likely to occur between them. Recombination can only be detected between two loci, both of which are heterozygous. The dominant/recessive relationships must allow for detection of recombinants. The most useful systems involve codominant alleles. Efficient mapping requires polymorphic loci, *i.e.* loci with two or more common alleles. Loci that have a single common allele are described as monomorphic. Any variations in DNA, whether in coding regions of genes or in noncoding regions, can be used as genetic markers, *i.e.* as a label for a particular point on a chromosome.

Forty-two di-nucleotide microsatellite, or simple-sequence repeat (SSR), markers were developed using CA and CT enriched genomic libraries of *Mangifera indica* L. Six cultivated mangoes and two wild species were tested for primer amplifications. Most loci could amplify *M. caloneura* Kruz and *M. foetida*. The average number of alleles per locus was 4.4. The average expected heterozygosity and the maximum polymorphism information content value were 0.57 and 0.53, respectively. The SSRs developed in this study together with 65 SSRs and 145 restriction fragment length polymorphism (RFLP) markers reported previously were used in the genetic linkage analysis. A partial genetic linkage map was constructed based on 31 F_1 progenies from a cross between 'Alphonso' and 'Palmer'. The map spanned a distance of 529.9 centi Morgan (cM) and consisted of 9 microsatellite markers (6 from this study) and 67 RFLP markers. The new SSR markers and the present map will be useful for mango genetic studies and breeding applications in the future (Chataporn *et al.*, 2015).

173 F_1 plants from a cross between *Mangifera indica* L. "Jin- Hwang" and *M. indica* L. "Irwin" and their parent plants were subjected to high-throughput sequencing and specific-locus amplified fragment(SLAF) library construction (Chun Luo *et al.*, 2016).After preprocessing, 66.02Gb of raw data containing 330.64M reads were obtained. A total of 318,414 SLAFs were detected, of which 156,368 were polymorphic. Finally, 6594 SLAFs were organized into a linkage map consisting of 20 linkage groups (LGs). The total length of the map was 3148.28 cM and the average distance between adjacent markers was 0.48cM.

Amount, regularity and low seed content of the crop are important properties of scion citrus cultivars. The genetic control of these traits was studied in a progeny derived from the cross *Citrus volkameriana* × *Poncirus trifoliate* using molecular marker analysis. Since the traits were not normally distributed, the Kruskal-Wallis non-parametric test was used for quantitative trait loci (QTLs) detection. Most of the QTLs detected correspond to the trait "number of fruits per tree", in agreement with its known physiological complexity. Related traits (fruit number, fruit size and seed number)are controlled by QTLs some of which are located in the same genomic regions, suggesting that undesired associations could be broken to some degree by recombination. QTL analysis over years revealed important effects of genotype-by-environment interaction on QTL detection. This result agrees with the differences found for the trait means among years, which was found to be related, among other causes, to the alternate bearing of some genotypes and the amount of rain before harvest. Most RAPD markers were scored as presence versus absence of a given band. Genetic interpretation of the isoenzymatic markers, GOT-2 and PGI-2; microsatellite markers, TAA1, TAA45, CAC23 and TAA52; and the RFLPs markers, pRLc3 and pRLc11.These markers were associated with QTLs for the traits fruit number, average fruit size, average fruit weight and average seed number. Their distribution through the genome is quite scattered. Two genomic positions(CAC23 at V7 and OPG19400) involve both fruit number and size QTLs. Marker alleles associated with increased number of fruits at V7 (*Fn9*) are also associated with more seeds per fruit (*Sn4*). On the contrary, the marker allele at OPG19400 (linked to *Fn14* and *Fs4*) associated with increased fruit number is also associated with a larger size of the fruit although these QTLs were detected only in *P. trifoliata*. Two of the three putative fruit weight QTLs are located in similar positions as 2fruit size QTLs. In these cases, the same marker alleles are associated with both increased fruit weight and fruit size.

Association Mapping

Also known as association genetics or linkage disequilibrium mapping is a method of mapping quantitative trait loci (QTLs) that takes advantage of historic linkage disequilibrium to link phenotypes (observable characteristics) to genotypes (the genetic constitution of organisms). Association mapping is based on the idea that traits that have entered a population only recently will still be linked to the surrounding genetic sequence of the original evolutionary ancestor, or in other words, will more often be found within a given haplotype, than outside of it. It is most often performed by scanning the entire genome for significant associations between a panel of SNPs (which, in many cases are spotted onto glass slides to create "SNP chips") and a particular phenotype. These associations must then be independently verified in order to show that they either (a)contribute to the trait of interest directly, or (b) are linked to/in linkage disequilibrium with a quantitative trait locus (QTL) that contributes to the trait of interest.

Advantages of AM over Linkage Mapping

1. Much higher mapping resolution.
2. Greater allele number and broader reference population.
3. Possibility of exploiting historically measured trait data.
4. Less research time in establishing an association

Genome Wide Association Mapping (GWAS)

also known as whole genome association study (WGA study, or WGAS), is an examination of many common genetic variants indifferent individuals to see if any variant is associated with a trait. GWAS typically focus on associations between single-nucleotide polymorphisms (SNPs) and traits like major diseases.

Why are GWAS possible now?

- ☆ Improved Genotyping Technology:
- ☆ Now have ability to type millions of SNPs in one reaction on a "SNP chip." The cost can be as low as $200-$300 per person.
- ☆ Two primary platforms: Affymetrix and Illumina. Design and analysis:
- ☆ Availability of SNP databases, HapMap, and other resources to identify the SNPs and
- ☆ improve chip design/capacity.
- ☆ Faster computers to carry out the millions of calculations make analyses more amenable.

Grapes

The cultivated grapevine (*Vitis vinifera* L.) is one of the first economically important fruit crops whose genome was sequenced. Having access to the genome sequence has facilitated the identification and development of genetic markers and has also enabled LD studies in grapes. The cultivated grapevine is primarily a self-pollinated perennial fruit crop. Most of today's cultivated grapes have been derived from controlled crosses among a few select cultivars, and the elite selections have been maintained by vegetative propagation. There are only a few studies describing LD patterns across the grapevine genome, as well as AM studies undertaken (Myles *et al.*, 2011). Integrating both QTL and AM strategies has proven highly effective in various studies. For example, Fournier-Level *et al.* (2010) combined QTL and AM to study the genetic patterns of anthocyanin content, a determinant of berry colour, in grapes. A strong QTL, accounting for 62 per cent of the variation in anthocyanin content, was mapped in an F_1 pseudo-testcross. Markers were identified in four Myb-type genes, selected based on metabolic profiles, within the QTL interval, and these were tested in a core collection of natural grape germplasm consisting of 141 individuals. Using Bayesian reconstruction of effective population size dynamics, the presence of extended LD in these genes was revealed. Furthermore, a multivariate regression analysis identified five polymorphisms in Vv MybA genes that accounted for 84 per cent of the observed variation. It was concluded that the observed variation in anthocyanin content in grape was mainly accounted for by a single cluster of three Vv MybA genes. Fruit development is an important economic trait of grapevine, but there is little known about the genetic and molecular controls of fleshy fruit development. Houel *et al.* (2010) have identified markers from a gene (flb) proposed to be linked to a fleshless berry mutation by resequencing genomic regions of two grape cultivars. They have also sequenced these gene fragments in a highly diverse set of both cultivated and wild *V. vinifera* genotypes to identify possible signatures of domestication in the cultivated *V. vinifera*. In this study, SNPs significantly associated to berry weight variation have been identified in the flb region. Moreover, eight gene fragments with significant deviations from neutrality of the Tajima's D parameter have been identified in the cultivated pool along with putative signatures of selection. Sweet- and floral-flavoured Muscat cultivars are highly regarded as table grapes and for winemaking. Using QTL analysis, Muscat flavour determination has been investigated and a major QTL on chromosome5 linked to the Muscat flavour has been found to co-localize with a VvDXS gene encoding 1-deoxy-D-xylulose 5-phosphate synthase (Emanuelli *et al.*, 2010). Upon resequencing of the Vv DXS gene in an ad hoc association population of 148grape cultivars, consisting of muscat-flavoured, aromatic, and neutral accessions, as well as muscat-like aromatic mutants and non-aromatic offsprings of Muscats, three SNPs in moderate LD that are significantly associated with muscat-flavoured cultivars have been identified, after accounting for population structure. They have also identified a putative causal SNP responsible for a predicted non neutral lsubstitution (Emanuelli *et al.*, 2010). Although LD rapidly decays in grapevine, LD could be extended in those genomic regions selected during domestication. Therefore, conducting a candidate gene-based study for traits important during domestication is feasible, although the availability of high densities of markers across the genome is required for genome-wide AM (Myles *et al.*, 2011).

Rosaseous Crops

For apples, Cevik *et al.* (2009) have used a pedigree-based QTL mapping approach and have identified markers (MdMADS2.1, MdMADS2.2, and MdMADS14) from two candidate orthologous FRUITFULL-like genes linked to fruit flesh firmness. Subsequently, they have used an association analysis to collect further evidence for association between MdMADS2.1 and flesh firmness in a population of 168 apple accessions. The association mapping population has allowed for confirmation that MdMADS2.1and fruit flesh are significantly associated. Additionally, recently developed high-throughput SNP genotyping assays, a 1536

SNP GoldenGate assay for apple (Khan *et al.*, 2012). and a 8000 SNP Infinium assay for apple, peach, and cherry(Chagne´ *et al.*, 2012), will provide more opportunities for candidate gene-based association mapping. Although the Infinium has 8000 SNPs, this is not sufficient for genome wide AM due to rapid LD decay in apple and in other rosaceous fruit crops. But, this can be used for whole genome scanning of marker-trait associations utilizing a joint linkage-association approach that makes use of both pedigree and diversity analyses. Presence of co-linearity between some rosaceous species should also be taken into consideration when drawing conclusions based on the economic value of a crop. Colinearity among species has been deemed acceptable by the Rosaceae research community as a unitary system for genetic analysis to overcome economic costs (Rikkerinket al., 2007). However, transferability of AM results across species may not always be feasible as there are also major biological differences, even within a single species, thus rendering synteny across species as a minor factor. However, it should also be noted that, unlike forest trees, where population structure is minimal, domesticated fruit trees can have substantial population structure that may confound AM inferences. Therefore, a large set of accessions/genotypes are required for reliable identification of associations by minimizing spurious associations due to population structure.

Conclusion

Ultimately, the goal of the breeder will be to assay the genetic makeup of individual plants rapidly and to select desirable genotypes in breeding populations. The construction of 'graphical genotypes' of each plant or progeny row would allow the breeder to determine which chromosome sections are inherited from each parent to facilitate the selection process and perhaps to reduce the need for extensive field tests. A logical extension of whole genome selection for the breeder would be to design the superior genotypes in silico, an approach described as 'breeding by design'. Thus, in the post-genomics era, high throughput approaches combined with automation, increasing amounts of sequence data in the public domain and enhanced bioinformatics techniques will contribute to genomics research for crop improvement. However, the costs of applying genomics strategies and tools are often more than is available in commercial or public breeding programmes, particularly for inbreeding crops or crops that are only of regional importance. Nevertheless, marker-assisted breeding or marker-assisted selection will gradually evolve into 'genomics assisted breeding' for crop improvement. Newly developed genetic and genomics tools will enhance, but not replace, the conventional breeding and evaluation process. The ultimate test of the value of a genotype is its performance in the target environment and acceptance by farmers.

Breeding of Perennial Horticultural Crops *Pages* **107–111**
Editors: S.K. Singh, V.B. Patel, A.K.Goswami, Jai Prakash and Chavlesh Kumar
Published by: **BIOTECH BOOKS, NEW DELHI**

***Chapter* 9**

Genome Editing

Shelly Praveen, T. Vinutha and Suneha Goswami

Division of Biochemistry,
ICAR-Indian Agricultural Research Institute, Pusa, New Delhi – 110 012

INTRODUCTION

Fruit development and ripening is a complex and genetically controlled process that is essential for acquiring quality attributes of the fruit. Fruit ripening is an irreversible developmental process that involves interplay of various metabolic, biochemical and physiological processes, leading to fruit softening, accumulation of sugars, volatile compounds and pigments. These alterations not only leads to changes in organoleptic properties but help in making the fruit more attractive for animal consumption, and therefore, facilitating seed dispersal. Despite the enormous number of reported studies using genome editing technology for gene functional studies in plants and crop improvement, only few studies have focused on improving or identifying key regulators of fruit development.

Parthenocarpic fruit develoment is a attractive trait of considerable importance for various industries related to agriculture as well as food quality and the development of parthenocarpic fruits is considered as a valuable goal especially for maintaining sustainable agriculture in the face of global warming. Parthenocarpy is the growth of ovary into a seedless fruit in the absence of pollination or/fertilization. In tomato, seeds are in general not considered as a negative trait of fresh market fruits since they contribute in a positive way to the fruit taste. However, Parthenocarpy improves the yield, quality and processing attributes of tomato. The vital processes of fruit development, including fertilization, are temporally and spatially controlled by phytohormones. It is an established fact that phytohormones play important role in fruit setting and their genetic manipulation can lead to seedlessness. Parthenocarpy may occur naturally or can be induced with phytohormones, particularly auxins, and is currently used to increase fruit production under adverse conditions for fruit set and growth. Several genes *viz.*, rolB (auxin signaling), CHS (auxin transport), ARFs and AUCSIA (auxin response) regulating auxin synthesis have been exploited to generate parthenocarpic tomato fruits through gene silencing technology (Schijlen *et al.*, 2007, Nancy *et al.*, 2005, Molesini *et al.*, 2009). Pat gene mutants of tomato have also been developed which are responsible for parthenocarpy. The pat gene enhance the growth rate of ovary during the first 10 days of anthesis, which correlates with a precocious onset of cell divisions in the pericarp and higher auxin, gibberellin and DNA contents in the ovaries (Mapelli *et al.*, 1978). Seedless eggplants and tomato fruits have obtained by expressing the auxin synthesising gene iaaM of Pseudomonas syringae pv. Savastanoi under the control of the ovule/placenta specific promoter from the DefH9 gene of Antirrhinum majus. In addition, silencing of a tomato floral organ-specification gene, TM29, a SEPELLATA-like MCM1, AGAMOUS, DEFICIENCE and SRF (MADS) box gene, also caused seedless fruits (Ampomah-Dwamena *et al.*, 2002). Recently, the MADS box gene AGL6 was shown to regulate parthenocarpy in tomato, and loss function of SlAGAMOUS-LIKE 6 resulted in tomato facultative parthenocarpy (Klap *et al.*, 2017). SlAGL6 encodes for a MADS-box protein belonging to the type II lineage MIKCC, subfamily AGL6, of the MADS-box transcription factor family (Smaczniak *et al.*, 2012). In tomato parthenocarpic fruits, SlAGL6acts as a key regulator of the transition between the state of 'ovary arrest' imposed towards anthesis and the fertilization-triggered fruit set. Mutations that enable fruit development in the absence of seeds development are of considerable importance in modern breeding of many crops, including tomato. In tomato seedlessness was reported to be associated with elevated content of total soluble solids (Falavigna *et al.*, 1978; Casas Diaz *et al.*, 1987; Ficcadenti *et al.*, 1999; Carmi *et al.*, 2003). Seedlessness can be of advantage in crops raised for processed products, such as tomato paste, paprika spice and canned melons, because the energy invested in separating the seeds from the processed product can be saved and parthenocarpy can be beneficial to seed companies as it assists in protecting their varieties. Here, we have employed CRISPR/Cas9 system to the produce parthenocarpic tomato plants via disruption of the tomato (Solanum lycopersicum) IAA9 (SlIAA9) and AGL6 (SlAGL6).

SlAGL6 belongs to MADS box gene family was shown to regulate parthenocarpy in tomato and loss of function of SlAGL6 resulted in tomato facultative parthenocarpy (Klap *et al.*, 2017). In tomato parthenocarpic fruits, SlAGL6 acts as a key regulator of the transition between the state of 'ovary arrest' imposed towards anthesis and the fertilization-triggered fruit set. we developed the technology *viz.*, "CRISPR/Cas9-AGL6 binary vector construct" by exploiting CRISPR/Cas9 tool to knockout the SlAGL6 gene of tomato on chromosome no. Solyc01g093960.

Role of SLAGL6 in Parthenocarpy

Parthenocarpy is the growth of ovary into a seedless fruit in the absence of pollination or/fertilization. AGAMOUS, DEFICIENCE and SRF (MADS) box gene, also caused seedless fruits (Ampomah-Dwamena *et al.*, 2002). Recently, the MADS box gene AGL6 was shown to regulate parthenocarpy in tomato, and loss function of SlAGAMOUS-LIKE 6 resulted in tomato facultative parthenocarpy (Klap *et al.*, 2017). SlAGL6 encodes for a MADS-box protein belonging to the type II lineage

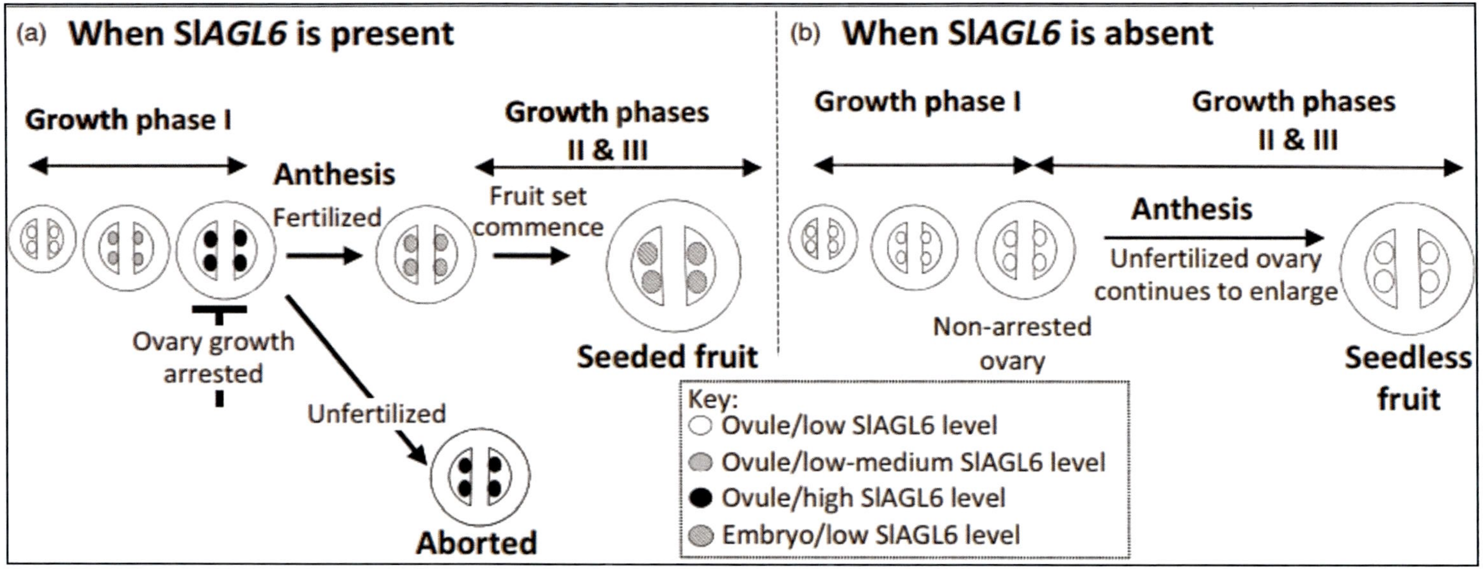

Figure 9.1: A Model Depicting the Key Role of SlAGL6 in the Regulation of Fertilization-Dependent Fruit Set. (a): under normal conditions, expression of SlAGL6 allows ovary growth at Growth phase I until it reaches higher level, during anthesis further growth of ovary is inhibited leading to "ovary arrest". Anthesis followed by fertilization induces downregulation of SlAGL6 expression in young embryo which enables growth and dproduction of fully developed seeded fruit.; (b): Loss of "ovary arrest" function in Slagl6 mutant s causes enlargement of ovary which continues even after anthesis at growth phase II and III allows fertilization independent development of seedless fruits.(*The figure Adopted from Klap et al., 2017*).

MIKCC, subfamily AGL6, of the MADS-box transcription factor family (Smaczniak *et al.*, 2012). In tomato parthenocarpic fruits, SlAGL6acts as a key regulator of the transition between the state of 'ovary arrest' imposed towards anthesis and the fertilization-triggered fruit set (Figure 9.1).

Structure of SlAGL6 and Development of CRISPR/Cas9-SlAGL6 Constructs

CRISPR/Cas9 technology was exploited to knockout the SlAGL6 gene (Solyc01g093960). *SlAGL6* consists of 8 exons and its encoded protein has two

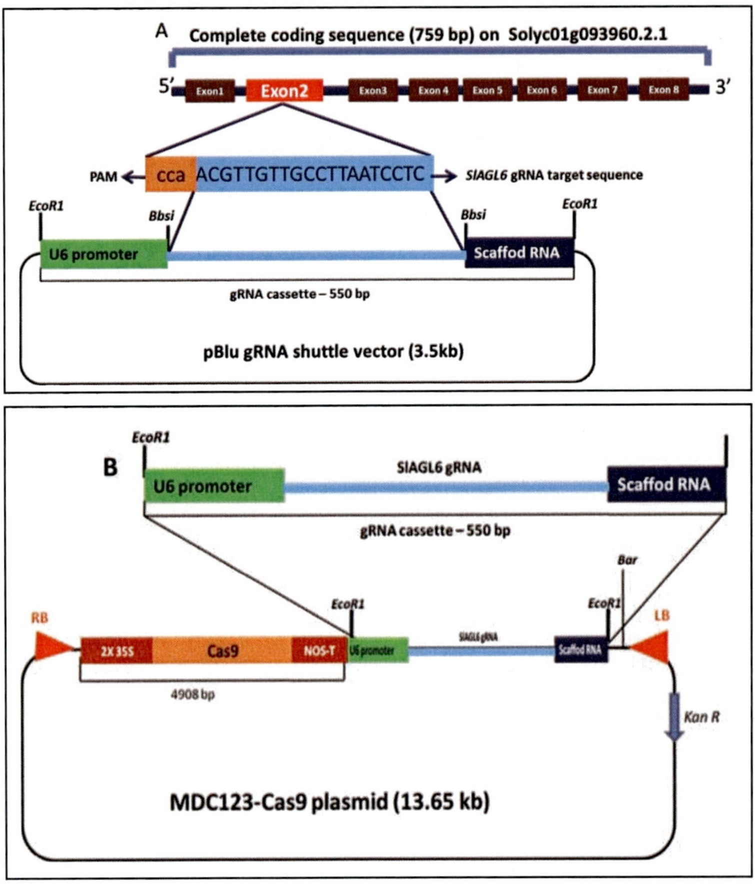

Figure 9.2: Overview of *SlAGL6* Gene Structure and *SlAGL6*Target gRNA Cloning into Shuttle Vector. A) Schematic of *SlAGL6* gene structure consisting of 8 exons with the complete CDS size of 759 bp. The chosen target guiding sequence(shaded in blue) for CRISPR/Cas9 modification in exon 2 of *SlAGL6* is presented, the PAM (its reverse complement) shaded in orange is depicted in lower case letters, and all the restriction sites needed for cloning are represented. B) Schematic of Cas9 construct harboring Cas9 (41139 bp) under 2x 35S promoter (768 bp) and Nos terminator (257bp) followed by gRNA cassette (U6-SlAGL6 gRNA-scaffold RNA). Bar gene resistance to glufosinate, T-DNA RB (right boarder) and LB (left boarder) are presented. Bacterial selection by Kanamycin resistance (Kan R) gene is also presented.

domains *viz.*, MADS-box domain (1-61 aminoacid) and K-box domain (85-193 aminoacid). Synthetic gRNA was designed to target the second exon of ***SlAGL6.*** It was incorporated into a Cas9 expressing binary vector and confirmed by restriction analysis and sequencing (Figure 9.2).

REFERENCES

AmpomahDwamena C, Morris B A, Sutherland P, Veit B, Yao JL. (2002). Downregulation of TM29, a tomato SEPALLATA homolog, causes parthenocarpic fruit development and floral reversion. *Plant Physiology*, 130: 605–617.

Carmi, N., Salts, Y., Dedicova, B., Shabtai, S. and Barg, R. (2003) Induction of parthenocarpy in tomato via specific expression of the rolB gene in the ovary. *Planta*, 217: 726–735.

Casas Diaz, A.V., Hewitt, J.D. and Lapushner, D. (1987) Effects of parthenocarpy on fruit quality in tomato. *J. Am. Soc. Hort. Sci.* 112: 634–637.

Falavigna, A., Badino, M. and Soressi, G.P. (1978) Potential of the monomendelian factor pat in the tomato breeding for industry. *Genetica Agraria*, 32: 159–160.

Ficcadenti, N., Sestili, S., Pandolfini, T., Cirillo, C., Rotino, G.L. and Spena, A. (1999) Genetic engineering of parthenocarpic fruit development in tomato. *Mol. Breeding*, 5: 463–470.

Klap C, Yeshayahou E, Bolger A M, Arazi T, Gupta S K, Shabtai *et al.* (2017). Tomato facultative parthenocarpy results from SlAGAMOUS-LIKE 6 loss of function. *Plant Biotechnol. J.* 15: 634–647. doi: 10.1111/pbi.12662.

Mapelli S, Frova C, Torti G and Soressi G P (1978) Relationship between set, development and activities of growth regulators in tomato fruits. *Plant Cell Physiology* 19: 1281-88.

Molesini B, Pandolfini T, Rotino G L, Dani V and Spena A (2009) Aucsia Gene Silencing Causes Parthenocarpic Fruit Development in Tomato. *Plant Physiology* 149: 534-48.

Nancy D (2015) Scrutiny of Strategies in developing Parthenocarpic Tomato (*Solanum lycopersicum*). *Indian Journal of Applied Research*, 5: 71-74.

Schijlen E G W M, de Vos R C H, Martens S, Jonker H H, Rosin F M, Molthoff J W, Tikunov Y M, Angenent G C, Tunen A J and Bovy A G (2007) RNA Interference Silencing of Chalcone synthase, the first step in the flavonoid biosynthesis pathway, leads to parthenocarpic tomato fruits. *Plant Physiology*, 144: 1520-30.

Smaczniak C, Immink, R G H, Angenent G C, and Kaufmann K (2012). Developmental and evolutionary diversity of plant MADS-domainfactors: insights from recent studies. *Development*, 139: 3081-3098.

Breeding of Perennial Horticultural Crops *Pages* **113–120**
Editors: S.K. Singh, V.B. Patel, A.K.Goswami, Jai Prakash and Chavlesh Kumar
Published by: **BIOTECH BOOKS, NEW DELHI**

Chapter **10**

Molecular Mapping

Rakesh Singh

Principal Scientist,
ICAR-National Bureau of Plant Genetic Resources, New Delhi

INTRODUCTION

The methods used to identify the locus of a gene and the distances between genes using molecular markers is called molecular mapping Genome mapping. It is used to assign short DNA sequences (molecular markers) or specific genes to particular regions of chromosomes and to determine their relative linear orders and distances. The essence of all genome mapping is to place a collection of molecular markers onto their respective positions on the genome. These molecular markers come in all forms.

Molecular Markers

DNA-based markers rely on differences in DNA sequences (polymorphisms) between two parental lines. Polymorphisms can result from various factors that lead to either nucleotide changes or differences in DNA segment lengths such as mutations, errors in DNA replication, and insertions, inversions, and deletions of DNA fragments. There are several established approaches for the detection of polymorphisms using molecular markers. A list of most commonly used molecular markers is given below:

PCR Based	*Non PCR Based Techniques*
RAPD	RFLP
AFLP	
CAPS	
SCAR	

PCR Based	Non PCR Based Techniques
SCoT	
SSR/Microsatellites	
EST-SSR	
SNP	

The advent of automated Sanger sequencing and especially recent advances in NGS technologies led to the development of a second generation of markers based on sequence information. SNPs differ by a single nucleotide A, T, C, or G at a given locus between different individuals, populations, and parental lines. If this variation occurs between the members of the same population, these variations are considered alleles (*e.g.*, A or T), and most SNPs have only two alleles. SNPs have emerged as the markers of choice because of their abundance and high-throughput detection capacities.

Types of Molecular Maps

1. Genetic Maps

Genetic maps are based on recombination frequencies between genetic markers and genes, and linked markers/genes form linkage groups showing their relative order. Genetic mapping relies on the fact that nuclear genomes are made up of chromosomes, which contain both genes and non-coding DNA. When homologous chromosomes pair at meiosis, they recombine at various positions along the chromosomes. Thus, recombination is the basis for genetic linkage mapping and determining the order of markers along the chromosome, that is, markers are separated by genetic distances calculated based on the amount of meiotic recombination that occurs between them.

Genetic Map Construction

Polymorphisms between Parents

In plants, most populations are derived from crossing two highly homozygous parents. The F_2 populations are commonly used and generally a good choice for chromosome mapping, other types of populations, such as backcross (BC), doubled-haploid (DH), and recombinant inbred (RI), are also commonly used. Each type of population has its advantages and disadvantages. F_2, BC, and DH populations can be developed very rapidly, while RI populations are developed by advancing each line by single-seed descent for many generations with the goal of selfing to homozygosity. The first step for molecular mapping is to study the parental polymorphism between the parents this makes the basis for study the segregation pattern in the population. The markers polymorphic between parents are the basis for genetic map construction.

Recombination

- ☆ Permits Mapping by providing:

- ❐ Linkage groups
- ❐ Distances
- ☆ During Meiosis
- ☆ Linkage Groups - markers that tend to remain together.
- ☆ Distance - the further apart two markers lie, the more often recombination will occur between those markers.
- ☆ Markers on the same chromosome can be so far apart that they appear in different linkage groups (Figure 10.1).

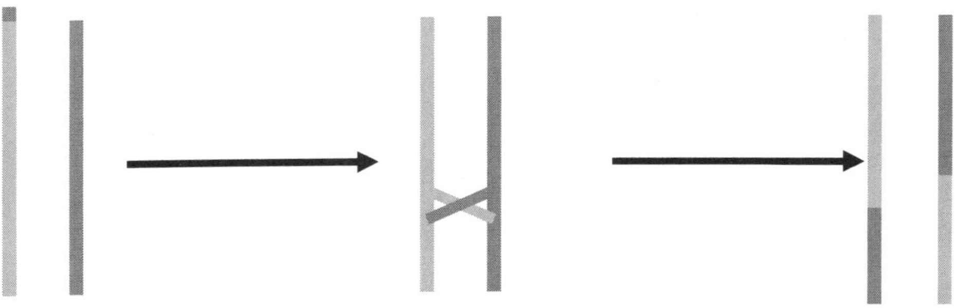

The development and analysis of genetic linkage maps lead to an abundance of information regarding genome structure. From a more applied perspective, they provide knowledge regarding the locations of genes and DNA markers associated with them (Figure 10.3). Genetic mapping is a great resource for trait mapping and map-based cloning studies; however, a genetic map is not sufficient for sequencing a genome. Polymorphism level is many times a limitation for genetic mapping; however, the advanced and low-cost NGS approaches have been a big boost to overcome this limitation. GBS has been widely accepted and is now being used to map target traits in various crops. In Table 10.1 a list of traits mapped in horticultural crops is given for reference.

Table 10.1: QTL Mapping in Important Horticultural Crops

Crops	*QTL/Gene Function*	*References*
Tomato	Fruit shape Fruit Sugar Content Fruit Weight	Liu *et al.* (2002), Fridman *et al.* (2002, 2004) Frary *et al.*, 2000; Cong *et al.*, 2002
Potato	Resistance to Ro1Globodera ostochiensis (Gro1-4) Quality trait cold-sweetening (invGE/GF) Flavonoid 3_,5_-hydroxylase (f3_5_h)	Paal *et al.* (2004) Li *et al.* (2005) Jung *et al.* (2005)
Apple	Scab resistance gens	Xu and Korban (2003)
Sugarbeet	Nematode resistance gene	Cai *et al.* (1997)
Pepper	Virus resistance gene	Ruffel *et al.* (2002)
Cauliflower	Orange gene	Lu *et al.* (2006)
Papaya	CpCYC b	Blas *et al.* (2010)

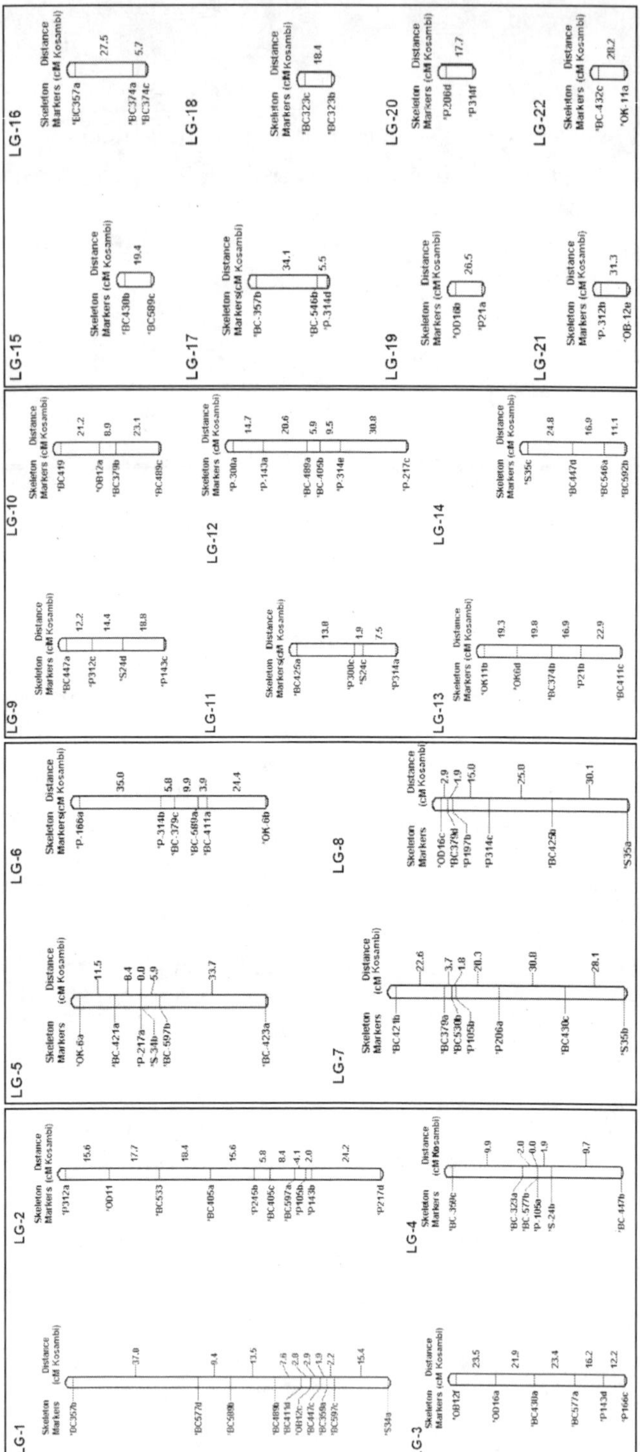

Figure 10.1: Linkage Map for *Cucurbita maxima* Based on Randomly Amplified Polymorphic DNA (RAPD) Markers [*Source*: Singh *et al.*, 2011, *Indian J of Horticulture* 68(1): 44-50].

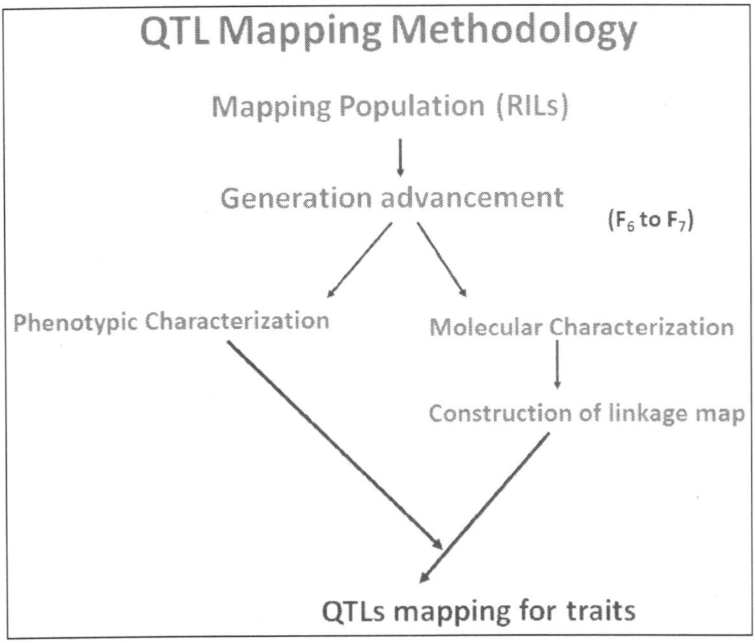

Figure 10.2: Strategy of QTL Mapping for Trait of Interest.

2. Physical Maps

A physical map of a given genome shows the physical locations of genes and other DNA sequences of interest, and distances are typically measured in base pairs (bp). The ultimate physical map is the complete genome sequence itself- based on actual distances in base pairs.

Physical Map Construction

In contrast to genetic mapping where distances between landmarks are calculated based on recombination frequency, physical mapping determines the actual physical distance. Physical mapping can be done cytologically by chemically staining and viewing whole chromosomes using techniques such as in situ hybridization (ISH) and C-banding. Such techniques have very low resolution in terms of physical mapping because chromosomes are viewed at the cellular level usually at metaphase. However, recent techniques, such as fiber-fluorescence in situ hybridization (FISH) where nuclear DNA is lysed on a glass slide and used for in situ mapping, can provide a much higher resolution. The highest-resolution physical mapping is obtained by sequencing the DNA itself. It is usually preceded by constructing local contiguous sequences (contigs) of large-insert DNA clones and anchoring the contig to a genetic map.

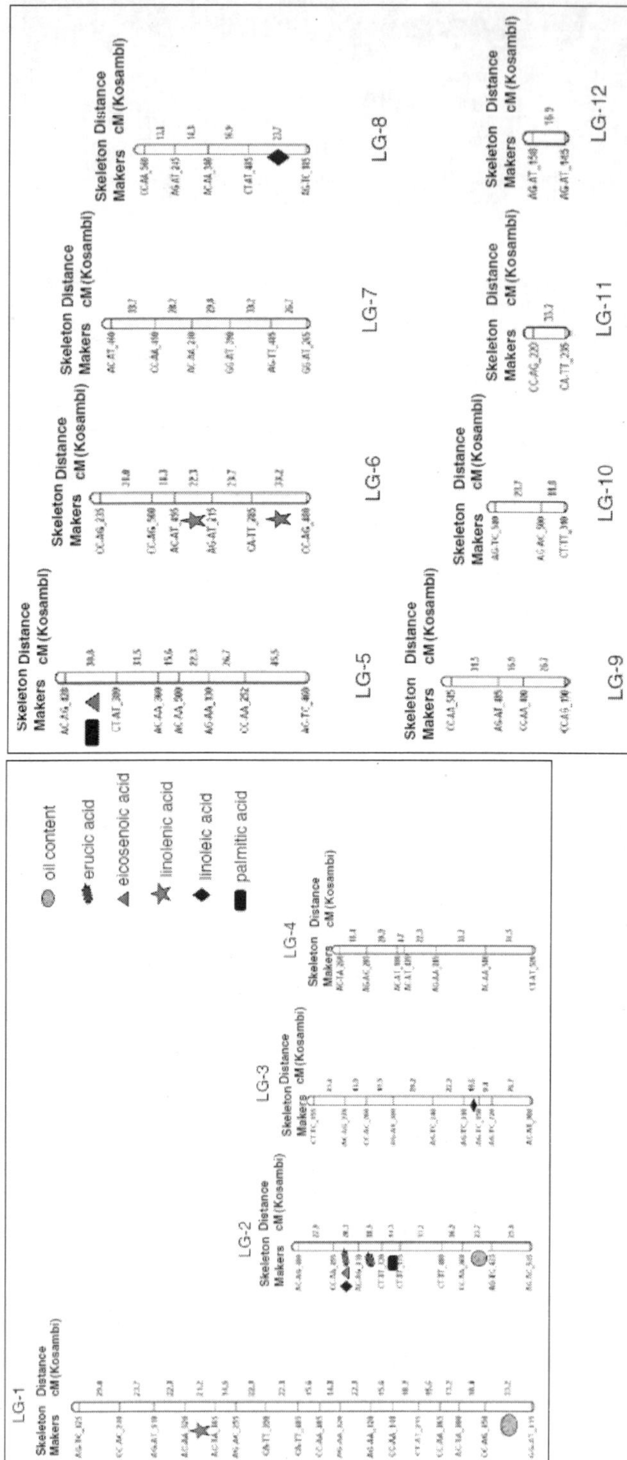

Figure 10.3: Mapping of QTLs for Fatty Acid Composition in Indian Mustard [*Brassica juncea* (L.) Czern. and Coss.] Based on AFLP Markers [Singh *et al.*, (2013), *J Plant Biochemistry and Biotechnology*, 22(1): 80-89.]

Hence a physical map of a piece of DNA tells us the location of certain markers

- ☆ A marker is a short sequence
- ☆ Given a sequence and a chromosome - try to find the place of the sequence

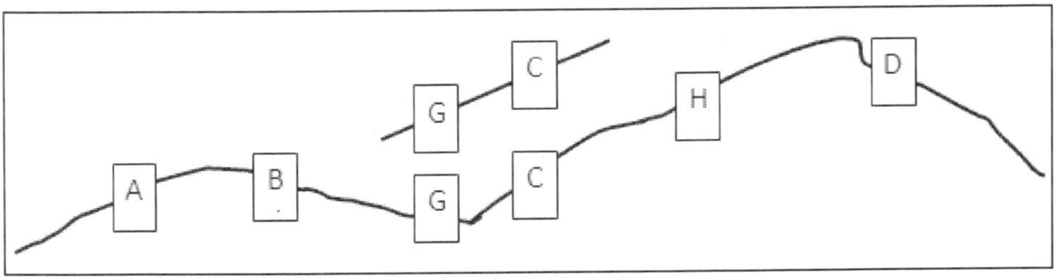

- ❒ Generally used to resolve regions much larger than 1 Mb (*e.g.* whole chromosomes)
- ❒ Map is created by fragmenting the DNA molecule using restriction enzymes and then looking for overlaps
- ☆ The pieces are too big to sequence, so this is not the same problem as fragment assembly!

Physical maps have led to a wealth of information regarding the physical locations of morphological traits and evolutionary translocation breakpoints and genome-wide structure and organization.

Association Mapping

Association analysis, also known as LD mapping or association mapping, is a population-based survey used to identify trait-marker relationships based on linkage disequilibrium.

Association Mapping Vs QTL Mapping

- ☆ Association Mapping can be conducted directly on the breeding material, therefore:
- ☆ Direct inference from research to breeding is possible
- ☆ Phenotypic variation is observed for most traits of interest
- ☆ Marker polymorphism is higher than in biparental populations
- ☆ Routine evaluations provide phenotypic data
- ☆ Association Mapping provides other useful information about:
- ☆ Organization of genetic variation
- ☆ Polymorphism across the genome

The ultimate goal in map construction is the deciphering of the linear DNA sequences of the full complement of chromosomes of an organism and the utilization of map information in trait mapping. The practical applications of the genome maps and reference sequences are best realized only when allelic diversity among diverse germplasm is better understood. In crops where sequence information is not available, comparative genomics-based tools can be very useful for providing a virtual gene order based on synteny.

Breeding of Perennial Horticultural Crops *Pages* **121–129**
Editors: S.K. Singh, V.B. Patel, A.K.Goswami, Jai Prakash and Chavlesh Kumar
Published by: **BIOTECH BOOKS, NEW DELHI**

Chapter 11

Linkage Mapping

Vignesh Muthusamy, Rajkumar Zunjare and Firoz Hossain

Division of Genetics,
ICAR-Indian Agricultural Research Institute, Pusa, New Delhi – 110 012

INTRODUCTION

Genetic mapping (also known as linkage or meiotic mapping) refers to the determination of the relative position and distances between markers along chromosomes based on the mean number of recombination events, involving a given chromatid, in that region per meiosis. Genetic map construction requires that the researcher develop appropriate mapping population, decide the sample size and type of molecular marker(s) for genotyping, genotype the mapping population with sufficient number of markers, and perform linkage analyses using statistical programs (Linkage1, GMendel, MapMaker, MapManager, JoinMap *etc.*). The construction of detailed genetic map with high level of genome coverage is the first step for localizing genes or quantitative trait loci (QTL) that are associated with economically important traits, marker assisted selection, comparative mapping between different species, a framework for anchoring physical maps, and the basis for positional cloning of genes. Hence, development of appropriate mapping population forms the basic platform for most of the molecular marker assisted plant breeding applications.

Mapping Populations

A population used for gene mapping usually obtained from controlled crosses is commonly called a mapping population. The first and most critical step in producing a mapping population is selecting two genetically divergent parents, which show clear genetic differences for one or more traits of interest at phenotype level. Sufficient variation for the traits of interest at the DNA sequence is equally important to trace the recombination events for the easier identification of polymorphic informative makers. The parents should be genetically divergent enough to exhibit sufficient polymorphism and at the same time they should not be genetically too distant to cause sterility of the progenies and/or very high levels of segregation distortion during analysis. Consideration must be given to the source of parents used in developing mapping population. Genetic divergence between parents influences marker-trait associations and hence economic significance of map.

Development of Mapping Populations

Progenies from the second filial generation (F_2), F_2 derived F_3 (F_2:F_3) populations, backcross (BC) populations, recombinant inbred lines (RILs), doubled haploids (DHs), near isogenic lines (NILs), and immortalized F_2 Population have been used for genetic mapping crop plants. F_2 populations are developed by selfing F_1 hybrids derived by crossing the two parents while BC population is produced by crossing F_1 back into one of the parents (the recipient or recurrent parent). $F_{2:3}$ population is generated by selfing the individual F_2 segregants for a single generation. RILs are developed by single-seed selections from individual plants of an F_2 population; such selections continue for 6-8 generations. If backcross selection is repeated at least for six generations, more than 99 per cent of the genome will be derived from recurrent parent. Selfing of selected individuals from BC_7F_1 will produce BC_7F_2 lines that are homozygous for the target gene, which is said to be nearly isogenic with the recipient parent (NILs). A DH population is produced by anther/pollen culture of F_1 plants. Plants will be regenerated using tissue culture techniques after induction of chromosome doubling from pollen grains or haploid embryos resulting from species crosses. Immortalized F_2 populations can be developed by paired crossing of the randomly chosen RILs derived from a cross in all possible combinations excluding reciprocals. The set of RILs used for crossing along with the F_1s produced, provide a true representation of all possible genotype combinations (including the heterozygotes) expected in the F_2 of the cross from which the RILs are derived. The RILs can be maintained by selfing and required quantity of F_1 seed can be produced at will by fresh hybridization. This population therefore provides an opportunity to map heterotic QTLs and interaction effects from multi-location data.

QTL Mapping

The term Quantitative Trait Locus (QTL) was first coined by Gelderman (1975) which defines QTL as "a region of the genome that is associated with an effect on a quantitative trait". Conceptually, a QTL can be a single gene, or it may be a cluster of linked genes that affect the trait. A quantitative trait exhibits continuous variation in its phenotypic expression. The trait expression may be controlled by many genes (a few hopefully of *large effect* and the others of *small effect*). Phenotypic expression of the trait is affected by genetic (*gene × gene interaction; epistasis*) and non-genetic factors [(*macro-* and *micro-environment variations* and *measurement errors* (*Noise*)] apart from their interaction with the environment (*gene × environment interaction*). A genetic marker that tends to co-segregate with the trait is likely to be close to a QTL controlling that trait. In QTL mapping, we seek an association between marker alleles (genotypes) and trait values (phenotypes). The major purpose of QTL mapping is primarily to estimate the locations, number, magnitude of phenotypic effects, and mode of gene action, of individual genetic loci (QTLs) for its application in marker-assisted breeding. Assessing the relative contribution of a QTL to the total genetic variance for a specific trait (R^2), analysis of QTL x E effects and identification of parental lines contributing the favourable QTL allele are the other important components of QTL analysis. The fundamental challenge in QTL analysis is to disentangle the genetic signal at any individual locus from the noise. With Polymorphic markers, QTL analysis can be done with or without a genetic linkage map. A linkage map needs adequate polymorphic markers and allows mapping of QTL, estimating their number, phenotypic effects, and mode of gene action. Without a linkage map, only QTL-harbouring markers can be identified; QTL location, number, effect and mode of gene action cannot be determined. QTL analysis can be done using a *mapping population* (Linkage analysis) or a *natural/breeding population* (Association analysis). Classical analysis of QTLs employed statistical techniques which were based on means, variances and co-variances provided negligible information on what these genes were, where they were located, and how they controlled the trait. Major advancements in PCR-based marker systems, applied statistics and quantitative genetics revolutionised the QTL research.

Genetic Mapping – Approaches

The main goal of genetic mapping is to detect markers in close proximity to the gene(s) controlling the complex quantitative traits. Genetic mapping can be done mostly in two ways:

1. Using the experimental populations (also referred to as "biparental" mapping populations) that is called QTL-mapping as well as "genetic mapping" and
2. Using the diverse lines from the natural populations or germplasm collections that is called LD-mapping or "association mapping".

Biparental Mapping of QTLs

Requirements

1. A suitable mapping population generated from phenotypically contrasting parents
2. A reasonably saturated linkage map based on molecular markers
3. Reliable phenotypic screening of mapping population
4. Powerful statistical packages to analyse genotypic and phenotypic information in combination.

Detailed Steps

1. Selection of phenotypically contrasting parents for target trait (and as many important traits as possible) and genetically divergent parental lines with due consideration of pedigree information.

2. Identification of a large number of polymorphic markers between parental lines with adequate genomic coverage. Different marker systems have different levels of resolution for detecting genomic variation. Select a system that allows experimental detection of heritable genomic variation among individuals of the mapping population. Co-dominant markers provide more power for QTL detection than dominant markers. For eternal RIL and DH, both types of markers are equally informative.
3. *Generation of a suitable mapping population*: Ability to detect QTLs or the information contained in F_2 or F_2-derived populations and RILs are relatively higher than others. $F_{2:3}$ populations are capable to measure the effects of additive and dominance gene actions at specific loci. Recombinant Inbred Lines (RILs) in general and Doubled Haploids (DHs) wherever available are the most preferred mapping populations for high resolution QTL mapping, though additive gene action can only be measured. Though, backcross (BC) population provides limited genetic information, marker-assisted introgression can be integrated with QTL mapping (AB-QTL Method). Mapping population of size 500-1000 is considered ideal for high map resolution and finer marker order. So choice of mapping population and its size should be, therefore, based upon the objectives of the experiment, the timeframe as well as resources available for undertaking QTL analysis.
4. *Partitioning of mapping population into different genotypic classes (based on genotypes at polymorphic marker loci)*: "Genotyping" refers to the process of determining the genotype (the genetic constitution) of an individual by the use of biological assays. For locating QTLs, a dense map is preferable as this allows greater precision of QTL location. QTL mapping, like linkage mapping, relies on the frequency of detectable recombination events, which, beyond a given marker density can only be increased by increasing size of the mapping population. So a possible approach is to initially use an evenly spaced sparse map to detect significant chromosomal regions to which more markers could be subsequently saturated for fine-scale localization of QTLs.
5. *Phenotyping of mapping population for target trait*: Phenotyping is as important as genotyping and is an often unrecognized, but the most critical part of QTL analysis. Robust and high resolution QTL mapping demands reliable phenotyping of traits of interest under defined target environmental conditions over locations and years. Phenotypic data that explains the effect of QTLs should come from multi-environment testing. Use appropriate intra-environment experimental designs and biometric tools to obtain accurate and precise estimate of genetic parameters.
6. *Apply correlative statistics to determine 'linkage disequilibrium' between a marker and the QTL*. Assess the probabilities of associations between phenotypic values and genotype scores at each marker locus, across the set of mapping population progenies. QTL Cartographer, MapQTL, PlabQTL, MultiQTL are the most commonly employed QTL mapping softwares. QTL analysis

approaches could be classified as follows according to the number of genetic markers used as *unit of analysis* in analysing the data.

Salient Features of most Commonly Employed QTL Detection Methods

Single point analysis analyses one marker at a time for its association with QTL employing linear regression analysis. It is computationally simple, quick and do not require molecular linkage map. But the magnitude of QTL detected is underestimated because of recombination and less likely to detect QTL further from a marker because of cross over effects between the loci. The method cannot determine whether the markers are associated with one or more QTLs.

Simple Interval mapping (Lander and Botstein, 1989) localizes QTL into marker intervals and takes full advantage of a linkage map. Presence of a putative QTL is estimated if the log of odds ratio (LOD) exceeds a critical threshold. Formulae available for calculating significance levels are appropriate for interval mapping when the genome size, number of chromosomes, number of marker intervals, and overall false positive rate desired are given. It is most widely used approach. By using tightly linked markers for analysis, it compensates for the recombination between markers and the QTL, thereby increases the probability of statistical detection of the QTL, and provides an unbiased estimate of QTL effect, but when multiple QTLs are segregating in a cross (which is usually the case), SIM fails to take into account genetic variance caused by other QTLs.

Composite interval mapping (Zeng, 1994 and Basten *et al.*, 2001) employs adjacent markers and/or other QTL as cofactors. It combines interval mapping for a single QTL in a given interval with multiple regression analysis on marker(s) associated with other QTL. It considers a marker interval plus a few other well-chosen single markers in each analysis, so that n-1 tests for interval-QTL associations are performed on a chromosome with n markers. By using linked markers as cofactors, the test is not affected by QTL outside the region, thereby increasing the precision of QTL mapping.

Advanced Backcross-QTL (AB-QTL) Strategy (Tanksley and Nelson, 1996) combines QTL analysis with variety development. It is particularly useful for transfer of valuable QTLs from un-adapted germplasm into elite breeding lines. QTL analysis delayed until the BC_2 or BC_3 generation. QTL-NILs derived from advanced BC populations in one or two additional generations are utilized to validate the QTL activity. QTL-NILs also represent improved commercial inbreds (over the original parental line) for one or more quantitative traits. So AB-QTL method enables simultaneous QTL mapping and its validation.

Multiple Interval Mapping (Zeng *et al.*, 1999) *is a* multiple QTL-oriented method that allows estimation of number, positions, effects, and epistatic interactions among significant QTLs simultaneously. ***Multiple-Trait Interval Mapping*** tests and estimates *pleiotropy* and *pleiotropy-vs-linkage, QTL x environment* interactions and QTLs affecting multiple traits.

The care and expense invested in generating marker and trait data should be accompanied by *equal care* in biometric analysis of data. A resampling technique to measure the quality of statistical estimates and models called **Bootstrapping** and statistical method to evaluate the predictive capability of a given set of models called as **Cross validation** are utilized to validate the QTL statistical analysis. Once mapped and validated, the QTL mapping information can be utilized for, genotype building/QTL pyramiding, QTL introgression, hybrid breeding, marker-assisted recurrent selection (MARS), and understanding basic mechanisms to resolve complex biological problems.

Association Mapping as an Alternative Approach

Turning the gene-tagging efforts from biparental crosses to natural population of genotypes (or germplasm collections), and from traditional QTL mapping to linkage disequilibrium (LD)-based association study became a powerful tool in mapping the genes of interest. This leads to the most effective utilization of ex-situ conserved natural genetic diversity of crop germplasm resources. 'Association genetic analysis' or Association mapping' makes use of genomic surveys of LD. LD refers to non-random association of alleles at different loci, either on the same or on different chromosomes. This is more appropriately referred to as "Gametic phase disequlibrium". Originally developed for human genetics, statistical methods and their derivatives for detection of LD are now increasingly being applied to crop plants like maize, leading to analyses of population genetic structure and QTL detection.

LD can be detected statistically, and has been widely applied to map and eventually clone a number of genes underlying the complex genetic traits. The main advantage of association mapping isthat it can explore all the recombination events or mutations in a given population, and with higherresolution compared with linkage mapping. Unlike conventional linkage analysis, LD mapping does not require family or pedigree information and can be applied to a range of experimental and non-experimental populations.

Steps

1. Selection of genotypes from the natural population or germplasm collections (preferably from core collections with wide coverage of genetic diversity)
2. Phenotypic evaluation of the selected genotypes for trait of interest (yield, quality *etc.*), preferably in multiple environments
3. Genotyping the mapping population with available molecular markers (either candidate-gene based approach or whole genome approach)
4. Quantification of the extent of LD of the population using molecular marker data
5. Assessment of the population structure (the level of genetic differentiation among the individuals within a sampled population) and kinship (coefficient of relatedness between pairs of individuals within a sample) in the population

6. Correlation of phenotypic and genotypic data by the use of an appropriate statistical tool, based on the information gained through quantification of LD and population structure, that reveals markers positioned within close proximity of targeted trait of interest.

Linkage Disequilibrium

Genetic linkage generally refers to coinheritance of twoloci that are located within a genetic distance on the chromosome. LD isa nonrandom association of alleles at different loci, describingthe condition with unequal (increased or reduced) frequency of the haplotypes in a population at randomcombination of alleles at different loci. LD is not the sameas linkage, although tight linkage may generate high levels ofLD between alleles.

The simplified explanation of the commonly usedLD measure, *D* or *D'*(standardized version of *D*), is thedifference between the observed gametic frequencies ofhaplotypes and the expected gametic haplotype frequenciesunder linkage equilibrium ($D = P_{AB} - P_A P_B = P_{AB} P_{ab} - P_{Ab} P_{aB}$).

There are many genetic and demographic factors that determines LD a crop species. Although mutation and recombinationare one of the strong impact factors influencing LD,generally, factors affecting LD can be grouped into twocategories: (1) factors that increasing LD, and (2) factorsthat decreasing LD. The increase of LD is observed withnew mutation, mating system (self-pollination), geneticisolation, population structure, relatedness (kinship), smallfounder population size or genetic drift, admixture, selection(natural, artificial, and balancing), epistasis, and genomicrearrangements.The decrease of LD is observedwithhigh recombination and mutation rate, recurrent mutations,outcrossing, and gene conversions.

Population Structure

Population structure due to adaptation, domestication or selection is the most important constraint to the success of association studies. Population structure can cause some allele frequencies to differ significantly between sub populations, which can create unexpected LD between unlinked loci across the genome. Several statistical methods have been used to control the effect of population structure in association analyses including: genomic control; structured association; principal components analysis (PCA); nonmetric Multidimensional Scaling (nMDS); and the unified mixedmodel approach. A twostage dimension determination approach using both PCA and nMDS has been demonstrated to be the best approach to capture the major structure of association panels in order to maximize the rejection of false positives while maximizing the statistical power to identify true associations.

Candidate Gene Based Association Analysis

The candidate gene method of association analysis is a hypothesisdriven approach for complex trait dissection. It involves genotyping or resequencing those genes considered to have a high probability of association with the phenotype(s) of interest within the germplasm being tested. There are two common strategies to this approach. In the first method, the gene of interest is sequenced as fully as possible across a limited number of diverse lines (typically 24 to 48) to identify the causal polymorphisms, such as SNPs causing amino acid changes or InDels in untranslated or translated regions. The selected polymorphisms can then be screened across a large germplasm collection using inexpensive simple PCRbased SNP/InDel genotyping assays (rather than sequencing) to confirm the associations between genotype and phenotype.

In the second method, the partial or entire gene is sequenced in all individuals of a germplasm panel to identify significant associations, either the causal polymorphism(s) or a polymorphism that is in LD with a causal polymorphism. Although this is more expensive than the first strategy, it may identify rare polymorphisms which may be missed by the first strategy.

Genome Wide Association Studies (GWAS)

Candidate genebased association analysis, as mentioned above, is a hypothesisdriven approach thatrequires detailed prior knowledge of potential candidate genes. The historical emphasis on thisapproach resulted from the low QTL detection power of association analyses based on random markers. However, the ability to do high throughput genotyping in crop plants is dramatically increasing, it should be possibleto identify markers within and closely linked to genes contributing to complex target traits. This genome wide association studies (GWAS) have been widely used in human and animal systemsand already in some plants with small genomes.The minimum number of markers needed for a successful GWAS depends on the genome size and therate of LD decay of the target germplasm.Determining which method to use has generally been based on the resource and the time availability.

Uses of Assocation Mapping in Crop Improvement

The genes found to have significant associations with target traits can be resequenced in a diverse panel of germplasm to identify the causal alleles. It will then be possible to identify the most favorable alleles for breeders' purposes and develop simple PCRbased markers for markerassisted selection (MAS).Genebased markers are more efficient (as they are more accurate) than linked markers, since the markertrait association will not be lost during segregation in the course of recurrent breeding selection cycles. In addition, and perhaps most importantly, the results from association analysis can be used to predict the best haplotype that breeders should be aiming for in order to obtain the optimum expression of their target trait.

In the first candidate-gene association mapping study in plants, DNA sequence polymorphisms within the *D8* locus were associated with flowering time in maize. Following these pioneer studies of association mapping in maize, several other traits such as phenotypic variation in flowering time, endosperm colour, starch production, maysin and chlorogenic acid accumulation, cell wall digestibility, and forage quality were associated using SNP markers of candidate genes. In barley, various traits such as yield, yield stability, heading date, flowering time, plant height, rachilla length, resistance to mildew and leaf rust were associated with different types of molecular markers. In rice, association mapping has not widely been applied yet due to highly structured population of rice (due to high selfing). Association mapping approach was also successfully applied in tetraploid potato where resistance to wilt disease, bacterial blight, *Phytophtora* that utilized a pedigree-based mixed model.

Limitations

To pyramid multiple genes with small effects from diverse genetic backgrounds into a single line remains a huge challenge due to the practical and unpredicted implications of epistasis and genotypebyenvironment effects. In addition, as the number of target traits and source lines increases, the size of populations required to capture the increasingly rare ideal segregants rapidly becomes unmanageable.

Conclusions

The association mapping methodology, initially developed by the human geneticists, has found its successive application in plant germplasm resources. The examples of association mapping studies performed in various plant germplasm resources including model plant Arabidopsis and extended to various crop germplasm largely demonstrate the flourish of crop genomics era with the utilization of powerful LD-based association mapping tool. This is also a good indicative of the potential utilization of this technology with the other crops and plant species in the future. The near-future completion of genome sequencing projects of various crop species, powered with more cost-effective sequencing technologies, will certainly create a basis for application of whole genome-association studies. This will provide more powerful association mapping tool(s) for crop breeding and genomics programs in tagging true functional associations conditioning the trait, and consequently, its effective utilization in crop improvement programmes.

REFERENCES

Acquaah G (2012). *Principles of plant genetics and breeding*. John Wiley and Sons. 740p.

Liu B H (1997). *Statistical Genomics*. CRC Press. 648p.

Lynch, M and Walsh B (1998). *Genetics and Analysis of Quantitative Traits*. Sinauer Associates. 980p.

Singh BD and Singh AK (2015). *Marker-Assisted Plant Breeding: Principles and Practices*. Springer India. 514p.

Xu Y (2010). *Molecular Plant Breeding*. CABI. 734p.

Breeding of Perennial Horticultural Crops *Pages* **131–142**
Editors: S.K. Singh, V.B. Patel, A.K.Goswami, Jai Prakash and Chavlesh Kumar
Published by: **BIOTECH BOOKS, NEW DELHI**

Chapter **12**

Mapping Populations for Gene Tagging

Gyan P. Mishra, H.K. Dikshit and Muraleedhar Aski

Division of Genetics,
ICAR-Indian Agricultural Research Institute, Pusa, New Delhi – 110 012

INTRODUCTION

The developments in the DNA based marker technologies have led to the hope that marker-assisted breeding programs will use marker–trait associations for marker-assisted selection of trait of interest. With the molecular breeding approach, many challenges also exist especially for the complex traits. A major limitation has been the limited ability to detect and quantify marker–trait relationships, especially for the traits influenced by the effects of gene-by-gene and gene-by-environment interactions (Podlich *et al.*, 2004). The development of molecular marker technology has caused renewed interest in genetic mapping. Genetic map construction requires selection of most appropriate mapping population(s), calculation of pairwise recombination frequencies using these populations, establish linkage groups and estimate map distances and determine the map order. Once mapped, the molecular markers are efficiently used in tagging several individual traits that are extremely important for a breeding programme. Since large mapping populations are often characterized by different marker systems, map construction has been computerized. Computer software package, such as Linkage 1, GMendel, Mapmaker, Mapmanager and Joinmap have been developed to aid in the analysis of genetic date for map construction. These programmes use data obtained from the segregating populations to estimate recombination frequency that are then used to determine the linear arrangement of genetic markers.

Mapping Populations

A population used for gene mapping is commonly called a mapping population. Commonly used mapping populations are usually obtained from controlled crosses. Selection of a population for genome mapping involves choosing of parents and determining a mating scheme. Decisions on selection of parents, mating designs and the type of markers should depend upon the objectives of experiments. Parents of mapping populations must have sufficient variation for the traits of interest at both the DNA sequence and the phenotype level. The variation at DNA level is essential to trace the recombination events. The more DNA sequence variation exists, the easier it is to find polymorphic informative makers. When the objective is to search for genes controlling a particular trait, genetic variation of trait between parents is important. If the parents are greatly different at phenotypic level for a trait, there is a reasonable chance that genetic variation exists between the parents, although uncontrolled environmental effects could create large phenotypic variation without any genetic basis for the effects.

Selection of Parents for Development of Mapping Population

Selection of parents for developing mapping population is critical to successful map construction. Consideration must be given to the source of parents (adapted *vs* exotic) used in developing mapping population. Chromosome pairing and recombination rates can be severely disturbed (suppressed) in wide crosses and generally yield greatly reduced linkage distances. Wide crosses will usually provide segregating populations with a relatively large array of polymorphism when compared to progeny segregating in a narrow cross (adapted x adapted). To have significant value in crop improvement programme, a map made from a wide cross must be collinear (*i.e.* order of loci similar) with map constructed using adapted parents.

Types of Mapping Populations

Different types of mapping populations that are often used in linkage mapping are F_2, F_2 derived F_3 (F_2:F_3), backcrosses, doubled haploids (DHs), recombinant inbred lines (RILs), and near isogenic lines (NILs). The characteristic features, merits and demerits of each of these populations are briefly presented below.

F_2 Population

An F_2 population is developed by selfing (or intermating for cross pollinated species) among F_1 individuals. These F_1 individuals are developed by crossing two parents that show significant polymorphism for whichever type of loci we are going to score. F_2 individuals are products of single meiotic cycle. Ratio expected for dominant marker is 3:1 and for co-dominant marker is 1:2:1. The advantage of this population is that, it is considered best population for preliminary mapping. It requires less time for development and can be developed with minimum efforts, when compared to other populations. But the main limitation is that the linkage established using F_2 population is based on one cycle of meiosis. These populations are of limited use for fine mapping. Quantitative traits cannot be precisely mapped using F_2 population as each individual is genetically different and cannot be evaluated in replicated trials over locations and years. Thus, the effect the G x E interaction on the expression of quantitative traits cannot be precisely estimated. It is not a long-term population and is impossible to construct exact replica or increase seed amount.

F_2 Derived F_3 (F_2:F_3) Population

F_2:F_3 Population is obtained by selfing the F_2 individuals for a single generation. Advantage of this population is that it is suitable for specific situations like mapping quantitative traits and recessive genes. This population is used for reconstituting the genotype of respective F_2 plants, if needed, by pooling the DNA from plants in the family. The main demerit is that like F_2 population, it is not *'immortal'*.

Backcross Population

Backcross populations are developed by crossing the F_1 with one of the two parents used in the initial cross. Usually in genetic analysis, backcross with recessive parent (testcross) is used. With respect to molecular markers, the backcross with dominant parent (B_1) would segregate in ratio 1:0 and 1:1 for dominant and co-dominant markers, respectively. However, backcross with recessive parent (B_2) or test cross would segregate in a ratio of 1:1 irrespective of the nature of marker. The merits of this population are like an F_2 population, the backcross populations require less time to be developed. However, the recombination information in case of backcrosses is based on only one parent (the F_1). The specific advantage of backcross populations is that, the populations can be further utilized for marker-assisted backcross breeding. While the major drawback of using F_2 or backcross populations is that the populations are not eternal. Therefore, our source of tissue to isolate DNA or protein will be exhausted at some point in time. We then would have to begin mapping again in another population.

Doubled Haploids (DHs)

Chromosome doubling of anther culture derived haploid plants from F_1 generates DHs. The suitability of doubled-haploid progenies for mapping project has been demonstrated in by Lefebvre *et al.* (1995) in pepper. DHs are also products of one meiotic cycle, and hence comparable to F_2 in terms of recombination information. The expected ratio for the marker is 1:1, irrespective of genetic nature of marker (whether dominant or co-dominant).

The major advantage of DHs are that they are permanent mapping population and hence can be replicated and evaluated over locations and years and maintained without any genotypic change. They are useful for mapping both qualitative and quantitative characters. It can be used for instant production of homozygous lines, thus saving time. The demerits of this population are that the recombination from the male side alone is accounted. Further, anther culture induced variability should be taken care of.

Near-Isogenic Lines (NILs)

NILs are generated either by repeated selfing or backcrossing the F_1 plants to the recurrent parents. NILs developed through backcrossing are similar to recurrent parent but for the gene of interest, while NILs developed though selfing are similar in pair but for the gene of interest (however, differ a lot with respect to the recurrent parent). Expected segregation ratio of the markers is 1:1 irrespective of the nature of marker. Like DHs and RILs, NILs are also *'immortal mapping population'*. They are suitable population for tagging the trait, wherever such population is available. They are quite useful in functional genomics. But the demerit is that it require many generations for development. It is directly useful only for molecular tagging of the gene concerned, but not for linkage mapping. Linkage drag is a potential problem in constructing NILs, which has to be taken care of.

Recombinant Inbred Lines (RILs)

Recombinant Inbred Lines (RILs) are developed by single-seed selections from individual plants of an F_2 population. Because of this procedure, these lines are also called F_2-derived lines. RILs are produced by continuous selfing or sib mating the progeny of individual members of an F_2 population until complete homozygosity is achieved. SSD method is best suited for developing RILs. Bulk method and pedigree methods without selection can also be used. RILs also equalize marker types like DHs, the genetic segregation ratio for both dominant and co dominant marker would be 1:1; RILs developed though brother-sister mating require more time than those developed through selfing. Merits of this population are that once homozygosity is achieved, RILs can be propagated indefinitely without further segregation (Table 12.1). Since RILs are immortal population, they can by replicated over locations and years and therefore are of immense value in mapping QTLs. RILs being obtained after several cycles of meiosis, are very useful in identifying tightly linked makers. RIL populations obtained by selfing have twice the amount of observed recombination between very closely linked markers as compared to population derived from a single cycle of meiosis. But the major demerit is that it requires many seasons/generations to develop and developing RILs are relatively difficult in crops with high inbreeding depression (Table 12.1).

Table 12.1: Relationship between Recombinant Inbred Population (level of inbreeding) and within-line Homozygosity at each Locus (per cent)

RI Population Level of Inbreeding	*Per cent Within-Line Homozygosity at Each Locus*
$F_{3:4}$	75.0
$F_{4:5}$	87.5
$F_{5:6}$	92.25
$F_{6:7}$	96.875
$F_{7:8}$	98.4375
$F_{8^{\cdot}9}$	99.21875

Genetics of Mapping Molecular Loci

Each of the mapping populations will give a specific segregation ratio at each locus. The knowledge of these ratios is important to determine if the population is expressing a skewed segregation ratio at any locus. The following are the ratios that we would expect at each locus for co-dominant and dominant makers segregating in the three types of populations (Table 12.2).

Table 12.2: Ratios Expected at each Locus for Co-dominant and Dominant Makers Segregating in the Three Types of Populations

Population	*Codominant Loci*	*Dominant Loci*
F_2 population	1:2:1	3:1
Backcross population	1:1	1:1
RI population	1:1	1:1
DH	1:1	1:1

First we must analyze our segregating population by various makers and determine that the segregation ratio of each locus is not deviating from the expected ratio, and then we can begin developing the map. One should normally include those loci with skewed segregation ratios in their mapping analysis. All of the segregation data is then compiled and used to derive the linkage relationship among the markers. This analysis is performed using computers and one program widely used is called MAPMAKER. This procedure is based on the maximum likelihood method. The output from this program is a linear relationship among the markers and the distance between the markers is measured in centi-Morgans.

Specific Mapping Topics

Log of Odds (LOD) Ratio

LOD is the log of the odds ratio between the two hypotheses: LOD = $\log_{10}$ [P $(H_1)_/$P (H_0)]. LOD simply means the probability of linkage over probability of no linkage of any trait of interest with any marker type. It means if LOD= 4 then the probability of linkage of the trait of interest is 10^4 with the marker type over probability of no linkage. In other words trait of interest is highly linked.

Kosambi Mapping Function

$r = 1/2 \tanh 2m$ (Where, r = recombination freq., m = map distance in Morgans). r itself is not really additive over long distances. The Kosambi function, like other mapping functions, is meant to give additive map distances (Paterson *et al.*, 1988).

Bulk Segregant Analysis (BSA)

The essence of this procedure is the creation of a bulk sample of DNA for analysis by pooling DNA from individuals with similar phenotypes. Besides the above-mentioned populations, Bulked segregant Analysis (BSA) approach, using any one of the above-mentioned populations (except NILs) is frequently used in gene tagging. BSA is based on the principle of isogenic lines. In BSA, two parents (say a resistant and susceptible), showing high degree of molecular polymorphism and contrast for the target trait are crossed and F_1 is selfed to generate F_2 population. For example, we may be interested in finding a molecular locus linked to a disease resistance locus.

In F_2, individual plants are phenotyped for resistance and susceptibility. Usually, the DNA isolated from 10 plants in each group is pooled to constitute resistant and susceptible bulks. The resistant parent, susceptible parent, resistant bulk and susceptible bulk, are surveyed for polymorphism using polymorphic markers. A marker showing polymorphism between parents as well as bulks is considered putatively linked to the target trait, and is further used for mapping using individual F_2 plants. Conceptually, the genetic constitution of the two bulks is similar, but for the genomic region associated with the target trait. Hence, they serve the purpose of isogenic lines in principle.

It has been observed over experiments that when 10 plants are sampled in each group for constituting the bulk, the probability of a polymorphic marker (between parents as well as bulks) not being linked to the target trait is extremely low (10-19). Hence, usually 10 plants are used for constituting the bulks. However, this number may vary depending upon the types of mapping populations used. In absence of isogenic lines, the BSA approach provides a very useful alternative for gene tagging (Michelmore *et al.*, 1991). The two DNA bulks were constituted in the polymorphism survey to eliminate from consideration those polymorphic markers that were less likely to be associated with disease resistance/susceptibility. Each of these bulk DNA samples will contain a random sample of all the loci in the genome, except for those that are in the region of the gene upon which the bulking occurred. Markers that showed a similar pattern of polymorphism between the parents and between the two bulks were considered as putative resistance/susceptibility related markers. These markers were used to genotype around 50-100 F_2 or >100 $F_{2:3}$ lines to test for their association with disease resistance/susceptibility. Therefore, any difference in RFLP, RAPD or STMS pattern between these two bulks should be linked to the locus upon which the bulk was developed. This is a powerful technique that has gained wide acceptance in the few years since it was first described.

Combining Markers and Populations

The genetic segregation ratio at maker locus is jointly determined by the nature of marker (dominant/co-dominant) and types of mapping populations (Table 12.3). Therefore, a thorough understanding of the nature of markers and mapping populations is crucial for any mapping projects. Markers such RFLPs, microsatellites and CAPS *etc.* are co-dominant in nature, while AFLP, RAPD, ISSR are often scored as dominant markers. Mapping populations such as RILs and DHs equalize marker type because of fixation of parental alleles at marker locus in homozygous condition. These population results in 1:1 segregation ratio at marker locus irrespective of genetic nature of markers, while an F_2 population segregates in 1:2:1 ratio for a co-dominant marker and in 3:1 ratio for dominant marker. Depending upon the segregation pattern, statistical analysis of marker data will vary.

Table 12.3: Genetic Segregation Ratio at Marker Locus in different Marker-Population Combinations

Marker	Nature	Genetic Segregation Ratio					
		F_2	RILs	DHs	NILs	Backcross Popn.	
						B1	B2
RLP	Codominant	1:2:1	1:1	1:1	1:1	1:1	1:1
RAPD	Dominant	3:1	1:1	1:1	1:1	1:0	1:1
AFLP	Dominant	3:1	1:1	1:1	1:1	1:0	1:1
Microsatellites	Codominant	1:2:1	1:1	1:1	1:1	1:1	1:1

Characterization of Mapping Populations

Precise molecular and phenotypic characterization of mapping population is vital for success of any mapping project. Since the molecular genotype of any individual is independent of environment, it is not influenced by G x E interaction. However, trait phenotype could be influenced by the environment, particularly in case of quantitative characters. Therefore, it becomes important to precisely estimate the trait value by evaluating the genotypes in multi-location testing over years using immortal mapping populations to have a valid marker-trait association.

Physical and Genetic Distances

All distances found on a linkage maps are the product of genetic recombination and therefore are considered to be genetic distances. Two pairs of loci that are genetically the same distance apart may not be physically the same distance apart because of suppressed recombination in the region between two of the loci. Those loci where suppression of recombination does occur will physically be farther apart. The Table 12.4 gives the relationship between the physical and genetic distance in three species.

Table 12.4: Relationship between the Physical and Genetic Distance in Three Species

Species	Kilobases/Centi Morgan
Arabidopsis	139
Tomato	510
Corn	2140

These values are estimates of a single region of the genome of each species, and it is quite common to see that two regions of the same species have different amount of DNA per genetic distance. Correlations between physical and genetic distances have also been performed using deletion stocks in wheat. A series of deletions stocks of a specific wheat chromosome were analyzed with probes known to hybridize to that chromosome. Each locus was assigned to the chromosome region between the breakpoint of the largest deletion where the band was present and the next larger deletion where the band was absent (Werner *et al.*,1992). The authors noted large discrepancies between the physical and genetic maps of chromosomes 7B and 7D. Two loci that are located near the centromere of 7B are 7 cM apart genetically, yet the distance between the two loci spans 25 per cent of the chromosome. Another region at the distal region of long arm of 7B which accounts for about 15 per cent of the chromosome is 91 cM long genetically. This points out the difference of recombination that can occur within a single chromosome.

Linkage Drag

One goal of plant breeding is to introduce a gene from a donor parent to improve a cultivar for a specific trait. For example, a wild germplasm could be used as a source of disease resistance or male sterility or fertility restoration *etc.* At the same time the breeder does not want to carry any of the other genes from the wild germplasm that might reduce the agronomic fitness of the cultivar. The backcross method of plant breeding is one manner in which the introduction of a specific gene is accomplished. One genetic feature though of backcross breeding is linkage drag. This refers to the reduction in fitness in a cultivar due to deleterious genes introduced along with the beneficial gene during backcrossing. Molecular makers offer a tool in which the amount of wild or alien DNA can be monitored during each backcross generation.

Quantitative Trait Locus (QTL) Mapping

Many traits do not fall into a discrete class but, when a segregating population is analyzed for these traits, a continuous distribution or normal distribution is found. The examples are yield, drought, heat or cold tolerance *etc.* in plants. These traits exhibit a bell-shaped distribution if we see the whole population as a whole. These traits are often referred to as quantitative traits and the loci controlling these traits are called quantitative trait loci or QTL. They are controlled by multiple genes, each segregating according to Mendel's laws. Traditional quantitative genetic research defined a quantitative trait in terms of variances. Mapping markers linked to QTLs identifies regions of the genome that may contain genes involved in the expression of the quantitative trait. Molecular makers associated with QTLs are identified by first scoring members of a random segregating population for a quantitative trait. The molecular genotype (homozygous parent A, heterozygous, or homozygous parent B) of each member of the population is then determined. The next step is to determine if an association exists between any of the markers and the quantitative trait.

Several methods have been described for estimating means of quantitative trait locus (QTL) genotypes using unreplicated progeny (Lander and Botstein, 1989; Simpson, 1989). The most common method of determining the association is by analyzing phenotypic and genotypic data by one-way analysis of variance and regression analysis. If the variance for the genotype class is significant, then the molecular marker used to define the genotype class is considered to be associated with a QTL. For those loci that are significant, the quantitative trait values are regressed onto the genotype. The R^2 value for the line is considered to be the amount of total genetic variation that is explained by the specific molecular marker. The final step is to take those molecular marker loci that are associated the quantitative trait and perform a multiple regression analysis. From this analysis, we will obtain an R^2 value which gives the percentage of the total genetic variance explained by all of the markers.

Types of Mapping Population

Random-mating populations are more difficult for QTL mapping, because linkage disequilibrium is a key to detecting QTLs with markers. It is essential to develop a suitable experimental mapping population using parental lines that are highly contrasting phenotypically for the target trait (*e.g.* highly resistant and susceptible lines). Another important requirement is that these parental lines should be genetically divergent; this is important to enhance the possibility of identifying a large set of polymorphic markers that are well-distributed across the genome. To fulfill the second criterion, one may have to carry out molecular polymorphism survey across a set of potentially useful lines so as to identify the most suitable ones for generation of mapping population.

As mentioned earlier, F_2-derived populations, RILs, backcross populations, NILs, DHs *etc.* could be employed for QTL mapping in various crop plants. The choice of a mapping population could vary based upon the objectives of the experiment, the time frame as well as resources available for undertaking QTL analysis. But, the ability to detect QTLs or the information contained in F_2 or F_2-derived populations and RILs are relatively higher than other. The primary advantage of F_2:F_3 families is the ability to measure the effects of additive and dominance gene actions at specific loci. Because RILs are essentially homozygous, only additive gene action can be measured. The advantage, though, of the RILs is the ability to perform larger experiments at several locations and even in multiple years. For many crops, it is not possible to generate enough seeds to perform a multiplications experiment with population of F_2:F_3 families. Modification of the genetic model is necessary to accommodate different types of populations.

Size of the Mapping Population

This depends on several factors, including the type of mapping population employed for analysis, genetic nature of the target trait, objectives of the experiment, and the resources available for handling a sizable mapping population in terms of phenotyping and genotyping. Usually the possible number of QTLs that can be detected with a population size of 25, 50, 100, 500, 1000 and infinity are 0, 1, 2, 3, 4 and 5, respectively. While analysis of a large number of individuals (~500 or more) would enable detection of even QTLs having small effects on the target trait, from the practical point of view (MAS), the basic purpose of QTL mapping would be largely served if one can detect the QTLs with major effects. This would require, in general, a mapping population of a size of 200-300 individuals.

Power of QTL Mapping

Statistical models and QTL parameter estimation methods have not been described for replicated progeny; however, Lander and Botstein (1989) have proposed using replicated progeny to increase power for mapping quantitative trait loci. The power of tests of hypotheses about means of QTL genotypes are partly determined by the genetic model (Lander and Botstein 1989) and the parameter estimation method (Simpson 1989). These factors do not determine how experiment design factors affect power; however, they do determine the sample size needed to achieve a given power.

The power of the test of a difference between means of QTL genotypes is increased by increasing the number of replications of lines (r) (Lander and Botstein 1989). They proposed using "progeny testing" ($r > 1$) to increase the "power of QTL mapping" and compared using replicated recombinant inbred (RI) and unreplicated backcross (BC) progeny. They also proposed, "recombinant inbred strains will thus typically be more efficient for QTL mapping than an equal number of backcross progeny" by "reducing the environmental variance through replicate phenotypic measurements within each recombinant inbred line" (Lander and Botstein 1989). Power is obviously increased by increasing r, but the effect of r has not been quantified. Furthermore, power for replicated progeny is determined by n and other factors in addition to r, and how these factors affect power has not been quantified. Increasing r decreases the standard error of a QTL genotype mean, and consequently increases power for testing hypotheses about means of QTL genotypes, but this standard error is determined by *r and n*. Beavis *et al.* (1991) analyzed four populations of maize and found molecular markers linked to plant height. No marker was consistently associated as a QTL with plant height in all four populations. Each of the ten maize chromosomes contained a marker linked to a QTL for at least one of the four populations. The statistical analysis of quantitative traits provided valuable information for the plant breeder.

Developments in Functional Genomics

Allele Mining

Allele mining is a research field aimed at identifying allelic variation of relevant traits within genetic resources collections. For identified genes of known function and basic DNA sequence, genetic resources collections may be screened for allelic variation by *e.g.* the "tiling strategy" using DNA chip technology. In that approach the basic DNA sequence of a gene is spotted on a chip in the form of large series of sequence overlapping probes consisting of 15–20 bases. Each base position in a fluorescently labeled sample is then interrogated for the presence of point mutations by monitoring hybridization signals with the spotted probes. Because the sequence of samples is determined in comparison with the primary composition of a gene, this method is also known as "re-sequencing". With this method new point mutations, in relatively large DNA fragments, can be detected.

Association Genetics

In allele mining studies as described in the previous section, allelic variation is analyzed for identified genes whose function and basic DNA sequence are known and whose map position in the genome will generally have been determined. On the contrary, in association genetic studies no prior information about the genes of interest is available, but associations between genetic markers and the considered traits are simply derived from observational research. Association genetics focuses on the identification of correlations between phenotypic traits and genetic markers with the aim to identify and locate the underlying genes in the genome (association mapping). Association genetics originated in human genetic studies focusing on the application of significant associations between marker data and diseases for mapping and diagnostic purposes.

Comparative Genome Mapping

Comparative genomics focuses on the integration of genome information derived from different species with the aim to obtain more insight in the genetic organization of traits through the identification of conserved mechanisms. This research field has emerged from previous findings of comparative mapping, by which conservation of tracts of collinear markers have been investigated not only among members of the same botanical family, but also among different species, and often led to better understanding of genome evolution. Comparative mapping has shown that a large proportion of the markers and genes are indeed located at comparable positions in the genome (synteny). Comparative mapping can identify inversions, translocation and duplications that have occurred. Genetic factors can also be assessed by comparing map distances of genes with conserved gene order in the two species.

Conclusions

Genetic map construction requires selection of most appropriate mapping population(s), calculation of pairwise recombination frequencies, establishment of linkage groups and estimation of map distances and finally determination of map order. Mapping of any trait involves a method of mapping a gene using some mathematical formulae and a mapping population. Today we have number of different mapping populations (*e.g.* DH, NIL *etc.*) with unique advantages of being 'immortal population'. Therefore we can grow these populations at different locations and at different places *i.e.* in different environments for tagging of both qualitative and quantitative traits. A large number of monogenic and polygenic loci for various traits have been identified in a number of plants, which are currently being exploited by breeders and molecular biologists together, so as to make the dream of marker-assisted selection come true.

Molecular analysis of quantitative traits now provides new tools, not only as selection tools for plant breeding, but as starting points for the cloning of these genes. These objectives could not have been realized without molecular markers. New developments in genomic research have given access to an enormous amount of sequence information as well as new insights on the function and interaction of genes and the evolution of functional domains, chromosomes and genomes. In this context, functional and comparative genomics can help in comparative genetic mapping and linkage analysis of useful agricultural traits. Future DNA marker techniques, such as the use of oligonucleotide arrays, are likely to be sequence based. Comparative analyses of sequence information in the growing databases now publicly available on the World Wide Web will be an invaluable resource for formulation of different strategies for molecular breeding. This will have an increasingly important role in the evolution of newer ways for molecular breeding.

REFERENCES

Beavis WD, Grant D, Albertsen M, Fincher R (1991). Quantitative trait loci for plant height in four populations. *Theo. Appl. Genet.*, 83: 141-145.

Lander S, Botstein D (1989). Mapping Mendelian factors underlying quantitative traits using RFLP linkage maps. *Genetics*, 121: 185-199.

Lefebvre V, Palloix A, Caranata C, Pochard E (1995). Construction of an intraspecific linkage map of pepper using molecular markers and doubled-haploid progenies. *Genome*, 38: 112-121.

Michelmore RW, Paran I, Kesseli RV, (1991). Identification of markers linked to disease resistance genes by bulked segregant analysis: A rapid method to detect markers in specific genomic regions by using segregating populations. *Proc. Natl. Acad. Sci.*, 88: 9829-9832.

Podlich DW, Winkler CR, Cooper M (2004). Mapping As you Go: An Effective Approach for Marker-Assisted Selection of Complex Traits. *Crop Sci.*, 44: 1560-1571.

Schachermayr G, Siedler H, Winzeler H, Keller B (1994). Identification and localization of molecular markers linked to *Lr9* leaf rust resistance gene of wheat. *Theor. Appl. Genet.*, 88: 110–115.

Simpson SP (1989). Detection of linkage between quantitative trait loci and restriction fragment length polymorphisms using inbred lines. *Theor. and Appl. Genet.*, 77: 815-819.

Werner JE, Endo TR, Gill BS (1992). Toward a cytogenetically based physical map of the wheat genome. *Proc. Nat. Acad. of Sci.*, 89: 11307-11311.

Singh, B.D., Singh, A.K. (2015). Marker-Assisted Plant Breeding: Principles and Practices. pp. 461. Springer-India, eBook DOI: 10.1007/978-81-322-2316-0

Singh RK, Singh R, Guoyou Ye (2010). *Molecular Plant Breeding: Principle, Method and Application*. Studium Press LLC, USA. pp. 467.

Breeding of Perennial Horticultural Crops *Pages* **143–148**
Editors: S.K. Singh, V.B. Patel, A.K.Goswami, Jai Prakash and Chavlesh Kumar
Published by: **BIOTECH BOOKS, NEW DELHI**

Chapter **13**

Functional Genomics

Jasdeep Chatrath Padaria
ICAR-National Research Centre on Plant Biotechnology, New Delhi – 110 012

INTRODUCTION

Augmentation of food production for the burgeoning global population requires optimum utilization of the available resources in a sustainable manner. Thus, at present the agricultural systems are under huge pressure to fulfill the increasing demand of food and feed in the context of global climate change with expanding population. Unfortunately, most of the land with agricultural potential is in areas with harsh environments which in scientific terminology is defined as stress. The term stress is derived from the Latin word *stringere*, which means a constraining or impelling force. In terms of agriculture production, stress is defined as "the external constraints which limit the rate of dry matter production of all or part of the vegetation below its genetic potential" (Grime, 1979). However, some researchers interpret stress with reference to economic yield, rather than dry matter production of the plant system. Any plant is considered to be under abiotic stress when it experiences a relatively severe shortage of an essential constituent or an excess of potentially toxic or damaging substance of abiotic origin. These abiotic soil stresses constitute a major limiting factor, which hamper plant productivity throughout the world. Over 60 per cent of all yield losses from crops that are currently grown worldwide are due to environmental stresses such as drought, temperature extremes or soil salinization. Furthermore, with the changing global climate these environmental stresses are becoming of paramount preposition especially in relation to crop productivity.

It is practically established fact that the climatic changes experienced by us in recent times, would increase tremendously in years to come which will lead to drastic reduction in agricultural productivity. Already these climatic variations such as increased global temperatures, annual rainfall, events of drought and flood are resulting in alteration in the edaphic attributes of soil. The change in the edaphic factors have resulted into the decreased soil fertility with water salinization and ultimately reduction in crop yield. So in the contemporary era of scientific advancement, it is of central significance to develop the mitigation strategies by the use of analytical and forward looking concepts to fulfill the rapidly increasing food demands with ecological sustainability.The development of crop plants that can tolerate different abiotic stresses would help combat the problem of reduced yields due to environmental stress. Currently traditional breeding strategies are limited by the restricted availability of beneficial genes. Therefore there is an urgent need to look out for genes especially for abiotic stress tolerance in plants growing under harsh environmental conditions. Crop plants in their natural environments are frequently exposed to different abiotic stresses like high salinity, extremes of both high and low temperatures and water deciency or drought. Whether plants are susceptible or tolerant to these adverse condition depends on expression of multiple stress responsive genes and their coordinated action, which also cross-talk with other modules of signal transduction involved in stress related pathways. Over long evolutionary scales, plants in response to different environmental stresses have attained mechanisms by which they can assess these stresses and modulate their physiological mechanisms accordingly. Thus it is important to look out for these genes and transfer them into abiotic stress sensitive important food crops using DNA recombinant technology. This strategy involves identification of abiotic stress tolerance gene from a hardy plant system, its isolation- characterization and functional validation in a model system for transgenic introgression into the desirable system.

What is Funcational Genomics?

Functional genomics refers to the development and application of experimental approaches to assess gene function by making use of information and reagents provided by structural genomics. It is a field of molecular biology that attempts to make use of the vast wealth of data given by genomic and transcriptomic projects (such as genome sequencing projects and RNA sequencing) to describe a gene (and protein) functions and interactions. It is characterized by high-throughput methodologies combined with statistical analysis of the results. Functional genomics uses data generated by RNA seq/Genome sequencing projects to describe gene, protein functions and interaction. It studies how genes and intergenic regions of the genome contribute to different biological processes. A researcher in this field typically studies genes or regions on a "genome-wide" scale (*i.e.* all or multiple genes/regions at the same time), with the hope of narrowing them down to a list of candidate genes or regions to analyse in more detail. The goal of functional genomics is to determine how the individual components of a biological system work together to produce a particular phenotype. Functional genomics focuses on the dynamic expression of gene products in a specific context, for example, at a specific developmental stage, under any kind of stress (biotic or abiotic). In functional genomics, we try to use our current knowledge of gene function to develop a model linking genotype to phenotype.

Functional genomics is used to study biochemical, cellular and physiological properties of each and every gene product.

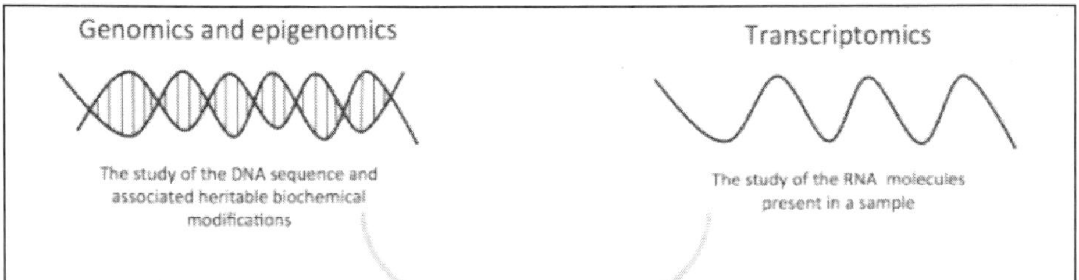

Figure 13.1: Functional Genomics is the Study of How the Genome, Transcripts (Genes), Proteins and Metabolites Work Together to Produce a Particular Phenotype. (https://www.ebi.ac.uk/training/online/course/functional-genomics-i-introduction-and-designing-e/what-functional-genomics)

Different between Classical and Modern Funcational Genomics

Classical functional genomics could examine only function of a single gene while modern functional genomics may target the expression analysis of 1000 to 10,000 genes at same time, expressed as a function of development. There are some approaches for functional genomics such as: SSH (Suppression Subtractive Hybridization), Microarrays, SAGE (Serial Analysis of Gene Expresssion).

Application of Functional Genomics

Functional genomics uses mostly multiplex techniques to measure the abundance of many or all gene products such as mRNAs or proteins within a biological sample

- **At the DNA level:** Genetic interaction mapping
- **At the RNA level:** Microarray, SAGE, RNA seq
- **At the protein level:** Yeast two hybrid, Affinity purification and Mass spectrometry
- **Loss of function techniques:** Mutagenesis, RNAi

The above mentioned techniques have been developed as component of functional genomics to elucidate the function of genes/sequences and can assist in understanding the genetic mechanism of growth, development and physiological responses of plant. Therefore, functional genomics can aid in discovery of alleles/QTLs, development of high-throughput markers, development of high-density linkage map and ultimately assigning the function of genes/QTLs related with yield and associated traits and biotic and abiotic stress tolerance to develop high yielding genotypes to mitigate the challenges of global food security.

Advantages, Disadvantages of Functional Genomics Technique

Functional genomics methods give an opportunity for accurate and comprehensive functional analysis that involves various fields of studies: genomics, epigenomics, proteomics, and interactomics. This is essential for filling the gaps in the knowledge about dynamic biological processes at both cellular and organismal level. However, each method has both advantages and limitations that should be taken into account before choosing the right method for particular research in order to ensure successful study.

Technique	*Advantage*	*Disadvantage*
RNA-Seq	Direct, quantitative and high throughput method. Does not require a priori knowledge about the genomic features. Appropriate for gene, transcripts (including alternative gene spliced transcripts) or allele-specific expression identification	High sequence similarity between alternative
SAGE	Direct and quantitative method. *A priori* knowledge about the gene sequences is not required. SAGE library requires a small amount of RNA as input. Simple data analysis	Low-throughput
cDNA microarray	Well-studied,-throughput and quantitative method Based on fluorescence (no need of radioactive probes)	Complicated data analysis
MS	High-throughput method that rightly identifies and quantifies proteins	The sample should be high-quality and homogenous Sometimes dissociation efficiency of complex protein is lower
Y2H	The two-hybrid technique is relatively simple. Appropriate for the first step in identifying interacting protein partners	The rate of false positive results is relatively high. The need of confirmatory test. Impossible interaction between two proteins at a time

Technique	*Advantage*	*Disadvantage*
Chemical mutagenesis (in animal model)	Mutation can be induced artificially and mutant phenotype can be recognized easily Genes can be cloned using standard procedures	Phenotype always reflects human beings
CRISPR-Cas9	The possibility to engineer the protein and RNA components of bacterial CRISPR system in order to recognize and cut DNA at desired locus	Work requires highly sterile conditions

Ziziphus nummularia: A Case Study

Ziziphus spp. (ber) belonging to the family Rhamnaceae, possess intrinsic and dual capability of drought avoidance and drought tolerance by means of various physiological mechanisms. These mechanisms include, stimulation of antioxidative metabolic pathways, osmotic adjustment and sensitive stomatal closure to minimize dehydration during adverse conditions. *Z. nummularia* (Burm.f.) Wight and Arn., a perennial herb commonly known as jhar ber is native to western India, southeastern Pakistan and southern Iran. *Z. nummularia* along with seventeen other species are wildly distributed in dry and hot climate sof the central India, northwestern plains and dry areas in the peninsular region. Both, the domestic and wild species of *Ziziphus* are found in the Indian deserts, but species *Z. mauritiana* and *Z. rotundifolia* are found only in the northwestern Indian desert along with the traditional species, *Z. nummularia. Ziziphus nummularia* responds to drought, salinity and temperature stress through a series of variations at physiological and developmental levels by inducing expression of a number of stress responsive genes.

Transcript Profiling in *Z. nummularia*

Two SSH libraries were constructed from one month old *Z. nummularia* seedlings which were subjected to drought stress through 30 per cent Polyethylene glycol (PEG 6000) treatment for 24 h and 48 h. A total of 672 ESTs were identified and annotated for gene ontology (GO) categorization. Fourteen drought responsive genes were validated by reverse transcription-quantitative PCR (RT-qPCR) at different duration of 30 per cent PEG 6000 treatment. Transcript pattern of genes showed significant upregulations with time in treated samples compared to control. The genes differentially upregulated in response to drought stress can serve as candidate genes for development of drought tolerant transgenics in food crops.

Transcriptome Sequencing of *Z. nummularia* under Drought

RNA-sequencing of one month old seedlings treated with PEG 6000 was performed using Roche GS-FLX454 Titanium pyrosequencing. A total of 367,176 raw sequence reads were generated, and upon adapter trimming and quality filtration 351,872 reads were assembled *de novo* into 32,739 unigenes. Further characterization of the unigenes indicated that 73.25 per cent had significant hits in the protein database. Kyoto encyclopedia of genes and genomes database (KEGG) identified 113 metabolic pathways from the obtained unigenes. A large number of drought-responsive genes were obtained and among them differential gene expression of 16 highly induced genes was validated by qRT-PCR analysis. To develop genic-markers, 3,425 simple sequence repeats (SSRs) were identified in 2,813 unigene sequences. The data generated shall serve as an important reservoir for the identification and characterization of drought stress responsive genes for development of drought tolerant crops.

Based on SSH library analysis *ZnAsr*1 gene was fully characterized for the first time from *Ziziphus nummularia*, which is one of the most low water forbearing plant. Full length *ZnAsr*1 gene was characterised and *in silico* analysis of *Zn*ASR1 protein was done for predicting its phylogeny and physiochemical properties. To validate transcriptional pattern of *ZnAsr*1 in response to drought stress, expression profiling in polyethylene glycol (PEG) induced Z. *nummularia* seedlings was studied by RT-qPCR analysis and heterologous expression of the recombinant *ZnAsr*1 in Escherichia coli. The nucleotide sequence analysis revealed that the complete open reading frame of *ZnAsr*1 is 819 bp long encoding a protein of 273 amino acid residues, consisting of a histidine rich N terminus with an abscisic acid/water deficit stress domain and a nuclear targeting signal at the C terminus. In expression studies, *ZnAsr*1 gene was found to be highly upregulated under drought stress and recombinant clones of *E. coli* cells expressing *Zn*ASR1 protein showed better survival in PEG containing media. *ZnAsr*1 was proven to enhance drought stress tolerance in the recombinant *E.coli* cells expressing *Zn*ASR1. The cloned *ZnAsr*1 after proper validation in a plant system, can be used to develop drought tolerant transgenic crops.

Breeding of Perennial Horticultural Crops *Pages* **149–157**
Editors: S.K. Singh, V.B. Patel, A.K.Goswami, Jai Prakash and Chavlesh Kumar
Published by: **BIOTECH BOOKS, NEW DELHI**

Chapter **14**

Cisgenesis: Its Potential in Fruit Crop Improvement

S.K. Singh, Megha R., Satyabrata Pradhan, Chavlesh Kumar, Jai Prakash, V.B. Patel and A.K. Goswami

Division of Fruits and Horticultural Technology,
ICAR-Indian Agricultural Research Institute, Pusa, New Delhi – 110 012

INTRODUCTION

Keeping in view of the exponential rise in world population, there is a need for breeding new fruit genotypes having higher quality and productivity with novel traits to sustain under the fluctuating climate. But the genetic improvement of the fruit crops through conventional breeding method is highly tedious due to its perennial nature, inherent long juvenile phase, highly heterozygous in nature, existence of self-incompatibility, linkage drag *etc.* (Petri and Burgos, 2005; Rai and Shekhawat, 2014; Limera *et al.*, 2017). The biotechnological tools offer several ways to solve the problem of conventional breeding in the fruit crops with greater extent. In this regard, transgenic approach for fruit crop improvement can be used as a strapping tool to complement and substitute the traditional breeding to meet the worldwide demand for high productive quality fruits. The release of genetically modified (GM) fruit crops into agricultural production has raised considerable debate, especially among the general public, politicians and bureaucrats. One approach is to address many of the ethical, religious and/or public concerns in the genetic engineering debate is the transfer of genes within the gene pools currently available to plant breeders, referred to as intragenic (Nielsen, 2003) or cisgenic (Schouten *et al.*, 2006). Fruit crop genotypes developed through cisgenic can not only be used for yield enhancement with higher nutritional quality but also to combat the climate change through development of various biotic and abiotic stress tolerances. However, such gene transfer still relies on vector systems based on DNA from prokaryotic origin that is concomitantly transferred to plants during genetic transformation. To ensure public acceptance of the transfer of genes within species, gene transfer must be achieved without the presence of any other DNA from foreign gene (Shiva Prasad *et al.*, 2003).

The Concept of Cisgenesis

Cisgenesis (literal meaning from "same" or "beginning") is one term for organisms that have been engineered using a process in which gene(s) are artificially transferred between organisms that could otherwise be conventionally bred. Unlike in transgenesis, gene(s) is only transferred between closely related organisms. The term cisgenesis was coined by Schouten and colleagues (Schouten *et al.*, 2006). Cisgenesis is the genetic modification of a recipient organism with a gene from a crossable sexually compatible organism (same species or closely related species). The gene (cisgene) includes its introns and is flanked by its native promoter and terminator in the normal sense orientation. Cisgenic plants can harbour one or more cisgenes, but they do not contain any part of transgenes or inserted foreign sequences. To produce cisgenic plants any suitable technique used for production of transgenic organisms may be used. Genes must be isolated, cloned or synthesized and transferred back into a recipient, where stably integrated and expressed. Presently, cisgenesis is described as the method of crop improvement, which includes specific alleles/genes in the breeder's gene pool to be introduced into new genotypes without the accompanying linkage drag (co-transfer of DNA sequences that are linked to the gene of interest), which generally occurs in conventional breeding. A diagram comparing procedure and the genetic changes achieved through conventional plant breeding, transgenesis and cisgenesis is given in Figure 14.1 (Anon, 2009).

Need for Cisgenesis

In case of conventional breeding introgression of foreign gene from related or unrelated species through backcross breeding needs almost 8-10 years. While, backcrossing there is probability of transfer of undesirable alleles along with desirable one due to tight linkage between them, which is known as linkage drag. Such limitations of classical breeding can be overcome by employing Cisgenesis or Intragenesis approach.

Cisgenic and intragenic derived genetically modified plants (GMP) are eco-friendly as classic bred plants and therefore exempted from GMP legislation.

In Europe, currently, this process is governed by the same laws as transgenesis but researchers at the Wageningen University in the Netherlands they feel that this should be changed and regulated in the same way as conventionally bred plants. In 2012, the European Food Safety Authority (EFSA) issued a report with their risk assessment of cisgenic and intragenic plants. The EFSA concluded that "similar hazards can be associated with cisgenic and conventionally bred plants, while novel hazards can be associated with intragenic and transgenic plants". Cisgenic

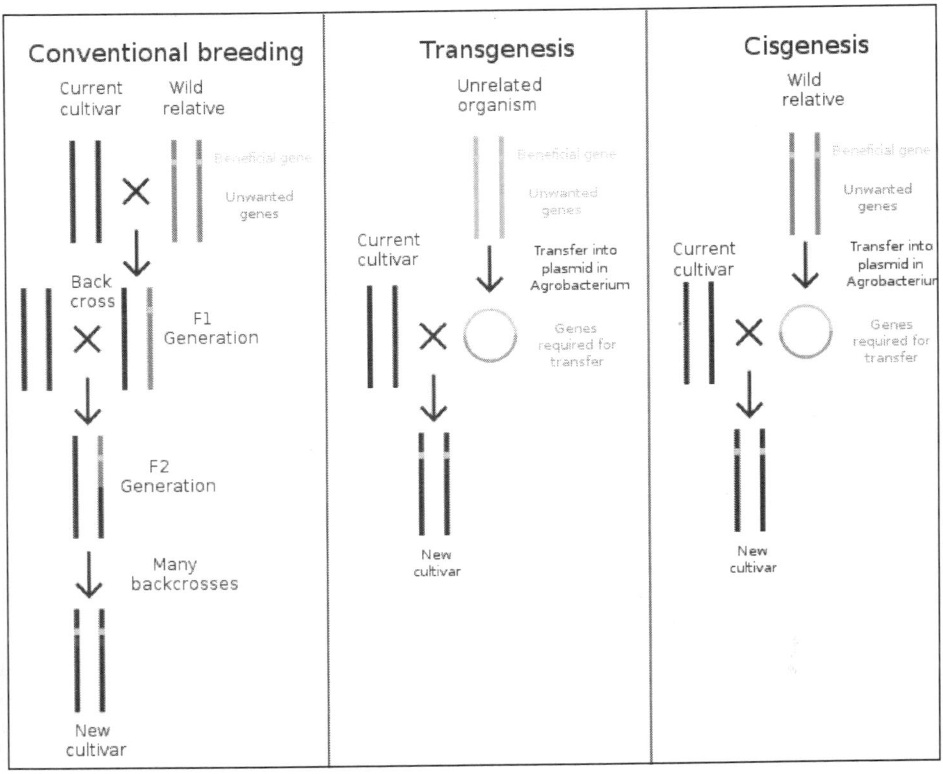

Figure 14.1: A Diagram Comparing the Genetic Changes Achieved through Conventional Plant Breeding, Transgenesis and Cisgenesis.

technique can help the fruit breeders to develop the desirable genotypes within a short span of time in comparison to conventional fruit breeding.

Transgenesis and Cisgenecis

In the case of transgenesis, the transferred gene usually is derived from an alien species that is neither the recipient species nor a close, sexually compatible relative. Transgenesis can extend the gene pool of the recipient species. Such a novel gene might provide the target plant with a new trait that neither occurs in the recipient species in nature nor can be introduced through traditional breeding. This novel trait might affect the fitness of the recipient species in various ways; a change in fitness can then spread through gene flow between a GM crop and its wild relatives, potentially creating shifts in natural vegetation. Consequently, lawmakers and regulatory authorities have paid much attention to the safety of deliberate releases of transgenic crops into the environment and have put in place the biosafety frameworks to control this risk (Jacobsen, 2007).

However, in the case of a cisgenic plant, the gene of interest, together with its promoter, has been present in the species or in a sexually compatible relative for centuries. Therefore, cisgenesis does not alter the gene pool of the recipient species and provides no additional traits. No changes in fitness occur that would not happen through either traditional breeding or natural gene flow. Similarly, cisgenesis carries no risks, such as effects on non-target organisms or soil ecosystems, toxicity or a possible allergy risk for GM food or feed-other than those that are also incurred by traditional breeding. This is the fundamental difference between cisgenesis and transgenesis. Consequently, the deliberate release and market introduction of cisgenic plants is as safe as the release and market introduction of traditionally bred plants. On the issue of safety, regulators could treat cisgenic plants the same as conventionally bred plants.

Cisgenics in Fruit Crops

Cisgenesis has been attempted only in a limited number of horticultural crops including fruit crops (Table 14.1). It has been applied to transfer the natural resistance genes to counter the devastating disease like scab (*Venturia inaequalis*) in apple (Vanblaere *et al.*, 2014), *Botrytis cinerea* in strawberries (Schaart *et al.*, 2004) and *Phytophthora infestans* in potato. Therefore, one can infer that cisgenesis to be a powerful alternative approach to transfer gene(s) of interest without much linkage drag in a single-step.

Apple

Most of the present day apple (*Malus* × *domestica*) cultivars are susceptible to scab, which is caused by the fungus (*Venturia inaequalis*). Fruit growers need to spray on average 15 times in one season to control the disease (Patocchi *et al.*, 2004). Later, Joshi *et al.* (2011) developed apple transformants carrying the apple-derived *HcrVf2* gene in a cisgenic as well as in an intragenic configuration. These transformants showed almost similar scab resistance levels as compared to the *Vf* resistant control cultivar obtained by classical breeding in cv. Santana.

Later on other researchers also developed cisgenic apple lines using various other genes from apple or related pome fruit species, *viz.*, *Rvi6* (Krens *et al.*, 2015; Wurdig *et al.*, 2015), and *HcrVf2* (Gessler *et al.*, 2014).

One of the major applications of cisgenesis in apple is to create blight resistant genotypes by transferring known resistance loci from wild genotypes into otherwise high yielding popular cultivars. The generation and selection of novel fire blight resistant apple cisgenic genotypes is expected to greatly improve the management of this devastating disease caused by *Erwinia amylovora*. Resistance sources to various diseases are often found in wild *Malus* genotypes, *e.g.* the *Rvi6* resistance gene from *M. floribunda* clone 821 for apple scab (Hough, 1994) or the *FB_MR5* of *M.* × *robusta* 5 (*Mr*5) for fire blight (Peil *et al.*, 2007). Kost *et al.* (2015) developed a cisgenic apple line C44.4.146 using the cisgene *FB_MR5* from *Mr*5 through *Agrobacterium tumefaciens*-mediated transformation in cultivar 'Gala Galaxy'. The line C44.4.146 was shown to carry only the cisgene *FB_MR5*, controlled by its native regulatory sequences and no transgenes were detected by PCR or Southern blot. This line was found to be phenotypically free from the blight infection (Figure 14.2).

Table 14.1: Application of Cisgenesis and Intragenesis in Various Fruit Crops

Fruit Crop	*Name of Gene*	*Source*	*Trait*	*Achievement*	*Reference*
Apple (*Malus domestica* Borkh.)	HcrVf2	*Malus domestica*	Apple scab resistance	Reduction (80 per cent) in fungal infection in cisgenic lines as compared to wild type Gala apple	Joshi *et al.* (2011)
	HcrVf2	Apple cv. Gala	Apple scab resistance	Cisgenic lines having the *HcrVf2* gene	Gessler *et al.* (2014)
	Rvi6	*Malus floribunda* 821	Apple scab resistance	Cisgenic lines had similar level of resistance to the *M. floribunda* wild type plants	Krens *et al.* (2015)
	Rvi6	*Malus floribunda* 821	Apple scab (Race 1) resistance	Two cisgenic lines resistant to apple scab	Wurdig *et al.* (2015)
	FB_MR5	Apple cv. Gala Galaxy	Fire blight resistance	Cisgenic line C44.4.146 having lower disease severity upon inoculation with *Erwinia amylovora*	Kost *et al.* (2015)
Grape (*Vitis vinifera* L.)	*VVTL*-1	*Vitis vinifera*	Powdery mildew resistance	Cisgenic lines showed delayed powdery mildew disease development	Dhekney *et al.* (2011)
Citrus (*Citrus paradisi*)	*C. clementina*-derived T-DNA-like region	*Citrus clementina*	Development of 'foreign DNA-free' intra-/cisgenic citrus cultivars	Transformation efficiency in 'Duncan' grapefruit was nearly 0.67 per cent	An *et al.* (2013)
Strawberry (*Fragaria* x *ananassa*)	*FaPGIP*	*Fragaria* x *ananassa*	Fruit rot (*Botrytis cinerea*) resistance	*Botrytis* resistance due to higher expression of *FaPGIP*	Schaart, (2004)

Figure 14.2: Results of fire Blight Inoculation on Grafted Apple Line C44.4.146 (left) and 'Gala Galaxy' *in vitro* (right) after 20 Days of Inoculation (*Source:* Kost *et al.*, 2015).

Grape

In grape, occurrence and damage caused by different pathogens result in heavy losses besides heavy application of plant protection chemicals. Hence, cisgenic is highly desired in grape breeding so that management of diseases like downy mildew, powdery mildew and anthracnose and Pierces' disease, nematode *etc.* Dhekney *et al.* (2011) developed grapevines engineered to express cisgenic *Vitis vinifera* thaumatin-like protein to produce fungal-disease resistance. The *V. vinifera* thaumatin-like protein (*vvtl*-1) gene was isolated from Chardonnay grape and reengineered for constitutive expression. Embryogenic cultures of cv. Thompson Seedless were initiated from leaves and transformed with *Agrobacterium* to regenerate cisgenic *VVTL*-1 plants. During greenhouse screening, two plants exhibited a 7-10-day delay in powdery mildew disease development and decreased severity of black rot disease was also observed in field tests. Berries exhibited a 42.5 per cent reduction in sour-bunch rot disease incidence compared to controls (non-transformed) after three week of storage at room temperature.

Citrus

In citrus, there is the development of a "foreign DNA-free" intragenic vector system, pUFCI (University of Florida Citrus Intragenic), by assembling T-DNA-like fragments with functional equivalents of T-DNA border sequences from *Citrus clementina* and recovery of positive intra-/cis-genic regenerants through PCR analysis. The transformation efficiencies obtained in both *Arabidopsis* and "Duncan" grapefruit indicate its great potential for citrus genetic improvement (An *et al.*, 2013).

Strawberry

It is a short duration very remunerative fruit crop. However, it suffers from different biotic stresses. Amongst them soft rot/fruit rot, caused by *Botrytis cinerea*, may result in significant losses in its production. Since *Botrytis* is proliferating particularly on ripe fruits, it may lead to pre- and post-harvest damages. Some polygalactouridinase (PG) gene inhibiting proteins (PGIPs) have been found in cell walls of a wide range of plants. Although the strawberry *PGIP* (*FaPGIP*) gene is naturally present in the strawberry genome, endogenous expression levels are expected to be too low for efficient inhibition of *B. cinerea*. Schaart (2004) developed cisgenic strawberry lines having higher *Botrytis* resistance due to more expression of *FaPGIP*.

Drawbacks

Although cisgenesis is having a great scope due to the ethical concern of transgenics, but it carries some of the probable theoretical drawbacks in its technology. A frequently mentioned drawback is the fact that cisgenes will be inserted at unknown places of the plant genome, which could bring unforeseen risks. Another theoretical drawback that is mentioned regarding cisgenesis is the mutation made at the insertion site of the cisgene and the unexpected accompanying phenotypic changes that might result from this. We just have seen that induced translocations and especially naturally occurring transposons can cause mutations in the crop genome, which even can be used. Usually, mutations are inherited recessively and represent loss of function. An important disadvantage of cisgenesis is food safety. Although still not reported from fruit crops but the use of wild *Solanum* sp. as source of genetic variation is bringing back different kinds of glycoalkaloids that have been removed during the breeding process in the past. The use of wild species as new source of genetic variation is accompanied by the re-introduction of these compounds.

Conclusion

The biotechnological tools are immensely help to the fruit breeders for the development of the new varieties of woody perennial fruit tree species. Further, development of the transgenics in fruit crops creates novel fruit cultivars while incorporating the genes from the other organism. The cisgenesis in fruit crops solved the problem of several bio safety issues while incorporating the genes from the crossable individuals or species. This technique is like classical breeding in the fruit crops but it substantially reduces the time duration and linkage drag. Although it has been used in a limited number of fruit crops but it have immense importance to the fruit breeders for increasing the efficiency and accuracy in the development of new varieties in the perennial woody fruit tree species.

REFERENCES

An, C., Orbovic, V. and Mou, Z. 2013. An efficient intragenic vector for generating intragenic and cisgenic plants in citrus. *American J. Plant Sci.* **4**(11): 2131-2137.

Anon. 2009. https: / /en.wikipedia.org/wiki/File: Breeding_transgenesis_cisgenesis.svg.

Dhekney, S.A., Li, Z.T. and Gray, D.J. 2011. Grapevines engineered to express cisgenic Vitis vinifera thaumatin-like protein exhibit fungal disease resistance. *In Vitro Cell. Devt Biol.-Plant*, **47**(4), 458-466.

Gessler, C., Vanblaere, T., Parravicini, G. and Broggini, G.A.L. 2014. Cisgenic 'Gala' containing the scab resistance gene from *Malus floribunda* 821 and the fire blight resistance genes from *M.* 'Evereste'. *Acta Hort.* **1048**: 43-50. doi: 10.17660/ActaHortic.2014.1048.4

Hough, L.F. 1944. A survey of the scab resistance of the foliage on seedlings in selected apple progenies. *Proc. American Soc. Hort. Sci.* **44**: 260-272.

Jacobsen, E. and Schouten, H.J. 2007. Cisgenesis strongly improves introgression breeding and induced translocation breeding of plants. *Trends Biotech.* **25**(5): 219-223.

Joshi, S.G., Schaart, J.G., Groenwold, R., Jacobsen, E., Schouten, H.J. and Krens, F.A. 2011. Functional analysis and expression profiling of *HcrVf1* and *HcrVf2* for development of scab resistant cisgenic and intragenic apples. *Plant Mol. Biol.* **75**(6): 579-591.

Kost, T.D., Gessler, C., Jänsch, M., Flachowsky, H., Patocchi, A. and Broggini, G.A. 2015. Development of the first cisgenic apple with increased resistance to fire blight. *PloS One*, **10**(12): e0143980.

Krens, F.A., Schaart, J.G., van der Burgh, A.M., Tinnenbroek-Capel, *I.E.*, Groenwold, R., Kodde, L.P. and Schouten, H.J. 2015. Cisgenic apple trees; development, characterization, and performance. *Frontiers Plant Sci.***6**: 286.

Limera, C., Sabbadini, S., Sweet, J.B. and Mezzetti, B. 2017. New biotechnological tools for the genetic improvement of major woody fruit species. *Frontiers Plant Sci.* **8**: 1418. doi: 10.3389/fpls.2017.01418.

Nielsen, K.M. 2003. Transgenic organisms-time for conceptual diversification? *Nature Biotech.* **21**(3): 227-228.

Patocchi, A., Bigler, B., Koller, B., Kellerhals, M. and Gessler, C. 2004. *Vr* 2: a new apple scab resistance gene. *Theor. Appl. Genet.* **109**(5): 1087-1092.

Peil, A., GarciaLibreros, T., Richter, K., Trognitz, F.C., Trognitz, B., Hanke, M.V. and Flachowsky, H. 2007. Strong evidence for a fire blight resistance gene of Malus robusta located on linkage group 3. *Plant Breed.* **126**(5): 470-475.

Petri, C. and Burgos, L. 2005. Transformation of fruit trees. Useful breeding tool or continued future prospect? *Transgenic Res,* **14**: 15-26. doi: 10.1007/s11248-004-2770-2.

Rai, M.K. and Shekhawat, N.S. 2014. Recent advances in genetic engineering for improvement of fruit crops. *Plant Cell, Tiss. Organ Cult.* **116**(1): 1-15.

Schaart, J.G. 2004. *Towards consumer-friendly cisgenic strawberries which are less susceptible to Botrytis cinerea*. Ph.D. thesis, Wageningen University: Wageningen, Netherlands.

Schouten, H.J., Krens, F.A. and Jacobsen, E. 2006. Cisgenic plants are similar to traditionally bred plants: international regulations for genetically modified organisms should be altered to exempt cisgenesis. *EMBO Rep.* **7**(8)**:** 750-753.

Vanblaere, T., Flachowsky, H., Gessler, C. and Broggini, G. A. 2014. Molecular characterization of cisgenic lines of apple 'Gala' carrying the *Rvi6* scab resistance gene. *Plant Biotech. J.* **12**(1)**:** 2-9.

Wurdig, J., Flachowsky, H., Sab, A., Peil, A. and Hanke, M.V. 2015. Improving resistance of different apple cultivars using the *Rvi6* scab resistance gene in a cisgenic approach based on the *Flp/FRT* recombinase system. *Molecular Breed.* **35**(3): 95.

Breeding of Perennial Horticultural Crops *Pages* **159–181**
Editors: S.K. Singh, V.B. Patel, A.K.Goswami, Jai Prakash and Chavlesh Kumar
Published by: **BIOTECH BOOKS, NEW DELHI**

Chapter 15

Phenomics: The Next Generation Phenotyping (NGP)

Viswanathan Chinnusamy[1], Rabi Narayan Sahoo[1], S. Naresh Kumar[1], S. Gopala Krishnan[1], S. Lekshmy[1], R. Dhandapani[1], Sudhir Kumar[1], Harikrishna[1], Ranjith Kumar Ellur[1], Swati Goel[1], Krishna K. Jangid[1], Rahul Tiwari[1], Anil Rai[2], A.R. Rao[2], Sudeep Marwaha[2], Sanjeev Kumar[2], Brejesh Lall[3], Santanu Chaudhury[3], Archana Chugh[3], Swati Bhugra[3], Padmini Swain[4], Sushanta K. Dash[4] and Jitendriya Meher[4]

[1]ICAR-Indian Agricultural Research Institute, Pusa, New Delhi – 110 012
[2]ICAR-Indian Agricultural Statistics Research Institute, New Delhi
[3]Indian Institute of Technology, New Delhi
[4]ICAR-National Rice Research Institute, Cuttack

INTRODUCTION

Enhancing agricultural production with environmental sustainability is the major goal of agricultural research. To meet the food demand by 2050, the growth rate of yield gain must be doubled. This needs the efficient use of available genetic diversity and use of modern biotechnology to genetically enhance the resource (water and nutrient mainly nitrogen) use efficiency and crop productivity. Phenotyping or characterization of plants is one of the earliest agricultural activities of humans started with domestication of crops plants. Later the phenotypic diversity within each species was exploited by the earliest crop improvement methods. Once the concepts of genetics were understood, it was established that the phenotype is the product of genotype x environment. Establishment of genotype-phenotype relationship is the key for modern genetic improvement methods of both plants and animals. The two pillars of analytical breeding are genotyping and phenotyping. Efficient use genomic information for crop breeding is potential solution to develop high yielding, resource use efficient and climate resilient crop varieties. Genomic technologies such as Next Generation Sequencing (NGS) and SNP arrays have enabled the plant scientist to obtain genotypic information of breeding material with relatively low cost and shorter time. However, the principal goal of identifying specific genotypes that are associated with phenotypes progressed only slowly as development in phenotyping has not kept pace with genomics. The wet chemistry and other actual measurement of growth and physiological processes based phenotyping is inherently low throughput, labor-intensive, costly, time consuming and often destructive due to organism-wide phenotypic data for same plant cannot be obtained, which lead to genotype-phenotype gap (Furbank and Tester 2011).

The *"Plant Science Decadal Vision for the decade 2015 to 2025"* on food, energy, environment, and health (Plant Science Research Summit 2013), as well as *"nine big ideas"* proposed by National Science Foundation (NSF), USA for solving pressing societal problems emphasizes the necessity of *"**understanding the rules of life: predicting phenotype and assemble plant traits in different ways to solve problems**"* (Mervis 2016). Thus, accurate phenotyping to obtain physiological information necessary for crop improvement is the key for further genetic improvement of crops. To address these necessities, the multi-disciplinary science of phenomics emerged recently.

Phenome is defined as expression of the genome as traits in a given environment. The human phenome project initiated in 1997 (Freimer and Sabatti 2003) led to the birth of phenomics (Bilder *et al.*, 2009). Phenomics is multidisciplinary science of sensor aided non-destructive high throughput automated acquisition and analysis of high-dimensional phenotypic data on an organism-wide scale. Phenomics, the Next Generation Phenotyping (NGP), offers solution to discover the inner workings of living plants and thus bridge the phenotype-genotype gap (Cobb *et al.*, 2013; Fiorani and Schurr 2013; Fahlgren *et al.*, 2015b; Großkinsky *et al.*, 2015). Phenomics involves 1) non-invasive sensors, 2) automated data processing to obtain phenotypic traits, 3) robotized delivery of plants to sensors or vice versa, 4) robotized plant culturing, and 5) automated analysis of processed data in a data management pipeline. Robotized delivery of plants to the imaging sensors is commonly used in controlled environment phenomics platform, while sensors are delivered to the plants in field phenomics platforms. Non-invasive sensors commonly used in non-destructive automated plant phenomics facility consists of various imaging cameras namely visual imaging, Hyperspectral imaging, IR thermography, NIR image analysis, Chlorophyll fluorescence imaging, bioluminescence imaging, fluorescence imaging, *etc.* Wide-range of phenotypic data on whole-plant during its entire life cycle can be acquired by using phenomics technologies that are not possible through conventional phenotyping methods (Kumar *et al.*, 2016). In addition, Light detection and ranging (LIDAR) and laser triangulation sensors are used for assessment of plant growth, shoot biomass, leaf angle distributions and canopy structure, while magnetic resonance imaging (MRI) is used for three-dimensional imaging of roots to obtain spatial information on the root system architecture of plants (van Dusschoten *et al.*, 2016). Phenomics is being employed in both controlled environment as well as in the natural field conditions (Figure 11.1). Different imaging methods, sensors used and phenotype data acquired are summarized in Table 11.1.

Phenomics Initiatives by Indian Council of Agricultural Reserach, New Delhi

Realizing the potential of phenomics, Australian government invested $51 million in 2007 and established the Australian Plant Phenomics Facility (APPF) in January 2010 (http://www.plantphenomics.org.au/about/). Since then, several government Institutes have established automated high throughput phenomics facility for crop plants. Soon, the Indian Council of Agricultural Research, New Delhi also initiated the establishment of phenomics facilities in India recently at ICAR-Indian Agricultural Research Institute, New Delhi; ICAR- Central Research Institute for Dryland Agriculture, Hyderabad; ICAR-Indian Institute of Horticultural Research, Bengaluru and National Institute of Abiotic Stress Management, Baramati, India. ICAR-IARI, New Delhi has established a *state-of-the*

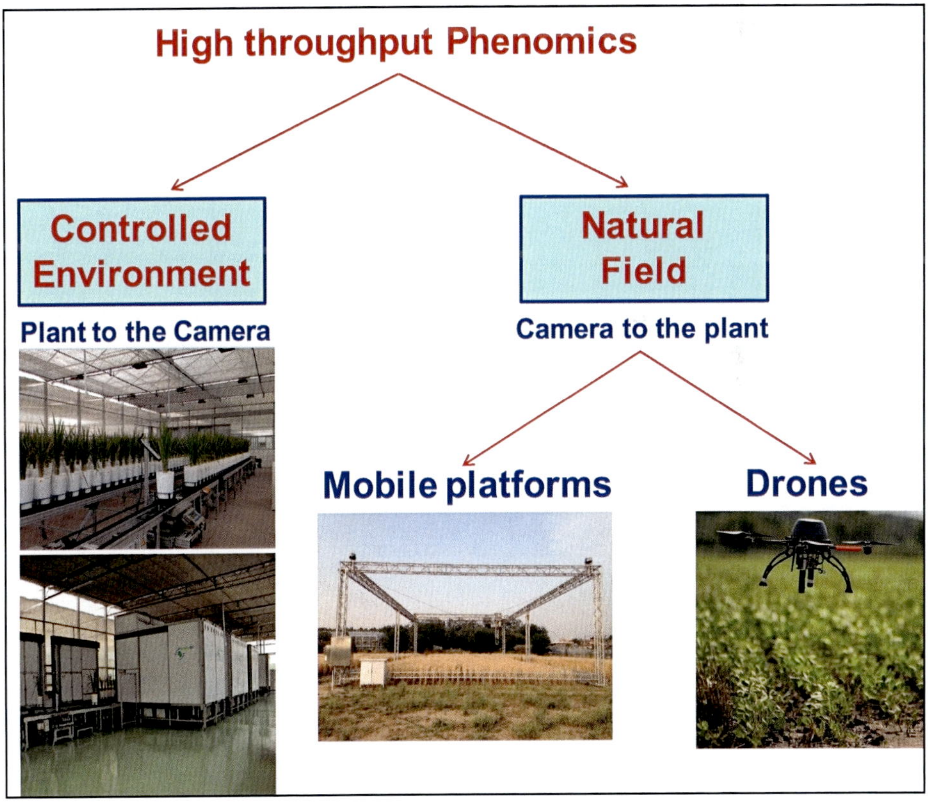

Figure 15.1: Phenomics Approaches Used in Controlled Environment and Field Conditions.

Table 15.1: High Throughput Non-invasive Sensors for Phenomics (Kumar *et al.,* 2016)

Sensor	*Principle of Trait Capture*	*Traits Measured*	*References*
Visible light imaging camera	Reflectance in the visible (400-700 nm) range depend upon plant growth, morphology, pigments, wax, health, *etc.* Image processing and segmentation in binary image, Hue Saturation Intensity (HSI) color model and L*a*b* color space are used to obtain phenotypic data.	Shoot/root growth, architecture, greenness, Leaf area, leaf rolling, senescence, growth rates, tillering, early vigor, plant height, phenology, biomass, convex hull, compactness, eccentricity	Golzarian *et al.* (2011) Das *et al.* (2016)
Fluorescence imaging camera	Light absorbed by short wave length is emitted as long wave fluorescence depending upon the composition of plant tissues with molecules with innate (auto) fluorescence characters. Chlorophyll molecules absorb light at shortwave length and emit fluorescence at red/far-red wavelength (680 and 735 nm). Nicotinamide (NAD) and flavin (FMN, FAD) coenzymes, pyridoxal phosphate, folic acid and secondary metabolites (phenolics, alkaloids and terpenoids) emit blue-green fluorescence when excited with UV light (340 to 360 nm).	Maximum quantum efficiency of PSII, photochemical quenching and non-photochemical quenching, which are highly sensitive to resource availability and stresses; Secondary metabolite mapping; Fluorescence can be used to detect metabolites and gene expression when tagged with non-native fluorophores such as transgenic plants expressing GFP/YFP fusion proteins.	Mishra *et al.* (2016)
Thermal imaging camera	The infrared energy (8 to 13 μm) emitted from object is converted into an electrical signal by the imaging (microbolometer) sensor. Tissue temperature is determined mainly by evapotranspiration.	Thermal images are used to infer stomatal conductance and plant health (biotic and abiotic stress).	Möller *et al.* (2007) Ludovisi *et al.* (2017)
Bioluminescence imaging camera	Emission of visible light from an enzymatic reaction inside the plan is captured using low-light imaging CCD cameras in darkness.	Transgenic plants expressing firefly LUCIFERASE or free calcium sensor AEQUORIN are useful to identify mutants and to assess physiological processes and stress responses.	Grant *et al.* (2000) Chinnusamy *et al.* (2002)
Near infrared imaging (NIR) and multispectral line scanning cameras	The reflectance of plants in the range of 900 to 1700 nm depends upon water content. Plants reflects large amount of 800 to 1400 nm light while soil reflectance is negligible.	Water content, leaf thickness, leaf area index; root soil moisture extraction pattern	Neilson *et al.* (2015)

Sensor	*Principle of Trait Capture*	*Traits Measured*	*References*
Hyperspectral Reflectance imaging camera (indium gallium arsenide sensors)	Spectral reflectance is imaged at nm resolution by VIS-NIR (visible-near infrared, 400–1000 nm) and SWIR (short wavelength infrared, 1000–2500 nm) cameras.	Several spectral indices are available to assess chlorophyll content, relative water content, nutrient status, chemical composition, plant health, photochemical reflectance index, genotype bar-coding	Romer *et al.* (2012) Sahoo *et al.* (2015) Wahabzada *et al.*, 2016
Stereo camera	Two RGB (red, green, blue) cameras to capture three-dimensional images	Shoot biomass and structure, leaf angle distributions, canopy structure	Biskup *et al.*, 2007
Light detection and ranging (LIDAR) and laser triangulation sensors	A laser light beam is projected onto plants and the energy scattered from the plant is captured for the computation of depth maps and 3D point clouds.	Shoot biomass and structure, leaf angle distributions, canopy structure	Kjaer and Ottosen (2015) Vadez *et al.* (2015)
NIR and Fourier transform infrared spectroscopy (FTIR) spectroscopy	NIR and FTIR measure chemical composition, respectively, from the NIR and long wave IR absorption and emission properties of plant tissues.	Quantification of water, protein, oil, sugars, starch, cell wall composition, lignin, and other larger molecules in seeds and other plant tissues	Chaerle *et al.* (2009) Bacıolu *et al.* (2017) Legner *et al.* (2018)

art automated high throughput plant phenomics facility for non-destructive and accurate characterization of a large number of germplasm and recombinant inbred lines under defined environmental treatment conditions (Funded by NASF, ICAR, New Delhi-110012). The phenomics facility has four hi-tech climate controlled greenhouses for cultivation of plants in defined environmental conditions. For plant cultivation, the facility is equipped with 1200 plant carriers with RFID chip tag. The plant carrier on moving field conveyer system randomizes plants within the greenhouse and carries plants for automated weighing and watering, and imaging at various imaging platforms. The facility has 5 automated weighing and watering stations for precise imposition of drought stress to plants and to measure transpiration and water use efficiency of plants. The facility has the following non-invasive image-based sensor platforms for measuring various plant traits (Figure 11.2):

1. ***Visual high resolution imaging:*** Reflectance in the visible (400-700 nm) range is captured by using high resolution camera from the top and side of the plants. Visual imaging is used to measure shoot/root growth, architecture, greenness, Leaf area, leaf rolling, senescence, growth rates, tillering, early vigor, plant height, phenology, biomass, convex hull, compactness, eccentricity, *etc.*
2. ***IR thermal imaging:*** The infrared energy (8 to 13 μm) emitted from plant is converted into an electrical signal by the imaging (microbolometer) sensor to measure tissue temperature. As tissue temperature is determined mainly by evapotranspiration, IR thermal images are used to infer stomatal conductance and plant health (biotic and abiotic stress).
3. ***Chlorophyll fluorescence imaging:*** Light absorbed by short wave length is emitted as long wave fluorescence depending upon the composition of plant tissues with molecules with innate (auto) fluorescence characters. Chlorophyll molecules absorb light at shortwave length and emit fluorescence at red/far-red wavelength (680 and 735 nm). This imaging system can measure chlorophyll fluorescence to calculate maximum quantum efficiency of PSII, photochemical quenching and non-photochemical quenching, which are highly sensitive to resource availability and stresses.
4. ***Near infrared (NIR) imaging:*** The reflectance of plants in the range of 900 to 1700 nm depends upon water content. Plants reflects large amount of 800 to 1400 nm light while soil reflectance is negligible. NIR shoot imaging system is used to measure water content and distribution in plants, leaf thickness and leaf area index, while NIR root imaging system is used to phenotype root soil moisture extraction pattern and root growth.
5. ***Visual-Near Infrared (VNIR) and Short wave Infrared (SWIR) - hyperspectral imaging systems:*** Spectral reflectance is imaged at nm resolution by VIS-NIR (400–1000 nm) and SWIR (1000–2500 nm) cameras. Several spectral indices are available to assess chlorophyll content, relative water content, nutrient status, chemical composition, plant health, photochemical reflectance index, genotype bar-coding.

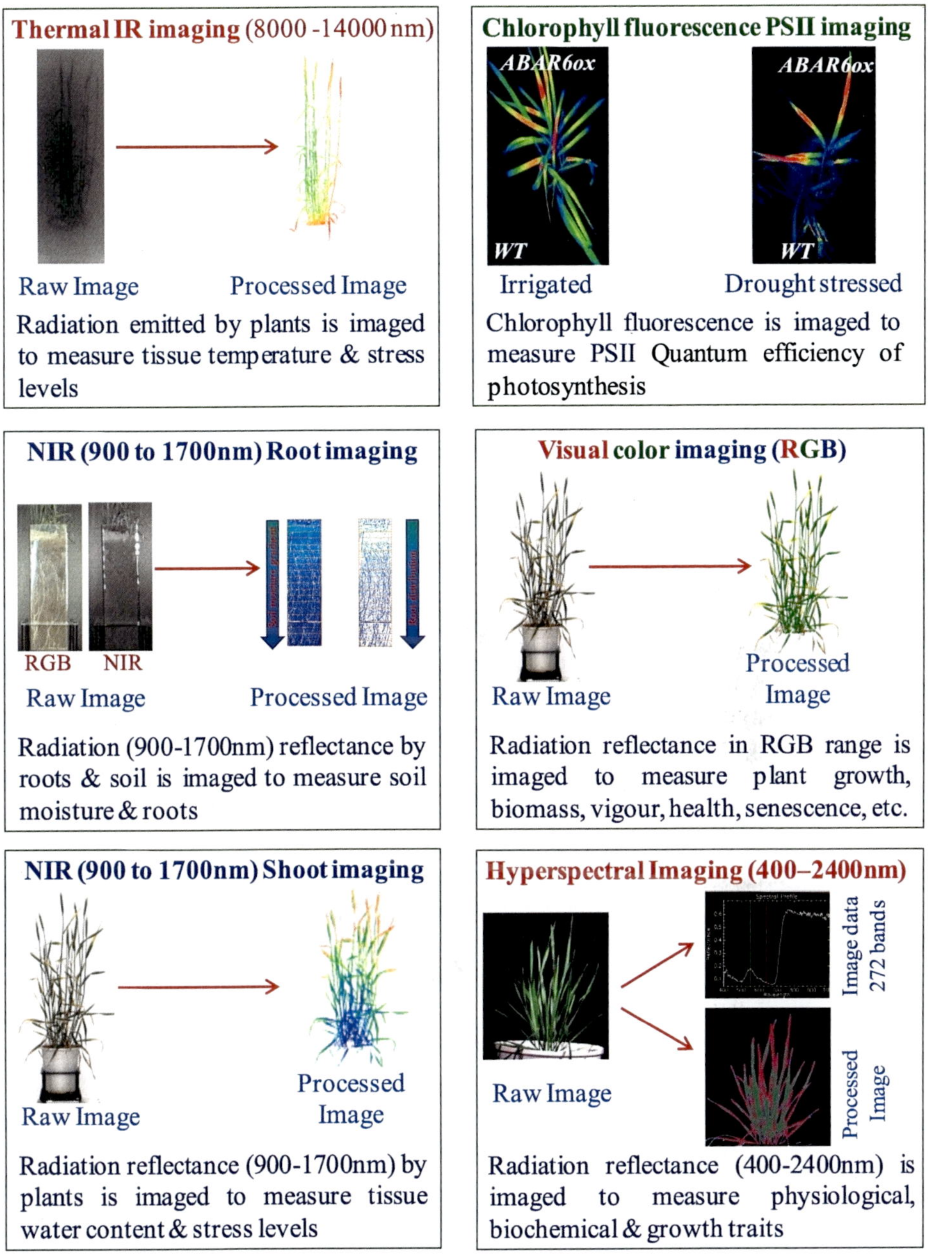

Figure 15.2: Examples of Image Analysis from different Sensors and the Physiological Traits Measured.

The automated weighing and watering stations will quantify the weight of pots before and after watering, in order to impose various drought/waterlogging/ nutrient deficiency stresses, and to assess input use efficiency. Thus, critical physiological traits contributing to the yield and stress tolerance can be measured by phenomics platforms with high throughput for a large set of plants at defined intervals during crop growth. The depth of component phenotypic traits and the spatio-temporal dynamic phenotypic data generated in phenomics are enormous and unparallel to the conventional phenotyping. Some of the utilities of phenomics facility are (Kumar *et al.*, 2016):

1. Dissection of complex traits into component traits
2. Germplasm screening to identify donors
3. Phenotyping of biparental population for Linkage mapping
4. Phenotyping of minicores for genome-wide association studies (GWAS)
5. Functional genomics, Forward and reverse phenomics
6. Gene function validation and selection of better transgenic events
7. Trait pyramiding in analytical breeding
8. Phenome-wide association studies (PheWAS)
9. Phenomic selection
10. Training of Genomic Selection models with deep phenotyping data
11. Development of ecophysiological crop simulation models for *in silico* phenotyping and ideotype design

Shri Narendra Modi, Hon'ble Prime Minister of India inaugurated and dedicated the "Nanaji Deshmukh Plant Phenomics Centre" to the Nation on 11th October 2017, on the event of the birth centenary celebration of Nanaji Deshmukh at IARI, Pusa, New Delhi. The major goals of this centre are:

1. To identify superior genotypes and novel genes useful for development climate resilient crop varieties.
2. To unravel the interaction of genes and the environment using big data analytics, the next step in expanding the boundaries of our knowledge in crop improvement and management.
3. To identify image features from different sensors that will be useful for UAV- and/or remote sensing-aided applications for resource and crop management in precision agriculture.
4. To develop globally competent scientific human resources in cutting edge research area of digital phenotyping, predicting plant behaviour in different environment and big data science useful for crop improvement and management.

Potential of Phenomics for Trait Dissection and Gene Mapping

Phenomics is being extensively used for establishing phenotype genotype relationship and QTL mapping. Some examples of biparental population based QTL mapping and genome-wide association (GWA) mapping using data from NGP phenomics are given in the Table 11.2. The relationship between QTL mapping under field conditions and controlled environment phenomics facility where plants were

Table 15.2: Examples of Phenomics Aided QTL Mapping

Crop/Model Plant	*Population*	*Phenomics Platform*	*QTLs Mapped*	*Remarks*	*Reference*
Arabidopsis	162 RILs and 92 NILs derived from a Cvi) × Ler cross	Visual image every 2 min for 8 hr; Controlled environment	QTLs for mean tip angle at each of the 241 time points	Time-dependent QTL were detected on chromosomes 1, 3, and 4	Moore *et al.*, 2013
Triticale	647 doubled haploid lines derived from four families; GWA mapping	A tractor pulled trailer equipped with two light curtains, three laser distance sensors, two 3D-Time-of-Flight cameras; Field conditions	23, 25 and 17 QTLs at 3 developmental stages; Two major QTLs	One major QTL on chromosome 5R is active throughout plant development while the other major QTL on chromosome 5A contributes strongly to biomass at the early stage	Busemeyer *et al.*, 2013
Rice	171 RIL and parental plants Bala × Azucena	Visual imaging of root systems growing in nutrient-enriched gellan gum at days 12, 14, and 16 Post planting	89 univariate QTLs across all days of imaging for various RSA traits	Many univariate and multivariate QTLs that we identified colocalized with previously identified root trait and drought resistance hotspots	Topp *et al.*, 2013
Barley	47 wild barley ILs of the S42IL library and the recipient parent Scarlett	RGB images; gravimetric measurement of water use, end point phenotypic data; Controlled environment	44 QTL for 11 traits;	Three QTL were identified for Absolute Growth Rate Integral (AGRI); Two QTL were detected for WUE	Honsdorf *et al.*, 2014
Triticale	647 DH lines derived from four families	phenomics data of biomass yield generated at three developmental stages; Controlled environment	10, 10, 9 QTLs were mapped for biomass at 3 stages respectively,	Of the several QTLs mapped, only 4 were common in all three stages, while 5, 4, and 4 were specific for biomass at satge1, 2 and 3 respectively	Liu *et al.*, 2014
Triticale	647 triticale DHs derived from 4 families	Visual image based plant height (PH) measurement at 3 developmental stages (PH1, PH2 and PH3); Field conditions	15 QTL for PH1, 18 for PH2 and 8 for PH3	Only 3 QTLs common for all three stages suggesting that the genetic control of plant height undergoes rapid temporal changes	Würschum *et al.*, 2014

Crop/Model Plant	*Population*	*Phenomics Platform*	*QTLs Mapped*	*Remarks*	*Reference*
Wheat	5000 RILs from a cross between Drysdale and Gladius	Conventional phenotyping in the field; Image based phenotyping in controlled environment	84 QTLs in Field; 21 QTLs for plant growth using the imaging platform	7 co-located QTLs were found for traits from the phenomics platform with that from the field	Parent *et al.*, 2015
Arabidopsis	324 accessions; GWA mapping	Visual top-view imaging and end-point fresh weight determination; Controlled environment	22 QTLs for fresh weight (endpoint), projected leaf area (at 12 different growth stages) and modelled parameters	Many of the growth QTLs would not have been identified with only endpoint fresh weight data	Bac-Molenaar *et al.*, 2015
Rice	378 diverse rice genotypes; Salinity tolerance; GWA mapping	Visible and Fluorescence imaging; Controlled environment	55 QTLs	Only 26 QTLs could be detected at one time point	Campbell *et al.*, 2015
Rice	553 genotypes phenotyped for salinity tolerance; GWA mapping	Visible image (one top and two side views); Controlled environment	Several QTLs for RGR, TR and TUE at different intervals	The QTL on chromosomes 11 is strongest in the first interval after salt stress (2–6 days after treatment), but not in the last interval (9–13 days after treatment)	Al-Tamimi *et al.*, 2016
Arabidopsis	324 accessions; GWA mapping	Visual top-view imaging, end-point fresh weight determination, gravimetric measurement of water use; Controlled environment	Twenty-one SNPs associate with FW, PLA over time, RWC and model parameters	Six time-dependent drought-QTLs; Of these, for five QTLs, most likely candidate genes could be identified	Bac-Molenaar *et al.*, 2016
Sorghum	97 RILs and the two parental lines BTx623, IS3620C)	RGB time-of-flight depth camera; Controlled environment	Five QTLs were mapped; alleles closely linked with the sorghum Dwarf3 gene, an auxin transporter, was found to play important role in shoot architecture.	Many of the QTLs identified via image-based phenotyping overlapped with QTLs for comparable traits discovered in prior field experiments	McCormick *et al.*, 2016

Crop/Model Plant	*Population*	*Phenomics Platform*	*QTLs Mapped*	*Remarks*	*Reference*
Maize	167 RILs with its parents (B73 and BY804)	106 traits across 16 developmental stages using the automatic phenotyping platform in controlled environment; also phenotyped under field conditions	988 QTLs were identified for 42 phenotypic traits across 16 time points; 42 to 82 QTLs at each time point	Several dynamic development QTLs were identified	Zhang *et al.*, 2017
Maize	252 diverse inbred lines; GWA mapping	Automated non-invasive phenotyping at 11 different developmental time points	Several QTLs mapped for different growth stages	Main effect loci detected show complex developmental phase-specific patterns of expression	Muraya *et al.*, 2017

grown in pots were studied. Phenomics approach was used to map QTLs in barley for growth under drought stress including growth rate and water use efficiency at seedling stage. Several QTLs showed co-localization with previously mapped QTLs under field conditions. A novel QTL that significantly increased biomass by about 36 per cent was identified (Honsdorf *et al.*, 2014). Further, in wheat by using phenomics approach, about 20 QTLs with strong effects, accounting for between 26 and 43 per cent of the variation were in a controlled environment showing that the G×E interaction could be reduced. Comparative analysis of QTLs mapped using phenomics approach with that are previously mapped under field conditions showed co-localization (Parent *et al.*, 2015). Combination of phenomics and genome-wide association studies (GWAS) in rice, 141 associated loci for 15 traits, 25 of which are previously known genes (Yang *et al.*, 2014). These performance evaluation studies demonstrated that phenomics approach is a suitable alternative to replace traditional laborious field-phenotyping for QTL mapping and positional cloning.

Superiority of Non-Destructive Phenomics over Conventional Field Phenotyping

Conventional phenotyping is often destructive and phenotypic data is obtained at few crop growth stages or at the end of the crop cycle. Automated NGP using phenomics technologies captures multiple phenotypic data throughout the crop growth stages and thus adds time-scale to the phenotypic data which is not available in the conventional phenotyping. Time-scale phenotypic data during different growth and development of crop is necessary for mapping the QTLs for component traits that contributes to crop development during specific growth stages. Plant growth models quantify 1) absolute growth rate (AGR), 2) relative growth rate (RGR), and Net Assimilation Rate (NAR), which require measuring biomass/leaf area at successive time points. However, raking these destructive measurements in field is limited due to space, time and cost limitations, and thus often only two point measurements are taken and fitted into simple logistic models. However, the results do not often fit with observations (Paine *et al.*, 2012). Phenomics is highly useful in measurement of plant growth and development on the organism-wide scale, and thus it is highly useful to measure dynamics of various component physiological traits that contribute to yield and stress adaptation. Automated phenomics enables the plant scientist to quantify traits that are difficult to measure under field conditions such as relative growth rate, transpiration, and water-use efficiency (WUE). Direct quantification of WUE requires gravimetric measurements of amount of water used for evapotranspiration and plant biomass. It is difficult to directly measure WUE for large number of germplasm lines and mapping populations. Hence only limited success has been achieved in identification of donors and QTLs for this important trait. Further, the physiological causes of the genotypic differences are not understood. Temporal measurements of water use and biomass in automated phenomics facility using *S. viridis* and domesticated *S. italica* revealed that both have similar biomass production, but *S viridis* maintained the water-use efficiency, while *S. italica* become less efficient growth under water-deficit. Conventional end point measurement could not have detected this temporal physiological response of genotypes in WUE as the soil available water changes

(Fahlgren *et al.*, 2015a). The dissection approach uses model-assisted methods to dissect complex phenotypes such as yield and drought tolerance into more simple and heritable traits. In barley, phenomics approach was used to identify novel traits, such as maximum growth rate and stress elasticity, associated with plant growth and drought tolerance. These traits are not measurable via traditional phenotyping approaches. In addition several image-based traits and model-derived parameters were identified which have potential for subsequent dissection of the genetic basis of complex agronomic traits (Chen *et al.*, 2014).

The genetic dynamics of plant traits were revealed by the introduction of time-axis by the use of automated phenomics to dissect the genetics of complex traits over the time-scale. Generally, heritability for a specific trait in a crop is considered stable, and traits with moderate to high heritability are given preference for genetic improvement. Automated image acquisition after every 2 min for 8 h of imposition of gravitropism and QTL mapping in Arabidopsis led to the mapping of time-dependent QTLs (Moore *et al.*, 2013).Leaf growth and development, a major determinant of photosynthetic capacity, is highly regulated by moisture and nitrogen availability. Genetic dissection of this trait was difficult as it needs measurement of this trait throughout a growing season. Using a time-lapse image analysis approach of phenomics, this complex trait was dissected and found to be highly heritable in Arabidopsis (Zhang *et al.*, 2012). The complexicity and plasticity of traits such as biomass and yield in triticale was studied with image-based phenotyping at three developmental stages. QTLs mapping identified some stage-specific QTLs and some QTLs common for two or more developmental stages, demonstrating a temporal contribution of these QTLs to the trait (Liu *et al.*, 2014). Phenomics of rosette growth in 324 accessions of Arabidopsis was compared with end-point weight measurement for GWAS. Use of temporal growth data detected time-specific QTLs which were undetected by endpoint measurement. Eleven of these time-specific candidate genes identified were annotated to be involved in the determination of cell number and size, seed germination, embryo development, developmental phase transition, or senescence. Of these eight genes have been previously demonstrate role with mutants and overexpression studies, suggesting the time-specific QTLs are true regulators of growth and development (Bac-Molenaar *et al.*, 2015). A recent study with non-destructive high throughput phenome of Arabidopsis accessions over spatial and temporal scale revealed that heritability for some traits is dynamic. The heritability of ÖPSII (Fq2/Fm2 , a useful proxy for the light use efficiency for CO_2 fixation) showed recurrent daily rise which was unaffected by the difference in light intensity, while that of chlorophyll reflectance index and projected leaf area (PLA, an indirect estimate of estimate of above ground biomass) gradually changed through time and responded strongly to light intensity. The heritability of PLA showed significant temporal flexibility ranging from 0.04 to 0.83 within the course of 6 h. This suggests the necessity of organism-wide spatial and temporal phenotyping in phenomics to understand the heritability of traits of agricultural importance (Flood *et al.*, 2016). Thus, spatial and temporal phenotyping of crops in phenomics facility will help understand and improve these important traits under water and nitrogen limited conditions.

In silico Phenotyping

Phenomics is highly useful for GWAS and linkage mapping of complex traits such as biomass and height in triticale (Busemeyer *et al.*, 2013; Würschum *et al.*, 2014), root architecture in rice (Topp *et al.*, 2013), yield component in rice (Yang *et al.*, 2014), root gravitropism in Arabidopsis (Moore *et al.*, 2013), *etc.* Phenomics was employed to map salinity tolerance using 378 diverse rice genotypes. Visual image based growth analysis led to the identification of a genomic region on chromosome 3 for the early growth response, while chlorophyll fluorescence imaging identified a region on chromosome 1 that regulate both regulates both the early growth rate and long term ionic stress effects under salinity stress (Campbell *et al.*, 2015). Rice genome is predicted to encode 37,544 genes. Functions of the some of the genes have been elucidated at molecular level, and their impacts on some phenotypes have been studied. However, effect of individual genes on whole plant phenome is critical ultimately to predict the plant traits from plant genome in different environmental conditions. An attempt has been made in yeast to study the phenotypes of essential gene mutations in yeast and PhenoM (Phenomics of yeast Mutants) database was developed (Jin *et al.*, 2012). Loss function mutants, transcriptome and phenomics data was used to elucidate the differential functions two stress responsive genes *AtRD22* and *AtUSPL1*belonging to BURP domain gene family in Arabidopsis (Harshavardhan *et al.*, 2014). However, such efforts are limited in important food crops such as rice and wheat.

Development of a gene based crop model to prediction of complex traits under diverse environmental conditions is an important area of research. For instance, an ecophysiological model predicts pre-flowering duration as affected by temperature and photoperiod was developed using barley RILs. Along with this, QTLs were mapped for the model input trait and values of the model-input traits predicted for the RILs from the QTL were fed back into the ecophysiological model. This model could predict the flowering time for eight field trial environments, and thus ecophysiological model was capable of extrapolating QTL information from one environment to another (Yin *et al.*, 2005).Similarly, wheat heading date could be predicted by using ecophysiological crop simulation model with QTL based parameter inputs (Bogard *et al.*, 2014). Using the marker-based parameter trait values, marker-based values of ILs for seven yield component traits were estimated and were fed to the GECROS model. This model could simulate yields of the ILs under well-watered and drought conditions, and identify virtual ideotypes which had 10–36 per cent more yield than those based on markers for yield per se (Gu *et al.*, 2014). Combining crop simulation models with genomic information and genetic modelling can accelerate delivery of future cereal cultivars suitable for different target environments. However, the robustness of model-aided ideotype design need to be further be enhanced through the inputs from phenomics and genomics and multi-model ensembles (Rötter *et al.*, 2015). In India such efforts are totally missing now. We need to introduce gene/QTL and genomics information into existing ecophysiological models, and improve crop models based on information for lower organizational levels for complex traits (Kumar *et al.*, 2016).

Image Processing, IAPS and Phenome Data Bank

Recently, significant progress has been made in development of image processing algorithms for quantification of phenotypic parameters (Li at al. 2014, Fahlgren *et al.*, 2015b, Singh *et al.*, 2016). Thermal imaging which measures plant temperature, a surrogate for stomatal conductance, can be combined with visual images to differentiate shaded area from leaf area that is fully sunlit and a thermal index based on temperature differences between the canopy and reference surfaces was developed to calculate stomatal conductance and stress monitoring in *Vitis vinifera* (Leinonen and Jones 2004, Möller *et al.*, 2007). A method for the segmentation and the automated analysis of time-lapse plant images from phenotyping experiments was developed using a plant appearance model implemented with Gaussian mixture models, which utilizes incrementally information from previously segmented instances (Minervini *et al.*, 2014).A fully automatic approach to image based 3D plant reconstruction was developed (Pound *et al.*, 2016).Hyperspectral imaging containing reflectance values of continuous wavebands of the electromagnetic spectrum from Visual (400–700 nm), near-infrared (NIR, 700–1000 nm) and short wave-infrared (SWIR, 1000–2500 nm) are influenced by physiology and biochemical composition of plants. To uncover latent hyperspectral characteristics of diseased plants reliably, the hyperspectral images are converted into a corpus of text documents, and probabilistic topic models were applied to identify content and topics of documents. By this approach, intuitive tool for hyperspectral imaging has been developed to automatically track the development of three foliar diseases of barley (Wahabzada *et al.*, 2016).Multimodal fusion methodology consisting of visual colour images and NIR reflectance images were used to study the dynamic phenotypic responses of a C4 cereal crop plant to nitrogen and water deficiency over time (Neilson *et al.*, 2015). Machine learning (ML) tools for extracting patterns and features from the image data and multimodal fusion of trait information from different kinds of images need to be developed to enable stress phenotyping (Singh *et al.*, 2016).Phenome database with curated and labelled phenome data will be useful to expedite both discovery and application in crop breeding (Krajewski *et al.*, 2015).

Although image analysis software is provided by LemnaTec in the phenomics facility, significant research work is needed to develop models to estimate relevant phenotypic traits values from sensor measured digital images. Image acquisition in phenomics is relatively simple but data analysis to retrieve required parameters is not. Several methods are being developed for data analysis. Leaf area measurement is often complicated by overlap of leaves. For precisely measure the leaf are, a novel computational model called HPGA (High throughput Plant Growth Analysis) was developed to measure leaf area individually by using parameters such as leaf tips and the short curvature areas around them, and length to estimate the area in Arabidopsis (Paine *et al.*, 2012). Since the leaf length-to-area model is genome specific, we need to develop such models for quantification of leaf area in rice and wheat. Some progress has been made in automated multimodal phenotyping using phenomics facility. The Plant Accelerator(®) at The University of Adelaide, Australia, analysed the phenome of sorghum under water-limited conditions with different levels of fertilizer. Using color RGB and NIR images, the study could identify accurately genotypic differences using R scripts based robust parsimonious models (Neilson *et al.*, 2015). Combination of machine learning (ML) algorithms and computer vision appears to be a promising approach. Testing of three different algorithms: k-nearest neighbour (kNN), Naive Bayes Classifier (NBC), and Support Vector Machine revealed that different ML algorithms for segmentation are required for different kind of images (Navarro *et al.*, 2016). A fully automated approach to image based3Dplantreconstruction was developed based modelling of complex architecture of leaf surfaces in wheat and rice (Pound *et al.*, 2016).Hyperspectral remote sensing is an automatic, quick and non-destructive method of assessing plant growth parameters, water and nutrient levels in crop plants. Team at IARI has employed different univariate and multivariate models to assess foliar N and relative water contents from captured crop spectra (Sahoo *et al.*, 2015). They also explored physical radiative transfer models for retrieval of plant biophysical parameters like LAI, chlorophyll content and water from hyperspectral Bidirectional Reflectance factor (BRF) data on crops like maize, soybean, wheat and mustard both at controlled field condition and regional scale (Mridha *et al.*, 2015, Tripathi *et al.*, 2012). BRF data were further explored to characterize plant geometry which can further be used for genotypic discrimination and retrieval of morphometric parameters. Visible RGB image was analysed and indexed for assessing leaf area and chlorophyll content (Das *et al.*, 2016). The hyperspectral reflectance data is used for genotypic discrimination, assessment of sugar levels in plants and water-deficit stress levels in rice and wheat crops identifying differential response of genotypes to water deficit stress (Das *et al.*, 2016, 2017, 2018). Team also initiated drone based field phenotyping of rice and wheat genotypes. Recent trends in plant phenomics have been to explore stand alone use of hyperspectral imaging sensors for retrieval different biochemical and biophysical parameters. Fusion and processing of multimodal data acquired through sensor or sensors is many times preferred for assessing parameters with better accuracy which otherwise not possible.

The phenome data from plant phenomics is growing exponentially. One of the major issues the determination of the quality of phenotypic data, *i.e.* to remove system errors in the data collection process from the biological responses. The Integrated Analysis Platform (IAP) is highly useful for complete end-to-end pipeline for large-scale image based phenotyping, starting from the image capture to image analysis, extraction of relevant phenotypic component traits based on relational models, data management, to the automated generation of experiment reports (Klukas *et al.*, 2014).A dynamic filter was developed to effectively identify the abnormalities in time-series quantum yield of photosystem II phenotype data (Xu *et al.*, 2015). The Plant Genomics and Phenomics Research Data Repository (PGP) was initiated by the Leibniz Institute of Plant Genetics and Crop Plant Research and the German Plant Phenotyping Network. PGP is an infrastructure to comprehensively publish voluminous data on phenomics and genomics, and fulfils the FAIR data principles-findable, accessible, interoperable and reusable. PGP is registered as research data repository at BioSharing.org, re3data.org and OpenAIRE as valid EU Horizon 2020 open data archive (Arend *et al.*, 2016). With a vision of transforming science through data-driven discovery, National Science Foundation, USA funded the establishment of CyVerse in 2015. It provides powerful computational infrastructure to handle huge datasets and complex analyses, thus enabling data-driven discovery in life sciences. CyVerse offers powerful extensible platforms, data storage, bioinformatics tools, image analyses, cloud services, APIs, *etc.* (http://www.cyverse.org/). With large number of Institutes in India are involved in genomics and availability of automated phenomics platform in four of the ICAR Institutes *viz.*, IARI, CRIDA, IIHR and NIASM warrant development of data repository similar to that of PGP in India.

Pheonomic for Crop Management

Besides the tremendous potential of phenomics in basic sciences and crop breeding, it has huge potential in crop management and precision agriculture. The sensors used for non-invasive image acquisition can be loaded in drones and Unmanned Aerial Vehicles (UAVs). Drones and UAVs with different kinds of sensors (RGB Visual, IR Thermal, Multispectral and Hyperspectral) can be used to fly over a large area of crop field to obtain phenotypic information such as phenological stage, crop health, water status, and nitrogen status, *etc.* (Vergara-Díaz *et al.*, 2016; Gracia-Romero *et al.*, 2017) This information can be useful for variable rate application or precision application of water, nutrients and agrochemicals in the necessary areas of the crop field (Figure 11.3).

Figure 11.3: Use of Drones for Phenotyping. Inset shows image obtained from the Drone clearly distinguishing crop with different irrigation and nitrogen treatments.

Conclusion and Perspectives

Recent advancements in use of NGP with phenomics platform enhanced the phenotyping capabilities as compared to few traits measured by conventional methods. Performance evaluation studies have shown that controlled environment as well as field phenomics is a suitable complementary approach, and in certain cases such as biotic stress, resource use efficiency and positional cloning phenomics can replace traditional laborious field-phenotyping. Besides GWA mapping, phenomics will be very useful in Phenome-wide Association Studies (PheWAS). Significant progress has been made in PheWAS to identify SNP-disease association in medical sciences. The availability of deep phenotypic data in spatial and temporal scale from NGP in phenomics is expected to accelerate PheWAS in plants. Besides, deep phenotypic data from phenomics will be very useful in training genomic selection models more accurately, and thus aid in genomic selection in crops. Further phenome features can also be used for phenomic selection (PS) in analogy with GS as complementary method (Kumar *et al.*, 2016). We need to develop human resource in the area of image analysis and big data science to effectively use the phenomics for accelerated analytical breeding for crop improvement.

Acknowledgement

The Nanaji Deshmukh Plant Phenomics Centre is funded by National Agricultural Science Fund, Indian Council of Agricultural Research, New Delhi (Grant No. Phen 2015/2011-12 and Grant No: NASF/CRISPR-Cas-7003/2017-18).

REFERENCES

Al-Tamimi N, Brien C, Oakey H, Berger B, Saade S, Ho YS, Schmöckel SM, Tester M, Negrão S. 2016. Salinity tolerance loci revealed in rice using high-throughput non-invasive phenotyping. *Nature Communications*, 7, 13342, doi: 10.1038/ncomms13342.

Arend D, Junker A, Scholz U, Schüler D, Wylie J, Lange M. 2016. PGP repository: a plant phenomics and genomics data publication infrastructure. *Database (Oxford)*. 2016. pii: baw033. doi: 10.1093/database/baw033.

Bac-Molenaar JA, Granier C, Keurentjes J J, Vreugdenhil D. 2016. Genome-wide association mapping of time-dependent growth responses to moderate drought stress in Arabidopsis. *Plant Cell and Environment*, 39, 88-102.

Bac-Molenaar JA, Vreugdenhil D, Granier C, Keurentjes JJB. 2015. Genome-wide association mapping of growth dynamics detects time-specific and general quantitative trait loci. *J Exp Bot.* 66: 5567–5580.

Baðcýoðlu M, Kohler A, Seifert S, Kneipp J, Zimmermann B. 2017. Monitoring of plant–environment interactions by highthroughput FTIR spectroscopy of pollen. *Methods in Ecology and Evolution* 8: 870-880.

Bilder RM, Sabb FW, Cannon TD, London ED, Jentsch JD, Parker DS, Poldrack RA, Evans C, Freimer NB. 2009. Phenomics: the systematic study of phenotypes on a genome-wide scale. *Neuroscience* 164: 30-42.

Biskup B, Scharr H, Schurr U, Rascher U. 2007. A stereo imaging system for measuring structural parameters of plant canopies. *Plant Cell Environ.* 30: 1299-1308.

Bogard M, Ravel C, Paux E, Bordes J, Balfourier F, Chapman SC, Le Gouis J, Allard V. 2014. Predictions of heading date in bread wheat (*Triticum aestivum* L.) using QTL-based parameters of an ecophysiological model. *J Exp Bot.* 65: 5849-65.

Busemeyer L, Ruckelshausen A, Möller K, Melchinger AE, Alheit KV, Maurer HP, Hahn V, Weissmann EA, Reif JC, Würschum T. 2013. Precision phenotyping of biomass accumulation in triticale reveals temporal genetic patterns of regulation. *Sci Rep* 3: 2442. doi: 10.1038/srep02442.

Campbell MT, Knecht AC, Berger B, Brien CJ, Wang D, Walia H. 2015. Integrating image-based phenomics and association analysis to dissect the genetic architecture of temporal salinity responses in rice. *Plant Physiol.* 168: 1476-89.

Chaerle L, Lenk S, Leinonen I, Jones HG, Van Der Straeten D, Buschmann C. 2009. Multi-sensor plant imaging: towards the development of a stress-catalogue. *Biotechnology J.* 4: 1152-1167.

Chen D, Neumann K, Friedel S, Kilian B, Chen M, Altmann T, Klukas C. 2014. Dissecting the phenotypic components of crop plant growth and drought responses based on high-throughput image analysis. *Plant Cell* 26: 4636-55.

Chinnusamy V, Stevenson B, Lee B.-H, Zhu JK. (2002). Screening for gene regulation mutants by bioluminescence imaging. Science's STKE 140: PL10.

Cobb JN, Declerck G, Greenberg A, Clark R, McCouch S. 2013. Next-generation phenotyping: requirements and strategies for enhancing our understanding of genotype-phenotype relationships and its relevance to crop improvement. *Theor Appl Genet.* 126: 867-87.

Das B, Sahoo RN, Biswas A, Pargal S, Krishna G, Verma R, Chinnusamy V, Sehgal VK, Gupta VK. 2018. Discrimination of rice genotypes using field spectroradiometry. *Geocarto International* doi.org/10.1080/10106049.2018.1506507.

Das B, Sahoo RN, Pargal S, Krishna G, Gupta VK, Verma R, Viswanathan C. 2016 Measuring leaf area index from colour digital image of wheat crop. *Journal of Agrometeorology* 18: 22-28.

Das B, Sahooa RN, Pargal S, Krishna G, Verm R, Chinnusamy V, Sehgal VK, Gupta VK, Dash SK, Swain P. 2017. Quantitative monitoring of sucrose, reducing sugar and total sugar dynamics for phenotyping of water-deficit stress tolerance in rice through spectroscopy and chemometrics. *Spectrochimica Acta Part A: Molecular and Biomolecular Spectroscopy* 192: 41-51.

Fahlgren N, Feldman M, Gehan MA, Wilson MS, Shyu C, Bryant DW, Hill ST, McEntee CJ, Warnasooriya SN, Kumar I, Ficor T, Turnipseed S, Gilbert KB, Brutnell TP, Carrington JC, Mockler TC, Baxter I. 2015a. A versatile phenotyping system and analytics platform reveals diverse temporal responses to water availability in Setaria. *Mol. Plant* 8: 1520–1535.

Fahlgren N, Gehan MA, Baxter I. 2015b. Lights, camera, action: high-throughput plant phenotyping is ready for a close-up. *Curr Opin Plant Biol.* 24: 93-99.

Fiorani F, Schurr U. 2013. Future scenarios for plant phenotyping. *Annu Rev Plant Biol.* 64: 267-91.

Flood PJ, Kruijer W, Schnabel SK, van der Schoor R, Jalink H, Snel JF, Harbinson J, Aarts MG. 2016. Phenomics for photosynthesis, growth and reflectance in

Arabidopsis thaliana reveals circadian and long-term fluctuations in heritability. *Plant Methods* 12: 14. doi: 10.1186/s13007-016-0113-y.

Freimer N, Sabatti C. 2003. The human phenome project. *Nat. Genet.* 34: 15–21.

Furbank RT, Tester M. 2011. Phenomics—technologies to relieve the phenotyping bottleneck. *Trends Plant Sci.* 16: 635–644.

Golzarian MR, Frick RA, Rajendran K, Berger B, Roy S, Tester M, Lun DS. 2011. Accurate inference of shoot biomass from high-throughput images of cereal plants *Plant Methods* 7: 2 https: //doi.org/10.1186/1746-4811-7-2.

Gracia-Romero A, Kefauver SC, Vergara-Díaz O, Zaman-Allah MA, Prasanna BM, Cairns JE, Araus JL. 2017. Comparative Performance of Ground vs. Aerially Assessed RGB and Multispectral Indices for Early-Growth Evaluation of Maize Performance under Phosphorus Fertilization. *Front Plant Sci.* 8: 2004. doi: 10.3389/fpls.2017.02004. eCollection 2017.

Grant M, Brown I, Adams S, Knight M, Ainslie A, Mansfield J. 2000. The RPM1 plant disease resistance gene facilitates a rapid and sustained increase in cytosolic calcium that is necessary for the oxidative burst and hypersensitive cell death. *Plant J.* 23: 441-50.

Großkinsky DK, Svensgaard J, Christensen S, Roitsch T. 2015. Plant phenomics and the need for physiological phenotyping across scales to narrow the genotype-to-phenotype knowledge gap. *J Exp Bot.* 66: 5429-40.

Gu J, Yin X, Zhang C, Wang H, Struik PC. 2014. Linking ecophysiological modelling with quantitative genetics to support marker-assisted crop design for improved yields of rice (*Oryza sativa*) under drought stress. *Ann Bot.*114: 499-511.

Harshavardhan VT, Van Son L, Seiler C, Junker A, Weigelt-Fischer K, *et al.*, 2014. AtRD22 and AtUSPL1, members of the plant-specific BURP domain family involved in *Arabidopsis thaliana* drought tolerance. *PLoS One* 9(10): e110065. doi: 10.1371/journal.pone.0110065.

Honsdorf N, March TJ, Berger B, Tester M, Pillen K. 2014. High-throughput phenotyping to detect drought tolerance QTL in wild barley introgression lines. *PLoS One* 9(5): e97047. doi: 10.1371/journal.pone.0097047. eCollection 2014.

Honsdorf, N., March, T. J., Berger, B., Tester, M. and Pillen, K. (2014). High-throughput phenotyping to detect drought tolerance QTL in wild barley introgression lines. *PLoS One*, 9, e97047. doi: 10.1371/journal.pone.0097047. eCollection 2014.

Jin K, Li J, *viz.*eacoumar FS, Li Z, Min R, Zamparo L, *viz.*eacoumar FJ, Datti A, Andrews B, Boone C, Zhang Z. 2012. PhenoM: a database of morphological phenotypes caused by mutation of essential genes in Saccharomyces cerevisiae. *Nucleic Acids Res.* 40 (Database issue): D687-94.

Kjaer KH, Ottosen CO. 2015. 3D Laser triangulation for plant phenotyping in challenging environments. *Sensors (Basel)* 15: 13533–13547.

Klukas C, Chen D, Pape JM. 2014. Integrated Analysis Platform: An open-source information system for high-throughput plant phenotyping. *Plant Physiol.* 165: 506-518.

Krajewski P, Chen D, Æwiek H, van Dijk AD, Fiorani F, Kersey P, Klukas C, Lange M, Markiewicz A, Nap JP, van Oeveren J, Pommier C, Scholz U, van Schriek M, Usadel B, Weise S. 2015. Towards recommendations for metadata and data handling in plant phenotyping. *J Exp Bot.* 66: 5417-27.

Kumar S, Raju D, Sahoo RN, Chinnusamy V. 2016. Phenomics: unlocking the hidden genetic variation for breaking the barriers in yield and stress tolerance. *Indian J Plant Physiol.* DOI 10.1007/s40502-016-0261-0

Legner N, Meinen C, Rauber R. 2018. Root differentiation of agricultural plant cultivars and proveniences using FTIR spectroscopy. *Front. Plant Sci.* https: //doi.org/10.3389/fpls.2018.00748.

Leinonen I, Jones HG. 2004. Combining thermal and visible imagery for estimating canopy temperature and identifying plant stress. *J Exp Bot.* 55: 1423-1431.

Li L, Zhang Q, Huang D. 2014. A review of imaging techniques for plant phenotyping. *Sensors* 14: 20078–20111.

Liu W, Gowda M, Reif JC, Hahn V, Ruckelshausen A, Weissmann EA, Maurer HP, Würschum T. 2014. Genetic dynamics underlying phenotypic development of biomass yield in triticale. *BMC Genomics* 15: 458. doi: 10.1186/1471-2164-15-458.

Ludovisi R, Tauro F, Salvati R, Khoury S, Mugnozza GS, Harfouche A. 2017. UAV-based thermal imaging for high-throughput field phenotyping of black poplar response to drought. *Front. Plant Sci.* 27 https: //doi.org/10.3389/fpls.2017.01681.

McCormick RF, Truong SK, Mullet JE. 2016. 3D sorghum reconstructions from depth images identify QTL regulating shoot architecture. *Plant Physiology,* 172, 823-834.

Mervis J. 2016. NSF director unveils big ideas. *Science* 352: 755-756

Minervini M, Abdelsamea MM, Tsaftaris SA. 2014. Image-based plant phenotyping with incremental learning and active contours. *Ecological Informatics* 23: 35-48.

Mishra KB, Mishra A, Novotná K, Rapantová B, Hodaòová P, Urban O, Klem K. 2016. Chlorophyll a fluorescence, under half of the adaptive growth-irradiance, for high-throughput sensing of leaf-water deficit in Arabidopsis thaliana accessions. *Plant Methods* 12: 46. doi: 10.1186/s13007-016-0145-3.

Möller M, Alchanatis V, Cohen Y, Meron M, Tsipris J, Naor A, Ostrovsky V, Sprintsin M, Cohen S. 2007. Use of thermal and visible imagery for estimating crop water status of irrigated grapevine. *J Exp Bot.* 58: 827-38.

Moore CR, Johnson LS, Kwak IY, Livny M, Broman KW, Spalding EP. 2013. High-throughput computer vision introduces the time axis to a quantitative trait map of a plant growth response. *Genetics* 195: 1077–1086.

Mridha N, Sahoo RN, Sehgal VK, Krishna G, Pargal S, Pradhan S, Gupta VK, Kumar DN. 2015 Comparative evaluation of inversion approaches of the radiative transfer model for estimation of crop biophysical parameters. *Int. Agrophys.*, 29: 201-212.

Muraya, M. M., Chu, J., Zhao, Y., Junker, A., Klukas, C., Reif, J. C. and Altmann, T. (2017). Genetic variation of growth dynamics in maize (Zea mays L.) revealed through automated non-invasive phenotyping. *Plant Journal*, 89, 366-380.

Navarro PJ, Pérez F, Weiss J, Egea-Cortines M. 2016. Machine learning and computer vision system for phenotype data acquisition and analysis in plants. *Sensors* (Basel).16(5). pii: E641. doi: 10.3390/s16050641.

Neilson EH, Edwards AM, Blomstedt CK, Berger B, Møller BL, Gleadow RM. 2015. Utilization of a high-throughput shoot imaging system to examine the dynamic phenotypic responses of a C4 cereal crop plant to nitrogen and water deficiency over time. *J Exp Bot.* 66: 1817-32.

Paine CE, Marthews TR, Vogt DR, Purves D, Rees M, Hector A, Turnbull LA. 2012. How to fit nonlinear plant growth models and calculate growth rates: an update for ecologists. *Methods Ecol Evol.* 3: 245-256.

Parent B, Shahinnia F, Maphosa L, Berger B, Rabie H, Chalmers K, Kovalchuk A, Langridge P, Fleury D. 2015. Combining field performance with controlled environment plant imaging to identify the genetic control of growth and transpiration underlying yield response to water-deficit stress in wheat. *J Exp Bot.* 66: 5481-92.

Plant Science Research Summit. 2013. Unleashing a decade of innovation in plant science: A vision for 2015–2025.

Pound MP, French AP, Fozard JA, Murchie EH, Pridmore TP. 2016. A patch-based approach to 3D plant shoot phenotyping. *Machine Vision and Applications* 1-13. doi.10.1007/s00138-016-0756-8

Ray DK Mueller ND West PC Foley JA. 2013. Yield trends are insufficient to double global crop production by 2050. *PLoS ONE* 8: e66428.

Römer C, Wahabzada M, Ballvora A, Pinto F, Rossini M, Panigada C, Behmann J, Léon J, Thurau C, Bauckhage C. 2012. Early drought stress detection in cereals: simplex volume maximisation for hyperspectral image analysis. *Funct. Plant Biol.* 39: 878-890

Rötter RP, Tao F, Höhn JG, Palosuo T. 2015. Use of crop simulation modelling to aid ideotype design of future cereal cultivars. *J Exp Bot.* 66: 3463-76.

Sahoo RN, Ray SS, Manjunath KR. 2015: Hyperspectral remote sensing of agriculture. *Current Science* 108: 848-859.

Singh A, Ganapathysubramanian B, Singh AK, Sarkar S. 2016. Machine learning for high-throughput stress phenotyping in plants. *Trends Plant Sci.* 21: 110-124.

Topp CN, Iyer-Pascuzzi AS, Anderson JT, Lee CR, Zurek PR, Symonova O, Zheng Y, Bucksch A, Mileyko Y, Galkovskyi T, Moore BT, Harer J, Edelsbrunner H,

Mitchell-Olds T, Weitz JS, Benfey PN. 2013. 3D phenotyping and quantitative trait locus mapping identify core regions of the rice genome controlling root architecture. *Proc Natl Acad Sci USA* 110: E1695–E1704.

Tripathi R, Sahoo RN, Sehgal VK, Tomar RK, Chakraborty D, Nagarajan S. 2012, Inversion of PROSAIL model for retrieval of plant biophysical parameters. *J. Indian Soc. Remote Sensing* 40: 19-28.

Vadez V, Kholová J, Hummel G, Zhokhavets U, Gupta SK, Hash CT. LeasyScan: a novel concept combining 3D imaging and lysimetry for high-throughput phenotyping of traits controlling plant water budget. *J Exp. Bot.* 66: 5581-5593.

Van Dusschoten, D., Metzner, R., Kochs, J., Postma, J. A., Pflugfelder, D., Bühler, J., Schurr, U. and Jahnke, S. (2016). Quantitative 3D analysis of plant roots growing in soil using magnetic resonance imaging. *Plant Physiology,* 170, 1176-88.

Vergara-Díaz O, Zaman-Allah MA, Masuka B, Hornero A, Zarco-Tejada P, Prasanna BM, Cairns JE, Araus JL. 2016. A novel remote sensing approach for prediction of maize yield under different conditions of nitrogen fertilization. *Front Plant Sci.* 7: 666. doi: 10.3389/fpls.2016.00666. eCollection 2016.

Wahabzada M, Mahlein AK, Bauckhage C, Steiner U, Oerke EC, Kersting K. 2016. Plant phenotyping using probabilistic topic models: uncovering the hyperspectral language of plants. *Scientific Reports* 6: 22482. DOI: 10.1038/srep22482.

Würschum T, Liu W, Busemeyer L, Tucker MR, Reif JC, Weissmann EA, Hahn V, Ruckelshausen A, Maurer HP. 2014. Mapping dynamic QTL for plant height in triticale. *BMC Genet* 15: 59. DOI: 10.1186/1471-2156-15-59.

Xu L, Cruz JA, Savage LJ, Kramer DM, Chen J. 2015. Plant photosynthesis phenomics data quality control. *Bioinformatics* 31: 1796-804.

Yang W, Guo Z, Huang C, Duan L, Chen G, Jiang N, Fang W, Feng H, Xie W, Lian X, Wang G, Luo Q, Zhang Q, Liu Q, Xiong L. 2014. Combining high-throughput phenotyping and genome-wide association studies to reveal natural genetic variation in rice. *Nat Commun.* 5: 5087. doi: 10.1038/ncomms6087.

Yin X, Struik PC, van Eeuwijk FA, Stam P, Tang J. 2005. QTL analysis and QTL-based prediction of flowering phenology in recombinant inbred lines of barley. *J Exp. Bot.* 56: 967–976.

Zhang X, Hause RJ Jr, Borevitz JO. 2012. Natural genetic variation for growth and development revealed by high-throughput phenotyping in *Arabidopsis thaliana. G3 (Bethesda)* 2: 29-34.

Zhang X, Huang C, Wu D, Qiao F, Li W, Duan L, Wang K, Xiao Y, Chen G, Liu Q, Xiong L, Yang W, Yan J. 2017. High-throughput phenotyping and QTL mapping reveals the genetic architecture of maize plant growth. *Plant Physiology,* pii: pp.01516.2016. doi: 10.1104/pp.16.01516.

Breeding of Perennial Horticultural Crops *Pages* **183–195**
Editors: S.K. Singh, V.B. Patel, A.K.Goswami, Jai Prakash and Chavlesh Kumar
Published by: **BIOTECH BOOKS, NEW DELHI**

***Chapter* 16**

Mutation Breeding

Awtar Singh and Kanhaiya Singh

Division of Fruits and Horticultural Technology,
ICAR-Indian Agricultural Research Institute, Pusa, New Delhi – 110 012

INTRODUCTION

Genetic improvement of fruit crops is essential for increasing production. The major problems facing tree fruit breeding are long juvenile periods, unavailability of suitable germplasm, and large tree size. In many fruit crops, breeding by controlled crosses is also hampered due to delayed flowering, unsuccessful fruit setting due to abortive embryos or massive fruit drop. For most fruit tree species, the popular/ desired varieties/cultivars have a range of agronomic and horticultural issues. The exploitation of genetic variation, natural or induced, assists greatly in the genetic improvement of fruit trees. When the desired variability is not found in the cultivated varieties or in the germplasm of cultivated species or when a high yielding variety has oligogenic defect such as susceptibility to a disease, mutation breeding is the best course of line. Further, when there is a tight linkage between desirable and undesirable characters, mutation breeding offers solutions. In those crops where sexuality is absent and generation of variability through recombination process is not possible or the breeding cycle is very long as in most fruit crops, induced mutations can help in creating new variability. The fruit growers are very reluctant to varietal changes, which limits the use of cross-breeding in fruits. Other specific problems (*e.g.* polyploidy, incompatibility, apomixis) can make obtaining useful recombinants a laborious task. In contrast, induced mutations change only one or a few specific traits of an elite cultivar and can contribute to fruit improvement without upsetting neither the requirements of the fruit industry nor the consumers.

Mutations are heritable changes in the phenotypes of organisms. These changes are the results of chemical changes at the level of genes. Such changes can bring about new and heritable character variations in plants, which can be selected and used for the establishment of crop varieties with new characters. The frequency of mutations can be increased with the help of certain mutagens and mutations induced in this way are called induced mutations.

A mutant variety is a new plant variety that is bred through either (Suprasanna *et al.*, 2015):

1. Direct use of a mutant line that is developed through mutagenesis or somaclonal variation.
2. Indirect use of a mutant line, which is used as a parental variety in cross-breeding.
3. Use of mutant gene/allele (trait).
4. Use of wild species' genes translocated into plant genomes through irradiation/mutagen derived translocations.

History of Plant Mutagenesis

The history of plant mutations could be traced back to 300 BC with reports of mutant crops in China. Mutations as a mechanism of creating variability were first identified by Hugo de Vries in the late nineteenth century. He considered this variability as heritable changes by mechanisms very distinctive from segregation and recombination and described this occurrence as swift changes in organisms, which were hereditary. He then coined the term 'mutation'. Radiation-induced mutations as a tool for generating novel genetic variability in plants advanced as a field after the discovery of the mutagenic action of X-rays demonstrated by Stadler. The first commercial mutant variety was produced in tobacco in 1934. As a breeding tool, mutagenesis became very popular from the 1950s onwards when a large range of crops were predominantly treated with irradiation to increase trait variation (Oladosu *et al.*, 2016).

Mutagenesis in Crop Improvement

Induced mutations play an important role in crop improvement for: (1) development of improved varieties, (2) induction of male sterility, (3) production of haploids, (4) creation of variability, (5) overcoming self-incompatibility, and (6) improvement in adaptation.

Induction of Mutations

Induced mutations are highly effective to enhance natural genetic resources, and have significantly assisted in developing improved cultivars of fruits. More than 80 mutant varieties have been registered in the Mutant Variety Database of FAO up to 2016 (FAO/IAEA, 2016). Table 16.1 depicts a list of mutant traits of different fruits resulting from physical mutagen treatments. Further, many fruit crops were considered for mutation induction or under various mutation induction programmes (Table 16.2). Many other varieties were also developed via chemical mutagenesis.

Table 16.1: Mutant Traits of different Fruits by Radiation Induced Mutations

Fruit	*Mutant Traits*
Apple	Early maturing, red fruit skin color, compact tree, russet-free fruit skin, variegated leaf
Pear	Disease resistance
Peach	Disease resistance, fruit size and yield
Citrus	Seedless, red color fruit and juice
Apricot	Early maturing
Plum	Early flowering
Pineapple	Spineless
Loquat	Fruit size
Pomegranate	Dwarfness
Papaya	Dwarfness
Japanese Pear	Disease resistance
Indian Jujube	Fruit morphology, earliness
Banana	Earliness, bunch size
Apricot	Earliness

Source: M. Jain, 2002.

Table 16.2: List of Fruit Crops Used/Considered for Induction of Mutations

Fruit Crop	*Potential for Improvement*
Mango	Disease resistance, *in vitro* technology, enhanced flavour, stress tolerance
Date Palm	Creation and stabilization of microclimates in semi-arid regions, provide food shelter and fuel, Bayoud disease
Coconut	Staple crop in maritime, insular and coastal regions
Cashew	High yield, stress tolerance and kernel quality
Avocado	Rootstock resistance to *Phytophthora* root rot
Papaya	Short rotation, papaya ring spot, delayed ripening
Citrus spp.	Less seeded or seedless mutants, molecular markers, disease resistance, increased water efficiency and use of less fertilizers
Carambola	Self compatible, compact, sub acid
Jujube	Fruit quality, disease resistance, early blooming, fruit colour and shape
Guava	Early flowering, seedless, abiotic and biotic stresses
Annona	Dwarf, shelf-life, reduced seed size or number
Litchi	Reduced seed size, extend harvest time
Banana	Nematode, Black sigatoka disease resistance

Source: M. Jain, 2002.

Procedure of Mutation Breeding

Mutation breeding employs three types of mutagenesis. These are induced mutagenesis, in which mutations occur because of irradiation (gamma rays, X-rays, ion beam, *etc.*) or treatment with chemical mutagens; site-directed mutagenesis; and insertion mutagenesis, which is due to DNA insertions, either through genetic transformation and insertion of T-DNA or activation of transposable elements (Oladosu *et al.*, 2016). The key point in mutation breeding is the process of identifying individuals with a target mutation, involving mutant screening and mutant confirmation. Mutant confirmation helps in detection of false mutants (Oladosu *et al.*, 2016).

Mutagenic Agents

Agents of artificial mutations are called mutagens. They are grouped into two broad categories, chemical and physical mutagens. Mutagenesis can be performed with all types of planting materials, *e.g.* whole plants, usually seedlings, and *in vitro* cultured cells. Nevertheless, the most commonly used plant material is seed. Recently, the induction of mutations in vegetatively propagated plants is becoming more efficient, using single cells and other forms of *in vitro* cultured plant tissues. The starting materials for the induction of mutations are vegetative cuttings, scions, or *in vitro* cultured tissues like leaf and stem explants, anthers, calli, cell cultures, microspores, ovules, protoplasts, *etc.* Chemical mutagens are preferably used to induce point mutations, but physical mutagens induce gross lesions, such as chromosomal aberrations or rearrangements. It is noteworthy that the frequency and types of mutations are direct results of the dosage and rate of exposure or administration of the mutagen rather than its type.

Physical Mutagenesis

The major advantage of using physical mutagens (Table 16.3) compared to chemical mutagens is the degree of accuracy and sufficient reproducibility, particularly for gamma rays, which have a uniform penetrating power in the tissue (Oladosu *et al.*, 2016). Mutagen doses that lead to 50 per cent lethality (LD_{50}) have often been chosen. It can be argued that LD_{50} is quite arbitrary and might lead to a high number of (mostly deleterious) mutations in plants. This could go to the extent that desirable mutations are either lost or overlooked due to either plant mortality or poor agronomic performance in the following generations. Therefore, a mutation rate targeting a lower LD (*e.g.* LD_{20}) with a survival rate of 80 per cent appears to be more suitable for mutation breeding.

Chemical Mutagenesis

The effect of most chemical mutagens (Table 16.4) on plant materials is generally considered milder. The chemical mutagens can be applied without sophisticated equipment or facilities. The ratio of mutational to undesirable modifications is generally higher for chemical mutagens than for physical mutagens. However, chemical mutagens are generally carcinogenic and therefore, extra care must be taken for health protection during the procedure. The concentration of the mutagen, the length of treatment and the temperature affect the efficiency of mutagenesis. As chemical mutagens are very reactive, it is important to use fresh batches of the chemical(s).

Table 16.3: Commonly Used Physical Mutagens

Mutagen	*Source*	*Characteristics*	*Hazard*
X-rays	X-ray machine	Electromagnetic radiation; penetrates tissues from a few millimetres to many centimetres	Dangerous, penetrating
Gamma rays	Radioisotopes and nuclear reaction	Electromagnetic radiation produced by radioisotopes and nuclear reactors; very penetrating into tissues; sources are 60Co (Cobalt-60) and 137Cs (Caesium-137)	Dangerous, very penetrating
Neutrons	Nuclear reactors or accelerators	There are different types (fast, slow, thermal); produced in nuclear reactors; uncharged particles; penetrate tissues to many centimetres; source is 235U	Very hazardous
Beta particles	Radioactive isotopes or accelerators	Produced in particle accelerators or from radioisotopes; are electrons; ionize; shallowly penetrating; sources include 32P and 14C	May be dangerous
Alpha particles	Radioisotopes	Derived from radioisotopes; a helium nucleus capable of heavy ionization; very shallowly penetrating	Very dangerous
Protons	Nuclear reactors or accelerators	Produced in nuclear reactors and accelerators; derived from hydrogen nucleus; penetrate tissues up to several centimetres	Very dangerous
Ion beam	Particle accelerators	Produced positively charged ions are accelerated at a high speed (around 20 per cent _80 per cent of the speed of light) deposit high energy on a target	Dangerous

Source: Oladosu *et al.*, 2016.

Table 16.4: Commonly Used Chemical Mutagens

Mutagen Group	*Example*	*Mode of Action*
Alkylating agents	MNU; ENU; MMS; EMS; DMS; DES; MNNG; ENNG; NDMA; NDEA	React with bases and add methyl or ethyl groups and, depending on the affected atom, the alkylated base may then degrade to yield a basic site, which is mutagenic and recombinogenic, or mispair to result in mutations upon DNA replication.
Azide	Sodium azide	Same as alkylating agents.
Hydroxylamine	Hydroxylamine	Same as alkylating agents.
Antibiotics	Actinomycin D; mitomycin C; azaserine; streptonigrin	Chromosomal aberrations also reported to cause cytoplasmic male sterility.
Nitrous acid	Nitrous acid	Acts through deamination, the replacement of cytosine by uracil, which can pair with adenine and thus through subsequent cycles of replication lead to transitions.
Acridines	Acridine orange	Intercalate between DNA bases thereby causing a distortion of the DNA double helix and the DNA polymerase in turn recognizes this stretch as an additional base and inserts an extra base opposite this stretched (intercalated) molecule. This results in frame shifts, *i.e.* an alteration of the reading frame.

Mutagen Group	*Example*	*Mode of Action*
Base analogues	5-bromo-uracil (5-BU); maleic hydrazide; 5-Bromo-deoxy-uridine; 2-amino-purine (2AP)	Incorporate into DNA in place of the normal bases during DNA replication thereby causing transitions (purine to purine or pyrimidine to pyrimidine); and tautomerization (existing in two forms which inter-convert into each other, *e.g.* guanine can exist in keto or enol forms).

Source: Oladosu *et al.*, 2016.

Types of Mutations

Mutations can be broadly divided into intragenic or point mutations (occurring within a gene in the DNA sequence), intergenic or structural mutations within chromosomes (inversions, translocations, duplications and deletions) and mutations leading to changes in the chromosome number (polyploidy, aneuploidy and haploidy). In addition, it is important to distinguish between nuclear and extranuclear or plasmone mutations (Pathirana, 2011). Loss or gain of one or more base pairs is referred to as deletions and insertions, respectively, and if they do not occur in multiples of three nucleotides in the sequence, it results in a frame-shift mutation. Therefore, deletion or insertion of a single base pair or two may result in a substantial effect, including the loss of function through a shift within the reading code.

i) Chromosome Rearrangements or Structural Mutations

Structural mutations are the result of chromosome breaks and their rearrangements. Ionizing radiation results mostly in such changes. Four categories of such rearrangements can be distinguished: deletions or deficiencies, duplications, inversions and translocations. About 90 per cent of the chromosome aberrations brought about by ionizing radiation are deletions and often are lethal. However, some deletions can block biochemical pathways. Duplications have been a cornerstone in crop evolution, particularly at diploid level. Inversions (180° rotation of blocks of nucleotides) and translocations are also common in crop plants (Pathirana, 2011).

ii) Changes in Chromosome Number – Genome Mutations

Mutations at the genome level, leading to changes in chromosome number are also important. These mutations are divided into allopolyploidy, autopolyploidy and aneuploidy. Allopolyploidy is the result of a combination of genomes of two or more species and commonly arises through inter specific or intergeneric hybridization followed by the doubling of chromosomes.

New Methods for Mutation Induction

Fast development of *in vitro* based methods in the past two decades has helped plant breeders to develop more efficient procedures using plant cell and tissue cultures. This has been particularly useful in developing novel variation in vegetatively propagated crops such as banana, and *Citrus* sp. (Pathirana, 2011). *In vitro* cultures have the advantage of producing large mutant populations in a limited space, and the possibility of screening for characters as tolerance to salt, toxic elements or fungal toxins in the laboratory environment. This allows the transfer of only putative mutants for field screening, and in the case of vegetatively propagated crops, removal of chimaera through several subcultures. *In vitro* mutation induction and screening methods have been established for widely cultivated crops such as banana (Pathirana, 2011).

Insertional mutagenesis is a useful method for producing mutants to investigate gene function because those mutations are tagged with DNA fragments of known sequence. There are two methods to construct such mutants: T-DNA insertions and transposon tagging. T-DNA insertions are considered genetically modified, carrying T-DNA sequence from *Agrobacterium tumefaciens*. Transposable elements are an internal genetic factor that has been harnessed recently for mutation breeding through their transcriptional activation.

Recent Advances in Mutation Breeding

Recent biotechnological techniques have the potential to provide sufficient methods of vegetative propagation, genetic improvement through improved mutation and screening techniques and genetic transformation. The advances holding significance for fruit trees are plant tissue culture and plant molecular biology, which can now be integrated with conventional techniques in generating new mutations (Jain, 2002). Plant tissue culture advances including somatic embryogenesis, somaclonal variation, microprogagation and micrografting, cryopreservation of embrygenic cell cultures, *in vitro* selection of cells and tissues to fungal toxins, somatic hybridization, embryo rescue have already been applied to fruits.

In vitro Mutagenesis

The use of *in vitro* techniques for mutation induction is important because they allow large number of propagules for induction of mutations in a small space, and several cycles of subculture can be carried out in a short time to dissociate mutated sectors from non-mutated ones, and increase the population to sufficiently large numbers for selection. Irradiation of somatic embryogenic cultures is advantageous to prevent chimerism, since their origin is from a single cell. However, somatic embryogenesis may cause rejuvenation that may lead to delay in flowering in some fruit trees. In that case, budwood or seed can be irradiated. Three to four times grafting of newly developed shoots from irradiated budwoods on rootstocks would help in overcoming the chimera problem (Jain, 2002). *In vitro* mutagenesis has been reported in many fruit crops (Table 16.5) and for many agronomic characters (Table 16.6).

In vitro Selection of Mutants

The selection pressure, such as salt, phytotoxins *etc.*, is applied on the irradiated material (callus, cell suspension, protoplasts) for the selection of mutants and plants can be regenerated and multiplied in large-numbers for further evaluation in the greenhouse and field. Regenerated mutant plants of strawberry exhibited tolerance against *P. cactorum* fungus under greenhouse conditions. Different types of molecular markers were tried for mutant identification (Table 16.7).

Table 16.5: *In vitro* Mutagenesis in Vegetatively Propagated Crops

Crop Species	*Treated Material*	*Mutagen and Dose (LD 50)*	*Plant Regeneration*	*Selected Mutants/Lines*
Banana (*Musa* spp.)	Shoot tips	Carbon-ion beam (0.5–128 Gy)	Direct regeneration	Disease-resistant lines
Banana (*Musa* spp.)	Shoot tips	γ-rays (60 Gy)	Direct regeneration	Mutant Novaria; earliness
Banana var. Lakatan,	Shoot tips Shoot tips	γ-rays (40 Gy) γ-rays (25 Gy)	Direct regeneration	Mutant variety Klue Hom Thong KU1
Pineapple var. Queen	Crowns	γ-rays	Axillary bud regeneration	Lines with reduced spines
Pear	*In vitro* shoots	γ-rays (3.5 Gy)	Microcuttings from shoots	Mutants for russeting, fruit shape and Size, Small tree size, wide branch angle and short internodes

Source: Suprasanna *et al.*, 2015.

Table 16.6: Agronomic Traits that can be Selectable in Cultured Plant Cells

Trait	*Selection Agents*
Disease resistance	Pathotoxin/crude toxin/culture filtrate (CF)
Herbicide tolerance	Herbicide
Salt tolerance	High salt (NaCl) concentration or sea water
Metal tolerance	High concentration of toxic metals
Cold tolerance	Cold temperature
Drought	Polyethylene glycol or other high osmoticum
Enhanced amino acid accumulation	Amino acid analogues or high concentrations of amino acids
Flooding tolerance	Anaerobic conditions

Source: Suprasanna *et al.*, 2015.

Table 16.7: Types of DNA Markers Used for Analyzing Genetic Variation

Type of Marker	*Markers Used*
1. Genomic DNA restriction site-based	Restriction fragment length polymorphism (RFLP)
2. PCR-based markers	Random amplified polymorphic DNA (RAPD)
	Sequence-characterized amplified region (SCAR)
	Simple sequence repeats (SSR)
	EST (expressed sequence tag)-SSR
	Inter-simple sequence repeats (ISSR)
	Single-nucleotide polymorphism (SNP)

Type of Marker	*Markers Used*
	Sequence-related amplified polymorphism (SRAP)
	Allele-specific associated primers (ASAP)
	Variable number tandem repeats (VNTR)
	Sequence-tagged sites (STS)
3. Combination of restriction and PCR	Amplified-fragment length polymorphism (AFLP)
	Cleaved amplified polymorphic sequence (CAPS)
4. Array-based molecular markers	Diversity array technology (DArT)
	Microarray-based marker Bead Array

Source: Suprasanna *et al.*, 2015.

Strategies for Mitigating the Drawbacks in Mutation Induction in Fruit Crops

The identified bottlenecks of mutation breeding in vegetatively propagated/ fruit crops can be ameliorated through a strategic use of biotechnologies. Cell and tissue biology techniques are useful for the efficient production of mutants while molecular biology techniques enhance efficiency in the rapid genotyping of the mutation events (Mba *et al.*, 2013).

i. Cellular and Tissue Biology Strategies

A critical bottleneck in the routine application of induced mutations in plants is the low level of efficiency of the processes; quality and quantity of induced mutant populations are sub-optimal. The problem of chimeras, is exacerbated in vegetatively propagated plants. A further significant hurdle is the need to have the mutated segment in a homozygous state so that the mutation, usually recessive, could manifest as a phenotype. Several *in vitro* techniques have been shown to significantly mitigate these bottlenecks to induced mutations, which are cell suspension cultures including somatic embryogenesis, and rapid *in vitro* multiplication (cf. Mba *et al.*, 2013).

ii. Somatic Embryogenesis

The added expense, in terms of time and resources, could be circumvented using single cells as starting materials for inducing mutations and subsequently regenerating whole plantlets from them. This process typically involves mass proliferation of undifferentiated cells – callus – and subsequently through the modification of culture media, the induction of embryogenic properties in the cells. Typically, the most common paths taken include the production of cell suspension cultures and friable embryogenic calli (FEC). The individual cell is then exposed to the mutagen, and taking advantage of the totipotency of each plant cell to regenerate into a whole plant, plantlets are regenerated. This ability to grow individual plant cells under aseptic conditions, and regenerating whole plantlets, permits the exposure of individual cells to mutagens. Arising from an individual cell, each plantlet is devoid of chimeras. The use of cell suspension cultures as starting materials in induced mutations has been successfully used in the generation of banana mutants (cf. Mba *et al.*, 2013).

iii. Rapid *in vitro* Multiplication

In instances where somatic embryogenesis is an impractical option, or where protocols have not been validated, *in vitro* techniques, especially rapid multiplication systems using meristematic regions of plants, could be used to achieve some level of efficiency in induced mutations. Plantlets are regenerated, but not at the same levels of homozygosity and incidence of chimeras as in somatic embryos. Such plantlets are then passed through several cycles of regeneration to dissociate chimeras. Through this method (irradiation and *in vitro* culture of shoot tips), an induced mutant banana variety, Novaria was developed (cf. Mba *et al.*, 2013).

iv. Molecular Biology Techniques for Enhancing Efficiency in Induced Mutations

Molecular biology strategies offer mechanisms for direct querying of target genes for changes. Molecular biology might not provide conclusive evidence of the desired mutation but such techniques will significantly obviate the need for field trials of large populations, enhancing the efficiency of the processes. Reverse genetics strategies, especially Targeting Induced Local Lesions IN Genomes (TILLING), through its inherent high throughput platform, promise to be indispensable tools for the efficient and rapid identification of mutation events. Molecular markers have been applied to the identification of germplasm in several crops (Jain, 2002).

Biotic Stress Resistance

Resistance to black spot disease (*Alternaria alternata* pv. Japanese pear), considered to be the most serious disease in Japanese pear (*Pyrus Serotine* var. *culta*), was induced in Japan in the 1960s through chronic irradiation of the cultivar 'Nijisseiki' and a resistant cultivar 'Gold- Nijisseiki' was released (Pathirana, 2011). The success of this programme led to radiation breeding of susceptible 'Shinsui' and 'Osa-Nijisseiki' cultivars resulting in resistant cultivars 'Kotobuki-Shinsui' and'Osa-Gold'. The 'Osa-Gold' mutant has the added advantage of being self-compatible, eliminating the need to grow pollinators. The breeding lines tolerant to salinity and water stress have also been developed, particularly in citrus fruits.

Achievements in Mutation Breeding

Important agronomic traits improved by induced mutations in some major fruits include disease resistance (Japanese pear and peach), seedlessness (citrus, guava), dwarfness (papaya, pomegranate), and earliness (banana, apricot, jujube, plum, apple). The selection of an appropriate technology is very important for mutation induction and the selection of mutants. Advances in plant tissue culture and plant molecular biology can be integrated with conventional techniques in generating new mutations (IAEA-TECDOC-1615, 2009). The application of mutation techniques has generated a vast amount of genetic variability and is playing a significant role in plant breeding and genetics and advanced genomics studies. The widespread use of mutation techniques in plant breeding programmes throughout the world has generated thousands of novel crop varieties in hundreds of crop species, and billions of dollars in additional revenue. The FAO/IAEA Mutant Variety Database or MVD collects information on plant mutant varieties (cultivars) released officially or commercially worldwide. Data on the mutagen and dose used, the characters improved, and agronomic data if available are among the information provided (Kole *et al.*, 2015).

In the context of MVD a mutant variety is a new plant variety that is bred through:

- Direct use of a mutant line that is developed through physical and chemical mutagenesis, or somaclonal variation and activation of endogenous transposable elements;
- Indirect use of a mutant line/lines, which is/are used as a parental variety/varieties in cross breeding (cross between mutant lines or with a commercial variety/varieties);
- Use of wild species' genes translocated into plant genomes through irradiation facilitated translocations, *e.g.* genes of wheat wild relative species.

Officially released cultivars according the FAO Mutant variety database accessed on 27/12/2016 includes 3246 varieties.

New Techniques for Mutation Induction, Screening and Utilization

The new developments for mutation induction include the use of space flight, ion irradiation, and transposable elements for mutation induction, the use of restriction endonuclease for site directed, homologous recombination; DNA markers linked to mutated genes for marker-assisted selection and tracing of the gene, and TILLING (Target Induced Local Lesions IN Ge-nome), as well as different variant versions for high throughput screening of mutated alleles (Chakraborty and Paul, 2012). TILLING is a high-throughput and low cost reverse genetic method for the discovery of induced mutations. The method has been modified for the identification of natural nucleotide polymorphisms, a process called Ecotill-ing.

Selection of the desired genotype is critical irrespective of the procedure used to create variation. The development of mo-lecular probes offers a tremendous opportunity to select desired mutants. Hence, mutation induction, molecular marking of useful and selected mutations, sequencing of the mutated genes and the developments of molecular probes will be critical in the continued and expanded use of induced mutations, mutants and cultivars. Induced mutations will also play an increasing role in creating crop cultivars with traits such as enhanced uptake of specific met-als, deeper rooting system, tolerance to drought and salinity, and resistance to diseases and pests as a major component of the environmentally sustainable agriculture (Janick, 2012).

Practical Considerations in Induced Crop Mutagnesis

The dose of a mutagen that achieves the optimum mutation frequency with the least possible unintended damage is regarded as the optimal dose for induced mutagenesis. For physical mutagens, this is estimated by carrying out tests of radio-sensitivity. Its predictive value, therefore, guides the researcher in the choice of optimal exposure dosage depending on the plant materials and the desired outcome. The LD_{50} for many fruit crops has been standardized under *in vitro* culture conditions (Predieri, 2001). Important factors influencing the outcome of mutagenesis using chemical mutagens include the condition of the mutagenic solution; inherent characteristics of the targeted tissue; the environment; concentration of mutagen; treatment volume; treatment duration; temperature; presoaking of seeds; pH (7.0); catalytic agents ($Cu_2C$and Zn_2C) and post-treatment handling. Factors influencing the outcome of mutagenesis using physical mutagens includes oxygen; moisture content; temperature; physical ionizing agents; dust and fibres; biological and infectious agents. In general, the steps required for inducing and detecting mutations vary among sexually and asexually propagated plants/crops but there are some basic principles that they share. The common practical considerations that need to be considered in mutation breeding as summarized by Mba (2013) include the following.

- ✰ A perfect understanding of the genetic makeup of the traits to be improved.
- ✰ Understanding the mode of reproduction of the target crop.
- ✰ The determination of the material that is to be used for the propagation prior to treatment.
- ✰ Knowledge of the number of sets of chromosomes in the nucleus of a cell (ploidy).
- ✰ Determinations of the genetic pedigree of the target crop for inducing mutations.
- ✰ Selection of an appropriate mutagen and dose.
- ✰ Identification of infrastructure for successful selection of desired mutants.
- ✰ Screening techniques for dissociation of chimeras from stable mutants.

REFERENCES

Chakraborty, N.R. and Paul, A. 2012. Role of induced mutations for enhancing nutrition quality and production of food. *International Journal of Bio-resource and Stress Management* **4**(1): 091-096.

FAO/IAEA. 2016. *Mutant Variety Database*. Cited on 24/12/2016.

IAEA-TECDOC-1615. 2009. *Induced Mutation in Tropical Fruit Trees* May, 2009.

Janick, J. 2012. Fruit Breeding: past, present and future. *Temperate Fruit Breeding, Fruit, Vegetable and Cereal Science and Biotechnology* **5** (Special Issue 1).

Jain, S.M. 2002. A review of induction of mutations in fruits of tropical and subtropical regions. *Acta Hort.* **575**: 295-302.

Kole, C. *et al.*, 2015. Application of genomic assisted breeding for generation of climate resilient crops: Progress and prospects. *Frontiers in Plant Science* **6**: pp. 16.

Mba, C., Afza, R., Jankowicz-Cieslak, J., Bado, S., Matijevic, M., Huynh, O. and Till, B.J. 2013. Enhancing genetic diversity through induced mutagenesis in vegetatively propagated plants. In: Q.Y. Shu (Ed.) *Induced Plant Mutations in the Genomics Era*. Food and Agriculture Organization of the United Nations, Rome, 2009, 262-265.

Oladosu, Y., Rafii, M.Y., Abdullah, N., Hussain, G., Ramli, A., Rahim, H.A., Miah, G. and Usman, M. 2016. Principle and application of plant mutagenesis in crop improvement: a review. *Biotechnology and Biotechnological Equipment*, **30**(1): 1-16.

Pathirana, R. 2011. Plant mutation breeding in agriculture. *CAB Reviews: Perspectives in Agriculture, Veterinary Science, Nutrition and Natural Resources* **6** (032): pp. 20.

Predieri, S.2001. Mutation induction and tissue culture in improving fruits. *Plant Cell, Tissue and Organ Culture* **64**: 185–210.

Suprasanna, P., Mirajkar, S.J. and Bhagwat, S.G. 2015. Induced mutations and crop improvement. In: B. Bahadur *et al.* (eds.), *Plant Biology and Biotechnology: Volume I: Plant Diversity, Organization, Function and Improvement*, Springer India, 593-617.

Breeding of Perennial Horticultural Crops *Pages* **197–202**
Editors: S.K. Singh, V.B. Patel, A.K.Goswami, Jai Prakash and Chavlesh Kumar
Published by: **BIOTECH BOOKS, NEW DELHI**

***Chapter* 17**

Nucleo-cytoplasmic Interactions in Hybrid Breeding

Naveen Singh[1], Rajendra Singh[1], Yashpal[1], Navinder Saini[1], A.K. Goswami[2], and Awtar Singh[2]

[1]Division of Genetics, [2]Division of Fruits and Horticultural Technology, ICAR-Indian Agricultural Research Institute, Pusa, New Delhi – 110 012

INTRODUCTION

Natural selection act upon coreplicons at a different rate in the alloplasmic lines and may cause genotypic and/or phenotypic modification(s) to confer adaptive or reproductive advantage to the gene(s) comprising a particular coreplicon. Phenotype in these lines is modified to propagate the genes comprising that set. Such a modification may or may not be consistent with the fitness of other coreplicons which comprise the total genome. Therefore, depending upon their contribution in determining the phenotype, two coreplicons may act in conflict or harmony. Conflict under such situation is due to difference in the inheritance in cytoplasmic and nuclear genes. Cytoplasmic genome have uniparental inheritance, whereas, it is bi-parental in case of nuclear genes. Therefore, selective interests differ for fitness and other morphological and physiological traits among the coreplicons. Natural selection during evolution, in general, leads to resolution of conflicting nucleo-cytoplasmic demands through alteration in nuclear genome. Whereas, conflict among the coreplicons can result from mutation or deliberate substitution of parental cytoplasm or some of its organelles. A classical examples of this nucleo-cytoplasmic conflict is cytoplasmic male sterility. Because of their maternal inheritance, cytoplasmic genes will not be selected for pollen production. Thus, a mitochondrial mutation causing male sterility will be selected only if it confers female fitness. On the other hand, natural selection acts on nuclear genes to restore male fertility through fertility restoration (*Rf*) genes. Genomic changes are, thus, expected when (i) native cytoplasm of a species is substituted by cytoplasm of a highly differentiated alien species and (ii) native genome is transferred to alien cytoplasm.

Nucleocytoplasm interactions, thus, cause:

1. Reciprocal differences due to maternal effect
2. Reduced fitness: Cytopalsmic male sterility (Banga and Banga 1998), reduced female fertility, pistilloidy (Kihara, 1951), changed floral morphology *etc.*
3. Changed pollination mechanism
4. Changed vegetative traits: leaf variegation (Katu and Tokumashu, 1980), leaf chlorosis, dwarfness, reduced nectarines *etc.*
5. Change in ploidy level: induction of haploids (Tsunewaki *et al.*, 1974)
6. Mutations in the alien genome: higher mutation rate leads to changes in genome to adjust into alien cytoplasm
7. Wider adaptations of allopolyploids to adverse climatic conditions.

Nucleocytoplasm interactions are responsible for creation of genetic variability through mutations and chromosomal substitutions in the alloplasmic lines. Alloplasmic lines having multiple genomes (amphiploids) has resulted from hybridization between two plant species where the chromosome set contributed by each parent undergo doubling and produces in the hybrid with chromosome number equal to sum of the diploid number of two parents. These allopolyploids have different set of genomes in a uniparental cytoplasm, thus, encountering variable degree of nucleo-cytoplasmic interactions. The level of interaction depends upon the evolutionary differences between the genomes and/or cytoplasm contributing species/genra.

Allopolyploids

The wide spread occurrence to polyploids has been attributed to the potential of a polyploid species to adapt to a wider range of habitat and survive better is unstable climate than their diploid progenitors. Natural allopolyploids are likely to be originated through chromosome doubling of hybrids produced through chance natural hybridization between two distinct species of the same or different genus. Such Fi's would be generally sterile since the chromosome of this parental species are expected to be differentiated and are not likely to pair during meiosis. Chromosome doubling may have occurred in somatic cells of F_1 hybrids due to irregular meiosis; this would give rise to doubled sectors either in apical meristems or auxiliary buds. Sexual progeny of such branches will be amphiploid *i.e.* having two copies of each genome. Genomic charges that occurred upon the formation of polyploid, particularly allopolyploids changes, might also lead to more harmonious behaviour and activity of the different constituent genomes. In many cases these change facilitated the establishment of newly formed polyploids as successful species.

Allopolyploidy generally combines the morphological and physiological features of the parent species, but it is very difficult to predict the precise combination of parental characteristics that would appear in the allopolyploid. However, allopolyploid are in general more vigrous then diploids, but this is also not true in all cases. These also appear to differ in their adaptability from the parental species, often they are more hardy. Many allopolyploids particularly in grasses, are apomictic which is an efficient mechanisms of countering the sterility of distant hybrids without doubling their chromosome number.

Production of Allopolyploids

Experimental production of allopolyploids is achieved through chromosome doubling with the help of colchicines of distant hybrids. Such allopolyploids are often knows as synthetic allopolyploids or synthetic amphiploids. As during allopolyploidization we are crossing two different species (interspecific) or two different genera (intergeneric), so naturally there will be difficulty is producing there distant hybrids which varies to great extent. The difficulties encountered in the production of distant hybrids are failure in zygote formation or seedling development. These problems can be resolved by adopting special crossing techniques such as bud pollination and embryo rescue techniques followed by chromosome doubling or protoplast fusion using polyethylglycol (PEG). Furthermore, both sexual and asexual (cybrids) development of alloplasmic lines has been successfully achieved by many workers. It has been learnt that

- ☆ More divergent the parents are, the greater the subsequent genomic change in the allopolyploid are expected
- ☆ Nuclear genome of maternal origin experiences less change than the paternal contribution than the one having paternal origin.

Abnormalities Associated with Alloplasmic Lines and their Rectification

Several abnormalities are reported in different crop plants due to changes enforced by Nucleocytoplasm interactions:

- ☆ Reduced fitness/male sterility
- ☆ Leaf chlorosis leading to lower photosynthetic activity
- ☆ Delayed maturity
- ☆ Poor floral morphology
- ☆ Female sterility
- ☆ Poor nectaries
- ☆ Yield penalty
- ☆ Dwarfness/low biomass *etc.*

Nucleo-cytoplasmic interactions in Citrus, on the other hand, has led to sexual incompatibility, nucellar embryogenesis, male sterility and female sterility *etc.* (Grosser and Gmitter 1990). Most of these abnormalities are corrected by different genetic interventions. Developmental abnormalities like chlorosis can be corrected through plastid substitution, whereas, mmitochondrial recombination helps improving floral morphology and fruit traits. Therefore, development of alloplasmic lines through somatic hybridization has been considered as better method for development of these lines. In addition to these approaches identification and introgression of corrective genes from donor species is considered as one of the options for overcoming abovementioned defects. Although some of the abnormalities like male sterility provides opportunity for development of seedless fruits.

Utilization for Allopolyploidy in Citrus Improvement

Somatic hybridization is being used to generate key allotetraploid breeding parents for use in interploid crosses to generate seedless triploids (Grosser and Gmitter 2005). These triploids can be developed by keeping allotetraploid somatic hybrids as pollen parents in interploid crosses followed by embryo rescue when necessary, to generate thousands of triploid hybrids for mandarin, grapefruit/ pummelo, and acid lime/lemon improvement (Figure 14.1). Monoembryonic allotetraploid somatic hybrids have also been used as females in interploid crosses to generate hundreds of grapefruit-pummelo type hybrids. With this latter approach, triploid hybrids can be recovered directly from seed without the need for embryo rescue; however, it has been more difficult to generate monoembryonic allotetraploid somatic hybrids because the embryogenic callus lines used as effective fusion partners have been derived from polyembryonic donors that produce nucellar-derived embryos. It is extremely difficult to obtain embryogenic callus from monoembryonic cultivars in citrus. Sedysheva and Gorbacheva (2013), on the other hand, have demonstrated development of triploid hybrids in apple by using autotertraploids.

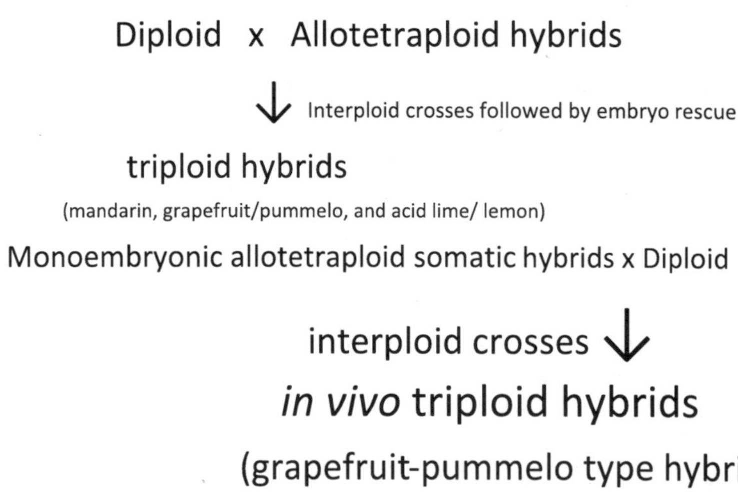

Figure 14.1: Development of Triploid Seedless Hybrids using Allotertaploid Citrus Lines.

Breeding Approach for Citrus Improvement using Allotertraploids

- ☆ *Breeding for scion improvement*: allotetraploid somatic hybrids as superior breeding parents in interploid crosses to generate triploids (Figure 14.1).
- ☆ *Breeding for rootstock improvement:* breeding at the tetraploid level using selected somatic hybrid rootstocks as parents provides an opportunity to recombine alleles from three or four proven diploid rootstocks, upon which to impose subsequent selection. This powerful approach can maximize the genetic diversity in zygotic progeny. Direct utilization of allotetraploids may ensure biotic and abiotic stresses resistance, creation of greater selectable genetic variation and induce dwarfness in root stocks.
- ☆ *Breeding for both scion and rootstock simultaneously:* Evaluation of allotetraploid somatic hybrids has indicated the possibility of producing new seedless cultivars at the tetraploid level as well (Grosser and Gmitter 2005, 2011).

Future Breeding Opportunities

In fruit crops, such as citrus, allopolyploids can be utilized for improvement of both scions and root stock traits for:

1. Abiotic stresses tolerance: Drought, Salinity and Cold
2. Biotic stresses resistance: Disease and nematode resistance
3. Fruit quality: Seedless fruits and improved fruit size
4. Plant architecture: Dwarf plant type and dense canopy leading to sunburn tolerance

REFERENCES

Bajaj Y.P.S. 1977. Protoplast isolation, culture and sometic hybridization. In: *Plant Cell Tissue Organ Culture* (eds). Reinert J and Bajaj YPS. Springer-verlag. Berlin Heidelberg, New York. pp. 467-96.

Gale, M.D. and Devos, K.M. (1998) Plant comparative genetics after 10 years. *Science,* 282, 656–658

Gill, B.S. 1991. Nucleo-cytoplasmic interaction (NCI) hypothesis of genome evolution and speciation in polyploid plants, in Proceedings of the Kihara Memorial International Symposium on Cytoplasmic Engineering in Wheat (Sasakuma, T. and Kinoshita, T., eds), pp. 48–53, Yokohoma, Japan

Grosser J. W. and Gmitter F. G. 2005. 2004 SIVB Congress symposium proceedings thinking outside the cell—applications of somatic hybridization and cybridization in crop improvement, with citrus as a model. *In vitro Cell Dev. Plant,* 41: 220–225

Grosser J. W. and Gmitter F. G. 2011. Protoplast fusion for production of tetraploids and triploids: applications for scion and rootstock breeding in citrus. *Plant Cell Tissue Organ Culture,* 104: 343–357

Gupta P.K. 2000. Numerical changes is chromosomes. In: *Genetics* (eds) PK Gupta pp - 240-51.

Katu M. and Tokumasu S. 1980. Nucleus substitution of *Brassica japonica* with *Raphanus sativus* L. and its resultant chlorophyll deficiency. *Euphytica*, 29: 97-106.

Kihara H. 1951. Substitution of nucleus and its effects on genome manifestation. *Cytologia*, 16: 177-193.

Kirti P.B., Narasimhulu S.B., Mohapatra T., Prakash S., Chopra V.L. 1993. Correction of chlorophyll deficiency in alloplasmic male sterile *Brassica juncea* through recombination between chloroplast genomes. *Genet. Res.*, 62: 11-14.

Leitch, I.J. and Bennett, M.D. 1997. Polyploidy in angiosperms. *Trends Plant Sci.*, 2: 470–476

McNauyhtor I.M. 1972. Synthesis and sterility of Raphanobrassica. *Euphytica*, 22: 70-88.

Neal C Stoskopff (1999) Interspecific and intergeneric hybridization. **In:** *Plant Breeding: Theory and Practices.* Scientific Publishers (India) pp. 345-72.

Pelletier G., Primard C, Vedel F, Chetrit P, Remy R, Rousselle P, Renard M. 1983. Intergeneric cytoplasmic hybridization in Cruciferae by protoplast fusion. *Mol. Gen. Genet.*, 191: 244-250

Pelletier G, Primard C., Vedel F., Chetrit P., Renard M.D., Mesquida J. 1987. Molecular phenotypic and genetic characterization of mitochondrial recombinants in rapeseed. Proc 7th Int Rapeseed Conference, Poznan, Poland, pp. 113-118

Sedysheva G. A. and Gorbacheva N. G. 2013. Estimation of New Tetraploid Apple Forms as Donors of Diploid Gametes for Selection on a Polyploidy Level, *Universal Journal of Plant Science*, 1(2): 49-54

Shyam Prakash 1973. Artificial synthesis of *Brassica juncea* Coss. *Genetica*, 44: 249-63.

Singh B.D. 2001. Pollyploidy. In: *Plant Breeding: Principles and Methods*, pp. 661-68.

Song K., Lu P., Tang K. and Osborn T. C. (1995) Rapid genome changes in synthetic polyploids of Brassica and its implication in polyploid evolution. In: *Proc. Natl. Acd. Sci., USA*, 92(72): 19-23.

Tsunewaki, K., Endo, T. and Mukai, Y. 1974. Further discovery of alien cytoplasms inducing haploids and twins in common wheat. *Theor. Appl. Genet.*, 45: 104-109.

Uprety, D.C., Prakash, S. and Tomar, V.K. 1990. Cytoplasm influences the photosynthetic efficiency in *B. carinata. Agron. and Crop Sci.*, 165: 207-210.

Breeding of Perennial Horticultural Crops *Pages* **203–208**
Editors: S.K. Singh, V.B. Patel, A.K.Goswami, Jai Prakash and Chavlesh Kumar
Published by: **BIOTECH BOOKS, NEW DELHI**

Chapter 18

Somaclonal Variations

Zakir Hussain and Pawan Yadav

Division of Vegetable Science,
ICAR-Indian Agricultural Research Institute, Pusa, New Delhi – 110 012

INTRODUCTION

Somaclonal variation, a term coined by Larkin and Scowkraft (1981) for plant variants derived from any form of cell or tissue cultures. Clonal propagation as well as in vitro culture of plants may generate genetic variability, *i.e.*, somaclonal variations as a result of gene mutation or changes in epigenetic marks. The variations have provided a new and alternative tool to the breeders for obtaining genetic variability relatively rapidly in annual as well as perennial fruit crops, which are either difficult to breed or have narrow genetic base. At present, micropropagated plants, in various crops, such as strawberry, papaya, banana, grapes, pineapple, citrus, *etc.*, are preferred over plants propagated through conventional means. However, several instances of somaclonal variations have been reported in different horticultural crops. The notable example could be banana in which occurrence of off-types from tissue cultured plantlets ranged from 6 to as high as 90 per cent (Smith 1988). Somaclonal variations may confer ecological advantages to clonal plants that could outweigh evolutionary costs. Relying to various degrees on vegetative reproduction, the capacity to conserve or reverse gene regulation changes over cell divisions has clear potential for optimization of plasticity and acclimation in response to environmental variation encountered. Clonal plants may be one of the best examples of organisms taking advantage of epigenetic acclimation as an alternative to the slower mechanisms of adaptation through natural selection. Many of these variations have contributed significantly in crop improvement. In clonally propagated horticultural crops, there is limited scope for genetic variability enhancement as no hybridization is involved. However, from the point of commercial micropropagation, variation of any kind, in particular, genetic variations may be considered obstructive and worthless; since, such variations may lead to loss of genetic delity. Use of plant tissue culture technology is not only helpful in multiplication of planting material, it can greatly improve the quality of planting material. However, it is of immense signicance to assure the genetic uniformity of *in vitro* raised plants at an early stage.

Somaclonal variation is a common phenomenon in both in vitro and in vivo propagated Musa plants. Such somaclones (off-types) are more common in bananas and plantain regenerated in vitro than in conventionally propagated Musa (Drew and Smith, 1990; Smith, 1988; Vuylsteke *et al.*, 1988; Vuylsteke and Ortiz, 1996). Certain genotypes are known to exhibit a high variation rate, while others rarely do (Cote *et al.*, 1993; Smith, 1988; Vuylsteke *et al.*, 1991). Dwarfism in the 'Cavendish' bananas and inflorescence variation in the plantains are the most commonly observed morphological changes. Changes in leaf size, shape and colour of the pseudostem and flower are also common. Somaclonal variation in bananas include intrinsic as well as culture-induced factors. Genetic changes induced during micropropagation can be influenced by the choice of explants, the composition of the media (nature and concentration of phytohormones), number of subcultures or length of time of in culture, and the level of dedifferentiation the tissues undergo in the culture. Intrinsic factors include the genetic stability of the cultivar or genotype used in micropropagation.

Despite the advantages somaclonal variations proffer, it has been unable to leave a significant impression on mango breeding via development of advantageous off-types of existing selection, which lack in some useful attributes.

Somaclonal Variations: A Potential Mechanics for Clonal Plant Success

Somaclonal variations may confer ecological advantages to clonal plants that could outweigh evolutionary costs. Clonal plants may be one of the best examples of organisms taking advantage of epigenetic acclimation as an alternative to the slower mechanisms of adaptation through natural selection. If epigenetic processes are important in matching organismal response to the environment, this may prove to be a mechanism that will buffer plants against the challenges of current and future rapid environmental changes.

Sources of Variations Detected in Plant Tissue Culture

The maintenance of genetic integrity is of primary importance in germplasm conservation. However, unexpected variations can occur during natural and *in vitro* propagation to generate off-types. Many times the variation observed is not due to *in vitro* protocols but is the resultant of mixoploidy and chimeric structures. The initial phenotypic variation may result from unknown residual tissue culture effects that persist. But this does not inherited in subsequent somatic generations. It is often reported that these observations are unstable, epigenetic events (Harding, 2004).The genetic events include alteration in chromosome number such as polyploidy and aneuploidy; alteration in chromosome structure like translocation, deletion and duplication, mutation in DNA sequence, while epigenetic changes can occur as a result of alterations in DNA methylation patterns, changes in histone modifications, or a combination of these mechanisms. Although frequent during *in vitro* propagation, genomic variations also occur naturally in plant somatic and reproductive tissues as plant response to environmental cues (Irie *et al.*, 2011).

Measures to Minimize Somaclonal Variations in *In vitro* Multiplied Plants

1. **Effect of explants:** It is generally recognized that phenotypic variation during plant regeneration is minimal when precautions are taken to avoid the dedifferentiated "callus" phase.
2. **Somatic embryos:** Meristem derived cultures without a significant dedifferentiation step give rise to normal or nearly normal plants due to the more stringent genetic requirements imposed by embryo formation. Somatic embryos have been considered a stable explant due to its origin from a single cell. However, when the origin of somatic embryos is multicellular, the chances of variations increased.
3. **Axillary buds:** The stimulation of axillary bud development is not expected to generate genetic instability, since this technique uses the normal ontogenetic route for branch growth by lateral meristems. The tea plants regenerated from enhanced axillary branching cultures produced plants from nodal explants showed a 7.43 per cent and 16.3 per cent level of polymorphism in the nuclear and the mt genome, respectively.
4. **Age of cultures/Number of subcultures:** An increase in the variability with an increase in culture age has been reported in date palm and banana.
5. **Genotypic differences:** The genotypic differences in the frequency of somaclonal variations have been reported in banana. Some species are inherently more unstable than others during propagation.
6. **Plant growth substances:** The minimum use of plant growth regulators is always recommended for conservation through tissue culture. The concentration of plant growth regulators has been found to affect the frequency of somaclonal variation in banana.

Techniques to Detect Somaclonal Variation

The characterization of *in vitro*-conserved and cryopreserved germplasm by traditional methods such as:

1. **Phenotypic/Morphological markers**: Assessments reflect whole genome interactions affecting the phenotype; plants are subject to phenotypic plasticity, mutations/deletions in non-coding regions may go undetected.
2. **Cytological markers**: Theseinclude chromosome number; detects ploidy changes: poly/aneuploidy; difficult in some species with small chromosomes; difficult to detect small chromosomal deletion events.
3. **Biochemical markers**: Isozyme/protein profiles are good indicators of gene expression; isozymes are a small fraction of the total protein in a

plant; patterns subject to environmental changes. Biochemical analysis by isoenzymes and electrophoresis techniques to separate molecules has been developed in order to detect genome modifications. However, these analytical approaches have serious limitations because describing only partially the genetic changes and there are difficulties in assess those changes if any, at the DNA sequence level. In addition the evaluation and interpretation of the results cover many years of research and probes are time dependent (Table 18.1).

Table 18.1: Occurrence of Somaclonal Variations as Affected by the Choice of Explants

Crop Species	*Explants/Explants Source*	*References*
Almond (*Prunus dulcis*)	Axillary branching (-)	Martins *et al.* (2004)
Chimeric 'Maricongo' banana	Vegetative and floral axis tip	Krikorian *et al.* (1993)
Cavendish group of bananas (*Musa* sp.)	Chimeric shoot tip (-)	Israeli *et al.* (1995)
Kiwifruit (*Actinidiadeliciosa* (Chev.)	Leaf blades and petioles	Prado *et al.* (2007)
Oil palm (*Elaeisguineensis* Jacq.)	Mature zygotic embryos	Rival *et al.* (2013)
	Immature zygotic embryo	Sanputawong and Te-chato (2011)
	Immature leaves	Lucia *et al.* (2011)
Papaya (*Carica papaya* L.)	Axillary shoot tips underwent cryopreservation	Kaity *et al.* (2009)
Sweet cherry (*Prunus avium*)	Shoot apical portions	Piagnani and Chiozzotto (2010)
Rootstock Mr.S 2/5, *Prunus cerasifera* Erhr	Leaf	Muleo *et al.* (2006)
*Vitis*spp.	Nodal segment (-)	Alizadeh *et al.* (2008)
Cocoa (Theobroma cacao)	Closed immature flower buds	Rodrı´guez Lo´pez *et al.* (2010)
Strawberry (*Fragaria* x *ananassa* Dutch.)	Shoot tips and nodal segments	Karim *et al.* (2015)
Bananas (Musa spp.)	Shoot apex	Sahijram *et al.*, 2010
Cultivated apple (*Malus domestica*)	Actively growing shoots	Noormohammadi *et al.*, 2015
Pineapple *Ananas comosus* (L.)	buds	Pérez *et al.*, 2012

4. **Molecular markers**: Although RFLP, RAPD, AFLP and RAF techniques are excellent tools for the detection of variations in primary DNA structure, other approaches are needed to detect epigenetic changes. Analysis of chromatin relies on detecting variations in its conformation which is dependent on nucleosomal structure. DNA methylation analysis focuses on the methylation of cytosine. It is important to note that techniques such as RAPD, AFLP and RAF screen small portions of the genome that are widely distributed, thus any detected variation is not representative

of the entire genome. The observed differences in marker profiles could potentially be caused by transposable elements or point mutations. Furthermore, detected variations may lie in non-coding regions and may not affect phenotypic characteristics and development (Table 18.2).

Table 18.2: *In vitro* Selection of Desirable Traits and Development of some Commercially Exploited Varieties through Somaclonal Variation in different Perennial Horticultural Crops

Horticultural Crop	*Characteristic of Somaclone*	*References*
Apple rootstock	Resistance to white root rot (*Dematophora necatrix*)	Modgil *et al.* (2012)
Banana (*Musa acuminata* L.)	Semi-dwarf and resistant to Fusarium wilt TC1-229	Tang *et al.* (2000)
	Fusarium wilt-resistant somaclonal variants of banana cv. Rasthali	Ghag *et al.* (2014)
	Var. CIEN-BTA-03, resistant to yellow Sigatoka	Gime´nez *et al.* (2001)
	superior horticultural traits and resistance to Fusarium wilt	Lee *et al.* (2011)
Grapevine (*Vitis vinifera* L.)	Resistant to *Botrytis cinerea* and *Plasmopara viticola*	Kuksova *et al.* (1997)
Kiwi fruit (*Actinidia deliciosa*)	5 somaclones, derived from cv. Tamuri, tolerant to NaCl	Caboni *et al.* (2003)
Mango (*Mangifera indica* L.)	Resistant to *Colletotrichum gleosporiensis*	Litz *et al.* (1991)
Myrobolan (*Prunus cerasifera* Erhr)	Water logging-tolerant clone variant (S.4) of myrobolan root stock Mr.S 2/5 for peach cv. Sun Crest	Lacona *et al.* (2013)
Olive (*Olive europea*)	Bush olive somaclone (BOS), columnar olive somaclone (COS)	Leva *et al.* (2012)
	Somaclone S 122-1 was found resistant to bacterial canker (*Pseudomonas syringae* pv. *syringae*)	Hammerschlag (2000)
Pear (*Pyrus* sp.)	Resistant to *Erwinia amylovora*	Viseur (1990)
Pineapple (*Anana scomosus* L., Merr.)	Spineless variant	Jaya *et al.* (2002)
	Cvs. P3R5 and Dwarf, variation in fruit color, growth habit, fruit size and length of plant generation cycle	Pe´rez *et al.* (2009, 2012)
Strawberry (*Fragaria* sp.)	Resistant to *Fusarium oxysporum* f. sp. *Fragariae*	Toyoda *et al.* (1991)
Sweet orange (*Citrus sinensis* (L.) Osb.)	Somaclone of OLL (Orie Lee Late) sweet orange; late maturing; suitable for fresh market or processing, exceptional juice quality and flavor	Grosser *et al.* (2015)

REFERENCES

Chaudhury R, Pandey R, Mandal BB, Sharma N, Gupta S, Malik, SK and Hussain Z. (2001) Germplasm Conservation: *In vitro* Repository and Cryobank In: Dhillon, B.S., Varaprasad KP, Srinivasan K, Singh M, Archak S, Srivastava U. and Sharma GD. (eds). *National Bureau of Plant Genetic Resources: A compendium of achievements*, NBPGR, New Delhi, pp. 209-223.

Engelmann F. (2010) Use of biotechnologies for the conservation of plant biodiversity *In vitro* Cell. Dev. Biol.—Plant DOI 10.1007/s11627-010-9327-2.

Harding K. (2004) Genetic integrity of cryopreserved plant cells: A review CryoLett. 25: 3-22.

Irie VB, Chinyere A, Sandra N and Abdou T. (2011) Genomic characterization of natural and somaclonal variations in bananas (*Musa spp.*) Plant Mol. Biol. Rep. 29: 440–448.

Kaity A, Ashmore SE, Drew RA and Dulloo ME. (2008) Assessment of genetic and epigenetic changes following cryopreservation in papaya Plant Cell Rep 27: 1529–1539.

Karim R, Ahmed F, Krishna Roy U, Ara T, Islam R and Hossain M. (2015) Varietal Improvement of Strawberry (Fragaria x ananassa Dutch.) Through Somaclonal Variation Using *In vitro* Techniques. J Agr Sci Tech 17: 977-986.

Krishna H, Mahdi A, Singh D, Singh U, Chauhan N, Eftekhari M, and Sadh RK (2016) Somaclonal variations and their applications in horticultural crops improvement. 3 Biotech. 6(1): 54 -72.

Martins M, Sarmento D, Oliveira MM. (2004) Genetic stability of micropropagated almond plantlets, as assessed by RAPD and ISSR markers. Plant Cell Rep 23: 492–496.

Noormohammadi Z, Fazeli S, Sheidai M, Farahani F. (2015) Molecular and genome size analyses of somaclonal variation in apple rootstocks Malling 7 and Malling 9. Acta Biol. Szeged 59: 139-149.

Pérez G, Yanez E, Mbogholi A, Valle B, Sagarra F, Yabor L, Aragón C, González J, Isidrón M, Lorenzo JC. 2012. New Pineapple Somaclonal Variants: P3R5 and Dwarf. Am J Plant Sci.3: 1-11.

Rodrý´guez Lo´pez CM, Bravo HS, Wetten AC, Wilkinson MJ. (2010) Detection of somaclonal variation during cocoa somatic embryogenesis characterized using cleaved amplified polymorphic sequence and the new freeware Artbio. Mol Breeding 25: 501–516.

Sahijram L., Soneji J. R., Rao M. N. (2010). Molecula and Genetic characterization of somaclonal variation in micropropagated bananas (*Musa spp.*). Genes genom genet, Special Issue 1: 9-17.

Sanputawong S, Te-Chato S. (2011) Analysis of somaclonal variation of callus, somatic embryo and plant regeneration of *in vitro* oil palm (*Elaeis guineensis Jacq.*). J Agric Technol 7(2): 531–545.

Breeding of Perennial Horticultural Crops *Pages* **209–225**
Editors: S.K. Singh, V.B. Patel, A.K.Goswami, Jai Prakash and Chavlesh Kumar
Published by: **BIOTECH BOOKS, NEW DELHI**

Chapter **19**

Breeding Priorities and Strategies

M.R. Dinesh

ICAR-Indian Institute of Horticultural Research
Hessarghatta, Bengaluru – 89, Karnataka

INTRODUCTION

Majority of the fruit crops are perennial in nature and owing to the outcrossing during their evolutionary stages they are highly heterozygous. Genetic improvement is extremely difficult due to the complex genetic nature. The germplasm collection, conservation and evaluation take long time and area. Although the concept of core collection is very much valid, there is always a question mark when it comes to highly heterozygous crops as to whether the sample is good enough to represent the population. Due to their propagation by seeds earlier the large variability present needs to be evaluated for various characteristics. The genetical studies carried out in many of these crops are not conclusive and the progeny performance cannot be predicted. The biometrical methods to estimate the parameters *viz.*, combining ability, heritability using the methodologies such as diallel have proven to be futile exercise to a very great extent. Parental selection, which is so vital for a successful breeding programme is very difficult as mainly selection has to be based upon the phenotype. Breeding for particular traits like biotic or abiotic stress has been difficult as resistance is not found within the spp and interspecific/intergeneric hybridization is difficult because of incompatibility barriers for *e.g.* the 'PRSV' breeding programme in papaya. In highly heterozygous crop like mango, there exists strong genotype x environmental interaction, which alters the performance of improved varieties.

Problems in Perennial Crop Breeding

The main problems in perennial breeding has been the highly heterozygous nature, they have long juvenile phase, complex taxonomy, nucellar embryony and cross/self-incompatible. The selection of parents in a breeding programme becomes difficult. Parents when they are selected based on the phenotype and used in the breeding programme, progeny performance or the manifestation of hybrid vigour becomes unpredictable. The progenies also are influenced to a great extent by the environment. However, one of the advantages of the heterozygosity is that in the F_1 generation itself segregation will be maximum and the desirable trait in the progenies can be fixed by vegetative propagation. Perennial crop breeding takes long time. The seedling progenies take more than eight to nine years to fruit thereby making evaluation of progenies a long duration programme. The absence of molecular markers and pre-selection indices in most of the crops is hampering the breeding programme. Several attempts have been made in crops like mango by vegetative propagation to induce early flowering through grafting (Kulkarni, 1991; Kulkarni, 1986 and Kulkarni, 1988 a)

In most of the fruit crops, taxonomy is complex and the nomenclature ambiguity in crops like mango due to the large unevaluated variability has hindered improvement. The chromosomal aberrations in the case of seed propagated crops generally get deleted. But in vegetatively propagated crops these aberrations get fixed ultimately resulting in undesirable recombinants. The presence of nucellar embryony in crops like Citrus and mango interferes in the identification of zygotic seedlings. Hence, there is an urgent need to develop markers to identify them at the nursery stage. In some of the crops cross and self-incompatibility comes in the way of selecting parents for the breeding programme. Raising a large population, which is the basic necessity for selection, is extremely difficult and the large area needed for the evaluation is another hindrance in the breeding programme.

Table 19.1: Crops and Centre of Origin

Sl.No.	*Crop*	*Ploidy Number*	*Place of Origin*
1	Mango	2n=40	Indo-Burma
2	Banana	2n=22,33,44	South East Asia
3	Citrus	2n=18	North western China
4	Grape	2n=38	Black Sea to Caspian Sea
5	Papaya	2n=18	Latin America
6	Guava	2n=22	Latin America
7	Sapota	2n=26	Mexico
8	Pomegranate	2n=16,18	Iran, and the highlands of Turkmenistan.
9	Fig	2n=26	West Asia
10	Avocado	2n=24	Mexico and West Indies
11	Aonla	2n=28	Tropical Asia; Indo-china
12	Custard Apple	2n=14	Tropical America
13	Pineapple	2n=50,75,100	Brazil
14	Jamun	2n=40	East India- Malaya

The conventional breeding strategies aimed at the improvement are similar to any other crop. However, the production of inbreds are totally ruled out barring a few seed propagated crops like papaya, in which also only 90 per cent homozygosity can be achieved through selfing. Priorities keep changing and we need to appropriately use modern day technologies to bring about improvement. Many of the fruit crops have its base in south-east nations (Table 19.1).

Breeding Priorities

Increasing productivity of perennial crops have always been a challenge. Except in crops like grapes, productivity has remained an issue in many of the crops. Perennial crop breeding has two components, rootstock and scion breeding. One of the ways of increasing the production is bringing the marginal soils under cultivation. We are in the era of high-density planting. Hence, there is a need to develop genotypes amenable for high density and also there is a need to develop genotypes resistant to new biotic and abiotic stresses. In addition to this breeder will have to concentrate on developing rootstock varieties resistant to abiotic stresses. Lack of understanding in the inheritance of traits is one of the most important hindrances in perennial crops. In fruit breeding, which is based on either conventional (hybridization, selection, mutation) or biotechnological methods employing embryo culture, regeneration from protoplasts, somatic hybridization, *in vitro* mutant selection, genetic transformation, production and doubling of haploids would help in understanding the inheritance of traits. Integrated approach with both biotechnological tools and conventional can help in obtaining good results in a short time.

The breeding priorities are:

1. Developing rootstocks for biotic and abiotic stresses
2. Scion genotypes suitable for high density planting
3. Scion varieties for biotic stresses-blight, mildew, canker *etc.*

Rootstock Breeding

Morphological resistance offered depends on root structure, length, width and root hairs, biochemically by creating osmotic environment thereby escaping from stress and physiologically by synthesis of toxins, tannins, phenolic acids *etc.* Schmid *et al.* (2003), evaluated the resistance of various populations against the Phylloxera revealed that the hypersensitive reactions on leaves and roots are linked. If plants showed resistance on their leaves, the roots were most of the time also resistance, but the resistance is inherited at different levels, depending on the other parent and the resistance originates from more than one gene. But the susceptible parent has an important influence on proportion of resistant hybrids in the population.

Pavel Pavlousek (2011) conducted a study on evaluation of drought tolerance of new grapevine rootstock hybrids: this study involved 20 new hybrids and four existing rootstock varieties. The scheme of this study involves three variant water supplies. The study showed that drought has a significant effect on shoot length, Chlorophyll content Index and visual observation. Here they have concluded that BORNER (*Vitis cinerea*) rootstock can be used as suitable genetic resource for the purpose of the breeding for drought tolerance. Accumulation of excess of Cl-

and Na+ can cause specific ion toxicities. By selecting rootstocks that restrict the uptake of these ions will serve as a beneficial tool for salinity for *e.g.*, Mandarin, Grapefruit, Rough lemon, Sour Orange, Rangpur lime *etc.* in Citrus, *Vitis champini*, Ramsey *etc.* in grapes and Borchard, Duke-7 and G755C in avocado. In mango, 3200 seedlings of 80 different mono and polyembryonic types were evaluated, among them 13-1 has showed highest level of tolerance. The rootstock 13-1 was observed to have less salt retention capacity than re-translocation into soil; therefore, it is regarded as salt tolerant rootstock in mango. Sathisha *et al.* (2007), characterized 10 grape rootstocks for their various morphological, physiological and biochemical parameters. Significant differences were observed for most of the parameters studied. Rootstocks in the group of *Vitis berlandierii* x *Vitis rupestris*, such as 110 R, 1103 P, 99 R and B2-56, had a significantly higher content of total phenols, flavonoids, proline and total protein.

Germplasm Collection

Many of the fruit crops *viz.*, mango and jackfruit have their primary centre of origin in India. However, collection, characterization and evaluation to solve the nomenclature ambiguity is necessary. Germplasm collection and recording the traditional knowledge is necessary to help in the selection of traits. Diversity studies need to be carried out, which could help in the selection of the parents.

Pre-selection Indices

Markers are to be developed for various trait selection at the early stage. Apart from the molecular markers, there are morphological markers and more recently biochemical markers are being used. The role of volatiles needs to be understood in a better way.

Breeding Strategies

The breeding strategies should to achieve our goals of developing improved varieties should be to combine the conventional breeding strategies with the modern technologies. However, at no time should we feel all that is conventional has not yielded results,

Diversity Studies

India being one of the most important diversity centres of many of the tropical fruits, it is necessary that it is exploited properly. In this direction, we need to promote 'on-farm conservation' of diversity. In addition, there is a need to identify 'custodian of genetic diversity' who can help in the conservation. Work in this direction carried out has resulted in good selections (Anon, 2016).

Introduction

It is one of the most important methods of creating variability in a crop. Introduction can be within a particular region or from outside essentially from outside the country. A good germplasm strength is a must for good improvement programme. One of the important aspects of such an introduction is the genotype x environmental interaction. If a variety has to perform in the place of introduction, then it should be least sensitive to the soil and climatic changes. A pointer to this is the introduction of the mango variety Dashehari in the southern regions of the country. Although the quality of the fruit has not changed, the size of the fruit has come down. However, even in such cases as well as the introduction from outside, care should be taken not to introduce new pest and diseases. Even if an introduction fails in its purpose, these can always be used as pre-breeding lines, which would help in the development of improved recombinants. Fruit crops like papaya and guava, which have been introduced from Latin American countries, have been able to establish well in India. Many of the mango varieties *viz.*, introduced from India to the American continent has performed well in these regions. Hence, introduction can play an important role in the improvement of crops.

Clonal Selection

Clonal selection is one of the important tools in the breeding programme that helps in creating variability in woody perennials especially in crops like mango. The exact basis for clonal variants is still not known, but spontaneous mutants may be one of the reasons for such variants. In crops like mango innumerable varieties have added to the confusion of nomenclature ambiguity by way of clonal variation. However, clonal selections have been reported in grapes, mango *etc.* In clarifying the nomenclature ambiguity biotechnological tools come in handy. Tissue culture can also be used for rapid multiplication of the selections made. Selections made out of clones can also be used as prebreeding lines as they may have some desirable traits other than that of the original mother clone. One of the main drawbacks of the clonal selection is that their susceptibility to diseases and pests. If the original mother clone is susceptible then the selections made also will be susceptible.

Open Pollinated Selections

As mentioned above the number of progenies raised is one of the most important steps in the breeding programme. In some of the perennial crops like mango, due to fruit drop the number of fruit recovery from a cross is very low. In many crops' flowers are delicate. In such cases there is a need to develop alternate strategies to raise large number of progenies. One of the methods would be to raise open pollinated progenies from a desired variety wherein trait improvement is sought after or in an otherwise ideal variety, which suffers from any one or two defects. In these progenies female parent is known but the male parent is not known. Validation of these half-sib progenies for their parentage can be carried out by using molecular markers once a hybrid progeny is selected. The evaluation of hybrid progenies can be carried out by planting them in high density so that the large number of progenies can be accommodated. Lavi *et al.* (2004) opine that due to the difficulty of performing controlled crosses, open pollination could be the strategy to generate the seedlings.

Hybridization

The success of a hybridization programme depends on the selection of parents and the number of recombinant progenies selected out of a particular cross. In the case of many of the fruit crops, the flower structure is complex, and the crossing is difficult. Intervarietal hybridization is quite common in almost all the fruit crops. However, interspecific or intergeneric hybridization is not common except barring few of the temperate crops like strawberry. In most instances' compatibility mechanism between the genera or species has not been worked out or there is incompatibility mechanism operating. In some of the fruit crops like mango use of wild spp holds promise (Iyer, 1991). Intergeneric hybridization has been used in papaya for development of tolerant lines to 'PRSV' (Dinesh *et al.*, 2007 and 2013). Interspecific hybridization has been adopted in developing rootstocks resistant to diseases *viz.*, wilt resistance in guava (Edward, 1960). Interspecific hybridization is very widely practiced in some of the fruit crops like Citrus, wherein intervarietal hybridization work is very limited as the desired traits are available across the species and there is no barrier. In the development of better-quality fruits, it is also used in custard apple breeding programme (Jalikop, 2011). Use of widely related species is also observed since long in banana, although, viable hybrids have not been generated. Multi-varietal orchards in some places have shown better yields, which needs to be probed with the pollen studies carried out.

Mutation

Most of the fruit crops are highly heterozygous. However, heterozygosity is one of the advantages when it comes to mutation. In highly heterozygous crops wherein there is a need for improvement in a single trait mutation have been of great use. One of main limitations is use of vegetative material, which due to the unlikelihood of uniform penetration and distribution of mutagenic agent can result in chimera. Mutation is one of the methods that can be relied upon to develop pre-breeding lines and to bring about change in the quantitative traits. Another very important fact is that there are crops like grapes wherein lot of spontaneous mutants have become established varieties *viz.*, Sonaka and Saritha to quote a few. It is also to be noted here that mutation breeding can be a good supplement for conventional breeding, more so with the emergence of tissue culture propagation, which often throws up spontaneous mutants. Mutation as a methodology can also be tried in some fruit crops spp where there is no disease resistance source within. It can also be adopted in cases where there is a need to break linkages. Seeds, pollen grains, buds/cuttings or complete plant can be used for mutagenesis. It depends on whether the crop is sexually or asexually propagated and type of mutagen to be used. Pollen grains are also infrequently used because, they are difficult to collect in large quantities, hand pollination (with treated pollens) is rather difficult and survival percentage of pollens is relatively low.

Although nearly all the parts of a plant can be induced to develop meristematic tissue, shoot tip meristems in leaf buds or in seeds are typically targets for mutation. In sexually propagated crops, seeds are commonly because seed can tolerate extreme environmental conditions. From treated seeds M_1 generation is developed. In M_1 generation large number of plants is grown with wider spacing. Dominant mutations are recorded if any (generally mutations are recessive and do not express in M_1).

Mutation breeding does hold promise in heterozygous perennials, as the first effect that is observed in the M_1 generation is the branching habit and dwarf statured plants. This will help in developing varieties for high density planting. Although there is every chance of mutants reverting back, raising large population and fixing the mutagenic changes through vegetative propagation would help in developing improved recombinants.

Ploidy Manipulation

Ploidy manipulation is also considered as part of mutation breeding by some. It has been employed in many of the crop improvement programmes. Ploidy manipulation can help fruit breeders especially the production of haploids and doubled haploids as they can show the way shortening the time needed to produce homozygous lines compared to conventional breeding and also in the production of triploids, which would be triploids. However, one of the basic needs for undertaking ploidy manipulation is that the crop should have lower chromosome number. In some crops another hindrance is noticed that is the conversion of tetraploids into diploid like behavior, which is termed as 'diploidization' (Ozkan and Feldman, 2009). Oritz*et al.*,(1997), through ploidy manipulations by utilizing the local triploid plantain landraces and diploid or cultivated wild bananas in interspecific hybridization in triploid-diploid crosses, followed by *in vitro* seed germination through embryo culture, and further rapid micropropagation of selected genotypes developed a tetraploid plantain type. In the case of Citrus, it has been observed that occurrence of duplication in the female gametes varies between less than 1 per cent to more than 20 per cent (Soost, 1987; Iwamasa and Nito, 1988).

In fruit crops, high heterozygosity, long generation cycle with long juvenile period, large size, and often self-incompatibility do not allow the haploidization easily through the conventional methods (Germanà, 2006). *In vitro* androgenesis is single-step to produce haploid or double haploid plant from heterozygous origins (Srivastava and Chaturvedi, 2008). Development of homozygous lines through gametic embryogenesis results in single step improvement of the crop. In a conventional breeding programme, a pure line is developed after several generations of selfing. However, in fruit crops, there is no way of obtaining haploids through conventional methods. Hence, fixation by doubling the chromosome complement presents a possibility using haploids. Haploids along with marker-assisted selection can help in shortening the time taken for fruit improvement.

Another fruit crop in which work has been carried out to produce haploids is in apple. Induction of maternal haploids using irradiated pollen for pollination is one of the methods. Pollen germination on the stigma stimulates embryo development and growth of the pollen tube within the style, although irradiated pollen is unable to fertilize the egg cell. It has been used successfully in several species (Jana Murovec and Borut Bohanec., 2012). Froelicher *et al.* (2007), have reported the same in mandarin.

Haploidy, doubled haploidy and gametic embryogenesis in fruit crop breeding can bring about the shortening of the breeding period along with the understanding of the inheritance pattern of some of the most important traits. A better understanding of the gametic embryogenesis process could make haploid production a powerful fruit crop breeding tool in the future. There is a need to improve the induction rate (the frequency with which gametes form embryos) and to increase the survival rate. Through haploid breeding the prediction of progeny performance be it disease resistance or any other trait in question can be addressed, which is of great importance in perennial crop breeding.

Markers in Fruit Breeding

Conventional methods of perennial fruit crop improvement have almost exclusively relied on traditional breeding techniques. These techniques are expensive, time consuming and space requiring, especially for tree species due to the need of a large population of seedlings from which only few promising ones are selected. The juvenile period of tree crops may last three to five years before onset of fruits so that 20 years or more are common before the successful introduction of a gene of interest (Baird *et al.*, 1996).Therefore, in this context, molecular tools recently made available for breeders have become tools of choice because they permit to reduce cost by fastening the selection process of woody fruit plants. Further, traits of importance like fruit quality, resistance to diseases, high yield *etc.* is polygenic in nature for which classical breeding is generally not suitable for detection and selection of traits. A quantitative trait locus (QTL) analysis with help of markers and linkage maps may be more appropriate for detection of such traits of interest. Also, molecular markers are being used for maintenance of core collections, for the assessment of genetic diversity between genotypes and fingerprinting(Kanupriya and Ravishankar, 2013).

Genetic Diversity and Fingerprinting

The assessment of genetic diversity of fruit crops is an important step in breeding since a diverse range of plant material including landraces, modern cultivars, crop wild relatives and other wild species are available as the starting material. DNA based fingerprinting has been implemented in the characterization of germplasm. A large number of molecular markers have been developed and applied for analysis of genetic diversity and relatedness, and all markers have their strengths and weaknesses (Schlötterer, 2004; Weising*et al.*, 2005).

Random Amplified Polymorphic DNA (RAPD) Markers

Comigration of non-homologous fragments can be a problem at higher taxonomic levels but is largely negligible at the intraspecific level or when the studied species are closely related (Rieseberg, 1986). RAPD markers have, despite the above-mentioned limitations, been widely used in plant genetics and breeding, *e.g.*, to estimate genetic diversity, detect phylogenetic relationships, and create genetic maps (Weising*et al.*,2005).

Simple Sequence Repeats or Microsatellites (SSR)

Microsatellites are short (1–6 bp) tandemly repeated sequences of DNA (Litt and Luty, 1989; Tautz, 1989). In microsatellite analysis, two primers are designed complementary to the sequences flanking a specific microsatellite DNA sequence. The main drawback of microsatellite analysis is the high expenses at the initial stages of primer development, when sequences for primer design must be isolated and cloned. However, some cross-species transferability of primer pairs (Yamamoto *et al.*, 2001, 2004) or software-designed primers based on publicly accessible sequence information from gene-bank accessions (Tonguc and Griffiths, 2004) can decrease the initial costs considerably.

Sequence-Characterized Amplified Regions (SCARs)

SCARs were originally developed from RAPD-derived markers (Paran and Michelmore, 1993), but have later also been developed from AFLP-derived markers (Mienie*et al.*, 2002).

Single Nucleotide Polymorphism (SNPs)

Single base differences between homologous DNA fragments are referred to as Single Nucleotide Polymorphism (SNPs). These markers represent most frequent polymorphism and are becoming more popular than SSRs as genetic markers in linkage analysis because these are more abundant and are suitable for automatic allele calling (Novelli *et al.*, 2006; Selmer *et al.*, 2009). A number of approaches have been developed to reduce the complexity of the genomes for cost efficient and simplified discovery of SNP markers using NGS, like, RNA-Seq, complexity reduction of polymorphic sequences (CRoPS), restriction-site-associated DNA sequencing (RAD-Seq), and genotyping-by-sequencing (GBS) (Daweyet al., 2011). Among these, RAD-Seq and GBS are proving to be particularly efficient in rapid and mass discovery of SNPs in highly heterozygous species. This is expected to help develop of ultrahigh density maps and association mapping. If each genotype of a mapping population has been characterized phenotypically for traits of interest, and has an individual barcode, then RAD and GBS markers can be used routinely for linkage mapping and QTL analysis (Rowe *et al.*, 2011).

Approaches for QTL Identification

The identification of QTLs is in most cases based on a technique called linkage mapping (or family mapping) for which a segregating population is generated from two parents which usually considerably differ for one or several trait(s) of interest. Two markers are linked if the recombination frequency is lower than 0.5, and this recombination frequency is then translated into a genetic distance unit (centimorgans, cM). On the other side, phenotypic data are scored in the population and generally on the parents. When a significant association is identified between the marker and the studied trait and after subsequent validation in different backgrounds or environments, the marker can be further used in a Marker-Assisted selection (MAS) program to introduce the trait of interest in commercial cultivars.

CRISPR/CAS (CRISPR-ASSOCIATED) Systems

Some bacteria have evolved a kind of immunity against viruses through a system called CRISPRs (Clustered Regularly Interspaced Short Palindromic Repeats). They consist of short sequences that originate from viral genomes and have been incorporated into the bacterial genome. Cas (CRISPR associated proteins) process these sequences and cut matching viral DNA sequences. In this technology, by introducing plasmids containing Cas genes and specifically constructed CRISPRs into eukaryotic cells, the eukaryotic genome can be cut at any desired position (Gaj *et al.*, 2013).CRISPR genome editing tools required two components *viz.* (1) the Cas9 endonuclease (2)single guide RNA (sgRNA). The sgRNA plays the major role in the target delivery since it is having guide sequence in thus, matches with the target DNA sequence (Khatodia *et al.*, 2016).

Brooks *et al.* (2014) reported the first successful transient of the CRISPR/Cas9 system in tomato. The protocol followed was primarily construction of CRISPR/Cas9 and its target delivery to the tomato homolog of the ArabidopsisARGONAUTE7 (SlAGO7) sequence that contained two sgRNAs with requisite binding region for Cas9 cleavage. In the same way, hairy root transformation could be an efficient way of introgression the mutant gene in tomato (Ron *et al.*, 2014). In addition, Wang *et al.* (2015) reported that, in DH (double-haploid) dry matter potato, gene knock out of targeting the StIAA2 gene, which encodes an AUX/IAA protein using CRISPR/Cas9 gene. Likewise, in sweet orange (*Citrus sinensis*) CsPDS gene was targeted by using CRISPR/Cas9 systems (Jia and Wang, 2014). There has been a continuous research accomplishment related to the application of the CRISPR/Cas9 technique in animal and plant systems (Khatodia *et al.*, 2016), proved that the CRISPR/Cas9 system is highly précised and efficient tool in targeting the genes in to the plant genomes thus, bring the non-characterized genes closer to the functional genomics studies in horticultural crops.

Biochemical Markers

Aroma compounds include a range of diverse low molecular weight, volatile chemicals that occur widely in nature. It plays an important role in fruit quality especially in fruits which are preferred for their aroma. In mango the major volatile aroma groups are monoterpenes, sesquiterpene, esters, alcohols, aldehydes, ketones, volatile fatty acids, some amount of phenols and some carotenoids. Aroma can be one of the important indicators for germplasm screening for fruit quality traits in fruit like mango. The composition and concentrations of aromatic volatiles in mango fruit are influenced by the cultivar [Pandit *et al.*, 2009 and Pino *et al.*, 2005]. Pandit *et al.* (2009) analyzed 27 mango cultivars, detecting 84 volatile compounds and the highest and lowest qualitative abundances of volatiles were detected in the 'Alphonso' and 'Pairi' cultivars, respectively. Johna *et al.* (1999), found differences in the compositions and concentrations of terpenoid compounds in the sap of seven Indian mango varieties, including monoterpenes β-myrcene, trans-/cis-ocimene, and limonene. Musharraf *et al.* (2016), identified seven terpenes, α-pinene, α-phellandrene, (+)-3-carene, sabinene, γ-terpinene, (-)-trans-caryophyllene, and α-humulene in the sap of nine Pakistani cultivars.

Sap burning is one of the traditional practices followed to select the best tender fruits for pickle purpose.Therefore, we also carried out sap burning test for surveyed fruit samples. Out of forty genotypes the mean time was 18.82 seconds, the maximum sap burning time was recorded in the genotype Kudige (28.25 seconds).

The compounds are related the pathway data, gene sequences and variations stored in the database allowing researchers to select a peak on a trace and have access to all other information related to that peak. By using these chemical fingerprints and their characteristic differences allows researchers to quickly find genes and variations within genes or their expression, that may be involved in the different sensory experiences that drive consumer preferences (Innes *et al.*, 2015).

Our objective in fruit breeding should be to develop varieties, with which we should be able to exploit the genetic potential to the fullest extent to increase quality production and productivity.

REFERENCES

Acquaah G., 2007. Principles of plant genetics and breeding Wiley-Blackwell, Malden.

Aleza, P., Juarez, J., Hernandez, M., Pina, J.A., Ollitrault, P. and Navarro, L. 2009.

Assani, A., Bakry, F., Kerbellec, F., Haicour, R., Wenzel, G. and Foroughi-Wehr, B. 2003, Production of haploids from anther culture of banana [*Musa balbisiana* (BB)]. *Plant Cell Rep*. 21: 511–516.

Baird, W.V., Ballard, R.E., Rajapakse, S. and Abbott, A.G. 1996.Progress in *Prunus mapping* and application of molecular markers to germplasm improvement. *HortScience* 31: 1099-1106.

Bouquet, A.B., Piganeau,B. and Lamaison, A.M. 1982, Influence du ge´notype sur la production de cals, d'embryons et de plantesentie'res par cultures d'anthe'res in vitro dans le genre Vitis. C.R. *Acad. Sci. Paris* 259: 569–574.

Bouvier L, Gue´ rif P, Djulbic M, Durel C-E, Chevreau E and Lespinasse Y 2002. Chromosome doubling of pear haploid plants and homozygosity assessment using isozyme and microsatellite markers. *Euphytica* 123: 255–262.

Bouvier, L., Zhang, Y.X. and Lespinasse, Y. 1993, Two methods of haploidization in pear, *Pyrus communis* L.: greenhouse seedling selection, in situ parthenogenesis induced by irradiated pollen. *Theor. Appl. Genet*. 87: 229–232.

Cersosimo, A., Altamura, M.M., Crespan, M. and Paludetti, G. 1990, Embryogenesis, organogenesis and plant regeneration from anther culture in Vitis. *Acta Hort.* 280: 307–314.

Chalak,L. and Legave, J.M. 1997, Effects of pollination by irradiated pollen in Hayward kiwifruit and spontaneous doubling of induced parthenogenetic trihaploids. *Sci. Hort*. 68: 83 93.

Chen, C., Lyon, M.T., O'Malley, D., Federici, C.T., Gmitter, J., Grosser, J.W., Chaparro, J.X., Roose, M.L. and Gmitter Jr., F.G. 2007. Origin and Frequency of 2n Gametes in Citrus sinensis × Poncirus trifoliata and their Reciprocal Crosses, Plant, Plant Science.

Christopher Brooks, Vladimir Nekrasov, Zachary B. Lippman*, and Joyce Van Eck 2014. Efficient gene editing in tomato in the first generation using the CRISPR/Cas9 system. *Plant Physiology* Preview. DOI: 10.1104/pp.114.247577.

Comai L., 2005. The advantages and disadvantages of being polyploid. *Nature Reviews Genetics*, 6: 836-846.

Davey JW, Hohenlohe PA, Etter PD, Boone JQ, Catchen JM, Blaxter ML, 2011. Genome-wide genetic marker discovery and genotyping using next-generation sequencing. *Nature Rev Genet.*, 12(7): 499–510.

De Witte, K. and Keulemans, J., 1994. Restrictions of the efficiency of haploid plant production in apple cultivar Idared, through parthenogenesis in situ. *Euphytica,* 77, No. 1-2: 141-146, ISSN 0014-2336.

Debnath, S.C., A. Jaime and T.D. Silva. 2007. Strawberry culture in vitro: Applications in genetic transformation and biotechnology. *Fruit, Vegetable and Cereal Science and Biotechnology*. 1: 1 -12.

Dineshhttp: //www.sciencedirect.com/science/article/pii/S0304423813003592 - aff0005 M.R., G.L. Veenahttp: //www.sciencedirect.com/science/article/pii/S0304423813003592 - aff0010, C. Vasugihttp: //www.sciencedirect.com/science/article/pii/S0304423813003592 - aff0005, M. Krishna Reddyhttp: //www.sciencedirect.com/science/article/pii/S0304423813003592 - aff0005, and K.V. Ravishankar., 2013.http: //www.sciencedirect.com/science/article/pii/S0304423813003592 - aff0005Intergeneric hybridization in papaya for 'PRSV tolerance. Scientia Horticulturae Volume 161: 357–360.

Dinesh, M.R., A. Rekha, K.V. Ravishankar, K.S. Praveen and L.C. Santosh, 2007. Breaking intergeneric crossing barrier in papaya using sucrose treatment. *Scientia Horticulturae*114: 33–36.

Edward, J.C. 1960. Wilt of Guava. Allahabad Farmers. 34: 289-293.

Esen, A. and Soost, R.K. 1972, Unexpected triploids in citrus: their origin, identification, and possible use. *J. Hered.* 62: 329–333.

Esen, A., R.K. Soost and G. Geraci. 1979. Genetic evidence for the origin of diploid mega- gametophytes in citrus. *J. Hered.*, 70: 5-8.

Evans, W. D., 1977. The use of synthetic octoploids in strawberry breeding. *Euphytica,* 26(2): 497-503

Froelicher, Y., Bassene, J.B., Jedidi-Neji, E., Dambier, D., Morillon, R., Bernardini, G.,Costantino, G. and Ollitrault, P. 2007. Induced parthenogenesis in mandarin for haploid production: induction procedures and genetic analysis of plantlets. *Plant Cell Reports*, 26(7): 937-944.

Frost, H.B. 1925, The chromosomes of Citrus. *J. Washington Acad. Sci.* 15: 1–3. Fei, K.W. and Xue, G.R., 1981, Induction of haploid plantlets by anther culture in vitro in apple. *Cv.* "Delicious". Sci. *Agric. Sin.* 4: 41–44.

Gaj T, Gersbach CA, Barbas 2013. CFZFN, TALEN, and CRISPR/Cas-based methods for genome engineering. *Trends Biotechnol.,* 31 (7): 397-405.

Germanà, M.A. 2006. Doubled haploid production in fruit crops. *Plant Cell, Tissue and Organ Culture,* 86: 131–146.

Germana' MA and Chiancone B 2001. Gynogenetic haploids of Citrus after in vitro pollination with triploid pollen grains. *Plant Cell Tiss. Org. Cult.,* 66: 59–66.

Gresshoff, P.M. and Doy, C.H. 1974, Derivation of a haploid cell line from *Vitis vinifera* and the importance of the stage of meiotic development of anthers for haploid culture of this and other genera. *Z. Pflanzenphysiol.* 73: 132–141.

Grosser, J.W., H.J. An, M. Calovic, D. Lee, C. Chen, M. Vasconcellos, and F.G. Gmitter Jr.2010. Production of New Allotetraploid and Autotetraploid Citrus Breeding Parents: Focus onZipperskin Mandarins. *HortScience*, 45(8): 1-4.

Hancook, J.F., D.H. Scott, and F.J. Lawrence. 1996, Strawberries. In: *J. Janick and J.N.Moore (Eds). Fruit Breeding:* Vine and Small Fruit. Vol. II: 419 -470. John Wiley and Sons, Inc., New York.

Hidaka, T. and Omura, M. 1989, Origin and development of embryos from microspores inanther culture of Citrus. *Jpn. J. Breed.* 39: 169–178.

Hirabayashi, T. and Akihama, H. 1982, In vitro embryogenesis and plant regeneration from the anther-derived callus of *Vitis.* In: *Fujiwara A (ed) Plant Tissue Culture* (pp. 547–548). Maruzen Co., Ltd, Tokyo.

Höfer, M and Hanke V., 1990. Induction of androgenesis in vitro in apple and sweet cherry. *Acta Hort.* 280: 333–336

Höfer, M and Hanke V., 1994. AntherenkulturbeiApfel und Kirsche: Einfluß des Pollenentwicklungsstadiums – KorrelativeBeziehungenzumorphologischen Blu¨ tenmerkmalen. *Gartenbauwissenschaft*, 59: 225–228

Höfer, M. and Grafe, Ch. 2003. Induction of doubled haploids in sweet cherry (*Prunus avium*L.). *Euphytica*, 130,-No. 2: 191-197, ISSN 0014-2336.

Höfer, M. and Lespinasse, Y. 1996. Haploidy in apple. In: *In Vitro Haploid Production in Higher Plants Vol. 3: Important Selected Plants*, Jain, S.M., Sopory, S.K. and Veilleux, R.E., pp261-276, Kluwer Academic Publishers, ISBN: 9780792335795, Dordrecht.

Höfer, M., 2004. *In vitro* androgenesis in apple – improvement of the induction phase. *Plant Cell Rep.*, 22: 365–370

Innes., David, Natalie.Dillon, Heather E. Smyth, Mirko Karan, Timothy A. Holton, Ian S.E. Bally, and Ralf G. Dietzen., 2015. Mamgomics information systems supporting advanced mango breeding in Genomics and Proteomics: Principles, Technologies, and Applications edited by DevarajanThangadurai and Jeyabalan Sangeetha Apple Press

Iwamasa, M., N. Nito and J. T. Ling., 1988. Intra- and intergeneric hybridization in the orange subfamily, Aurantioideae. International Citrus Congress (6th: 1988: Tel Aviv, Israel): Margraf. **1:** 123-130.

Iyer, C.P.A., 1991. Recent advances in varietal improvement in mango. *Acta Horticulturae, 291: 109-132.*

Jalikop, S.H. 2011. Breeding of Pomegranate and Annonaceous fruits. Proc. IInd IS on Pomegranate and Minor, including Mediterranean fruits. *Acta Hort.*, 890: 191-196.

Jana Murovec and BorutBohanec, 2012, Haploids and Doubled Haploids in Plant Breeding. www.intechopen.com

Jia, H. and Wang, N. (2014), "Targeted Genome Editing of Sweet Orange Using Cas9/sgRNA," Public Library of Science One, vol. 9, no. 4, April 7.

John, K.S., L.J.M. Rao, S.G. Bhat and U.J.S.P. Rao, 1999. Characterization of aroma components of sap from different Indian mango varieties. *Phytochemistry*, 52: 891-894.

Kadota, M., Han D-S and Niimi, Y., 2002, Plant regeneration from anther-derived embryosof apple and pear. *Hort Science* 37(6): 962–965.

Karasawa, K. 1971, On the occurrence of haploid seedlings in *Citrus natsudaidai*Hayata. Bull. Sakushingakuin*Junior Col. Woman*. 1: 1–2.

Kerbellec, F. 1996, Establissementd_une technique d_androgene' se pour l_ame´liorationge´ne´tique du bananier (Musa ssp.). *Ph.D. Thesis*, Ecole NationaleAgronomique de Rennes, France.

Khatodia, S., Bhatotia, K., Passricha, N., Khurana, S. M. and Tuteja, N. (2016), "The CRISPR/Cas Genome-Editing Tool: Application in Improvement of Crops," *Frontiers in Plant Science*, vol. 7, no. 506, April 19.

Kulkarni, V.J., 1986, Graft induced off season flowering and fruiting in mango (*Mangiferaindica* L.) *J. Hort. Sci.*, 1: 141-145.

Kulkarni, V.J., 1988 a. Further studies on graft induced off season flowering and fruiting in mango (*Mangifera indica* L.). *J. Hort. Sci.*, 63: 361-367.

Kulkarni, V.J., 1991, Physiology of flowering in mango studied by grafting. *Acta. Hort.*, 291: 95-104.

Lavi, U., K. Kashkush, D. Sa'ada, H. Shats, U. Ravid and E. Tomer. 2014. Mango Breeding and the Potential of Modern Biology. *Acta Horticulture*, 341: 146-151.

Litt, M. and Luty, J.A. 1989. A hypervariable microsatellite revealed by in vitro amplification of a dinucleotide repeat within the cardiac muscle actin gene. *American Journal of Human Genetics* 44, 397–401.

Litz. R.E and Conover. R.A., 1978. Recent advances in papaya tissue culture. *Proc. Fla State Hort. Soc.*, 91: 180–182.

Luro F, Maddy F, Jacquemond C, Froelicher Y, Morillon R, Rist D, Ollitrault P 2004. Identification and evaluation of diplogyny in clementine (*Citrus clementina*) for use in breeding. *Acta Hortic.*, 663: 841–847.

Luro, F., Maddy, F., Ollitrault, P. and Rist, D. 2000. Identification of 2n gamete parental origin and mode of nuclear restitution of spontaneous triploid Citrus hybrids. In: Proceedings of the Ninth InternationalCitrus Congress. Orlando, Florida, pp. 168–169.

Masaki Yahata, HisatoKunitake, Tsutomu Yabuya, Yukiko kashihara and Haruki Komatsu, 2005, Production of a double haploid from a haploid Pummelo using colchicine treatment of axillary buds. *J. Amer. Soc. Hort. Sci.*, 130(6): 899-903.

Mauro, M.C., Nef, G. and Fallot, J. 1986, Stimulation of somatic embryogenesis and plant regeneration from anther culture of *Vitis vinifera* cv. Cabernet Sauvignon. *Plant Cell Rep*. 5: 377– 380.

Mienie, C.M.S., Fourie, H., Smit, M.A., Van Staden, J., Botha, F.C. 2002. Identification ofAFLP markers in soybean linked to resistance to *Meloidogyne javanica* and conversion tosequence characterized amplified regions (SCARs). *Plant Growth Regulation* 37, 157–166.

Musharraf SG, Uddin J, Siddiqui AJ, Akram MI (2016) Quantification of aroma constituents of mango sap from different pakistan mango cultivars using gas chromatography triple quadrupole mass spectrometry. *Food Chem.*, 96: 1355–1360

Naess, S.K., Swartz, J.H. and Bauchan, G.R. 1998. Ploidy reduction in blackberry. *Euphytica*. 99: 57-73, ISSN 0014-2336.

Niemirowicz-Szczytt K. Zakrzewska Z. Malepszy S. Kubicki B 1983. Characters of plants obtained from *Fragaria* X *ananassa* anther culture. Acta Hortic 131: 231–237

Novelli VM, Cristofani M, Souza AA, Machado MA. 2006. Development and characterization of polymorphic microsatellite markers for the sweet orange (*Citrus sinensis* L. Osbeck). *Genet Mol Biol.*, 29(1): 90-6.

Oiyama I, Okudai N 1986. Production of Colchicine-induced Autotetraploid Plants through micrografting in monoembryonic Citrus Cultivars. Jpn J Breed 36: 371–376

Oiyama, I. and Kobayashi. 1993, Haploids obtained from diploid triploid cross of Citrus. *J.Jpn. Soc. Hort. Sci*. 62: 89-93.

Ortiz, R. and D.Vuylsteke, 1997. Developing new plantain varieties for Africa. Plant Varieties and Seeds 1997 Vol. 10 No. 1 pp. 39-57

Owen, H.R. and A.R. Miller. 1996. Haploid plant regeneration from anther cultures of three North American cultivars of strawberry (*Fragaria x ananassaDuch*).*Plant Cell Reports* 15: 905-909.

Ozkan H., Feldman M., 2009. Rapid cytological diploidization in newly formed allopolyploids of the wheat (Aegilops-Triticum) group. Genome 52: 926-934.

Pandey, K.K., Przywara, L. and Sanders, P.M. 1990, Induced parthenogenesis in kiwifruit (*Actinidiadeliciosa*) through the use of lethally irradiated pollen. *Euphytica*, 51: No. 1,: 1-9, ISSN 0014-2336

Pandit SS, Chidley HG, Kulkarni RS, Pujari KH, Giri AP, Gupta VS. 2009. Cultivar relationships in mango based on fruit volatile profiles. *Food Chem.*, 114(1): 363–372.

Paran, I. and Michelmore, R.W. 1993. Development of reliable PCR-based markers linked to downy mildew resistance genes in lettuce. *Theoretical and Applied Genetics*, 85: 985– 993

Pavel Pavlousek, 2011.Evaluation of drought tolerance of new grapevine rootstock hybrids. *J. Environ. Biol*, 32: 543-549.

Pavel Pavlousek, 2011, Evaluation of drought tolerance of new grapevine rootstock hybrids. *J. nv., Biology*. **32:** 543-549.

Pino JA, Mesa J, Munoz Y, Marti MP, Marbot R. Volatile components from mango (*Mangifera indica* L.) cultivars. J Agric Food Chem. 2005; 53(6): 2213–2223.

Rajasekaran, K. and Mullins, M.G. 1979, Embryos and plantlets from cultured anthers of hybrid grapevines. *J. Exp. Bot.* 30: 399–407.

Ramirez, C., Chiancone, B., Testillano P. S., Garcia-Fojeda, B., Germana,' M.A. and Risueno, M.C. 2003, First embryogenic stages of Citrus microspore-derived embryos. *Acta Biol. Cracov. Bot.* 45: 53–58.

Ravishankar, K.V., Kanupriya, A. Rekha, A Upadyaya, C. Vasugi, N. Vijayakumari, PoojaKishnani and M.R. Dinesh.2013. Omics approaches in tropical fruit crops: In OMICS Application in Crop Science-(Ed D. Barh), CRC Press Boca Raton, Florida, USA, 285-324.

Recovery and characterization of a *Citrus clementina*Hort. ex Tan. 'Clemenules' haploid plant selected to establish the reference whole *Citrus* genome sequence. *BMC PlantBiology*, 9: 110, ISSN 1471-2229.

Ron M, Kajala K, Pauluzzi G, Wang D, Reynoso M a, Zumstein K, Garcha J, Winte S, asson H, Inagaki S, *et al.* (2014) Hairy root transformation using *Agrobacterium rhizogenes* as a Tool for Exploring Cell Type-Specific Gene Expression and Function Using Tomato as a Model.Plant Physiol. doi: 10.1104/pp.114.239392

Rowe HC, Renaut S, Guggisberg A.,2011. RAD in the realm of nextgeneration sequencing technologies. Mol Ecol; 20(17): 3499-502.

Sathisha, J., Ramteke, S.D. and Karibasappa, G.S., 2007, Physiological and biochemical haracterization of grape rootstocks.*S. Afr. J. Enol. Vitic.,**Vol. 28,** No. 2:* 163-168.

Schlötterer, C. 2004. The evolution of molecular markers – just a matter of fashion? *Nature reviews/Genetics* 5, 63–69.

Schmidt, J., Manty, F. and Ruhl, E. H., 2003, Utilizing the complete phylloxera resistance of *itiscinerea* Arnold in rootstock breeding. *ActaHort* **603**: 393-410.

Selmer KK, Brandal K, Olstad OK, Birkenes B, Undlien DE, Egeland T. Genome-wide linkage analysis with clustered SNP markers. J Biomol Screen 2009; 14: 92–6.

Soost, R.K., 1987. Breeding Citrus-Genetics and nucellar embryony (In) improving vegetatively propagated crops (Abbott, A.J and Atkin, R.K-eds) Academic Press, New York, pp 83-110.

Srivastava, P. and R. Chaturvedi. 2008. In vitro androgenesis in tree species: an update and prospect for further research. *Biotechnol Adv* 26: 482-491.

Tautz, D. 1989. Hypervariability of simple sequences as a general source for polymorphic DNA markers. *Nucleic Acids Research* 17: 6463–6471.

Tonguc, M. and Griffiths, P.D. 2004. Genetic relationships of *Brassica* vegetables determinedusing database derived simple sequence repeats. *Euphytica* 137, 193–201.

Truong Xuan Nguyen, Ye-Su Song1 and Sung Min Park, 2012, Haploid plant production through anther culture in Day-neutral strawberry (*Fragaria x ananassa* Buch) cv. Albion. *J. ISSAAS Vol. 18, No. 1: 173-184.*

Tsay, H.S. and C. Y. Su. 1985. Anther culture of papaya (Carica papaya L). Plant Cell Rep4(1): 28-30. (SCI)

Verdoodt, L., Van Haute, A., Goderis, I.J., De Witte, K., Keulemans, J. and Broothaerts, W.

Viloria, Z., D.L. Drouillard, J.H. Graham, and J.W. Grosser. 2004. Screening triploid hybrids of'Lakeland' limequat for resistance to citrus canker. Plant Disease. 88: 1056-1060.

Wang S, Zhang S, Wang W, Xiong X, Meng F, Cui X. Efficient targeted mutagenesis in potato by the CRISPR/Cas9 system. Plant Cell Rep. 2015; 34: 1473–1476. doi: 10.1007/s00299-015-1816-7.

Weising, K., Nybom, H., Wolff, K. and Kahl, G. 2005. *DNA fingerprinting in plants: principles, methods and applications*. Taylor and Francis/CRC press.

Xue GR and Niu JZ 1984. A study on the induction of apple pollen plants. *Acta Hort. Sin.* 11: 161–164.

Yamamoto, T., Kimura, T., Sawamura, Y., Kotobuki, K., Ban, Y., Hayashi, T. and Matsuta,N. 2001. SSR isolated from apple can identify polymorhism and genetic diversity in pear. *Theoretical and Applied Genetics* 102, 865–870.

Yamamoto, T., Kimura, T., Soejima, J., Sanada, T., Ban, Y. and Hayashi, T. 2004. Identification of quince varieties using SSR markers developed from pear and apple. *Breeding Science* 54, 239–244.

Zarsky, V., Tupy, J. and Vagera, J. 1986, In-vitro-culture of apple anther. In: Moet-Henessy *Conference Fruit Tree Biotechnology*, Moet Henessy, Paris, October, p. 75.

Zhang, Y. X., Boccon-Gibod, J. and Lespinasse,Y. 1987, Obtentiond'embryons de pommierapre' s culture d'anthe' res. C.R. *Acad. Sci. Paris Ser.* III 305: 443–448.

Zhang, Y.X., Lespinasse,Y. and Chevreau E., 1991, Induction of haploidy in fruit trees. *Acta*

Zou, C.J. and Li, P.F. 1981, Induction of pollen plants of grape (Vitis vinifera L.). Acta Bot. Sin. 1: 79–81.

Breeding of Perennial Horticultural Crops *Pages* **227–239**
Editors: S.K. Singh, V.B. Patel, A.K.Goswami, Jai Prakash and Chavlesh Kumar
Published by: **BIOTECH BOOKS, NEW DELHI**

***Chapter* 20**

Parent Selection Criteria and Shortening the Breeding Cycle

Awtar Singh

Division of Fruits and Horticultural Technology,
ICAR-Indian Institute of Agricultural Research, Pusa, New Delhi – 110 012

INTRODUCTION

Perennial horticultural crops play a crucial role in human diets and as a commercial commodity in trade. Genetic improvement in these crops is hampered by several limitations including large size of the plant, long juvenile phase, high heterozygosity, inbreeding depression and environmental problems (*e.g.* fruit drops due to natural calamities). Besides, many important genetic traits in these crops exhibit polygenic inheritance and thus making breeding programme complicated. The long time between generations and the high per-unit cost of testing, makes it imperative that established and sound principles of breeding be followed. Breeding fruit crops relies on knowledge of traits being selected for. These traits need to be identified and quantified in such a manner as to enable an investigation of the genetic control of their expression (inheritance), and genetic and environmental factors that influence their expression, because there is strong genotype x environment interaction in the expression of these traits.

The main goals of research on fruit breeding are: to obtain new varieties with a shorter juvenile non-fruiting period, an increased yield, a longer ripening season, regular bearing, seedlessness and improved external and internal quality of the fruits. Another important aim in fruit tree improvement research is to make available new scions and rootstocks selected for resistance or tolerance to biotic and abiotic stresses. Fruit species breeding is based on either conventional (hybridization, selection, mutation) or biotechnological methods employing embryo culture, regeneration from protoplasts, somatic hybridization, *in vitro* mutant selection, genetic transformation and haploid production. Using an integrated approach with both biotechnological tools and conventional breeding methods, it is possible to obtain good results in a short time (Germana, 2006).

Artificial hybridization is the deliberate crossing of selected parents (controlled pollination). There are specific methods for crossing different species, which differ according to factors including floral morphology, floral biology, genetic barriers, and environmental factors. Hybridization is the most common conventional method of generating a segregating population for selection in breeding flowering species. In some breeding programs, the hybrid (F_1) is the final product of plant breeding and in others, further generations are must for getting the full expression of genetic traits. Hybridization is done for specific purposes, within the general framework of generating variability. However, hybridization is used for many purposes in the fruit crops improvement.

Gene Transfer

Sometimes, only a specific gene (or a few) needs to be incorporated into an adapted cultivar. Crossing is used for the gene transfer process, followed by back crosses to retrieve the desirable genes of the adapted cultivar. The procedure of backcrossing is different for the traits controlled by dominant and recessive alleles.

Recombination

Genetically diverse parents may be crossed to recombine their desirable traits. The goal of recombination, which is a key basis of plant breeding, is to forge desirable linkage blocks. Examples are many scion and rootstock varieties in several fruit crops. In general, the complementary parents are used in the hybridization programme.

Break Undesirable Linkages

Whereas forging desirable linkage blocks is a primary goal of plant breeding, sometimes crossing is applied to breakage of undesirable linkages, because the crossing over taking place is responsible for the breaking of the linkage and there is larger segregation for the trait under consideration.

Heterosis

Specially developed parents are crossed in a predetermined fashion to capitalize on the phenomenon of heterosis for cultivar development.

Maintenance of Diversity in a Gene Pool

Plant breeders may use a strategy of introgression (crossing and backcrossing selected entries with desired traits into adapted stocks) and incorporation to develop dynamic gene pools for breeding programmes.

Genetic Analysis

Geneticists make planned crosses to study the inheritance and genetic behaviour of traits of interest. Different methods of hybridization like Diallel and Line x Tester or other mating designs are used to get an understanding of General Combining Ability and Specific Combining Ability.

Hybridization is a straightforward and conventional approach of combining different traits available in the germplasm. The desirable hybrid progeny is selected in the F_1 or later generations, either it can result in a new variety or the selected genotype can serve as a breeding line in future breeding programmes. Most of the perennial fruit crops are propagated through vegetative means and if any improvement is achieved in the F_1, it is fixed by vegetative propagation. Therefore, only the F_1 is produced through hybridization and subsequent generations are produced through vegetative multiplication and this process fixes the genes in one generation only, whereas in seed propagated crops, many generations are required for fixing of the traits. The hybridization in perennial fruit crops involves: i). selection of parents, ii). Production of hybrid (F_1) progeny and iii). selection of superior genotypes. The fruit crops which reproduce by sexual means can only be hybridized and in others, other means of hybridization as somatic hybridization need to be applied.

Choice of Parents for Hybridization

The first step is to choose parents, usually with complementary strengths and weaknesses. For example, a very cold hardy grape with small clusters might be crossed with a large clustered variety lacking hardiness to combine both major qualities in the offspring. Selection of appropriate parents is critical as success of new varieties developed by hybridization depends on the parents used in the initial crossing programme. A breeder must discrete in selection of parents, making the right combination of crosses and raising a nursery by growing a large hybrid population, which requires land, expensive labour and other resources; in addition, evaluation of numerous hybrids is time consuming. According to the breeding objectives appropriate donor parents can be located from the vast collection. The value of F_1 progeny depends on the parents involved in the breeding hybridization programme. Therefore, the parents are selected based on their known performance, which depends on their general combining ability. It will be highly desirable to know the relative values of general combining ability and specific combining ability in the crop to be improved. If the GCA is high, a small number of parents is sufficient, but in case where SCA is important, many parents are to be used in the hybridization programme so that a large F_1 progeny can be produced. One or two cycles of inbreeding in parents will be desirable in hybridization programme. Traditionally, hybridization involves the selection of two parents likely to transmit high fruit quality and other important characteristics to the offspring (progeny). Complementary parents are normally chosen based on the desirable traits they contain and are likely to pass on to their offspring. There is a need to have a clear understanding in the choice of parents and of their positive and negative attributes through developing knowledge of the progeny. Normally cultivars with commercial potential are included in crosses whereas seedlings are used as the second parent. The ideal is to identify a range of parents to choose from with traits that will be in line with the latest market trends. Parents should not be used repeatedly in the breeding program, except in cases where certain traits are needed in a specific family and where selected seedlings are then 'back crossed' to one of the parents. The combinations of a parent can enhance the quality of the progeny or contribute to its faults. Studying variation within and between families therefore aids the selection of ideal parents and parental combinations. For transgressive segregation in some traits, preliminary hybridization is required to be taken before the full breeding programme. This is done through the quantification of variation of desirable and undesirable traits. In parent selection, the following points are considered:

1. The parents should be unidentical/complementary but reproductively compatible. Further, parents to be crossed are usually obtained from the same species or a close relative as the crosses among distant relatives are generally nonviable. Exceptions are there as in citrus where intergeneric hybridization is common. More over many intergeneric crosses can be attempted using distant hybridization and somatic hybridization and good examples are available in citrus. Many intergeneric crosses were attempted through these techniques for specific traits like greening bacterium, fungal and viral diseases and abiotic factors including salinity and water stress.
2. The parents together should supply the critical genes needed to accomplish the breeding objective. They should have complementary characteristics. Examples are again found in the hybrid programmes of citrus and trifoliate orange, where trifoliate orange contributes the genes for biotic stress control and citrus for abiotic stress control. If the complementary parents ae available, gene pyramiding is possible in one generation and that too through normal breeding procedures of sexual hybridization.
3. One parent is usually designated as female. Whereas some breeding methods may not require this designation, breeders usually select one parent to be a female and the other a male (pollen source). It is required as some of the traits are controlled by cytoplasmic genes, which exhibit maternal inheritance in general. In some cases, selected parents of cross-pollinated species may be isolated and allowed to randomly cross-pollinate each other (*e.g.* mango).
4. If some characters are governed by non-nuclear genes, then the male and female parents are selected with great care as cytoplasm is generally contributed by female parent only.
5. The female parent usually needs some special preparation. In complete flowers (with both male and female organs), the flowers of the parent selected to be female are prepared for hybridization by emasculation. Emasculation is eliminated in some crossing programs by taking advantage of male sterility (renders pollen sterile) when it occurs in the species, *e.g.* in Novel oranges and some Clementine mandarins. The emasculation is also not required in self-incompatible plant species like pummelo and only bagging and pollination will work. Pollen is often physically or manually transferred. Artificial hybridization often includes artificial pollination, whereby the breeder physically deposits pollen from the male parent onto the female stigma. The phenomenon of dichogamy,

non-synchronous flowering, *etc.*, need to be considered before planning the hybridization programme.

Hybridiation for Production of F_1 Progeny

The parents are crossed in the set combinations or in two way/reciprocal manners. The hybridization programme is followed as per the anthesis, dehiscence and receptivity of the flower. The emasculation is done at proper time or just before the opening of the flower with a minimum disturbance to stigma. The opened and immature flowers must be removed. After emasculation, the flower (s) is covered with a butter paper bag. In the next morning, the pollination is done and again the flowers are covered. After the fruit set, the bags are removed. The fruits are harvested at the proper maturity stage and the seeds are sown to raise the F_1 progeny. Anthesis, dehiscence and receptivity of the flower are the important criteria for deciding the hybridization programme. The process of emasculation can be avoided if the parents are self-incompatible and only bagging can serve the purpose.

Selection among the F_1 Families

Following hybridization, large segregation takes place in the F_1 progeny itself due to the high heterozygosity of the parents and selection for the desired genotypes can be practiced in this generation. Use of molecular and morphological markers, if available, significantly reduces progeny population to be planted for evaluation. Undesirable seedlings are discarded based on set pre-selection criteria. The hybrid seedlings are raised and planted in the field as such or they are grafted on the rootstocks or on the grown-up plants for flowering and fruiting. The hybrid plants are selected for the desirable attributes and selected plants are multiplied for multi-location testing. Progeny selection for the traits of interest should be carried out by the breeder with most care since some of the traits like fruit size, rind and juice colour, ripening date, sugars and acidity in the juice may be influenced by the environment. Knowledge on the inter-relation and inheritance of various fruit and plant traits is useful in making discrete selection of appropriate genotypes in the segregating population. Once superior selections are made, they must be multiplied clonally (so that the genetic integrity of the selected progeny is not lost), and tested against ruling variety in a replicated trial before identifying for release. Multi-location evaluation is preferred before recommending for commercial cultivation. Vegetative propagation allows for simultaneous testing of seedlings as well as their maintenance in the field.

Following hybridization, many varieties of scions and rootstocks in citrus, apple, pear, plum, mango, guava and other fruits have been developed and released all over the world and are being cultivated commercially. In some fruits, inter-specific hybridization is also possible like many varieties of apple, strawberry and citrus are inter-specific varieties. In case of *Citrus*, many intergeneric hybrids were also produced and are being used commercially, *e.g.* Citranges (*Citrus sinensis x Poncirus trifoliata*) and Citrumelo (*Citrus x paradisi* Macf. x *P. trifoliata*).

Flower and Flowering Issues in Hybridization

The flower has a central role in hybridization. The success of a crossing program depends on the condition of the flower regarding its overall health, readiness or receptiveness to pollination, maturity, and other factors. The actual technique of crossing depends on floral biology (time of pollen shedding, complete or incomplete flower, self- or cross-pollinated, flower size and shape).

Flower Health and Induction

It is important that plants that are to be crossed be in excellent health and be properly developed. This is especially so when flowers are to be manually emasculated. Once successfully crossed, adequate amount of seed should be obtained for planting the first generation. The parents to be mated should receive proper lighting, moisture supply, temperature, nutrition, and protection from pests. In some species, a special temperature treatment (vernalization) is required for flower induction. Furthermore, temperature affects the pollen shedding in flowers. Consequently, extreme temperatures may cause inadequate amounts of pollen to be shed for successful artificial pollination. Pollen quantity and quality are influenced by the relative humidity of the growing environment. Extreme moisture conditions should be avoided. Parents should be fertilized with the proper amounts of nitrogen, phosphorus, and potassium for vigorous plant growth to develop an adequate number of healthy flowers.

Synchronization of Flowering

In artificial pollination, the breeder should be familiar enough with the species to know its flowering habits regarding time from planting to flowering, duration of flowering, mechanisms of natural pollen dehiscence and fertilization, and time of peak pollen production, to take advantage of the window of opportunity of anthesis (pollen shed) for best crossing outcomes. To ensure that parents in a crossing program will have flowers at the same time, some adjustments must be made in crossing programme or pollen storage must be done, particularly in trifoliate orange, which flowers early or late in the season. Some adjustments in flowering time can be induced artificially by giving some abiotic stress as these can advance or delay the initiation of flowering.

Selecting Female Parents and Suitable Flowers

After selecting parents for a cross, it is necessary in artificial crosses to designate one parent as female (as previously stated), as well as identify which type of flowers on the parent would be most desirable to cross. Because the pollen or male gamete is practically without cytoplasm, and because certain genes occur in the extranuclear genome (such as CMS), it is critical that parents to be treated as female plants are selected judiciously. Some markers may be used to distinguish between selfed and hybrid seed on the female plant. For example, trifoliate leaf character in crosses of citrus and trifoliate orange. Another critical aspect of flower physiology is the age of the flower when it is most receptive to pollination. The breeder usually determines the optimal stage of flower maturity by examining its physical appearance. Telltale signs are variable among species. Usually, fully opened flowers would have already been pollinated by undesirable pollen. Furthermore, flowers in the same inflorescence have different maturity levels. The first inflorescence is more suitable for crossing than later ones; also, flowers at the base and middle of the inflorescence give better results than those at the top. Reciprocal crosses can help in deciding the order of parents *i.e.* female or male parent.

Factors to Consider for Success for Hybridization

It is critical to know the duration of stigma receptivity and pollen viability. The maximum time between emasculation and pollination that can be tolerated varies among species. It is convenient to emasculate flowers and pollinate later, either during the same day or even later. The caution to observe is that prolonged delay between the two operations increases the chance of contamination from undesirable pollen. To reduce this risk, emasculated flowers may be covered with bags (*e.g.*, glassine bag, cloth bag). Pollen quality and quantity varies with the weather and time of day. Because emasculation is done before anthers are mature, pollination is done the next day or 2–3 days later, when the stigma is receptive. Before making crosses, the pollen viability needs to be estimated, particularly, when stored pollen is to be used in the crossing programmes.

Breeding Strategy

Establishment of a high quality, but genetically variable foundation stock is the first step in the breeding process. From this stock, selections are made and used as next generation parents, enabling genetic improvement by selecting and mating genetically superior trees in each generation. Hybridization among the best selections is performed, followed by the selection of superior offspring for use as cultivars and as parents of the next generation. Parents of the new generation should always be selected offspring from the last cross to produce the advanced cycle. Once high-quality breeding families have been established, breeders are in a much better position to be able to develop new cultivars and to meet consumer demands. It is essential that breeding families be large enough to permit the combination of superior traits and to increase the possibility of generating individuals of superior quality.

Heritability and Variation Estimates

Heritability determines the degree of resemblance between relatives and expresses the extent to which the phenotypes are determined by the genes transmitted from the parents. Knowledge about how characteristics are inherited will increase breeding efficiency. The most important function of heritability is therefore its predictive role in selection. In addition to heritability estimates, some further evidence on the nature of the genetic make-up in traits can be ascribed to the effect of genes, *i.e.* major effects (qualitative) or minor interactive genes of quantitative nature. The latter implies that the trait is then seen as a complex, genetically determined trait. Most of the traits that fruit breeders are interested in are controlled by more than one gene. The quantitative nature of traits is complicated by environmental factors such as temperature, day length, drought and nutrient availability and their interactions. Progress in breeding programmes is conditioned by the magnitude and the nature of the genetic and non-genetic variation in the traits. A study concerning quantitative genetics includes statistical analyses of data and provides a scientific base for modern fruit breeding. Quantitative genetics helps in the understanding of how traits are passed from parents to offspring and in explaining the influence of the environment on the expression of the traits. Offspring performances can be predicted by the quantification of variation, thereby maximizing the likelihood of obtaining the desired characteristics. The breeder must not only understand whether the trait sought after is qualitative or quantitative, but also how the variation within a quantitatively determined trait is defined and how large the environmental impact on that character is. Thus, the variation of specific traits needs to be quantified to determine appropriate breeding strategies for family and breeding stock improvement. Now a days software is available for statistical analysis and predictability of results. Accurate heritability estimates can be determined following suitable mating designs as Diallel and Line x Tester design.

Marker-Assisted Selection

Molecular markers are used as tags for specific genes located near them. In modern breeding programs, markers are used to screen seedlings for the presence or absence of genes of interest. Markers are necessary to detect seedlings in progenies that carry more than one resistance gene and to pyramidize various genes for a specific type of resistance in one individual. With the help of these markers, desirable hybrids can be identified at an early stage and at least putative desirable hybrids can be selected and there is less load in the final selection population of hybrids. The final field evaluation is always compulsory in the field.

Somatic Hybridization

The hybridization is season dependent and its applications are possible in sexually compatible parents. Somatic hybridization offers the possibility of crossing sexually-incompatible parents also and therefore offers great scope for improvement. It can also combine the characters of both the parents which are governed even by non-nuclear genes. Somatic hybridization via protoplast fusion has become a relevant biotechnological technique that facilitates ploidy manipulation in efforts to develop improved plant cultivars. Initial successes have been primarily to access disease resistance genes from related wild species that are difficult or impossible to hybridize by conventional breeding methods. Although not apparent at the onset, the most important application of the technique may be the production of novel breeding parents for use in unique conventional breeding strategies. In the CREC citrus breeding program, somatic hybridization is having a profound impact on fresh citrus cultivar development through production of superior tetraploid breeding parents via symmetric fusion of elite diploid cultivars, followed by interploid crosses to generate seedless triploids for mandarin, grapefruit/pummelo and acid lemon/lime improvement. Due to the heterozygous nature of allotetraploids obtained via somatic fusion, greater genetic and phenotypic diversity can be found from interploidy conventional crosses using the somatic hybrid allotetraploid as compared to an autotetraploid parent (Grosser and Gmitter, 2011). The evaluation of numerous interploid crossing combinations can lead to the identification of superior allotetraploid parents that should contribute to rapid progress in the development of improved seedless fresh fruit citrus cultivars.

For rootstock improvement, feedback from the field evaluation of somatic hybrid rootstocks has identified superior parents for use in a unique tetraploid rootstock breeding program that maximizes genetic diversity in tetraploid zygotic progeny. Continued research should result in the release of improved, widely adapted "tetrazyg" rootstocks that exhibit all required disease tolerance/resistance and tree size control.

In woody plants, generally characterized by a long reproductive cycle, a high degree of heterozygosity, large size, and, sometimes, self-incompatibility, it is not possible to obtain haploidization through conventional methods. The absence of pure lines in woody plants makes genetic studies rather difficult to conduct. The great potential of employing haploidy, doubled haploidy and gametic embryogenesis in fruit crop breeding is clear (Germana, 2006). Haploids can improve the efficiency and the speed of the usually cumbersome, time-consuming, laborious and sometimes rather inefficient conventional breeding methods. Although *in vitro* culture of gametes is a standard tool for plant breeders in many crops, this has yet to be achieved in fruit crop breeding since the deployment of gametic embryogenesis in fruit crops improvement is still hampered by low frequencies of embryo induction, albinism, plant regeneration, plant survival and the genotype dependent response.

Strategies to Shorten the Breeding Cycle

Fruit trees have long generation intervals as depicted in Table 1. The juvenility is defined as the period elapsing between seed germination and first flowering of the seedling. During this time a plant cannot induce flower and the meristems must acquire reproductive competence becoming able to sense and respond to signals that induce flowering (Hanke *et al.*, 2007). The length of the juvenile period is important for fruit breeders from two perspectives: non-flowering seedlings may not be evaluated in terms of flowering and fruiting traits characteristic of the adult tree, and non-flowering seedlings are not suitable as crossing partners in backcross breeding programmes. Tree development from the juvenile to the adult stages requires several years and is a disadvantage for breeders. To accelerate the onset of flowers by reducing the juvenile phase is an important aim for research. Detailed understanding of genetic factors controlling the process of flower formation during tree development is basic to develop appropriate techniques to shorten the juvenile period. The acquisition of reproductive competence in plants is both under genetic and environmental control. Various strategies have been developed to shorten the generation time which can be adopted by the breeders based on certain modifications in the genetic and environmental factors (Zhang *et al.*, 2010). To be fully effective for rapid cultivar development, shortening the breeding cycle must be linked to a reduction in the number of breeding cycles. For tree fruits and nuts, the past few years have witnessed the initiation of a transition from traditional breeding techniques, based largely on phenotype, to genome-assisted breeding approaches, based on previously cataloged trait-locus associations, resulting in fewer breeding cycles. In addition, efficient genotype selection is most advantageous where a reduction in the length of breeding cycle is possible. The following strategies are generally used for shortening the breeding cycle:

i) **Selection of mutants from the induced variability**: In grapevine, a mutant dwarfed plant was identified that flowers early, takes less space in the green house and enables rapid cycling of generations (Boss and Thomas, 2002).

ii) **Grafting seedlings onto dwarfing rootstocks**: Grafting apple seedlings onto dwarfing rootstocks can stimulate earlier flowering, but juvenile phases often persist for more than two years (Flawchowsky *et al.*, 2009). The seedlings can also be grafted on the bearing plants, where they can flower early.

iii) **Microbudding**: Microbudding has reduced the total time taken from selection to final fruit setting and in crops like citrus. Following this technique, commercial yields can be obtained within two years from budding to harvesting.

iv) **Manipulation of cultural conditions:** In woody perennial plants, the length of the juvenile period is influenced by environment and is inversely correlated with vigor (Nocker and Gardiner, 2014). Accordingly, environmental conditions that reduce vigorous growth, such as mineral deficiency, low light, water stress, defoliation or cold stress, tend to delay the transition from the juvenile to adult phase, whereas the conditions that allow for vigorous growth can shorten the period of juvenility. In apple (*Malus x domestica*), where field-grown seedlings typically do not flower until they are at least 5 years old, plants can be promoted to the adult reproductive phase after as little as 10 months under optimal growth conditions. The apical portion of the plant, having attained the adult state, can be grafted to a rootstock for further growth and maintenance. A drawback of such an approach is that plants can grow very tall and become difficult to manage in a controlled environment or greenhouse setting (Zimmerman, *et al.*, 1985). The following situations are considered: (a) Embryo rescue or chemical treatment of the seed to break seed dormancy offers a shortcut to flowering. (b) Maintaining seedlings under optimal growth conditions can greatly abbreviate the juvenile phase. Typically, plants may become very tall, and the apex can be grafted to a rootstock for further growth and maintenance. (c) Chilling may be required for effective floral development.

v) **Use of virus induced gene silencing**: This technique exploits the enodogenous antiviral defence mechanism, which launches RNA-mediated attacks against viral genome. VIGS is being applied in apple breeding, where flowering is induced by inoculating seedlings with an Apple latent spherical virus (ALSV) that expresses a flowering gene (FT) from *Arabidopsis thaliana* (Yamagishi *et al.*, 2011). ALSV does not induce disease symptoms, does not spread horizontally to other trees and vertical transmission of the virus via seed occurs at very low rates. Since the use of VIGS does not result in a GM cultivar, it can be used as a breeding tool without resulting crops being labeled as GMOs.

vi) **Induction of flowering using genetic engineering**: Two ways to induce early flowering by transgenic approach are constitutive expression of genes promoting flowering, such as LFY, API and FT or the suppression of genes that delay flowering, such as TFLI have been utilized in apple (Hanke and Flachowsky, 2012) and both approaches induce early flowering in the transgenic apple plants.

The knowledge gained on flowering from the studies on *Arabidopsis thaliana* can be utilized in understanding similar processes in perennial crops/fruits, which generally have long generation time and are not amenable to genetic analysis. FT and/or SOCI genes are responsible for promoting floral meristem identity genes and floral organ identity genes. The LFY gene of *Arabidopsis* were expressed in citrus (Pena *et al.*, 2001) and many other genes from *Arabidopsis* and other sources were expressed in tree species. Flachowsky *et al.* (2007) created transgenic apples containing a gene from silver birch (*Betula pendula*) that induces early flowering. Using this rapid flowering trait, breeders are cycling through a generation each year, crossing early flowering transgenic offspring with elite cultivars or wild relatives with desirable traits like resistance to apple's primary pathogens, apple scab and fire blight (Flachowsky *et al.*, 2011). The same technology holds good promise in other perennial plants (*e.g.* citrus and pear) that have been genetically modified for precocious flowering (Endo *et al.*, 2005; Matsuda *et al.*, 2009) (Table 20.1).

Table 20.1: Length of the Juvenile Phase under Field Conditions for some Tree Fruit/Nut Crops

Crop	*Duration (Years)*	*Crop*	*Duration (Years)*
Almond	3–4	Lemon	5–10
Cherry	3–5	Mandarin	5–10
Peach	≥3	Tangerine	5–10
Pistachio	4–10	Pecan	≥5
Walnut	5–9	Apple and Pear	6–12
Orange	5–10	Avocado	15+

Conclusions

Breeding of perennial tree fruit crops is transitioning from lengthy, brute-force approaches, where large populations are generated, maintained for many years under field conditions and finally evaluated purely on phenotype, to highly intensive and selective approaches involving the genomics assisted identification of valuable germplasm, and cycling through generations as quickly as possible. To be more effective in rapid breeding, further advancement of genomics technologies and genetic maps should be accompanied by further reduction in generation time. The use of marker-assisted selection can help in shortening the breeding cycle and also permits the use of parents which were earlier thought non-crossable owing to the inheritance related problems for some of their desirable traits.

REFERENCES

Boss, P.K. and Thomas, M.R. 2002. Association of dwarfism and floral induction with a grape green revolution mutation. *Nature* **416**: 847-850.

Endo, T., Shimada, T., Fujii, H., Kobayshi, Y., Araki, T. and Omura, M. 2005. Ectopic expression of an FT homolog from citrus confers an early flowering phenotype on trifoliate orange (*Poncirus trifoliata* L. Raf.). *Transgenic Research* **14**: 703-712.

Flachowsky, H., Hanke, M.V., Peil, A., Strauss, S.H. and Fladung, M. 2009. A review on transgenic approaches to accelerate breeding of woody plants. *Plant Breeding* **128**: 217-226.

Flachowsky, H.,Le Roux, P.M., Peil, A., Patocchi, A., Richter, K. and Hanke, M.V. 2011. Application of a high-speed breeding technology to apple (*Malus x domestica*) based on transgenic early flowering plants and marker-assisted selection. *New Phytologist* **192**: 364-377.

Flachowsky, H., Peil, A., Sopanen, T., Elo, A and Hanke, M.V. 2007. Over-expression of BpMADS4 from silver birch (Betula pendula Roth.) induces flowering in apple (*Malus x domestica* Borkh.). *Plant Breeding* **126**: 137-145.

Germana,M.A. 2006. Doubled haploid production in fruit crops. *Plant Cell, Tissue and Organ Culture* **86**: 131–146.

Grosser, J.W. and Gmitter, F.G. Jr. 2011. Protoplast fusion for production of tetraploids and triploids: applications for scion and rootstock breeding in citrus. *Plant Cell Tiss. Organ Cult.* **104**: 343–357.

Hanke, M.V. and Flachowsky, H. 2012. Biotechnological approaches to shorten the juvenile period in fruit trees. *Acta Hort.* **929**: 309-314.

Hanke, M.V., Flachowsky, H., Peil, A. and Hattasch, C. 2007. No flower no fruit-genetic potentials to trigger flowering in fruit trees. *Genes, Genomes Genomics* **1**: 1-20.

Matsuda, N., Ikeda, K., Kurosaka, M., Takashina, T., Isuzugawa, K., Endo, T. and Omura, M. 2009. Early flowering phenotype in transgenic pears (*Pyrus communis* L.) expressing the CiFT gene. *Journal of the Japanese Society for Horticultural Science* **78**: 410-416.

Nocker, S. van and Gardiner, S.E. 2014. Breeding better cultivars, faster applications of new technologies for the rapid deployment of superior horticultural tree crops. *Hort. Res.* **1**: 1-8.

Pena, L., Martin-Trillo, J., Jaurej, J.A., Pina, L. and Martinez-Zapater, J.M. 2001. Constitutive expression of *Arabidopsis* LEAFY and APETALAI genes in citrus reduces their generation time. *Nat. Biotechnol.* **19**: 263-267.

Yamagishi, N., Sasaki, S., Yamagata, K., Komori, S., Nagase, M., Wada. M., Yamamoto, T. and Yoshikawa, N. 2011. Promotion o flowering and reduction of a regeneration time in apple seedlings by ectopical expression of the *Arabidopsis thaliana* FT gene using *Apple latent spherical virus* vector. *Plant Molecular Biology* **75**: 193-204.

Zhang, H., Harry, D.E., Ma, C., Yuceer, C., Hsu, C.Y., Vikram, V., Shevchenko, O., Etherington, E. and Strauss, S.H. 2010. Precocious flowering in trees: the FLOWERING LOCUS T gene as a research and breeding tool in *Populus*. J. Exp. Bot. **61**: 2549-2560.

Zimmerman, R.H., Hacket, W.P. and Pharis, R.P. 1985. Hormonal aspect of phase change and precocious flowering. *Encycl. Plant Physol.* **11**: 79-115.

Breeding of Perennial Horticultural Crops *Pages* **241–249**
Editors: S.K. Singh, V.B. Patel, A.K.Goswami, Jai Prakash and Chavlesh Kumar
Published by: **BIOTECH BOOKS, NEW DELHI**

Chapter **21**

Banana Improvement

Vikramaditya Pandey and S.K. Singh

Division of Horticultural Sciences,
Krishi Anisandhan Bhavan-II, Indian Council of Agricultural Research HQ,
Pusa, New Delhi – 110 012

INTRODUCTION

Banana is a monocotyledon, giant herb not a tree. Most of the cultivars are sterile, parthenocarpic clones. In the wild, no fruits without a seed (and only in last decade seed (and only in last decade for oranges and grapefruit) and Introduced with farming and domesticated, along with all other major crops and animals, 8000-10000 years ago. It is 1-7 year plantation,vegetatively propagated (exclusively) 85 per cent used as local staple 20-30kg fruit bunch >100Mt/year.

It is cultivated banana cultivated banana it is origin from two species in Asia: *Musa acuminata* (the A genome) and Musa balbisiana (B genome). Bananas and plantains are difficult crops to breed because most of the important and popular varieties are highly sterile and therefore do not produce seeds. Furthermore, compared to many other important food crops, there is a relative lack of knowledge on Musa genetics and cytogenetics. Despite these constraints, important progress has been made in the genetic improvement of Musa in recent years and new varieties are now becoming available from breeding programmes.

The varieties that are used for commercial or subsistence production are predominantly triploid (2n =3x =33) cultivars that are evolutionary derived from crosses within and between diverse accessions of two diploid ancestor species, *Musa acuminata* Colla (A genome) and *M. balbisiana* Colla (B genome). The existing cultivars are usually classified into three genome groups, *i.e.*, predominantly AAA for dessert and highland bananas, AAB for plantains, and ABB for cooking bananas.

Genetic Improvement of Banana

The main objective for the genetic improvement of banana is to breed sterile triploid hybrids through the recombination of fertile cultivars and species that meet farmers' needs and consumers' demands (Ortiz and Vuylsteke, 1996). *Musa* breeding started about one century ago in Trinidad, aiming to develop disease- and pest-resistant cultivars for the export trade. Even so, all banana export cultivars grown today are still selections from somatic mutants of the group Cavendish and have a very narrow genetic base (Perrier *et al.,* 2011). A broad-based, improved *Musa* germplasm with pest/disease resistance will be a major component to achieve sustainable production of this vegetatively propagated, perennial crop. Such germplasm can be produced through conventional cross-breeding, enhanced by the utilization of innovative methods for the introduction of additional genetic variation. Also, the increased use of molecular markers will accelerate the process of recurrent selection of improved *Musa* germplasm and, hence, facilitate the development of new hybrids.

Banana breeding remains a technically challenging endeavor where several obstacles impede rapid breeding progress such as low reproductive fertility, triploidy, slow propagation, space requirement (6 m^2/plant) and time (2 years from seed to seed) needed. These problems have been alleviated by screening *Musa* germplasm for fertility, manipulating ploidy and the intensive application of tissue culture techniques

Introduction of some cultivators of banana was made with resistance to biotic stresses *e.g.* Lady Finger (EC 160160) resistant to bunchy top virus introduced from Australia and is being evaluated at IIHR, Bangalore and TNAU, Coimbatore. Further, cultivars Naine MS (EC 27237) from France and Valery from West Indies were introduced for utilization in improvement programme (Singh and Rana, 1993).

Crossing Technique

Pollinations are generally carried out between 7.00 and 10.00 A.M. Undehisced anthers from male flowers are collected and twisted gently to force them to dehisce. Using a soft sable hair brush, pollen grains are prised out and smeared gently over the stigmatic surface of those female flowers that open on the day of pollination. Pollinated flowers are covered with a soft-cloth bag. Pollen fertility is an important factor for a clone to be used as the male parent in a hybridization programme. Male fertility in banana clones/hybrids is assessed by pollen output per anther, pollen viability and pollen grain size.

Hybridization

The majority of edible banana and plantain cultivars are triploids (2n =3x =33), justifying the aim of breeding programs to develop high yielding and disease resistant hybrids in the triploid background, usually through crosses between diploid and tetraploid accessions. We established that such crosses produce predominantly triploid progeny in the 4x - 2x direction and predominantly diploid progeny in the 2x - 4x direction. Although success in banana breeding depends on production of seed upon crossing, hybrids should be parthenocarpic to find acceptance as edible banana. In India, breeding work was started at Central Banana Research Station, Aduthurai (Tamil Nadu) in 1949 (Sathiamoorthy and Balamohan, 1993). Afterwards breeding programme was also initiated at TNAU, Coimbatore and Kerala Agricultural University, Trichur. Technique of hybridization in banana is different from other crops. Pollination is best carried out in the morning. The bunches of female parent are bagged at shooting and each successive hand is pollinated as it is exposed. At maturity and ripening the bunch is cut and seeds are extracted. Seeds are sown at once in the green house.

Evaluation of hybrid progenies from seedlings to harvest may not be the correct phase instead, evaluation of the same under next vegetative phase *i.e.*, sucker to harvest stage will be ideal as full expression of yield potential could be observed only in the second crop of the F_1 progeny. The first crop (seedling to harvest) takes more than 15-19 months, where most of the energy of the plants is needed for corm formation. The hybrid H-212, a diplod (AB), resembles Ney Poovan (AB) and possesses resistance to nematodes and sigatoka leaf spot. Hence, this hybrid has been tested with Ney Poovan and found to be promising.

Approaches in Banana Breeding

Hybridisation between triploid cultivars and dipoloid cultivars often results in many primary tetraploids. As tripoids are preferred over tetraploids because of superior vigour and yield potential, these primary tetrapoids are subjected to further crossing with potential diploids to develop secondary triploids, These tetraploids need to be further improved by crossing again with diploids to improve bunch weight. The aim is to produce seedless varieties, preferably in the triploid background that is common to the majority of existing cultivars. This usually involves crossing 3x cultivars to 2x accessions that are donors of resistance genes, selecting 4x and 2x primary hybrids from the 3x - 2x progenies, and crossing 4x - 2x hybrids to produce secondary 3x hybrids.3n x 2 n superior diploid= yielding tetraploids, 4 n bred tetraploids hybrids x 2n superior diploids producing 'Natural triploids'. 2n meiotic restitution clones x 2n superior diploids producing 'Natural triploids. In many banana growing countries, initially wild diploid bananas (AA) were utilized as male parent and as a result, the resultant tetraploids had inherited many undesirable traits. Hence, it has been felt by banana breeders that the primary objective is to synthesize a good male parent. An ideal male parent must be highly resistant to Panama and Sigatoka diseases, must have vertical and compact bunch and fruits as large as the diploidy can allow and must be parthenocarpic having sufficient pollen to permit its use as a male parent.

Musa acuminata subsp. *burmannica* and its hybrids offer a good source of resistance to black Sigatoka. One such diploid developed in Honduras is SH 2989. Other male diploids worthy to be mentioned are SH 3142 for nematode resistance and SH 3176 evolved through multiple crosses for resistance to Black Sigatoka with desirable horticultural traits.

Since 1971, extensive inter-diploid crosses were made to synthesize new diploid forms at the Tamil Nadu Agricultural University, Coimbatore using the following parents:Matti (AA) is a diploid cultivar commercially grown in the southern most part of India. It exhibits a strong resistance to Sigatoka disease but is highly susceptible to nematodes. Its bunches weigh 12 to 19kg with 9 to 10 hands containing fairly long fingers. It sets seeds when pollinated, though it is highly male sterile. This cultivar is extensively used as female parent in the diploid breeding programmes. *M. acuminata* subsp. *burmannica* has been shown to have resistance to fusarium wilt Races 1 and 2, sigatoka diseases and nematodes. Other diploid clones involved in the diploid male parents synthesis at Coimbatore are the indigenous cultivars Anaikomban (AA) and Namarai (AA). Anaikomban is resistant to nematodes and fusarium wilt but susceptible to yellow sigatoka. It has long fingers (15 to 18 cm) and usually produces a smaller bunch weighing 6 to 8 kg. Namarai is a small slender plant, grown in Pulney and Sirumalai hills of Tamil Nadu. With small fruits having piquant flavor and pleasant acid sweet taste. It has very short pedicel. It is susceptible to both Sigatoka disease and nematodes but no incidence of Panama disease is known so far. The introduced diploids are Pisang lilin (AA) and Tongat (AA), known for their resistance to Panama disease and nematodes. Many synthetic hybrids (diploids) have been developed which have good horticultural characters including resistance to Sigatoka, Panama wilt and burrowing nematodes. These hybrids are now used as the male parents to cross with local triploid varieties or inter crossed to synthesise new triploid hybrids. It is a Pome group of banana of the genome AAB and closely resembles Virupakshi (AAB), a pome type banana popular in the hills of Tamil Nadu. Presently three pre-release cultures *viz.*, H.96/7 (akin to Karpooravalli)

At Kerala Agricultural University, two hybrids *viz.*, BRS-1 (Agniswar x Pisang lilin) and BRS -2 (Vannan x Pisang lillin) have been developed. BRS -1 (AAB) is 100 days earlier than Rasthali with significant differences in bunch weight. It has been released for homestead cultivation in Kerala, as it is resistant to sigatoka leaf spot. BRS-2 (AAB) is a medium statured hybrid, tolerant to leaf spot and panama disease, rhizome weevil and nematodes. The average bunch weight is 14 kg with 8 hands and 118 fruits in crop duration of 314 days.

Because of biotic stresses such as black Sigatoka (caused by Mycosphaerella fjiensis Morelet), considerable efforts have been made to breed for resistance, as well as to broaden the genetic base of these crops. PITA-9: A Black Sigatoka Resistant (BSR) hybrid from the "False Horn" plantain, a tetraploid hybrid having black Sigatoka resistance has been developed at International Institute of Tropical Agriculture (IITA), Nigeria.'BITA-3' is a tetraploid starchy banana hybrid with low partial resistance to black Sigatoka disease developed at IITA High Rainfall Station in Onne (Southeastern Nigeria), where both (Banana streak virus) and cucumber mosaic virus (CMV) have been observed. 'BITA-3' is a hybrid from the interspecific cross 'Laknau' x 'Taju Lagada', 'Laknau' is a female –fertile AAB starchy banana that closely resembles plantains. 'Taju Lagada' is an AA diploid Banana having a long bunch with many hands. BITA-3' produces heavy bunches.

Banana breeding programme has been taken up in Honduras by the Fundacion Hondurena De Investigation Agricola (FHIA) with the aim of developing superior diploid plantations combining desirable agronomic traits with resistance, which is then used for production of primary tetraploids. This organization has developed many FHIA hybrids, which possess resistance to nematodes, Fusarium wilt *etc.* Introduction and testing of these hybrids in India in various centers revealed superior performance of FHIA-1, FHIA-3, FHIA-21 and FHIA-25.

Mutation Breeding

Bud mutation in Indian banana is very common perhaps due to spontaneous rearrangement of chromosomes in somatic meristem and structural re-assortment. A great majority of edible bananas are triploids, a condition that interferes with normal equilibrium of plants and may provide the requisite stimulus to structural rearrangement of chromosomes, leading ultimately to the evolution of a new gene complex. Several natural sports of well established commercial clones have been recognized *e.g.*) High gate (AAA) is a semi-dwarf mutant of Gros Michel (AAA), Motta Poovan (AAB) is a sport of Poovan (AAB), Ayiranka Rasthali a sport of Rasthali (or Silk), Barhari Malbhog is a sport of Malbhog, Krishna Vazhai is a natural mutant of Virupakshi (or Pome), and Sambrani Monthan (ABB), a mutant of Monthan (ABB). In Nendran, more than six mutants have been recognized. One of these, Moongil, has undergone such a radical change that there is no male phase and a bunch has only one or two hands with biggest size fruits. Attu Nendran, Nana Nendran, Myndoli, Velathan and Nenu Nendran are a few mutants which have been selected for one or the other desirable character. Similarly, Ambalakadali and Erachi vazhai are mutants of Red Banana. The Kunnan variety of Malabar has provided a few mutants known as Thattilla Kunnan (male phase absent), Veneetu Kunnan, Adakka Kunnan and Thaen Kunnan. From cv.Monthan, Sambal Monthan, Nalla Bontha Batheesa, Sambrani Monthan, Pidi Monthan and Thellatti Bontha have been recognized as sports. At INIVIT, Cuba induced mutations of ABB cooking banana, Burrow Cemsa, was obtained (Rodriguez Nodals *et al.*, 1992). At TBRI, Taiwan, Tai Chiao 1 and GCTV, triploid bananas with Fusarium wilt resistance are obtained as a result of clonal variation of AAA Cavendish banana. The early flowering FATOM 1 was developed as a result of in vitro gamma irradiated meristem culture of cv.Grand Naine has been released in Malaysia.

Improvement of commercial triploids like cvs. Robusta and Rasthali through sexual hybridisation is very difficult as these are female sterile. Therefore, mutation breeding was initiated in 1995 with cultivars like Robusta (AAA) and Rasthali (Silk –AAB).

Biotechnology

Plant tissue culture and molecular biology techniques are applied to enhance the handling and improvement of banana. Important application of a cell biology are micro propagation for rapid multiplication and germplasm exchange, embryo culture/rescue for *in vitro* seed germination, cryopreservation of germplasm and genome manipulation through genetic engineering using cell suspensions or protoplast culture. Although, Vylsteke *et al.* (1996) reported that somaclonal variation through micropropagation is of limited use in plantain breeding, it has been successfully applied in Taiwan for the development of improved Cavendish bananacultivars with resistance to Fusarium wilt and acceptable fruit quality (Hwang 1991, Hwang and Ko, 1989).

An overall similarity of 63 per cent was found among the diploid cultivars. Greater diversity was seen within the AA group than within the AB group. Many natural mutants, such as 'Ambala Kadali', 'Erachi Vazhai', 'Thattila Kunnan', 'Vennetu Kunnan' and 'Rasa Kadali' showed significant variation from their parental genotypes

Molecular Markers

For example, DNA markers were useful for identifying putative diploid ancestors of both triploid Cavendish and Gros Michel cultivars (Raboin *et al.*, 2005). Furthermore, the level of genetic divergence within morpho-groups based on bunch type suggests that True Horn plantains are derived from False Horn plantains which in turn are derived from French plantains (H.K. Crouch *et al.*, 2000). However, most of the DNA marker-based research in *Musa* relied on the A genome-derived microsatellite markers, despite the importance of the B genome in plantains and cooking bananas. Hence, Buhariwalla *et al.* (2005) targeted the development of B genome microsatellite markers in *Musa*. These microsatellite markers can help fill an important gap for diversity assessment and linkage mapping in this crop. Recently, Youssef *et al.* (2011) showed the value of the sequence-related amplified polymorphism (SRAP) technique for analyzing genetic variation and relationships among *Musa* accessions. SRAP has several advantages over other molecular marker systems such as simplicity, reasonable throughput rate, disclosure of numerous co-dominant markers, ease of isolation of bands for sequencing, and, last but not least, targeting of open reading frames.

RAPD fragments specific for the A and B genomes of Musa have been identified, isolated and cloned. The RAPD markers are useful in preliminary genome classification, but they are not subspecies specific markers. Hence, more robust markers were developed using AFLP analysis of total DNA and PCR-RFLP of the ribosomal DNA internal transcribed spacer (ITS) regions, allowing for quantitative determination of genome composition and inferences on phylogenetic relationships. Thus, we established that *M. acuminata* may comprise 3 genetic subspecies (*burmannica, malaccensis*, and *microcarpa*) while *M. balbisiana* contains 2 forms (Singapuri, I-63), suggesting that there might be at least 3 A genomes and 2 B genomes. Furthermore, it seems that the A genomes in the plantains come from the subspecies *microcarpa* while the B genome comes from the I-63 form of *M. balbisiana*. Of practical importance to breeding is that the genome composition of all breeding lines and native accessions available at IITA has now been determined. The development of molecular markers to assist selection of traits that are expressed late in the plant cycle was pursued. A large mapping population of approximately 400 individuals was developed from the cross *M.a. subsp. microcarpa* 'Borneo' x M.a. subsp. Banksii. 'SF247'. So far, a putative AFLP marker for fruit parthenocarpy (distinguishing between low and high pulp content) and seedlessness was identified, subject to confirmation when phenotyping for three crop cycles is completed at Onne and at Ibadan.

In gene transfer methods, Sagi *et al.* (1995), from Katholieke University, Leuven. Belgium reported that the transgenic triploid cooking banana showing transient expression of GUS marker gene in pot growing in the green house from DNA particle bombardment on ABB cooking banana. The molecular markers are providing tools for phylogenetic investigations and cultivar identification, basic genetic research, marker assisted selection and diagnostics in pathogen identification (Table 21.1).

Table 21.1: Source of Resistance

Clone/Cultivars	*Biotic and Abiotic Stress*
Musa balbisiana	Drought
Calcutta-4	Black sigatoka
Pisang Lilin	Panama wilt (Race1)
SH3142 (Diploid hybrid)	Race 1 of Fusarium
Musa acuminata sp malaccensis	Race 1 and Race 2 of Fusarium
Musa acuminata sp burmannica	Bacterial wilt race 2, Moko disease
Pisang Jari Buaya (PJB)	Burrowing nematode
Tongat and Anaikomban	Nematodes

The genepool has the diversity there which can meet these challenges which can meet these challenges:

- ☆ Breeders need to get better and faster Breeders need to get better and faster
- ☆ Banana, has extra challenges – Staple food Staple food – Major income source in many communities – Sterile plant Sterile plant. Breeders need to get better and faster Breeders need to get better and faster
- ☆ Banana, has extra challenges – Staple food Staple food – Major income source in many communities – Sterile plant

Transgenics

It is essential to clearly define realistic targets for genetic engineering that can be included within innovative *Musa* breeding programs. Crouch *et al.* (1998) suggest putting unique, important genes into elite germplasm for subsequent use in conventional *Musa* breeding. Host-plant resistance to burrowing nematode (*Radopholus similis*) (Atkinson *et al.*, 2004), Xanthomonas wilt (Tripathi *et al.*, 2010), and Fusarium wilt (Paul *et al.*, 2011), as well as enhancing pro-vitamin A content (http://www.ausfoodnews.com.au/2010/03/18/gm-bananas-first-harvest-shows-promise.html) are among the most promising transgenically introduced traits for bananas today. Dessert bananas are also candidates for delivering a bite-sized vaccine for hepatitis B virus (HBV) to millions of people in the developing world (Kumar *et al.*, 2005). Plant-based production of an oral HBV has economic and other advantages over the existing vaccine delivery by injection. However, public acceptance of transgenic *Musa* cultivars will be the most important hurdle to overcome. Nonetheless, a recent survey in Uganda shows a high willingness to purchase transgenic banana among consumers (Kikilwe *et al.*, 2011). Furthermore,

Falck-Zepeda *et al.* (2008) indicate that Uganda foregoes potential annual benefits ranging between US$ 179 million to US$ 365 million by delaying the approval of transgenic bananas.

Genomics-Led Breeding

DNA markers have a wide range of applications for improving the efficiency of *Musa* breeding (J.H. Crouch *et al.*, 2000). Oligonucleotide primers were constructed from genomic regions bordering *Musa* microsatellites in order to detect polymorphisms in the number of simple sequence repeats (SSR; Crouch *et al.*, 1997). They have been used to distinguish diverse germplasm and full-sib hybrids in both tetraploid and triploid plantain hybrid families (Ortiz, 2011 and references therein). They have also been used for demonstrating the occurrence of recombination during the formation of 2*n* gametes from triploid plantain and the heterozygosity of a banana accession commonly used as a true breeding line in genetic research and *Musa* improvement programs (H.K. Crouch *et al.*, 1998). The analysis of variable numbers of tandem repeats (VNTR) of microsatellites was also able to identify those hybrids which may be most useful in subsequent breeding of secondary triploid hybrids (H.K. Crouch *et al.*, 1999). J.H. Crouch *et al.* (1999) found that amplified fragment length polymorphisms (AFLP), random amplified polymorphic DNA (RAPD) and VNTR detected a high level of polymorphism between parents and within their offspring. AFLP assays had by far the highest multiplex ratio whereas VNTR analysis detected the highest levels of polymorphism. However, Tenkouano *et al.* (1998b) indicated that hybrid performance was not significantly correlated with genetic similarity indices based on both pedigree and molecular data. The genetic differences among offspring families were essentially due to differences in additive effects of the parents. Further research shows that DNA markers can provide a more accurate description of genetic relatedness but pedigree-based analysis may be more useful for the selection of prospective parental combinations in *Musa* breeding (Tenkouano *et al.*, 1999a). Tenkouano *et al.* (1999b) noticed that prediction of offspring's bunch weight was best using genealogical distance and equal parental contribution, whereas predicted fruit size was most accurate when DNA marker data were used considering an unequal parental contribution.

Candidate markers for important attributes have been identified. For example, RAPD markers has been successfully used to distinguish A and B genome sequences in *Musa* (Pillay *et al.*, 2000) and for linking them to Fusarium wilt resistance (Javed *et al.*, 2004; Zambrano *et al.*, 2007), whereas an AFLP fragment was able to identify a dwarf variant (Engelborghs *et al.*, 2000). Likewise, one restriction fragment length polymorphism (RFLP) was strongly associated to host-plant resistance to black leaf streak and a second RFLP accounting for this host-plant resistance at a lower significance level was also mapped onto another linkage group (Carreel *et al.*, 1999). Lheureux *et al.* (2003) also found AFLP linked to the BSV expressed locus (*BEL*). As noted by Geere *et al.* (2005), some *Musa* species and cultivars contain a diverse array of endogeneous viruses.

Diversity Arrays Technology (DArT) markers displayed high polymorphism in *Musa* (Risterucci *et al.*, 2009). A total of 836 DArT markers were identified and used for genotyping. DArT markers reveal genetic relationships consistent with those provided by the other DNA markers but at a significantly higher resolution and speed, and for reduced costs. DArT and SSR markers have been further used for developing the first saturated map of *M. acuminata* (Hippolyte *et al.*, 2010). This map contains 489 markers (167 SSRs and 322 DArTs) covering 1197 cM in the 11 linkage groups. Ecotilling also seems to provide a robust and accurate platform for the discovery of *Musa* polymorphisms (Till *et al.*, 2010). Ecotilling can be therefore considered as tool for selecting banana germplasm for further breeding.

Conclusion

The success of breeding bananas should be measured by the extent to which new *Musa* cultivars are used profitably and sustainably by farmers. Access to bred germplasm is therefore very important, especially in a crop with a few active breeding programs but with a great demand from farmers needing improved germplasm worldwide. Breeding programs should be also monitoring changes in the production systems and markets to target new traits according to end-user needs. New challenges include epidemics, drought and temperature extremes due to climate change, which will affect banana farming. For example, gradual increases of temperature, intense precipitation followed by dry periods and high foliar humidity, favor Mycosphaerella leaf spot incidence and severity (Torrado-Jaime and Castaño Zapata, 2008). The genetic knowledge and the genomic tools accumulated in the last two decades will facilitate and speed breeding host-plant resistance to pests affecting the *Musa* crop. Changes in annual rainfall patterns and water scarcity also affect banana yields. The B genome seems to be a valuable source of alleles for adapting cooking bananas to mild drought environments (Ekanayake *et al.*, 1994, 1998). Genomics tools can also assist for identifying *Musa* putative drought- or temperature-responsive genes (Davey *et al.*, 2009; Santos *et al.*, 2009). Likewise, they will be useful for identifying plant stress-tolerant genes for further marker-aided breeding.

REFERENCES

Kumar, N. (2006). Problems and prospects of banana breeding in India. *J. Hort. Sci.*, 1(2): 77-94.

Pillay, Michael and Tenkouano, Abdou (2011). Banana Breeding: Progress and Challenges. CRC Press Taylor and Francis Group 6000 Broken Sound Parkway NW, Suite 300 Boca Raton, FL 33487-2742.

Pillay, M., A. Tenkouano, and J. Hartman. 2002. Bananas and Plantains: Future challenges in *Musa* breeding. Chapter 8. *In*: Kang M.S. (ed.). Crop improvement challenges in the twenty-first century. The Haworth Press, New York.

R. Ortiz (2013). Conventional Banana and Plantain Breeding. *Acta Horticulture*, 986.

Tenkouano, A. and Swennen, R. L. (2004). Progress in breeding and delivering improved plantain and banana to African farmers. *Chronica Horticulturae*, 44(1): 9-15.

Breeding of Perennial Horticultural Crops *Pages* **251–268**
Editors: S.K. Singh, V.B. Patel, A.K.Goswami, Jai Prakash and Chavlesh Kumar
Published by: **BIOTECH BOOKS, NEW DELHI**

***Chapter* 22**

Citrus Improvement

A.K. Dubey, Nimisha Sharma and R.M. Sharma

Division of Fruits and Horticultural Technology,
ICAR-Indian Institute of Agricultural Research, Pusa, New Delhi – 110 012

INTRODUCTION

Citrus, (oranges, mandarins, lemons, lime, grapefruits, tangerine *etc.*) is some of our planet's highest-yielding fruit crops, and distributed throughout the tropical and subtropical regions of the world. The top ten citrus-producing countries are China, India, Nigeria, Brazil, Mexico, USA, Spain, Egypt, Italy and Argentina (Zhang *et al.*, 2018). According to FAOSTAT, in 2014, the worldwide cultivation of citrus occupies more than 9.08 million hectares' areas and yielded more than 139.79 million tons, with a gross production value of 66.44 billion US$ dollars. For sustainable growth of citrus industry, attention is required to develop highly prized scion varieties as well as multi stresses tolerant rootstock hybrids. Various environmental stresses such as soil or water salinity, drought and the major biotic constrains include diseases such as CVC, citrus canker, leprosis, tristeza, black spot, sudden death, and more recently, huanglongbing, in addition to many pests that challenge the citriculture growth of citrus industry in many commercially citrus growing areas. These environmental factors deserve attention, which are associated to characteristics of individual citrus groups, expressing the necessity of better understanding to support future work on plant breeding, for instance, fruit quality (juice colour, seeedlessness, nutritional composition). In this crop, rootstocks play a preeminent role in improving yield, quality and protection against environmental stresses. Moreover, most of the environmental factors that limits the growth of citriculture may be combat through use of suitable resistant rootstocks. Successful scion and rootstock programme in citrus depend on the thorough knowledge of centre of origin/diversity, taxonomy, mechanism of evolving of species, cytology, breeding techniques and methods *etc.*

Origin

Name of the Species	*Origin Place*
Centre of origin	N E India and Northern Burma and Yunnan in south centre China (Recent study)
Citron (*Citrus medica*)	South China to India
Limes (*C. aurantifolia*)	East India
Lemon (*C. limon*)	Unknown (Cross citron x lime)
Sour orange (*C aurantium*)	South eastern Asia, India
Sweet orange	South China or Indonesia
Washington Navel (*C. sinensis*)	Brazil
Pummelo (*C. grandis*)	Malaysia and India
Grapefruit (*C. paradisii*)	West Indies
Mandarin (*C. reticulata*)	Indo-China

Taxonomy

The genus citrus and its close relative are member of the family rutaceae subfamily Aurantioidea. This sub family is of main concern because it represented 28 genera in the tribe citrae. Six of these genera like eremocitrus, microcitrus, citrus, poncirus, fortunella and clymenia comprise the true citrus fruits which are by and large evergreen trees, the leaves are simple and persistent except in poncirus where they are deciduous and trifoliate. Clymania is known only from the pecific island of New Ireland. It has not been hybridized with others (Figures 22.1 and 22.2).

Cytology

The basic chromosome number of *Citrus* has been recognized as x = 9 since the studies conducted in the 1920s (Frost 1925, Nakamura 1929), and this largely applies to species from the *Rutaceae* family. The great majority is diploid, only few polyploids have been identified in nature, including tetraploid Fortunella hindissi and Triphasis. The genome size was estimated as 0.76 or 0.82 μg DNA/2C, equivalent to 367 or 396 Mb/1C, in *C. sinensis* L. (Arumuganathan and Earle 1991), but varied among species, ranging from the largest citron (*C. medica* L., average value 398 Mb/haploid genome) to the smallest mandarins (*C. reticulata* Blanco, 360 Mb). The pummelo (*C. grandis* L.) had an intermediate genome size of 383 Mb (Ollitrault *et al.*, 1994)

History of Improvement

The pioneer systematic experiments planned by W.T. Swingle and H.J. Webber in 1892-93. Systematic work was planned and carried out by Swingle and Webber for diseases resistance in citrus. In India, attempts of Hybridization were made at Modibag garden, College of Agriculture in Poona 1911. After 1942, citrus hybridization work was carried out at Fruit Research Station, Kodur. Development of suitable rootstocks for sweet orange in Java was begun in 1928.

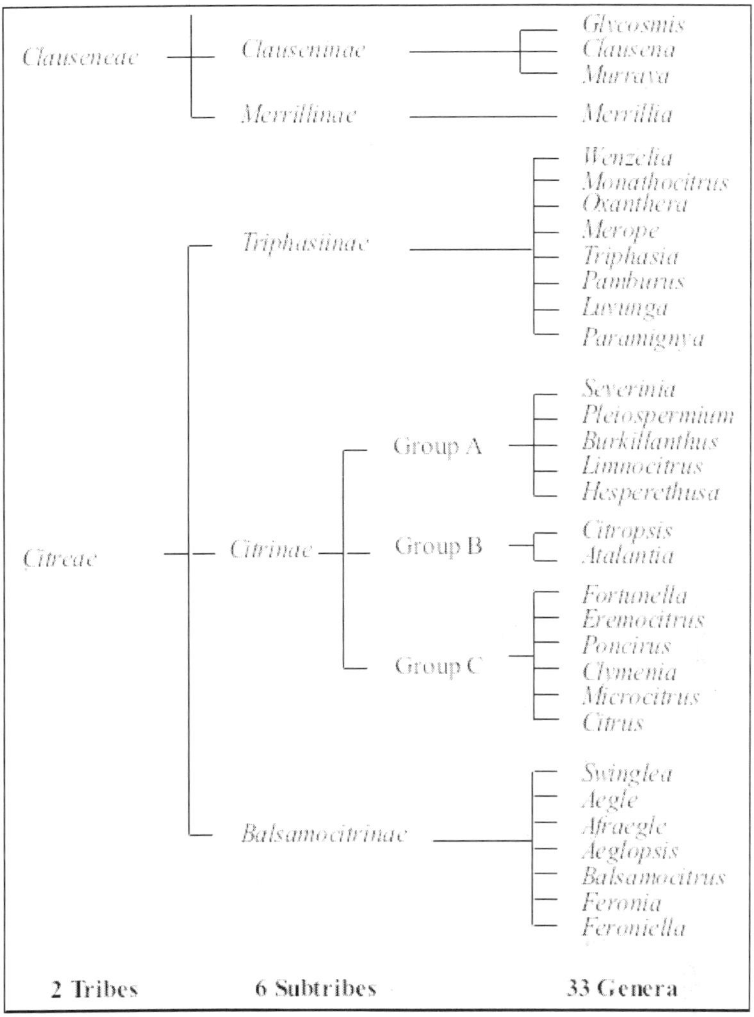

Figure 22.1: Aurantioideae Subfamily (after Swingle, 1967).

Achivements in Citrus Scion Varieties

Some Important Citrus Hybrids Developed from USDA

Tangelos

Samson Thoron Williams, Orlando (Tangerine x grapefruit)

Tangelolo: Wekiwa (Samson x grapefruit)

Tangor: Hybrid (Satsuma x Sweet orange)

Limequats; Eustis, Lkeland, Tavares (Maxican lime x kumquats)

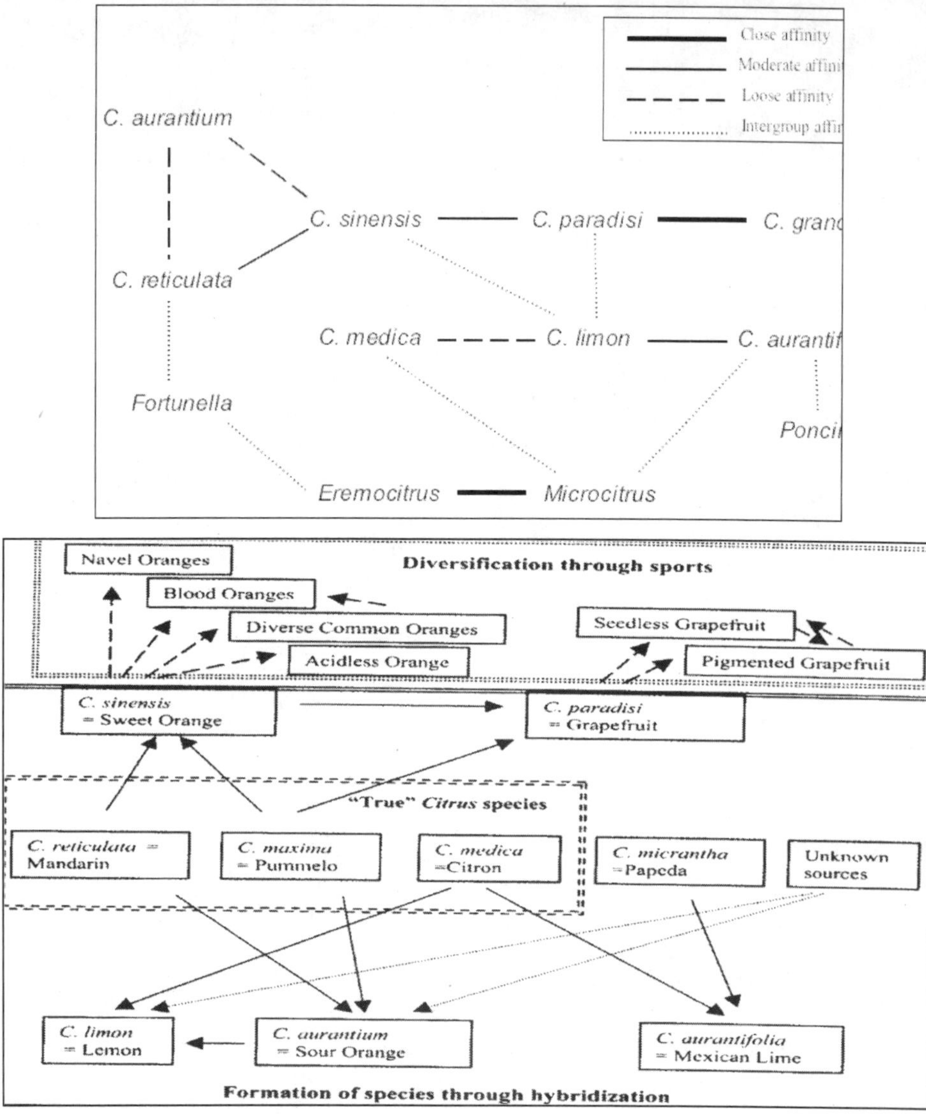

Figure 22.2: Affinity Relationships between some Citrus Species and Relatives after Swingle (1943) and Barrett and Rhodes (1976).

Mandarin

Robinson, Osceola, Lee, Nova, Bower (Clementine x Orlando); Page (Minneola x Clementine)

Hybrids of University of California

Tangelos: Pear tangelo (Imperial grpefruit x Willow leaf),

Tangors: Mency Tangor (Maltese Oval x Dancy), Frua (King x Dancy), Ruddy (King x Maltese Oval)

Mandarins: Kara (Satsuma x King) Kinnow (King x Willow leaf), Wilking (King x Willow leaf) Kency (King x Dancy), Honey mandarin (King x Willow leaf) Gold Nugget (Wilking x C. nobilisx Kincy (King x Dency tengerine)

Lemon: Owarban (Sarsuma x Lisbon)

Triploid: Thahiti (Not explained by Frost) Orangelo (Maltese oval x Marsh seedy), Gold Nugget Wilking (C. deliciousa willow leaf x C. nobilis King) x Kincy (King x Dancy tangerine)

Romelos: Chandler (Pommelo2240 x 2241)

Hybrids of Italy

Orange/Mandarin

Primosole (C. *unshiu* miho x C.reticulata carvathais), Sirio (C. *unshiu* miho x C. clementina Comune), Cami (C. clementina Comune x C. deliciosa Avana), Tacle (C. clementina Monreal x C. sinensis Tarocco), Clara (C. clementina Monreal x C. sinensis Tarocco).

Hybrids of Japan

Mandarin: Youkou (Kiyomi tangor (Miyagawa Satsuma x trovita ornage) x Ponkan mandarin), Hayaka (Imamura Satsuma x C. reticulata Nakano 3)

Tangor: Amakusa (Kiyomi tangor x okitsu wase) x Page orange)

Hybrids of Israel

Mandarin: Winola (Wilking x *C. nobilis*) x Minneola tangelo); Rishon (Wilking x C. reticulata Mikhal); Shani (Wilking x C. reticulata Mikhal)

Hybrids of India

Work at Kodur

Orange/Mandarin/Lemon

Hybrid: Sweet orange x pummelo; pumello x lemon, pumello x lime, lime x lemon, *C. sinensis* x Feronia limonia, Atlanta monophylla x Fortunella **(nothing significant achieved)**

Work at Akola

Acid Lime

N-52 (Acid lime x lemon or citron/kumquat/Rangpur lime)

Hybrid-2

Hybrid-4

Work at Bangalore and Chethali

Lime

ALH-77 (Kagzi lime x Nepali round lemon) resistant to bacterial canker

Work at IARI

Orangelo

CSH-11-11, CSH11-13, CSH-11-15, CSH-19-4 (Pemmelo white fleshed x sweet orange cv Mosambi)

Rootstock Hybrids

CRH-18-1, CRH-18-2, CRH-18-3, CRH-18-5, CRH-18-7 (Rough lemon x poncirus trifoliata cv Pomeroy), CRH-20-1 (small fruited mandarin x Troyer citrange).

Breeding Goals

- ☆ Scion- fruit color, size, shape, flavor, yield
- ☆ Disease resistance- CTV, Phytophthora, CVC, Alternaria, Scab, Greening, *etc.*

Modern Breeding Objective

- ☆ Vigor, yield potential, environmental stress tolerance
- ☆ Fruit type- mandarin, grapefruit, orange
- ☆ Fruit quality- flavor, size, seediness, appearance, shelf-life
- ☆ Salt and drought tolerance
- ☆ Cold and heat tolerance
- ☆ Fresh market fruit- size, sugars, low acid
- ☆ Increase emphasis on fruit quality and earliness for fresh market expansion

Procedures to be Adopted fro Varietal Development in Citrus

- ☆ Generation of more of hybrids through embryo rescue
- ☆ Hybridity confirmation by molecular markers (SSR)
- ☆ Selection based on stakeholder's needs and first evaluation of promising ones at experimental site.
- ☆ Second evaluation of selected promising hybrids in different citrus growing areas.
- ☆ Optimization of managemental practices and estimation of commercial value¸
- ☆ Production of saplings and commercial plantings

Breeding System

1. Introduction

2. Selection
3. Clonal selection
4. Cross-pollination- combine genes from different parents in hybrid progeny
5. Self-pollination- fix genes of interest in one line to stabilize phenotype (inbreeding)
6. Mutation- natural or induced genotypic modification
7. Polyploidy breeding
8. Biotechnological approaches

Introduction

This method involves the introduction of types from another region or country. It is usually attempted when the existing germplasm is felt unsuitable or lacks some desirable characters or merely enriches the existing germplasm. If introduction found suitable for commercialization, they can be recommended for cultivation. There are several cultivars introduced in India in different species such as in sweet orange, Jaffa, Valencia, Washington Navel, Pineaple, Early Gold, Hamlin, Westin, Itaborai *etc.* in Mandarin Kinnow, in tangerine: Dancy, WT Murcott, Daisy, in lemon: Eureka, Lisbon, in grapefruit: Duncan, Foster, Marsh Seedless, Redblush, Imperial *etc.*

Clonal Selection

A clone is a group of individuals resulting from sexual multiplication as originally defined by Webber (1903) or a clone is any plant variety (or strain), which is propagated vegetatively or asexually. A particular character difference may be due to a single gene difference or to several or many gene differences, on other hand one gene may affect several or many characters. A genetic character is a heritable characteristic of an organism or some part of an organism, such as particular shape of fruit or colour of rind or quality of fruit, which is more or less arbitrarily selected for special consideration. Clonal selection is easiest method for identification of improved cultivar. Clonal selection has been very effective within cultivar groups. This method involves the survey of different areas, collection of germplasm, raising of germplasm at a place, studying their performance for considerable time and finally selecting superior types (Table 22.1).

Table 22.1: Selection of Citrus Varieties from India

Crop	*Selection*	*Parent*	*Institution*
Orange	Yuvraj blood red	Blood Red	MAU, Parbhani
	Pusa Sharad	Malta	IARI, New Delhi
	Pusa Round	Malta	IARI, New Delhi
Mandarin	Madhukhed Seedless	Nagpur santra	MAU, Parbhani
Acid lime	Pramalini	Acid lime	MAU, Parbhani
	Vikram	Acid lime	MAU, Parbhani
	Jaydevi (PKM-1)	Acid lime	MAU, Parbhani
	Chakradhar lime	Kagzi lime	Periakulum, (T.N.)

Crop	Selection	Parent	Institution
	Tenali	Acid lime	Tirupati
	Saisarbati	Acid lime	Rahuri
	Pusa Udit	Kagzi lime	IARI, New Delhi
	Pusa Abhinav	Kagzi lime	IARI, New Delhi
Lemon	PAU selection	Lemon	PAU Ludhiana
	Dabwali baramasi	Lemon	PAU Ludhiana
	Pant Lemon	Lemon	Pant Nagar
	Kagzi Kalan	Lemon	Saharanpur

Hybridization

Hybridization followed to create new types by reshuffling the genes in the existing germplasm. Hybridization essentially involves the controlled crossing of two genetically different parents obtain a desirable type. In general, much of the hybridization in citrus has been interspecific rather than intraspecific.

Selection of Parents

For scion cultivar breeding, cultivar which are entirely zygotic are to be preferred as seed parents. The pummelo are such a group. If the available monoembryonic cultivars are not suitable as parent, cultivar, which produces some zygotic embryos, must be used. The good taste qualities of important characters of citrus fruits that can be used in breeding programme are as fallows. Outstanding easter principle, predominating sensation of sweetness bled with slightly perceptible tartness qualities of sweet orange and pleasant tang which is due to ester in loose skin orange are preferable characters to be incorporated in new hybrids. The peel oil in lemon and limes also gives desirable qualities in citrus improvement. Citron is a preferred parent to develop hybrids for preserve due to quality of firm pulp (Table 22.2).

Table 22.2: Some Important Citrus Genotypes for Breeding Programme

Genotypes	Positive Traits	Negative Traits
Scion improvement		
Pummelo	Deep red colour, highly nutritive with naringene and limonine, hardy plant	Thick peel, highly seedy with less juice content
Grapefruit	Seedlessness, deep red colour, nutritive with, naringene, limonine	Bitter juice, susceptible to sun scalding and canker
Lemon	Thornless, resistant to canker, summer month fruiting under subtropical region	Susceptible to measophyll collapse physiological disorder, less acidic and prone to fruit cracking
Lime	High citric acid, juicy, excellent fruit quality, tolerant to fruit cracking, highly polyembryonic	Highly susceptible to canker
Tangerine	Easy peeling, heavy yielding, bearer harvesting	Highly susceptible to granulation

Genotypes	Positive Traits	Negative Traits
Rootstock improvement		
Attani	Na and Cl excluder, dwarf	Highly susceptible to phytophthora
Rangpur lime	Cl excluder	Susceptible to Citrus Blight and citrus Sudden Death
Cleopatra mandarin	Cl excluder, good cold tolerance, unaffected by CTV and CVd, deep root system, is quite well adapted to dry conditions and drought tolerant	susceptible to citrus nematode and blight
Rough lemon	Vigorous, tolerant to CTV and CVd	Susceptible to citrus nematodes and Phytophthora root rot
Webster trifoliate	Na excluder, resistance to the citrus nematode, some species of Phytophthora, and to CTV, dwarf	susceptible to Cl and CVd, deciduous in nature
Rubidaux	Na excluder, *Phytophthora* and CTV resistant, dwarf	susceptible to Cl, deciduous in nature
Pomeroy	Na excluder, *Phytophthora* and CTV resistant, dwarf	susceptible to Cl, deciduous in nature
Troyer citrange	Na excluder, *Phytophthora* tolerant	Susceptible to Cl and Troyer and Carrizo inherited their susceptibility to CVd
Sour orange	Cl excluder, *Phytophthora* tolerant	Highly susceptible to CTV

Approaches to Enhance Hybrid Seedling Recovery in Polyembryonic Species

Within the perspective of high polyembryony, immature embryo culture is important (Hu and Ferrira, 1998) but selecting suitable embryo age, media and adequate *in vitro* procedures (Jaskani*et al.*, 2005) along with molecular markers which would expedite the recovery and recognition of hybrid seedlings (Bastianel *et al.*, 1998; Ruiz *et al.*, 2000) are needed for better recovery of hybrid seedlings (Table 22.3).

Embryo Rescue and *in vitro* Culture

The germination capacity of citrus embryos can be affected by their genetic structure and the stage of development (Viloria *et al.*, 2005). Carimi *et al.* (1998) found that zygotic embryos of sour orange could grow well *in vitro* when fruits were harvested at 105 days after pollination. Further they reported that the ability to form plants *in vitro* strongly increased with increasing embryo developmental stage. The zygotic seedlings were only derived from large- to medium-sized embryos. In Interploid sexual hybridizations, Viloria *et al.* (2005) found that germination of globular, heart and torpedo shaped embryos (defined as small embryos) was significantly affected by medium and genotype. Gamborg's medium induced 82.89 per cent germination. Further, Turgutoglu *et al.* (2015) found the highest germination rate from 2 mg/L GA_3 containing MT media which were harvested 115 days after pollination in Cleopatra mandarin x Carrizo citrange and Cleopatra mandarin x Flying Dragon were obtained with 65 per cent and 30 per cent, respectively.

Table 22.3: Embryony in Citrus and its Relatives

Common Name	*Species*	*Embryony Types*
Kumquat	*F. hindsii*	Polyembryony
	F.cassifolia	Polyembryony
	F.margarita	Monoembryony
	F.japonica	Monoembryony
Trifoliate	*Poncirus trifoliata*	Polyembryony
	Hybrids *of P. trifoliata*	Polyembryony
Papeda	*C. inchangensis*	Monoembryony generally
	C. junos	Mixture
	C. hongheensis	Monoembryony
	C.macropera	Monoembryony
	C.hystrix	Monoembryony or 2 embryo occasionally
Citron	*C. medica*	Monoembryony
Acid lime	*C. aurantifolia*	Polyembryony
lemon	*C. limon*	Polyembryony
	Hybrid *of C. limon*	Mixture
	Hybrid of C. medica	Mixture
Pummelo	*C.grandis*	Monoembryony
	Hybrid *of C. grandis*	Mixture
Grapefruit	*C. paradisi*	Polyembryony
Sour orange	*C. aurantium*	Polyembryony
	C. sinensis	Polyembryony
Mandarin	*C. mangshanensis*	Monoembryony
	C. reticulate	Polyembryony mostly

Culture media also plays important role for *in vitro* embryo germination and subsequent plantlet growth and survival. In citrus, many researchers have tried to alter the varied factors such as pH, Agar, sucrose, Malt, GA_3 concentrations (Carimi *et al.*,1998, Ribeiro *et al.*, 1998, 2000), macro and micro nutrient, vitamins of the MT medium (Morais, 1997), developmental stages (Carimi *et al.*,1998, Ribeiro *et al.*, 1999, Kurt and Ulger, 2014). Malt extract appears to play a vital role in embryo cultures of *Citrus* species (Molnár *et al.*, 2011). High germination rates of aborted embryos obtained from monoembryonic seed progenitors while inoculated on MS (Starrantino, 1992) or MS supplemented with malt extract and adenine sulfate (Starrantino and Reforgiato-Recupero, 1981). However, mandarin embryos did not germinate well on identical basal medium supplemented with giberellic acid (GA) (Khan *et al.*, 1996). Increasing the sucrose concentration to 6 per cent, and adding GA helped to enhance germination rates (62–90 per cent) of mandarin embryos (Sykes and Lewis, 1996). Chagas *et al.* (2003) found the best development of embryos at the globular stage using 50 per cent and 100 per cent of the MT medium plus 60 g L^{-1} and 90 gL^{-1} of sucrose, respectively, supplemented with 0.01 mg L^{-1} of GA_3. Further they observed that addition of activated charcoal or vitamins in the MT medium has shown to be unnecessary to be of no use to the development of globular embryos.

Identification of Hybrid Seedlings

One of the unusual features of citrus is the development in one seed of two or more embryos which is referred to as polyembryony. It often happens that the sexual (zygotic) embryo does not develop in which case almost all the embryos and subsequent seedlings are nucellar. Though some Citrus species produce only zygotic embryos (for example, pummelo, citron, Clementine, Temple, Persian lime, Kiyomi).

Though in some instance, hybrid seedlings can be identified from the relatively uniform nucellar seedlings by difference in size, vigour, growth habit, cotyledon colour, and leaf shape (Teich and Spiegel-Roy, 1972) or such other morphological characters but this is not certain. Spiegel-Roy and Teich (1972) reported thorn as a possible genetic marker for distinguishing zygotic from nucellar seedlings. In crosses of Eureka with Poncirus trifoliata, thorns were as good a criterion for distinguishing zygotes as trifoliation. Isozyme analysis is useful and reliable for distinguishing hybrids from nucellar seedlings. Isozyme techniques provide a reliable method to distinguish between nucellar (true-to-type) and zygotic (off-type) citrus seedlings. This method is quite accurate for identifying hybrids including most interspecific hybrids but unable to distinguish between closely related cultivars. Now more sophisticated DNA marker technique as RELF is used to develop maps of the citrus genome that may be useful in locating genes that have specific function.

Application of Molecular Approaches for Hybridity Confirmation

The use of molecular technologies will help to overcome the obstacles of juvenility and polyembryony in citrus breeding. DNA markers are genomic based, and considered to be the best of the biochemical markers. They have been widely used in citrus hybrid identification, such as SSR (Tan *et al.*, 2007), RFLP (Carimi *et al.*, 1998), and RAPD (Andrade-Rodrígue *et al.*, 2004). Further single nucleotide polymorphism markers have been emerged as the best markers for that they are the most abundant in the genome and the easiest to analyze. The citrus genomic sequence information (Xu *et al.*, 2013) and the increasingly accumulated citrus EST data are the very useful resources for the development of SNP markers. Zhu *et al.* (2013) found single nucleotide polymorphism based allele specific PCR (AS-PCR) was successfully employed as a handy tool to confirm and identify zygotic seedlings of Shantou-Suanju x trifoliate oranges crosses. SNP based AS-PCR is a rapid, simple, reliable, low cost method, and thus has a great potential in citrus hybrid identification. SSR markers have been screened for different parents for their polymorphism has been given in Tables 22.4 and 22.5.

Mutation Breeding

Spontaneous Mutation

- ✰ This is as evidenced by sudden change in heritable character occur frequently in citrus. Mutation is often observed as limb sport or sectors in fruits (Table 22.6).

Table 22.4: Suitable Primers of Identification of Hybrids of different Cross Combinations

Cross Combinations	*Primers/Marker Type*	*Reference*
Mandarin X pummelo	RAPD	Nageswara Rao *et al.*, 2007
Fremont x Midnight Valencia Fremont x Rio Red	SSR (AG14 and TAA03)	Ildiz *et al., 2013*
Robinson x Rhode red Valencia	SSR (CAT01, TAA01 and TAA45)	Ildiz *et al., 2013*
Yashar x Page	ISSR (GA)8YG)	*Golein et al., 2011*
Yashar x Ponkan	ISSR (GA)8YG, (AG)8YT, (AC)8YA)	Golein *et al.*, 2011
Valencia' orange x Femminello' lemon	SSR (CAC39)	Scarano *et al.*, 2003

Table 22.5: Summary of SSR Marker Screening for Polymorphism for Rootstock Parents

Cross/Primer name	*CAC-39*	*CgEMS-138*
Sour orange X Morton	M	P
Rangpur lime X Troyer	P	M
Sour orange X Troyer	P	M
Soh sarkar X Morton	M	P

M=Monomorphic, P=Polymorphic.

Dubey *et al.* (2018).

Table 22.6: Varieties Identified as Spontaneous Mutant

Mother Cultivar	*Mutant*	*Type of Mutation*
Washington Navel orange	Navel late Marrs, Leng, Autamn Gold, Powel summer, Willer Red	Limb sport
Pongan mandarin	Pongan 86-1	Bud sport
Owari mandarin	Clausellina	Bud sport
Walter grapefruit	Foster (Pink flushed)	Limb sport
Foster (Pink flushed)	Hudson (deep Red flush)	Bud sport
Thompson (light pink)	Red blush (deep red)	Limb sport

☆ They are also detected occasionally in nucellar seedlings or their budded progeny.

☆ As one of examples, Satsuma mandarin (*Citrus unsui*)

Induced Mutation

To induce mutation in citrus by gamma irradiation, X rays and other mutagent is known as induced mutation. Budwood or seeds are treated for a short period with gamma irradiation to induce mutation without killing the cells. In general, seeds (LD 50- 0.1 to 0.15Gy are more tolerant to gamma irradiation than buds (LD50- 0.05 to 0.09 Gy). The resulting seeds are buds are planted or propagated by tissue culture techniques. At low concentrations (0.01 and 0.05 per cent) for 1.5-3 h, ENU and MNU increased the percentage of pollen germinating on an artificial medium, percentage fruit set and seed germination percentage (Table 22.7).

Table 22.7: Varieties Developed through Induced Mutation

Mother Cultivar	*Mutant*	*Methods*	*Year*
Mandarin (China)	Fuyu	Irradiation of Ponggan	2000
	Tay-1	Ponkan spontaneous mutation	1997
Clementine (Italy)	Rubino	C. clementina Oroval bud mutantation	1994
	Spinoso	C. clementina Comune bud mutantation	1997
Clementine (Spain)	Marisol	C. clementina Oroval bud mutation	1985
	Esbal	C. clementina Fina bud mutation	1985
Tangor (Israel)	Mor	Irradiated budwood of Mucott Tangor	1991
Mandarin	Kousyun Ponkan	Nucellar seedling	1991

Polyploidy

In citrus, occurrence of polyploidy has been reported in many cultivars. Hybrid tetraploid seedlings seldom occur from crossing diploid, but several spontaneous tetraploids in citrus and poncirus have occurred as nucellar seedlings. Therefore, it appears that nucellar tetraploid can be derived from cultivars that produce moderate to high percentage of nucellar seedlings. Tetraploid hybrids from crossing diploid parents have rarely been found. These hybrids can be obtained in high frequencies when monoembryonic diploid seed parent were crosses with tetraploid pollen parents. These occur because of the frequent production of diploid megagametophytes as well as the failure of most triploid embryos. Failure is related to the ratio of the chromosome number of the endosperm to the number of chromosome of the embryo. Thus the tetraploid embryo accompanied by hexaploid endosperm giving on embryo an embryo to endosperm chrosome number ratio of 2:3. Oiyama, (1992) reported that tetraploid plants could be produced by colchicine treatment of axillary buds (0.5 mm) and subsequent micrografting onto a decapitated young seedling of Poncirus trifoliata. Monoembryonic cultivars, including Ellendale, Clementine and polyembryonic Mediterranean sweet and Kanton Orange were suitable seed parents for the production of hybrid tetraploids when crossed with tetraploid cultivars (Oiyama, 1992). The development of seedless varieties is one of the important goals for citrus breeding because seedlessness is very desirable from the viewpoint of the consumers. The production of triploid citrus plants, by crossing tetraploids with diploids, provides an important method of producing essentially seedless varieties. Triploid embryo result from the union of the usual haploid egg with diploid male gametes from the tetraploid pollen parent. The triploid can also be recovered from diploid x diploid crosses if a cultivar that produces enough diploid megagametophyte is used as seed parent. Two seedless tetraploid pummelo - grapefruit hybrids, Oroblanco and Melogold, which originated from diploid x tetraploid, have been released. Oiyama and Kobayashi (1991) reported pentaploids from small seeds of diploid Miyauchi iyokan (Citrus iyo) X diploid crosses using cv. as the seed parent. Diploid Miyauchi iyokan also produced hexaploid seedlings from small seeds in addition to triploid and tetraploid seedlings from normal seeds when crossed with tetraploid Funadoko (C. funadoko) and Trovita (C. sinensis). The seedlings produced from four monoembryonic diploid (2n = 18) citrus cultivars and triploid (2n = 27) hybrid derived from a cross between diploid Citrus medioglobosa X tetraploid C. funadoko and C. natsudaidai cv. Kawano exhibited several ploidy levels. Occurrence of a spontaneous octoploid in apomictic seedlings of a tetraploid Citrus hybrid was reported by Barrett and Hutchison (1982).

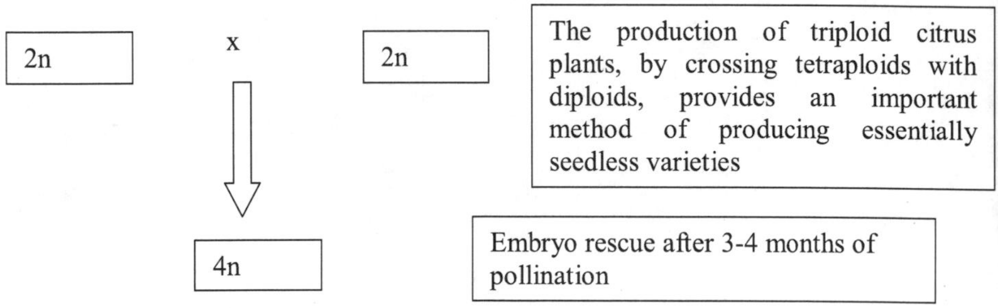

Production of Triploid.

1st Selected Triploid Genotype by Crossing

Oroblanco, Melogold (4n x 2n)

Other Tripoids

Alkantara: Oroval Clementine x Tarocco (2n x 4n)

Tacle: Monoreal Clementine x Tarocco (2n x 4n

Mandared: Monoreal Clementine x Tarocco (2n x 4n)

Citrus Genomics

Citrus species have small genomes. The size of the *Citrus sinensis* haploid genome is estimated to be 372 Mb (Arumuganathan and Earle 1991). Ollitrault *et al.* (1994) found significant genome size variation between citrus species. The largest and smallest genomes were respectively *Citrus medica* L. (the citron, with an average value of 398 Mb/haploid genome) and Citrus reticulata Blanco (the mandarin, with an average value of 360 Mb/haploid genome). The pummelo (*Citrus maxima* Burm.), was intermediate with an average 383 Mb/haploid genome. Interestingly, the secondary species presented intermediate values between their putative ancestral parental taxa: 370, 368, 381 and 380 Mb for *C. sinensis* (L.) Osb., *Citrus aurantium* (L.), *Citrus paradisi* Macf. and *Citrus limon* (L.) Burm. haploid genomes, respectively.

Biotechnological Intervention for Citrus Improvement

The development of biotechnological tools such as protoplast fusion has provided a means of bypassing natural barriers related to reproductive biology of citrus found in conventional breeding systems. Somatic hybridization in citrus is suggested as a useful technique for overcoming barriers of sexual hybridization and for improving citrus scion and rootstock (Ohgawara and Kobayashi, 1991). Somatic hybrids between Redblush and Duncan grapefruit (*Citrus paradisi*) with Avana, Tardivo di Ciaculli mandarin (*C. deliciosa* Tenore) and Fortune mandarin (*C. reticulata* Blanco) were obtained by symmetric protoplast fusion. Grosser (1986) used protoplasts from embryogenic suspension cultures of *Citrus sinensis* cv. Hamlin and fused chemically with leaf protoplasts of (sexually compatible) *Poncirus trifoliata* cv. Flying Dragon. Hybrid plants exhibited the dominant trifoliate leaf character of Flying Dragon. Fusion of Key lime protoplasts, isolated from an embryogenic suspension culture, with Valenica leaf protoplasts resulted in the regeneration of interspecific allotetraploid somatic hybrid plants. Protoplasts of the *C. sinensis* cv. Hamlin (sweet orange), isolated from nucellus-derived embryogenic callus, were fused with leaf protoplasts of *A. ceylanica.* Cybrid is obtained by the donor-recipient protoplast fusion method by which the nuclear division of the donor-protoplasts was arrested by gamma irradiation and the metabolism of unfused recipient protoplasts was transiently inhibited by iodoacetate. Transgenic plants of grapefruit (*C. paradisi*) cv. Rio Red have been obtained by Agrobacterium tumefaciens-mediated gene transfer using seedling-derived epicotyl segments as explants and kanamycin as the selective agent (Yang *et al.*, 2000).

DNA Marker Development and Application

Till now, abundant DNA markers, such as restriction fragment length polymorphism (RFLP), randomly amplified polymorphic DNA (RAPD) and cleaved amplified polymorphic sequences (CAPS) were developed and applied to cultivar identification (Matsuyama *et al.*, 1992, Omura *et al.*, 2003, Sugawara *et al.*, 2002, Ueda *et al.*, 2003). Numerous EST-SSR and genomic SSR markers have been developed from citrus genome sequences and assigned to the genetic map of the Clementine mandarin (Luro *et al.*, 2008, Ollitrault *et al.*, 2012). Among these markers, CAPS markers are generally less polymorphic than the other molecular markers but represent a convenient method that does not require special analysis instruments for genetic diagnosis. herefore, CAPS markers were utilized as the principal DNA marker for cultivar identification and for phylogenetic and linkage analyses. Based on "Minimal Marker", which is a computer program for the identification of minimal sets of discriminating DNA markers for efficient cultivar identification (Fujii *et al.*, 2013a), nine CAPS markers were sufficient to discriminate examined 33 cultivars.

Linkage Mapping and Marker-Assisted Breeding

Genetic maps are useful tools for breeding program to select hybrids with economically important traits, and high density genetic maps provide information on quantitative trait loci (QTLs) associated with pathogen resistance and fruit quality. Numerous molecular markers have been developed and used to construct citrus genetic maps over the past decade. The initial citrus genetic maps were constructed from RFLPs and isozymes (Durham *et al.*, 1992, Jarrell *et al.*, 1992). Subsequently, SSR markers were adopted for mapping (Kijas *et al.*, 1997). Thereafter, genetic maps have been extended to localize important traits, such as cold acclimation (Cai *et al.*,

1994), citrus tristeza virus (CTV) resistance (Cristofani *et al.*, 1999, Fang *et al.*, 1998, Gmitter *et al.*, 1996), fruit acidity (Fang *et al.*, 1997), apomixes (Garcia *et al.*, 1999, Kepiro and Roose 2010), nematode resistance (Ling *et al.*, 2000) and *Phytophthora gummosis* resistance (Siviero *et al.*, 2006). Ollitrault *et al.* (2012) constructed a genetic map of the Clementine mandarin using a SNP and EST-SSR marker, and Shimada *et al.* (2014) constructed a framework genetic map of mandarin progenies using CAPS and SNP markers. These maps were constructed using co-dominant genetic markers associated with a rather on served nd functional genome coding (Table 22.7).

Table 22.8: Target of Functional Metabolites Aimed to Enrichment in Fruit

Compounds	*Breeding Resources (Latin name)*	*Target Gene or Metabolites*
Limonoids	Satsuma mandarin	UDP glucose: limonoid glucosyltranferase
Polymethoxyflavonoids	Nou-6 (King x Mukaku kishu)	Flavonoid O methyltrasferase (nobiletin and tangertine from sinensetine)
Hesperisin	Satusuma mandarin	Conversion of rhamoside
Monoterpenes	Ponkan, sweet orange	Monoterpene biosynthetic gene

Problem in Citrus Breeding

Heterozygosity

Many citrus and related species are heterozygous in nature and hybrids in F_1 tends to exhibit variability.

Incompatibility

Many varieties of citrus either self or cross incompatible and can not be used in breeding programme. Hensz *et al.* (1975) reported that no fruits were set when Minneola and Orlando were self-pollinated. All tested varieties of *Citrus grandis* and their hybrids are self-incompatible. Thus Clemntine, Orlando Minneola, Sukega, Robinson, Nova, Page, Kagzi Kalan and Fairchild are self-incompatible. Some hybrids like Robinson, Nova and Page are cross incompatible also. Hybrids of self-incompatible are also self-incompatible. Washington Navel and Satsuma are pollen sterile.

Nucellar Embryony

One of the unusual features of citrus is the development in one seed of two or more embryos, which is referred to as polyembryony Some species like *Citrus grandis, C. medica* and *C. reticulata* variety Nagpur mandarin produced only zygotic seedlings. Earlier it was very difficult to identify zygotic and nucellar seedlings in the hybrid progeny at early stage. But now a day many techniques like isozymes markers and molecular marker are available which will help in early identification of hybrids.

Quantitative Inheritance

Most of the undesirable characters are dominant, therefore it is very difficult to obtain desirable characters. Most of economic importance characters rarely show single gene inheritance due to which inheritance studies are not easy. Furthermore, transfer of such characters to other varieties is also difficult.

Juvenility

The seedlings show a very long period of juvenility and come in bearing after many years, which require long time for evaluation of a hybrid.

REFERENCES

Barrett, H.C and Rhodes, A.M.(1976). *Systematic Botany*, 1(2) 105-136.

Barrett, H.C. and Hutchison, D.J. (1982) Proceedings of the International Society of Citriculture, 1981. Volume I., 29-30.

Batygina TB, Vinogradova GY. (2007). *Russian J. Dev. Biol.*, 38: 126–151..

Cameron JW, Soost RK (1979). *J. Am. Soc. Hortic. Sci.*, 104: 408–410.

Carimi, F., De Pasquale, F., Crescimanno, F.G. (1999). *Plant Cell Report*. 18: 935-940.

Chagas, E.A., Pasqual, M., Ramos, J.D., Pio, L.A.S., Dutra, L.F. and Cazetta, J.O. (2005). *Ciencia e Agrotecnologia*, **29**(6): 1125–1131.

Cheng, F.S.Z. and Roose, M.L., 1995. *Journal of the American Society for Horticultural Science*, 120(2): 286-291.

Chen, L.G., Omura M. and Hidaka, T., 1991. *Acta Horticulturae Sinica*, 18(1) 27-32.

Driver, J.A. and Kuniyuki, A.H. (1984). *Hort Science*, **19:** 507–509.

Fang, D.Q., Zhang, W.C. and Xiao, S.Y., 1993. *Acta Phytotaxonomica Sinica*, 31(4): 329-352.

Gamborg, O.L., Miller, R.A. and Ojima, K. 1968. *Experimental Cell Research*, **50:** 15-158.

Geraci, G., Esen, A. and Soost, R.K., 1975. *Journal of Heredity*, 66(3): 177-178.

Grosser, J.W., Gmitter, F.G. Jr., Tusa, N., Recupero, G.R. and Cucinotta, P., 1996. *Plant Cell Reports*, 15(9): 672-676.

Jaskani, M.J., Khan, I.A. and Khan, M.M. 2005. *Scientia Horticulturae*, **107**: 51–57.

Kurt, S. and Ulger, S. 2014. *International Journal of Fruit Science*, **14**: 42–48.

Malik, M.N., Scora, R.W. and Soost, R.K., 1974. *Hilgardia*, 42(9): 361-382.

Murashige, T. and Tucker, P.H., 1969. *1st International Citrus Symposium*, **3**: 1155–1161.

Niedz, R.P., Bausher, M.G. and Hearn, C.J., 1992. *Hort. Science*, 27(1): 43-44.

Oiyama, I. and Kobayashi, S., 1991. *Hort. Science*, 26(3): 292-293.

Oiyama, I., 1992. *Bulletin of the Fruit Tree Research Station*, No. 3, Extra, 68 pp.

Roose, M.L. and Traugh, S.N., 1988. *J. Am. Soc. Hort. Sci.*, 113: 100-105.

Samaan, L.G., 1982. *Euphytica.*, 31(1): 167-173.

Scarano, M.T., Abbate, L., Ferrante, S., Lucretti, S. and Tusa, N., 2002. *Plant Cell Reports*, 20(12): 1162-1166.

Scora, R.W., Kumamoto, J., Soost, R.K. and Nauer, E.M., 1982 *Systematic Botany*, 7(2): 170-177.

Scora, R.W., 1983. *Phytochemistry*, 22(1); 12.

Spiegel, Roy P. and Teich, A.H., 1972. *Euphytica*, 21(3): 534-537.

Spiegel, Roy P., 1979. *Euphytica*, 28(2): 361-365.

Tan, M., Song, J. and Deng, X., 2007. *Euphytica*, 157: 155–160.

Turgutoglu, E., Kurt, S. and Demir, G. 2015. *Ekin Journal of Crop Breeding and Genetics*, 1: 17-19.

Viloria, Z., Grosser, J.W. and Bracho, B. 2005. *Plant Cell Tissue and Organ Culture*, 82: 159–167.

Wang, J.F., Chen, Z.G. and Lin, T.X. 1999. *Biology of Reproduction*, 8(2): 57–63.

White, P.R., 1943. The Jacques Catlell Press, Lancaster.

Breeding of Perennial Horticultural Crops *Pages* **269–282**
Editors: S.K. Singh, V.B. Patel, A.K.Goswami, Jai Prakash and Chavlesh Kumar
Published by: **BIOTECH BOOKS, NEW DELHI**

***Chapter* 23**

Cucurbits Improvement

A.D. Munshi, T.K. Behera, A.K. Sureja, Jogendra Singh and B.S. Tomar

Division of Vegetable Science,
ICAR-Indian Institute of Agricultural Research, Pusa, New Delhi – 110 012

INTRODUCTION

Cucurbits are important group of vegetables in whole of the world and are consumed as salad (cucumber, gherkin and long melon), sweet (ash gourd, pointed gourd), pickles (gherkin, bitter gourd, ivy gourd), desserts (melons) and as vegetables (all except melons). The family *Cucurbita*ceae consists of about 118 genera and 825 species. The major cucurbits grown in India are cucumber, watermelon, pumpkin and squashes, muskmelon, ash gourd, snake gourd, ridge gourd, smooth gourd, bottle gourd, round gourd, bitter gourd, pointed gourd, ivy gourd and Gherkin. The characteristic features of cucurbits are:

- ☆ Viny growth habit (except barring summer squash)
- ☆ Annuals (except pointed gourd, ivy gourd, *kakrol* and chayote)
- ☆ Frost-sensitive
- ☆ Monoecious (except pointed gourd, ivy gourd and *kakrol*)
- ☆ Cross-pollinated
- ☆ Propagated by seed (except pointed gourd, ivy gourd and *kakrol*)
- ☆ Multi seeded fruit (except chayote)
- ☆ Mainly fruit *i.e,* 'pepo' is the edible part (exceptions: chayote all plant parts edible, pumpkin besides immature and mature fruit, flower is also edible)
- ☆ Mostly old world origin (except squashes, pumpkin and chayote)

Floral Biology and Pollination Behaviour

Vegetable	*Anthesis Time*	*Pollen Dehiscence Time*	*Duration*	
			Pollen Fertility	*Stigma Receptivity*
Muskmelon (*Cucumis melo*)	5.00-8.00 hr	5.00-8.30 hr	5.00-14.00 hr	2 hr before to 2-3 hr after anthesis
Watermelon (*Citrullus lanatus*)	6.00-8.30 hr	6.00-8.30 hr	7.00-11.00 hr	2 hr before to 3 hr after anthesis
Bottle gourd (*Lagenaria siceraria*)	11.00-16.00 hr	13.00-14.30 hr	On the day of anthesis till next morning	36 hr before to 60 hr after anthesis
Bitter gourd (*Momordica charantia*)	6.30-10.30 hr	7.30-10.30 hr	6.00-12.00 hr	One day before and one day after anthesis
Snake gourd (*Trichosanthes anguina*)	18.00-21.00 hr	Shortly before anthesis	10 hr before to 40 hr after dehiscence	7 hr before to 50 hr after anthesis
Ridge gourd (*Luffa acutangula*)	15.00-18.00 hr	15.00-18.00 hr	On the day of anthesis to till 2-3 days after anthesis in spring season and 1 day in rainy season	6 hr before to 80 hr after anthesis
Sponge gourd (*Luffa aegyptica*)	5.00-8.00 hr	5.00-8.00 hr	On the day of anthesis	10 hr before to 100 hr after anthesis
Cucumber (*C. sativus*)	5.30-7.30 hr	4.30-7.00 hr	Up to 14 hr	12 hr before to 6-7 hr after anthesis
Summer squash (*Cucurbita pepo*)	4.00-10.00 hr	4.00-11.00 hr	16 hr after anthesis	2 hr before and 10 hr after anthesis
Pumpkin (*Cucurbita moschata*)	4.00-8.00 hr	4.00-8.00 hr	16 hr after anthesis	2 hr before and 8 hr after anthesis

Sex Forms

- **Hermaphrodite**: Primitive form where only bisexual or perfect flowers are produced. As for example 'Satputia' cultivar of ridge gourd, cucumber and muskmelon.
- **Monoecious**: Where staminate and pistillate flowers are separately produced in the same plant. As found in cucumber, bottle gourd, bitter gourd, watermelon, pumpkin, squash, ash gourd, ridge gourd, sponge gourd, snake gourd *etc.*
- **Andromonoecious**: Staminate and perfect flowers produced separately in the same plant. For example, watermelon (in some cultivars), muskmelon (dessert type).
- **Gynomonoecious**: Where pistillate and perfect flowers are produced separately in the same plant. As found in cucumber.

- ☆ **Gynoecious**: Where only pistillate flowers are produced in a plant. Example, genetic stock of cucumber, it has been evolved from gynomonoecious sex.
- ☆ **Trimonoecious or gynoandromonoecious**: Where staminate, pistillate and perfect flowers are produced in the same plant in varying proportion (genetic stock of bitter gourd).
- ☆ **Androecious**: Where only staminate flowers are produced (pointed gourd).
- ☆ **Dioecious**: Where staminate and pistillate flowers are produced in separate plants, *e.g.*, pointed gourd, ivy gourd and *kakrol* (*Momordica dioica*) among cultivated cucurbits.

Sex Modification

Besides environmental factors, exogenous application of plant regulators can alter the sex ratio and sequence. if applied at 2-4 true leaf stage, the critical stage at which the suppression or promotion of either sexes is possible.Hence, modification of sex to desired direction has to be manipulated by exogenous application of growth regulators once,twice or even thrice at regular intervals. To promote greater number of pistillate flowers and alter the sex ratio in desired direction, several chemicals have been tried and found useful. In cucumber, Maleic hydrazide (MH) at 50 to 100 ppm, GA_3 5 to 10 ppm, 2- chloroethylphosphonic acid (commercially called ethrel, or ethephon or CEPA) 150 to 250 ppm; in watermelon, tri-iodobenzoic acid (TIBA) 25 to 50 ppm, boron 3 ppm; in bottle gourd, boron 3 ppm and calcium 5 ppm and in sponge gourd, ethrel 250 ppm have been found useful in bitter gourd. These bring about increased yield in terms of number of fruits per plant, but the individual fruit set is slightly reduced, especially in watermelon. This kind of sex modification is more useful and practicable, in crops like cucumber and bottle gourd, where continuous and simultaneous flowering, fruit set and fruit picking take place. Gynoecious lines of cucumber are maintained by induction of male flowers through sprays of GA_3/GA_7 at 1500 to 2000 ppm. Now-a-days silver nitrate at 300-400 ppm and silver thiosulphate at 250-300 ppm has been found to bring about this same modification.

Bitterness in Cucurbits

Most cucurbit species have bitter compounds due to presence of oxygenated tetra cyclic triterpene compounds namely *Cucurbita*cins in their foliage. *Cucurbita*cins are thought to be toxins produced by these plants as a defense against insects and herbivores. The cucurbit cultivars have been selected by plant breeders to have low *Cucurbita*cin content in the fruits except for ornamental gourds where higher level of *Cucurbita*cin is desirable. Bitter fruits in cucumber are occasionally produced when plants are exposed to drought during fruiting. Another way of producing bitter fruit is when bitter pollen grains fertilized the non-bitter ovule, the resulting fruit will be bitter in taste because the gene *Bt* (bitter fruit) is dominant over *bi* (bitter free) gene and the phenomenon is known as metaxenia. Generally, during development, varieties are tested to verify that the leaves and fruits are not bitter-tasting, and if maintained by self pollination (or sib-pollination in an isolation block), there will be no problem of bitterness

Breeding Objectives

The objectives of cucurbits breeding are highly crop specific but in general following aspects are taken into consideration during breeding program.

- **Growth habit:** Preference should be given to medium vine characteristics.
- **Maturity:** The number of node at which first pistillate (or hermaphrodite) flower appears, gives a fair indication of earliness, especially in those kinds where immature fruits are edible.
- **Sex ratio:** A high female to male ratio desirable.
- **Fruit shape:** Spherical or oblong in watermelon; spherical, flat or oblong in muskmelon; round, long, cylindrical, club shaped in bottle gourd; round, oblong, spherical shaped in pumpkin. A shape having high flesh recovery per volume, especially in muskmelon and pumpkin would be desirable. In salad or slicing cucumber, uniform long cylindrical shape without neck is desirable.
- **Fruit surface:** Attractive dark green with minimum of prickles in slicing cucumber. Netting and thick skin are useful in transportable quality in muskmelon and toughness of skin helps in watermelon for long distance haulage. In bitter gourd, continuous ridges and blunt tubercles are preferred. A non-ridged fruit skin is attractive in pumpkin. Sparse hairs persisting on the skin of bottle gourd and round melon indicate tenderness of fruit and edible maturity.
- **Flesh colour and thickness:** The flesh colour in muskmelon may be white or greenish or salmon-orange while in watermelon it can be pink or pale pink or deep pink and in pumpkin yellow, yellowish orange or deep yellow colour is met with. The flesh should be thick with small seed cavity in muskmelon and pumpkin. In cucumber, carpel separation showing hollow spots at edible maturity is undesirable. In most of the gourds where immature fruits are harvested like bottle gourd, ridge gourd, sponge gourd, round melon, snake gourd, *etc.*, the flesh should not become fibrous quickly in optimum sized fruits at edible maturity. Carotenoids produce pigmentation which results in the range of whites, creams, yellows, and oranges in the flesh color of squash and pumpkin. The expression of this flesh color is conditioned by the particular carotenoid type and concentrations which are influenced by genetic and environmental factors. The goal of the cucurbit breeding should be to investigate the various levels of carotenoids present and identify molecular markers associated with these carotenoids within pumpkin, summer and winter squash.
- **Sweetness**: The sweetness in melon is judged as total soluble solids content read by squeezing the juice in a hand refractometer, subject to varying conditions of environment, 10 per cent TSS may be taken as a minimum standard for selection of fruits to meet consumer acceptance.

- **Bitterness**: Bitterness in fruits of cucumber, bottle gourd, ridge gourd, *etc.*, should be eliminated.
- **Flavor**: High aroma in muskmelon is preferred by consumers.
- **Higher early yield and total yield**: The yield in terms of weight and number of fruits of marketable quality will be the deciding criterion in selection. Misshapen and poorly developed fruits should be rejected. Early yield, especially of higher percentage of total yield is a desirable character.
- **Seed**: Fewer and smaller seeds are desirable in watermelon. In cucumber, fruits should not produce mature seeds at edible maturity.
- **Wider adaptability**
- **Photo insensitivity**
- **Dwarf growth habits**
- **Resistance/tolerance** against biotic and abiotic stresses.
- **Parthenocarpic gynoecious and indeterminate varieties** for protected cultivation especially in cucumber
- **Bunching fruit habit** producing multiple pistillate flowers on individual nodes for harvesting finger size fruits to suit whole fruit canning for export especially in case of gherkin

Commonly Adopted Methods for Cucurbits Breeding

Indian subcontinent is the centre of origin for a number of wild and cultivated cucurbits. It is primary centre of origin for a number of wild and cultivated cucurbits like cucumber, snake gourd, ridge gourd, smooth gourd, pointed gourd, bitter gourd and bottle gourd. By utilizing different selection methods like mass selection, Individual plant selection, inbred selection, hybridization followed by selection, clonal selection and back cross methods, several high yielding varieties and F_1 hybrids have been developed in cucurbit vegetables.

Mass Selection

The best individual plants are selected on the basis of phenotypic characters and their open pollinated seeds are mixed for raising the succeeding generations. Since mass selection is made exclusively on the phenotypic performance of the mother plant without any progeny test, the success of selection depends largely on the heritability of the character under selection. This method is effective in improving simply inherited and highly heritable qualitative characters like sugar content in watermelon and muskmelon, disease resistance *etc.*

Individual Plant Selection

The species under the family *Cucurbita*ceae, though cross-pollinated in nature, generally do not show significant loss in vigour due to inbreeding. In these crops, inbreeding and individual plant selection through pedigree breeding can be practiced as effectively as in self-pollinated crops. The plants are selected only on individual basis and unlike mass selection; the seeds of the selected plants are not bulked together even if they are equally good and look similar. In this method, the progenies of individual plant for evaluation are developed by selfing because they do not show loss of vigour due to inbreeding. Homozygosity for the concerned characters can also be attained in the individuals of the progeny by such selfings. After necessary evaluation, the best selection may be treated as a new variety. It is most important breeding method followed in India for developing several varieties in different cucurbits.

Recurrent Selection

This population improvement scheme is designed to concentrate favourable genes scattered among number of individuals in the base population. This is based on selection of superior plants followed by controlled mating to produce the new base population. This method is effective in improving yield and other quantitative characters by enhancing additive genetic variance in the population. Even though inbreeding depression would be negligible or nearly absent in the parental lines, interference of natural self- and sib-pollination is major disadvantage of handicap in these crops for this method.

Inbred Selection

Inbreeding is the mating of closely related individuals, either by selfing or sib mating. Inbreeding increases homozygosity and reduces the proportion of heterozygosity in the population. Recessive, lethal and deleterious genes usually remain concealed in the heterozygous system of the cross pollinated crops which are increased in proportion upon rendering to homozygosity by inbreeding, thus resulting reduction in vigour-called inbreeding depression. *Cucurbita*ceous vegetable crops do not show loss of vigour due to inbreeding. For this reason, individual selection of inbreds can be practiced and maximum uniformity can be attained in several cucurbits namely muskmelon, pumpkin *etc.* Example, Muskmelon (Hara Madhu), Pumpkin (Arka Chandan,), Summer squash (Punjab Chappan Kaddu).

Hybridization followed by Selection

Controlled hybridization between selected parental lines followed either by selfing and pedigree breeding or open pollination and selection is also effective in developing improved varieties because as a result of gene recombination in the progeny following a cross, it is possible to possible to select desirable segregates. The hybridization aims at combining in a genotype the useful characters inherent in each parent separately. That is why; it can be called combination breeding. Generally parental lines with specific characters and belonging to the same species are utilized for the purpose of developing desirable segregates. Good number of varieties has been developed in India following this method of breeding. Example. Muskmelon (Pusa Sharbati: Kutana x PMR-6; Punjab Sunheri: Hara Madhu x Edisto; Hisar Madhur: Pusa Sharbati x 75-34; Punjab Rasila: WMR-29 x Hara Madhu), Watermelon (Arka Manik: IIHR-21 x Crimson Sweet), Cucumber (Himangi: Poinsett x Kalyanpur Ageti), Bottle gourd (Punjab Komal: LC-11 x LC-5), Ridge gourd (Arka Sujat: IIHR 54 x IIHR 18; Arka Sumeet: IIHr 54 x IIHR 24).

Practical Difficulty of this Breeding Method

- ☆ Huge requirement of space in the field for the evaluation of a reasonably large segregating population to identify the desirable one becomes the main practical difficulty of this breeding method.
- ☆ In distant hybridization, segregates with unacceptable horticultural traits usually predominate in the F_2 generation of the inter-specific crosses.
- ☆ The undesirable traits of the wild parental species are often dominant and consequently desirable segregates in the F_2 population may often be small.

Breeding for Resistance

Cucurbits are highly susceptible to several biotic and abiotic stresses. In India, resistance breeding has not been given the required importance, partly because cucurbits are considered as low value crops. India has large genetic resources in cucurbits and the river bed farmers have been unconsciously maintaining a large reservoir of germplasm of these crops, which have not been unfortunately documented, conserved and utilised in resistance breeding programmes.

At IARI, New Delhi (India) in muskmelon powdery mildew, downy mildew, *Fusarium* wilt and CGMMV were tackled noting sources of resistance, mode of inheritance. Two varieties namely DMDR-1 and DMDR-2 were found to be highly resistance to downy mildew and CGMMV, having TSS 10-12 per cent and average yield 170 q/ha. On the basis of their superior performance the varieties DMDR-1 DMDR-2 were identified as resistance source against (CGMMV) and downy mildew and DMDR-2 resistant to (CGMMV) and downy mildew was further identified for release. Muskmelon variety DMDR-1(IC-557706) was registered with NBPGR, New Delhi as resistant source against CGMMV and downy mildew(INGR -08044).

At Punjab Agricultural University, Ludhiana (India), downy mildew resistance from 'Phoot' (*C. melo var. momordica*) was transferred to MR-12 (Punjab Rasila). At IIHR, Hessarghatta, the cultivar 'Arka Manik' carries resistance to anthracnose, powdery mildew and blossom end rot bred from Crimson Sweet of U.S.A. Unfortunately, studies on breeding for insect resistance to red pumpkin beetle and fruit fly did not receive sustained support and sources of moderate resistance in bitter gourd and muskmelon for fruit fly and in bottle gourd for red pumpkin beetles, were identified in IIHR, Bangalore (India) and not pursued further. Powdery mildew caused by *Sphaerotheca fuliginea* is a major limiting factor in *Cucurbita* (squash and pumpkin) production in subtropical and tropical regions of the world In pumpkin,(*Cucurbita moschata*}, enough awareness for resistance breeding to powdery mildew and viruses, was not felt even though known sources of resistance and easy crossability with, *Cucurbita lundelliana* were available. Resistance to downy mildew (*Pseudoperonospora cubensis*) is reported in snap melon (*C. melo* var. *momordica*) while resistance to fruit fly is reported in *C. callosus etc.* Most of the resistant varieties in cucurbits have been developed by simple selection. In respect of multiple resistance, vegetable breeders of USA like Walker and Peterson, have developed open pollinated varieties and hybrids in cucumber carrying resistance to more than five or six diseases. In fact cucumber line WI. 2757 carries resistance to nine diseases.

Breeding by Alteration of Ploidy

Seedless Watermelon

a) *Production of Tetraploid Line*

Application of colchicines @ 0.2 to 0.4 per cent to growing point of young seedlings at 1- to 2- true leaf stage for 2-3 days successively under controlled condition avoiding direct sunlight. The tetraploid will be slow and terminal bud should be carefully promoted to grow. The 4X plants are characterized by thick and broad leaves and bigger stomata. The size of pollen grains is also bigger than diploid plants. The seeds are also broad and bigger in 4X plant but less in number (30-40 seeds/fruit) due to high pollen sterility. Maintenance of 4X lines should be done through selfing.

b) *Production of Triploid Seeds*

$$\text{♀ } 4X \times 2X \text{ ♂}$$
$$\downarrow$$
$$3X$$

It is done by hand pollinating the female flowers of tetraploid line with pollen from the male flowers of diploid parent or by pinching the all male flowers from the tetraploid line and then allow for insect/natural cross pollination, where 4X and 2X plants are grown in the same plot at the ratio of 4:1.

c) *Production of Ttriploid Fruits*

3X seeds have poor germination due to hand seed coat. So, seeds will have to be sown in flat on peat cups with moisture carefully maintained at a temperature of 24°-29°C in the first 72 hours. There will be no fruit set in triploid plants, if there are only 3X plants in the filed. Even few tetraploid plants occurring as a mixture would not be enough for good fruit set. Stimulation of fruit development has to be brought about by 2X pollen through bee pollination. Hence, a pollinizer diploid cultivar has to be planted in the field. The 3X plants are of long duration, fruits maturing in 120 to 130 days in contrast to 85-90 days in diploid cultivars. The firm texture of the rind is an attractive feature, but there may be few empty ovular cavities in the fruit. Sweetness will be uniformly distributed from the rind to the centre. Triploids are sterile both on male and female sides and unequal chromosomal distribution in meiosis causes pollen and ovular abortion. However, few ovules may develop into a hardened seed-coat and these are called ex-embryonate or 'false' seeds. Example. Triploid Watermelon (Pusa Bedana: Tetra-2 x Pusa Rasal).

Heterosis Breeding

Among the vegetable crops, cucurbits are distinct group where sex mechanism is unique and can be easily manipulated for production of F_1 hybrids. Further, low in breeding depression, high heterosis percentage and large number of seeds per fruit (per pollination) and low seed rate requirement per unit area, has distinct advantages in commercial exploitation of heterosis in this group of vegetables. The phenomenal success in heterosis breeding in cucumber and to a lesser extent in other cucurbits is entirely due to manipulation of sex expression to the desired direction. The genetic control of sex mechanism in cucumber, especially of the gynoecious sex form, has made it possible to exploit heterosis in cucumber.

Hybrid cucumber and squash are important not only for yield and earliness but also for external attributes of uniformity of size and shape, especially in slicing cucumber. F_1 hybrids have attractive colour, flesh texture and other quality traits. Most of the hybrids carry multiple resistances to diseases due to the fact that dominant genes control resistance to some diseases.

In muskmelon, heterosis breeding has different requirements. Being dessert fruits, quality characteristics of the hybrids should be superior in respect of uniformity and stability over pure bred cultivar. Earlier maturity is another major advantage in the acceptance of melon hybrids. It has been reported in cantaloupe hybrids that additive genes more than non-additive factors are responsible for the heterosis of most of the characters.

In summer squash (*C. pepo*), earliness and high number of fruits per plant were the advantages in hybrid. Application of ethephon or ethrel to young plants at 2-3 leaf stage would suppress male flower formation for 2-3 weeks and complete suppression can be achieved at higher concentrations of 600 ppm applied twice at 2- and 4-leaf stage has made hybrid squash seed production comparatively easier.

Likewise, heterosis breeding has potentialities in Asiatic cucurbits as well, like bottle gourd, bitter gourd, pumpkin, Indian squash melon, ridge gourds, *etc.* At IARI, New Delhi (India), bottle gourd hybrid Pusa Hybrid-3,bitter gourd hybrid Pusa Hybrid-2 and pumpkin hybrid Pusa Hybrid-3 showed distinct advantage in yield, earliness and other desirable character over their respective commercial checks and released for commercial cultivation. Earliness and uniformity in quality characteristics have been the advantages of watermelon hybrids extensively produced in Japan.

In most of the heterosis studies in different cucurbits, non additive gene action has been noted to be predominant for several yield components and quality traits, which point out the possibility of exploitation of heterosis breeding for the development of F_1 hybrids.

In cucurbits, hybrid seed production has been simplified by exogenous manipulation of sex mechanism and sex expression. Techniques were evolved to reduce the cost of hybrid seed production *viz.* use of male sterile lines, gynoecious lines and use of plant growth regulators. In case of monoecious cucurbits, special attention should be given to standardized techniques of artificial pollination and suppression of male flowers. A technique of easy and economical method of hybrid seed production by Pinching of male flower buds from female line was recommended by IARI, New Delhi (India). In this method, F_1 hybrid seed can economically be produced on large scale by pinching all the male flower buds from the female parent before opening and allow the male parent to grow side by side of female parent for natural cross-pollination. One row of male parent can be sown after every three rows of female parent as suggested by Choudhury and Sirohi (1971) for producing F_1 hybrid seed on large scale in bottle gourd. Since the male or staminate flowers in bottle gourd, pumpkin and Squash are quite big, showy, having long pedicels and less in number; the pinching operation of unopened flowers can easily be performed.

Techniques of Hybrid Seed Production

Bagging of Female Flowers and Hand Pollination

A day prior to anthesis, the female flowers before they open on the plants of female parent are covered by butter paper bags or buds of female flowers are tied with rubber band or clip specially where flower is large like pumpkin. In the afternoon of the same day, unopened male flowers on the plants of male parent are protected from pollen contamination by bagging or tying the petals. In the next day morning (7-11 a.m.), pollens are applied on the stigma of the protected female flowers. After hand pollination, the female flowers are again bagged or petals are tied. This practice is applicable in most of the cucurbits except muskmelon where andromonoecious is predominant. F_1 seed is collected from the mature fruits harvested from the plants of female parent.

Emasculation and Hand Pollination

This method is practiced in muskmelon where andromonoecious sex form is predominant. In the plants of female parent, the hermaphrodite flowers are emasculated a day prior to anthesis and protected by bagging. The male flowers on the plants of intended male parent are also protected. Pollination is done in the early morning of next day by taking pollen from protected male flowers of male parent. Hybrid seed is collected from the mature fruits harvested from the plants of female parent. A female and male planting ratio of 4:1 is followed to get maximum F_1 seed yield.

Pinching of Male Flowers

Production of hybrid seeds by this method is the most simple and economical and can easily be adopted by the growers who can identify male or female flowers. F_1 hybrid production in gourds can economically be done on large scale by pinching all the male flowers before opening from the female parent and allow the male parent to grow side by side of female parent for natural cross-pollination. One row of male parent can be sown after every three or four rows of female parent for producing F_1 hybrid seeds on large scale in bottle gourd, pumpkin, ash gourd and summer squash. All the fruits set in the plants of female parent would be necessarily through crosspollination by insects. For this, as a precaution, there should not be a single male bud left open in the plants of female parent as it will promote self or sib-pollination within the female parent. Anthesis of the flowers of bottle gourd taken in afternoon, the pinching operation should therefore be done in the forenoon. Isolation distance between different varieties should be kept about 400 metres. As the male flowers in bottle gourd, pumpkin, ash gourd and squash are quite big, showy, having long pedicels and less in number, the pinching operation can easily be performed. In *Luffa,* where male flowers are produced in racemes, pinching off male buds will not be complete and effective. For maximum fruit set and seed yield, availability of pollinator is pre-requisite. One medium sized bee colony per hectare would be enough in seed production block.

Use of Gynoecious Sex Form

Gynoecy condition where all the flowering nodes in the primary, secondary and tertiary branches bear pistillate flowers in the leaf axils, is by far the most important sex form which has made phenomenal exploitation of hybrid vigour possible in cucumber and also in muskmelon. It is observed that gynoecious sex forms are stable only under moderate regimes of temperature and photoperiodic conditions. When temperature exceeds more than 30ºC, the stability of gynoecious sex expression affects seed production.

The commercial production of gynoecious cucumber seed was made possible only when it was discovered that gynoecious inbreds could self reproduce if a growth regulator is applied to induce male flower formation.In the beginning GA3 ((1500-2000 ppm) was to be used for induction on male flower in cucumber. How-ever, due to erratic male flower induction by use of gibberellic acid, application of silver compound such as silver nitrate (250-400 ppm) is done to induce male flowers. Silver ions inhibit ethylene action and thus promote male flower formation in gynoecious cucumber plants. However, due to phytotoxic ef-fects such as burning of plants, silver thiosulphate (400 ppm) is now widely used by seed producers for the maintenance of gynoecious cucumber lines. It induces male flowering of cucumber plants over a longer period and is less phytotoxic com-pared to silver nitrate. When the temperature exceeds beyond 30°C the stability of gynoecious sex expression is affected. Unfortunately, the temperate gynoecious lines are unstable for gynoecy under high temperature and long photoperiodic conditions because of their thermo-specific response for gynoecious stability. That is why the gynoecy in cucumber did not receive much attention in the tropical countries.

Keeping in view the facts efforts have been directed during recent past towards developing gynoecious sex forms under our tropical and indigenously adapted back ground and three three gynoecious lines DC-103, DC-103 having stable gynoecious sex even at temperatures around 40°C have been developed at IARI. These lines are being further exploited in heterosis breeding programme. Hybrids of cucumber are produced mainly by crossing gynoecious lines with monoecious lines. Though, other systems of producing gynoecious hybrid seed such as gynoecious X gynoecious have been proposed but gynoecious X monoecious hybrids are still the most widely grown. Most of the commercial hybrids of cucumber based on gynoecious lines are a blend of gynoecious hybrid and monoecious seed. About 10 per cent blending is of monoecious genotype seed with gynoecious hybrid seed has been suggested by the author. In addition, homozygous gynoecious hybrid seed has been produced by crossing two gynoecious lines after one has been treated with a growth regulator to induce male flowers.

Development of Gynoecious Parthenocarpic Varieties of Cucumber for Protected Cultivation

Development of suitable Ideotype novel genes along with biotic and abiotic stress resistance is the need of the hour for breeding cucumber varieties/F_1 hybrids suitable for protected cultivation. The extension of protected cultivation and breeders' work has been linked by a close relationship. The increase in return in protected cultivation has made a greater effort in plant breeding possible, which has encouraged the expansion of crops. Further, plant breeding efforts have permitted to solve specific problems of varieties suitable for protected cultivation as well as general problems which also benefit the open field cultivation. Ideotype breeding iwith incorporation of useful genes like indeterminate plant type, gynoecy and parthenocarpic can be utilised for large scale fruit production under polyhouse.

Gene 'PP' responsible for parthenocarpy, Pp delayed parthenocarpy and pp non parthenocarpic traits (Incomplete dominance)' while F (Dominance)' responsible for gynoecy can be utilised through pure line and back cross breeding methods for the development of gynoecious parthenocarpic cucumber varieties/F_1 hybrids.Marker assisted breeding is also being carried out at IARI to identify markers for gynoecy and parthenocarpy traits for introgression to the cultivated varieties

The North Indian conditions permit a very short growing period for cucumbers as maximum yield reali-sation could not be obtained under open fields. So polyhouse technology with proper training, pruning and nutritional support can be exploited for off-season breeding of cucumber. Polyhouse production technology has shown four-fold realisation of yield in commercial hybrid during off season as compared with cultivation under open conditions.

Parthenocarpic gynoecious cucumber varieties are suitable for polyhouse cultivation as these varieties develop fruits automatically without any pollination. Now a days many of the private seed companies are selling varieties/F_1 hybrid of parthenocarpic cucumber at a very high price as their seeds are being sold on per seed basis.

Keeping in view the above facts a breeding programme on development of parthenocarpic gynoecious varieties of cucumber for protected cultivation was undertaken during 2008-2009 at Division of Vegetable Science IARI, New Delhi. The initial material was an exotic collection which was broken by using silver thio sulphate (3Mm) for induction of male flowers and individual plant selection was carried out and generations were advanced during spring summer and *kharif* season respectively, under insect proof net house. In winter season of 2012-13, seventeen breeding lines which showing true gynoecious and parthenocarpic behaviour were advanced to F_8 and were evaluated under low cost polyhouse.from 2012-13 to 2016-17.On the basis of five trials the line DPaC-6 was observed as most promising as it expressed 32.2 per cent, 29.8 per cent and 21 per cent.5 higher yield than check Pant Parthenocarpic Cucumber–2, Aviva and Asma, respectively. The yield obtained by DPaC–6 (126.0 t/ha, 1260kg/100sqm polyhouse) can be considered as quite high since it was obtained during off-season (winter season) under low cost polyhouse without using any energy. On the basis of it consistent superior performance DpaC-6 was released as Pusa Seedless Cucumber-6 by Delhi State Seed Committee 2018

Utilization of Male Sterility in Muskmelon

Five recessive nuclear male sterile genes (ms-1, ms-2, ms-3, ms-4 and ms-5) are known. The genetic male sterile line ms-1 (genotype ms-1) of muskmelon was introduced in India more than 12 years ago and still the only line available in India. The development of the superior F_1 cultivar of Punjab Hybrid (ms-1 x Hara Madhu) is the result of research and breeding work using ms-1. The male sterile line (ms-1, ms-1) is to be identified from the female and female plants to be removed. For this purpose, in the early morning, anthers of anthesis flowers from each plant of the female parent are pierced with needle. Male fertile flowers give light yellow powdery mass, while in male sterile plants, anthers produce sticky green immature pollen grain. Identified male sterile plants are tagged every day.

Use of Monoecious Line in Muskmelon

Considering the problems associated with the use of male sterile gene(s) for production of muskmelon hybrid (maintenance of single recessive gene in heterozygous condition and problem of identifying and rouging of 50 per cent male fertile plants from the female row at the time of flowering and also to overcome the andromonoecious forms), the scientists at IARI, New Delhi took keen interest in developing true breeding monoecious lines *viz.*, M1, M2, M3 and M4. Of various combinations developed, M3 x Durgapura Madhu *i.e.* Pusa Rasraj was found to be outstanding performance and has been released for commercial cultivation.

Chemical Suppression of Male Flowers and Open Pollination

There is yet another method of hybrid seed production by the use of chemicals. With the advent of growth regulating substances, very significant results have been obtained with regard to sex modification in cucurbits. It has now been possible to prove that the two trueleaf stage is the most responsive stage for application of chemicals for sex modification. Specific chemicals are known to induce femaleness or maleness as desired. In cucurbits like bottle gourd, pumpkin and squash female flowers can be increased by the application of ethrel (2-chloroethylphosphoric acid) at the rate of 200-300 ppm at two true-leaf and four-leaf stage. Ethrel helps in suppressing the staminate flowers and initiating pistillate flowers successively in the first few flowering nodes on the female parent. The row of male parent is grown by the side of female parent and allows natural cross pollination. In the absence of insect pollinators, hand pollination is possible when two sexes are separate. Precaution should be taken that during natural cross pollination; there should not be any other variety except the parents of hybrid. Four to five fruit-sets at initial nodes containing hybrid seed would be sufficient seed yield. Complete suppression of male flowers in squash can be achieved at higher concentration of 400-500 ppm ethrel applied twice has made hybrid seed production comparatively easier.

Seed Yield

Average hybrid seed yield of different cucurbits range from 300-600 kg/ha.

Future Thrusts

1. Asia, Africa and America are the centre of origin of *Cucurbitaceous* vegetables. High genetic variability present in these crops has not been fully utilized. This genetic variability can be successfully utilized for the development of high yielding and resistant (biotic and abiotic stresses) varieties and F_1 hybrids of cucurbits.
2. Riverbed areas cover around 60 per cent total cultivable area of cucurbits in India. Efforts should be made to develop suitable varieties and F_1 hybrids which can exploit its full potential under the sandy riverbed conditions with low input availability and subsistence farming conditions in India as well as in other countries.
3. Male sterile lines and gynoecious lines in different cucurbits reported so far have poor stability and they generally breakdown under high temperature conditions. Efforts should be made to develop stable tropical gynoecious and male sterile lines linked with marker and their proper utilization in heterosis breeding programme so that hybrid seed becomes cheap and it should be within the reach of average vegetable growers.
4. Cucumber, summer squash, bitter gourd and gherkin are important low volume, high value vegetables for protected cultivation. Efforts should be made to develop gynoecious parthenocarpic varieties and F_1 hybrids of these above vegetables so that they can be profitably exploited under protected conditions or greenhouse of vegetable farming.
5. *Cucurbitaceous* crop like bitter gourd, bottle gourd, cucumber and gherkin are important from export point of view. Therefore, efforts should be made to develop varieties and F_1 hybrids which can suit the requirement of international market.
6. Efforts should also be made to develop varieties and F_1 hybrids of pumpkin, ash gourd, gherkin and bitter gourd, which can suit the requirement of value addition and processing industry. Especially in case of gherkin, attempt should be made to develop varieties with bunching fruit habit producing multiple pistillate flowers on individual nodes for harvesting finger size fruits to suit whole fruit canning.
7. Lines capable of performing well during winter or in low temperature (*e.g.* off-season) needs to be developed. These lines can further be utilized in hybrid breeding programme so that their availability can be ensured in winter season or low temperature also.
8. Attempts should be made to develop multiple disease resistant varieties and hybrids, more particularly in muskmelon and cucumber which are highly susceptible to several diseases.

Breeding of Perennial Horticultural Crops *Pages* **283–314**
Editors: S.K. Singh, V.B. Patel, A.K.Goswami, Jai Prakash and Chavlesh Kumar
Published by: **BIOTECH BOOKS, NEW DELHI**

Chapter 24

Grape Improvement

M.K. Verma, S.K. Singh, V.B. Patel and Chavlesh Kumar

Division of Fruits and Horticultural Technology,
ICAR-Indian Agricultural Research Institute, Pusa, New Delhi – 110 012

INTRODUCTION

Grapevine is one of the most widely grown crops in the world and covers about 6,931 million hectares and produced to the tune of 7,428 million tonnes during 2016-17 (FAO, 2017). In India, 2683.26 thousand tonnes of grapes were harvested from an area of 136 thousand ha (NHB, 2017). Grapes occupies the premier position in exports with 232.9 thousand tonnes valued at Rs. 2,08,835 lakhs (APEDA, 2017). It is anticipated that worldwide 8,000 to 12,000 grapevine cultivars exist, mainly used for wine production (56.8 per cent) but also for table grapes (27.0 per cent), a mixed utilization for both wine and table grapes (7.3 per cent), and dried fruits (0.7 per cent). It is generally believed that *Vitis vinifera*, originated in the Middle East and that cultivation of the grapevine began during the Neolithic era along the eastern shores of the Black Sea in the region known as Transcaucasia in the areas south of the Caspian and the Black Sea. From its centre of origin, the grapevine travelled with man to other parts of the world and in ancient Egypt mention was made of grapevine growing and winemaking. The domesticated grapevine reached European and North African Mediterranean countries, first with the Phoenician and Greek trade routes and later with the Romans, who spread it through their Empire. From Europe, the domesticated grape spread globally with settlers to the more temperate climates in the Americas, South Africa, Australia and New Zealand. *Vitis vinifera* grows in temperate climates and flourishes in Mediterranean climates, can also be cultivated in tropical climates and with careful management and pruning may produce two crops a year.

Geographical Distribution

European grape (*Vitis vinifera* L.) is considered to have originated primarily between Caspian and Black Sea region. American grapes belonging to a large number *Euvitis* and *Muscadinia* species have originated in North America. Grape growing is an age old practice. Wild grapes were present in many places of Eurasia during Neolithic period. It was later spread from Transcaucasia, western Mediterranean counties, Asia Minor, Nile delta, and Balkan Peninsula. Islamic countries promoted for fresh consumption of grapes and its cultivation in Middle East and North Africa. This led to the introduction of table grapes into Spain, France, Italy and New World. In 1525, grapes were planted in Mexico followed by Peru, Chile and Argentina in 1550. Viticulture started in South Africa (1616), California (1697), Australia (1788) and New Zealand (early 19th Century).

Taxonomy and Species Diversity

Of the 17 genera belonging to the family Vitaceae, that developed its own identity inside the Eudicots at the end of the Cretaceous period about 100 million years ago, it is only the genus *Vitis* that produces edible fruits. Planchon (1887) divided the *Vitis* species between two sub-genera namely *Euvitis* (bunch grapes) and *Muscadinia* (Muscadine grapes). *Muscadinia* has a genome of 2n = 40, similar to the one of several other genera of Vitaceae (Ampelopsis, Ampelocissus, Parthenocissus), while for *Euvitis* it is 2n = 38. Though some researchers (Olmo 1986; Mullins *et al.*, 1992) argued that Muscadinia should be elevated to generic level as is supported by morphological, anatomical and karyological characteristics, it has not achieved general acceptance. Muscadinia is endemic to the southeastern states of the USA and among the three species known, only *Vitis rotundifolia* Michx. is of commercial value (Owen 1990).

Taxonomic Classification of *Vitis* and *Muscadinia* Species

A. Genus *Vitis* Subgenus *Euvitis*

Series: Aestivales

Vitis aestivalis Michx. – summer grape

Vitis aestivalis Michx. var. *aestivalis* – summer grape

Vitis aestivalis Michx. var. *bicolor Deam* – summer grape

Vitis aestivalis Michx. var. *lincecumii* (Buckley) Munson – long grape

Series: Cinerescentes

Vitis cinerea (Engelm.) Engelm. ex Millard – graybark grape

Vitis cinerea (Engelm.) Engelm. ex Millard var. *baileyana* (Munson) Comeaux – graybark grape *Vitis cinerea* (Engelm.) Engelm. ex Millard var. *cinerea* – graybark grape

Vitis cinerea (Engelm.) Engelm. ex Millard var. *floridana* Munson – Florida grape

Vitis cinerea (Engelm.) Engelm. ex Millard var. *helleri* (L.H. Bailey) M.O. Moore – Heller's grape

Vitis cinerea (Engelm.) Engelm. ex Millard var. *tomentosa* (Planch.) Comeaux – graybark grape

Series: Cordifoliae

Vitis monticola Buckley – sweet mountain grape

Vitis palmata Vahl – catbird grape

Vitis vulpina L. – frost grape

Series: Labruscae

Vitis labrusca L. – fox grape

Vitis x *labruscana* L.H. Bailey

Vitis mustangensis Buckley – mustang grape

Vitis shuttleworthii House – calloose grape

Series: Ripariae

Vitis acerifolia Raf. – mapleleaf grape

Vitis riparia Michx. – riverbank grape

Vitis rupestris Scheele – sand grape

Series: Hybrids

Vitis x *champinii* Planch. (pro sp.) *[mustangensis* x *rupestris]*, Champin's grape

Vitis x *doaniana* Munson ex Viala (pro sp.) *[acerifolia* x *mustangensis]*, - Doaniana grape

Vitis x *novae-angliae* Fernald (pro sp.) – pilgrim grape

Series: Non Grouped Species

Vitis arizonica Engelm., Canyon grape

Vitis x *bourquina* Munson ex Viala (pro sp.)

Vitis californica Benth. – California wild grape

Vitis girdiana Munson – desert wild grape

Vitis tiliifolia Humb. and Bonpl. ex Schult. – West Indian grape

Series: Viniferae

Vitis vinifera L.

Subspecies

Vitis vinifera L. ssp. *sylvestris* (C.C. Gmel.) Hegi – wild grape

Vitis vinifera L. subsp. *Vinifera* – Table and wine grape

Section: *Labruscoideae*

Vitis bashanica P.C. He

Vitis bellula (Rehd.) W.T. Wang

Vitis bellula var. *pubigera* C.L. Li

Vitis heyneana subsp. *ficifolia* (Bunge) C.L. Li

Vitis hui Cheng

Vitis longquanensis P.L.Qiu

Vitis menghaiensis C.L. Li.

Vitis pentagona Diels et Gilg

Vitis retordii Roman. Ex. Planch.

Section: *Sinocineriae*

Vitis sinocinerea W.T. Wang

Section: *Vitis*

Series: *Vitis*

Vitis amurensis Rupr., - Amur grape

Vitis amurensis Rupr. var. *dissecta Skvorts*

Vitis betulifolia Diels et Gilg

Vitis balanseana Planch.

Vitis chunganensis Hu

Vitis chungii Metcalf

Vitis fengqinensis C.L. Li

Vitis flexuosa Thunb.,

Vitis flexuosa var. *parvifolia* (Roxb.) Gagnep (Syn. *Vitis flexuosa* Thunb., *V. indica* Thunb., *V. parvifolia* Roxb., *V. wallichii* DC) – Creeping grape, Parvifolia

Vitis luochengensis W.T. Wang

Vitis luochengensis var. tomentoso-nerva C.L. Li

Vitis hekouensis C.L. Li

Vitis mengziensis C.L. Li

Vitis piloso-nerva Metcalf

Vitis pseudoreticulata W.T. Wang

Vitis wilsonae Veitch

Vitis yunnanensis C.L. Li

Series: *Piasezkianae*

Vitis lanceolatifoliosa C.L. Li

Vitis piasezkii Maxim.

Vitis piasezkii var. *pagnucii* (Planch.) Rehd.

Series: *Davidianae*

Vitis davidii (Roman.) Föex

Vitis davidii (Roman.) Föex var. *ferruginea* Merr. et Chun

Vitis davidii (Roman.) Föex var. *cyanocarpa* (Gagnep.) Gagnep.

Series: *Adstrictae*

Vitis bryoniaefolia Bunge

Vitis bryoniaefolia var. ternate (W. T. Wang) C.L. Li

Vitis zhejiang-adstricta P.L. Qiu

Series: *Romanetianae*

Vitis romanetii Roman. ex Planch.

Vitis romanetii Roman. var. tomentosa Y.L. Cao et Y.H. He

Vitis shenxiensis C.L. Li

Series: *Wuhanenses*

Vitis coignetiae Pulliat ex Planch

Vitis erythrophylla W.T. Wang

Vitis hancockii Hance

Vitis jinggangensis W.T. Wang

Vitis ruyuanensis C.L. Li

Vitis silvestrii Pamp.

Vitis tsoii Merr.

Vitis wenchouensis C. Ling ex W.T. Wang

Vitis wuhanensis C.L. Li

B. Genus: Muscadinia

Muscadinia rotundifolia Michx. var. *rotundifolia* – Muscadine grape

Muscadinia rotundifolia Michx. var. *munsoniana* (*Simpson ex Munson*) M. O. Moorei – Munson's grape

Muscadinia rotundifolia Michx. var. *popenoei* Fennell

The genus *Vitis* L. (Vitaceae) includes about 70 woody climber species, spread mostly in the temperate regions of the northern hemisphere. Two subgenera are commonly recognized: subg. *Vitis* (2n = 38; comprising the vast majority of species) and subg. *Muscadinia* (2n = 40; 2–3 species). The current distribution of the subg. *Vitis* includes northern South America, Central and North America, Asia, and Europe. In contrast subg. *Muscadinia* is restricted to the southeastern United States, northeastern

Mexico, Belize, Guatemala, and the Caribbean. The Angiosperm Phylogeny Group placed Vitaceae in the rosids. *Vitis* was found to be monophyletic and placed in a clade with the tropical Ampelocissus and Pterisanthes.

Centers of grapevine diversity are found in the southeastern US and East Asia. Up to 30 species are native to a vast area in eastern Asia, China, Japan and Java, two species across middle Asia and Europe, and up to 28 species across the eastern and southwestern US and Mexico. The genus *Vitis* is divided into two subgenera: Muscadinia Planch. (2n=40, one or two species) and Vitis Planch. (2n=38, the remaining species). Additional divisions within *Vitis* are "series" which are subgeneric groupings that have been used historically in the systematics of Vitis. They rank below "sections" that are more commonly used in plant systematics for groups of species. Although all wild species are considered diploid, there is evidence of hexaploidization in their distant past that is shared with all rosids. The two subgenera are nearly reproductively isolated while the species within subgenus *Vitis* are interfertile. All species are dioecious except *V. vinifera* which has hermaphroditic flowers, and *V. rotundifolia* Michx. which segregates for this trait. Many species have overlapping distributions, thus natural hybridization would occur were it not for ecological and phenological barriers. The classification of *Vitis* is surprisingly very confused in part due to the lack of agreement among systematic botanists as to what constitutes a true species and because of extreme morphological variation within the species. This has led to many extraneous species names. The systematics of *Vitis* is based primarily on morphology and molecular methods have only recently been used to study this taxonomic problem.

Most cultivated grapevines belong to the Euvitis sub-genus. These fall under three groups: i) The American group is made up of more than 20 species including *V. berlandieri, V. riparia* and *V. rupestris,* which were used as rootstock to address the phylloxera crisis. ii) The East Asia group consists of around 55 species, which are currently considered to be of limited importance to viticulture. iii) The Eurasian group is made of one single species, *Vitis vinifera* L., which accounts for most of the world's Vitis varieties. There are two sub-species of *Vitis vinifera: sylvestris,* which corresponds to the wild form of the vine, and vinifera, referring to the cultivated form. Over the years the cultivation of grapevines has led to a significant increase in genetic diversity via sexual reproduction.

History of Grape Breeding

The first cultivar successfully grown in the New World was 'Alexander', a native grape originating from *Vitis labrusca* L. It was selected around 1800 by the Frenchman Peter Legaux. First documented cultivars and defined crossings are 'Sage', 'Cunningham', 'Isabella', 'Catawba', and 'Flowers'. These and other cultivars are well known as American hybrids. In European countries and first in France major breeding activities emerged as a consequence of the introduction of powdery mildew (*Erysiphe necator*), anamorph: *Oidium tuckeri*, Berk.), phylloxera (*Daktulosphaira vitifoliae* Fitch), and downy mildew (*Plasmopara viticola* (Berk. and Curt. ex. De Bary)). These pathogens changed dramatically the many thousand years old tradition of viticulture in Europe. The use of sulphur and copper as first

Pusa Seedless

Pusa Urvashi

Pusa Navrang

Pusa Aditi

Pusa Trishar

Pusa Swarnika

found to possess useful fungicide activity in the Bordeaux mixture became inevitable to combat the mildew fungi, and still in our days an extraordinarity intense plant protection is necessary. In 1878 Millardet suggested to combine the fruit quality of *V. vinifera* L. subsp. *vinifera* and the resistance against powdery and downy mildew found in American wild species. A biological trick was found rather soon against phylloxera, which nevertheless took decades to be acceptable for the market: the use of grafted vines (scions of traditional cultivars (with leaf-resistance to phylloxera) on phylloxera root-tolerant rootstocks). An acceptable solution of the mildew problem by breeding took about 120 years to become reality and first cultivars showing good field resistance and high wine quality were introduced at the turn of the millennium

In addition to the activities initiated at public institutions in France at the end of the 19th century to combat the pests also various dedicated private viticulturists started their own breeding programmes in order to combine "European wine quality" with "American resistance". The resulting hybrids were called "direct producers" indicating that they could be grown on their own roots. Private French breeders like Albert Seibel (1844-1936), Georges Couderc (1850-1928), Eugene Kuhlmann (1858-1932), Bertille Seyve (1864-1939), Seyve-Villard (1895-1959) and others made thousands of crosses resulting in tens of thousands of seedlings from which the best grape genotypes where selected. Some of these showed quite mediocre wine quality combined with a high expression of resistance characteristics. They were recognized as the so-called "French Hybrids". In 1929 the plantation surface of these French Hybrids covered about 2,50,000 hectares and it reached its peak in 1958 with about 500,000 ha. Due to the limited wine quality and political decisions their area decreased later on. Nowadays the "French Hybrids" are almost totally removed from production. In retrospective, the bad image of the French Hybrids prevented any continuation of the breeding programmes in France. While the breeding efforts stopped in France, countries like Germany, Hungary, or others used the valuable French material for their own pur- suing breeding activities.

To introduce resistances into the gene pool of *V. vinifera* L. subsp. *vinifera* breeders generated F_1-plants by interspecific crosses. This strategy was quite successful for rootstock breeding, but for wine grapes it yielded only unacceptable genotypes. Consequently, Erwin Baur (1922) suggested to create in a first step a small number of interspecific hybrids between *V. vinifera* L. subsp. *vinifera* and a wild species as a resistance donor to generate an F_1 generation selected for resistance, vigour, and yield (10-12 plants). Following multiplication of these F_1 plants, in a second step the selection should be performed at the level of large populations (about 1,00,000 plants) of the F_2 generation generated from sister pollination. The outline was the consequent application of Mendel's laws re-discovered in 1900. To generate large numbers of seeds derived from defined crosses always remained a challenge. It finally turned out that it requires more than two generations from the wild to select acceptable genotypes and even more crosses to obtain really elite lines and new quality cultivars. The huge efforts in France prepared the ground for the break through though the "French Hybrids" failed. In Germany for example where resistance breeding was initiated in the early 1920th the development took a different direction. Husfeld was the first who proved that resistance and quality can be combined. His cultivars 'Aris' ((Oberlin 716) F_1 x 'Riesling', cross 1937) and 'Siegfriedrebe' ((Oberlin 595) F_1 x 'Riesling', cross 1936) showed a convincing wine quality and high mildew resistance. Unfortunately, these two cultivars could not satisfy the wine growers due to insufficient yield and virus susceptibility. A next generation cultivars like 'Phoenix' ('Bacchus' x 'Villard Blanc', cross 1964) or 'Regent' (('Silvaner' x 'Müller-Thurgau') x 'Chambourcin', cross 1967) was developed by Alleweldt. Husfeld and Alleweldt used a breeding scheme similar to MAS, which is a recent development. 'Regent', 'Phoenix', and other cultivars gained access. The numerous cultivars selected at various breeding stations in Germany are the outcome of continuation and the use of step-wise improved breeding material. They are today's basis of prosperous breeding, which will result in further improvements in regard to pathogen resistance and quality of grapevines.

In 1868 phylloxera (introduced in 1863) was identified as the devastating pest destroying the vineyards in France. Its rapid spread throughout France eliminated within 15 years about 8,00,000 ha of vineyards. Its subsequent spread throughout Europe was a serious threat for the survival of viticulture. No treatment whatsoever (*e.g.* removal of vines and/or various chemical treatments or flooding of vineyards with water) could stop the pest from dissemination, which was spread rapidly by planting material, wind, and surface water. Observations in the grape collection in Bordeaux showed that some American hybrids exhibited a certain resistance against phylloxera on their roots. In 1869, Laliman first suggested to use phylloxera resistant American vines as rootstocks for the traditional European grapevine varieties. In 1872 Bazille performed the first successful graftings.

American cultivars like 'Clinton', 'Jaquez' and others were recommended as rootstocks. But the degree of resistance of these cultivars proved to be not high enough. Hence Millardet recommended in 1878 to use pure American *Vitis* species like *Vitis riparia* Michx., *Vitis rupestris* Scheele, *Vitis cinerea* Engelm. var. *cinerea*, *Vitis vulpina* L., or *Vitis aestivalis* Michx. However, soon it became evident that the tolerance of these species to lime soils is rather poor. In 1887 Viala conducted an expedition through North America. In Texas he found *Vitis berlandieri* Planch. (today called *V. cinerea* Engelm. var. *helleri*) this grows very well on calcareous soils. But because of the poor rooting ability of this species crosses with other *Vitis* species, mainly with *V. riparia* Michx., were performed in several research institutes in France. This was the beginning of a target oriented rootstock breeding leading in the end to a series of rootstock cultivars with good rooting ability and good adaptation to calcareous soils.

A major impact came from the Hungarian wine grower Zsigmond Teleki when he received about 10 kg of seeds of open-pollinated *V. cinerea* Engelm. var. *helleri* in 1896 from Rességuier, a French viticulturist. Teleki grew about 40,000 seedlings and selected them first according to their morphology. Later he tested them in various calcareous soils. The best growing genotypes were propagated and multiplied. Some of the most promising genotypes were transferred to Franz Kober in Austria for further selection and finally distributed to various locations in Europe where very important rootstock genotypes like 'Kober 5 BB' could be selected.

There is no doubt about the vital importance of the development of rootstocks to rescue viticulture from phylloxera crisis. It is the greatest success breeders could have achieved. One of the most important objectives for rootstock breeding was the resistance against phylloxera. Therefore, great emphasis was given to elucidate the genetics of phylloxera resistance. Almost all of the material developed shows rather tolerance than resistance. Since rootstocks became available at the beginning of the 20th century and brought the solution of the phylloxera disaster, rootstock breeding activities declined. Nevertheless rootstock breeding programmes are continued and research is directed to elucidate the genetics of certain traits.

Genetic Variations in Grapevines

Three processes impacted development of cultivated grapevines: sexual reproduction, vegetative propagation and somatic mutations. New genotypes are produced by sexual reproduction, either by crossing or self-fertilization. Because individual grapevine plants have highly heterozygous genotypes any progeny produced from seed is a novel combination of parental alleles, resulting in phenotypic variation and segregation of traits in a progeny population. The selection of a particular phenotype, particularly a berry trait, can be a long process considering the juvenile period (three-to-five years) of grapevine plants and the additional time necessary for evaluation of a trait important for wine production. Furthermore, many generations might be necessary to recover the desired traits. Once identified, vegetative propagation (asexual) by cuttings is a method of maintaining and multiplying a highly desirable genotype so that a vineyard can be planted with a single cultivar. Cuttings are also a convenient method of transporting cultivars from one region to another. Cultivars grown today are maintained by vegetative propagation. Although clonal propagation should ensure that all plants grown from cuttings have the same genotype, the occurrence of a somatic mutation in one cutting and no other cuttings might eventually lead to plants of the same cultivar having a slightly different genotype and sometimes a different phenotype, referred to as clonal variation. This clonal variant thus appeared rapidly over one cycle of vegetative reproduction (three-to-four years). This clonal variation is more complex if the mutation is maintained in only one cell layer of the plant, resulting in genetic chimerism. If the mutation is maintained in only the L_1 layer (epidermis) then it will not be passed onto progeny by sexual reproduction.

Biology and Grape Breeding

Domestication caused morphological shifts that include larger berry and bunch sizes, higher sugar content, altered seed morphology and a shift from dioecy to a hermaphroditic mating system. There is interest in identifying the genes that contribute to these morphological shifts. For example, several papers have attempted to identify the gene(s) that are responsible for the shift to hermaphroditism, which were mapped to an ~150-kb region on chromosome 2.

Even without an understanding of grapevine floral biology and genetics, early farmers had the ability to select the best vines and plant new vineyards with vines propagated from their best selections. Today, the science is much more precise, but grapevine breeding is still based on hybridization schemes designed to develop a single elite individual vine combining the best traits of both carefully-chosen parents. Because most grapes are propagated from cuttings, and not grown from seeds, a single vine with good characteristics can be readily turned into many more vines, all with the same genes and thus the same characteristics. The earliest varieties were propagated from choice wild seedlings, the products of chance crosses between vines. The process of breeding, however, goes one step further, by controlling pollination and choosing which varieties will act as parents for seed production. Parents must be carefully, chosen based on the traits desired in the progeny, and what might already be known about a parent's ability to impart those desired traits to its seedlings.

Making crosses is all about flowers. Like many other organisms, grapes reproduce sexually. In other words, each seedling is a unique combination of the genes from a female and the genes from a male. Pollen from a male fertilizes the female ovary, and a seed develops. Depending on the individual vine, flowers might be female, male, or hermaphroditic (also known as "perfect", which means that each flower contains both male and female structures). By far, the majority of cultivated varieties are hermaphroditic. Because fruit does not result without fertilization, hermaphroditic vines are preferred because they are capable of fertilizing themselves, and every vine planted is capable of bearing fruit. The male parts of the flower are the stamens, each consisting of a pollen-bearing anther supported by a filament. Generally there are five anthers arranged around the ovary, but some varieties may have more or less. Pollen are shed as the anther ripens and are primarily transported by wind, though rarely more than 20 feet (7 meters) from the source. Pollination occurs when pollen lands on the female parts of a flower. Each pollen grain grows a long tube towards the eggs within the ovary, and sperm cells move down this tube. The female parts of the flower consist of the ovary, the stigma, and the style. The stigma and style provide the entrance to the ovary for pollen. When a sperm cell in the pollen tube unites with an egg cell in the ovary, an embryo (a new seedling plant) is produced. The embryo grows within the developing seed, while the entire ovary grows to become the grape berry itself with seeds contained within.

A few cultivated varieties, and many wild vines, are female. These plants have flowers which contain a well-formed ovary, stigma, and style, but have poorly formed male parts. The anthers produce pollen that are generally sterile, and the filaments supporting them are often short, and reflexed back away from the ovary. Such varieties require a male or hermaphroditic vine nearby to provide pollen in order to set fruit. Since male vines are incapable of bearing fruit, they are not used in commercial agriculture, though a few are used as rootstocks. The flowers of these vines contain functional anthers, but the center of the flower has only a small, rudimentary ovary and no stigma or style. These vines reproduce through fertilization of other vines by wind-blown pollen. Many wild vines found throughout North America and Asia is male.

a. Conventional Breeding Techniques

Pollen Collection

If the variety to be used as a male happens to be in flower at the same time as the female, then this step can be easily accomplished by collecting a few of the flowering clusters for pollen. If the male is significantly earlier, it may become necessary to collect and store pollen. This can be done by drying the clusters (1-2 days) and shaking out the pollen onto a sheet of glass or other smooth surface, then collecting it into a vial by scraping the glass with a razor blade. If, however, the male flowers after the female, then the pollen must be collected, dried, and stored in the freezer (at -2°C) and the cross made next year. Stored pollen usually remains viable, but it is often not as effective as fresh pollen. Another option where the male flowers before the female is to reverse the cross - use the male as a female and use the female as a male. This only works if both parents have perfect flowers.

Preparing the Female

The guiding principle in preparing female flowers is to prevent accidental pollination before the application of pollen from the selected male parent. Before the flowers open, each one is covered by a cap, short for calyptra. Though this cap eventually dries and comes off on its own, it must be removed a day or two before the cross is to be made to prevent self-pollination with perfect-flowered varieties. In most cases, the anthers can, with some care and practice, be removed at the same time as the calyptra, using forceps with very fine tips. This practice is called emasculation. After the cluster has been emasculated, you must wait one to four days before proceeding (more time in cool weather and less time in warm weather) so that the female parts of the flower mature further and attain peak receptivity. This is often indicated by the appearance of a bead of moisture on the stigma. During this time, the cluster is covered with a bag to prevent fertilization by external pollen sources. If the female variety produces female (imperfect) flowers, no emasculation is necessary. The flowers are covered with a paper bag until they are ready for pollen application.

Hybridization

Crossing is a matter of brushing the pollen onto stigmas of the emasculated cluster. If both parents are flowering simultaneously, the pollen-bearing cluster can be gently brushed against the emasculated female. However, if pollen was collected earlier and stored, application with a soft paint brush is recommended. Care must be taken to sterilize the brush (and your hands) with 70 per cent ethanol between pollinations. After pollen has been thoroughly applied to a cluster, taking care to brush every flower, the paper bag is replaced over the cluster, both to prevent fertilization by other pollen sources, and to protect the developing fruit from birds and the elements. The bags also help to identify the cluster locations at harvest time, at which time the seeds are removed from the individual berries.

Planting the Seeds

Over the course of the season, berries on the cluster mature and ripen within the bags. The clusters are collected and the seeds extracted from within the berries. These seeds contain embryos of new plants with an equal amount of genetic material from each parent, and each is genetically distinct. In order to separate inviable seeds from the viable, the seeds are dropped into a small beaker of water. Inviable seeds tend to float, and are skimmed off and discarded. The remainder are cleaned of any remaining pulp, dried, and stored at -2°C. Grape seeds require stratification in order to germinate well. Stratification is provided by placing the seeds, along with a dilute solution of fungicide on filter paper, in plastic bags and placing them in the refrigerator (5°C) for approximately three months. Without this chilling period, seed germination rates would be very low and uniformity in growth among those germinating would be poor. There are many variations on this stratification technique. Some researchers have utilized hydrogen peroxide and or gibberellic acid to reduce the time required for stratification or to increase germination percentage. In March, the seeds are planted in small peat pots and germinated in the greenhouse. After danger of frost has passed, they are moved outdoors to a sheltered "cold frame" location, where they can be covered if sub-freezing temperatures are expected. In June, our seedlings are planted outdoors into an irrigated nursery site.

Making Selections

A long term commitment is necessary to find the best selections once you've successfully made crosses and produced seedlings. The more seedlings planted and the longer spent evaluating them, the better the chance of finding the exceedingly rare selection with potential. However, the reality is that there are always limits in terms of time, space, and manpower. The transplanted seedlings grown in the nursery into the vineyard. Many seedlings have been eliminated in the nursery, where they are grown without disease control. Only vines appearing to be disease resistant in the nursery are then transplanted to a permanent vineyard location for further evaluation. It is more difficult to select for some other characteristics such as wine quality or cold hardiness, and several years of growth in the vineyard required for proper evaluation. The criteria used depends on your goals, persistence, creativity, knowledge of grape genetics, careful evaluation and some amount of luck are the keys to success.

Expected Period for Developing Varieties through Breeding

Through Clonal Selection

In this technique asexual reproduction method is used for phytosanitary selection for keeping cultivars healthy and stable in yield, selection of variations within a cultivar considering the important economic traits like seedlessness, colour, shape, aroma, sugar content, loose cluster and early maturity. It takes about 10-15 years and commonly practiced in *Vitis vinifera*.

Through Hybridization

Rootstock development takes about 30-50 years, table grapes require about 15-20 years and wine grapes takes 25-30 years involving sexual reproduction.

Breeding though Marker-Assisted Selection

It involves the use of pre-selection criteria for eliminating highly susceptible vines under greenhouse followed by MAS for traits difficult to evaluate prior to planting in the vineyard. The various stages of testing, seedlings- (1 vine), pre- (10 vines), intermediate- (50 vines) and main testing (500 vines), with increasing numbers of vines are followed by trials in viticulture practice. Usually developing a new cultivar requires 25 to 30 years through traditional breeding tools. Acceleration of the breeding process for up to 10 years is expected by the use of MAS and by merging pre- and intermediate testing to one testing phase as the planting material becomes available in bulk.

Breeding Objectives

Grapevine breeding is time consuming due to a long generation cycle, the requirement of several repetitions caused by environmental impact on the traits to get sufficient evaluation data for selection, limited plant material and slow propagation rates through hardwood cuttings. Furthermore, breeding goals need to be redefined and diversified according to the grapes/plants uses:

General Breeding Objectives

- ☆ Clonal selection is performed within existing cultivars in order to keep the cultivar phytosanitary healthy and morphologically stable. Clonal selection makes use of the limited genetic variation given within a vegetatively propagated genotype (a cultivar) to select for variants (mutants) of certain traits. These may be loose clusters, higher sugar accumulation, aroma variants *etc.* Sometimes clonal variants have become independent cultivars. For example berry colour mutants of 'Pinot Noir' are 'Pinot Gris', 'Pinot Blanc' and a mutant with earlier ripening time is 'Pinot Précoce Noir'

- In contrast to clonal selection controlled sexual reproduction is required for hybridization allowing genetic segregation through meiosis and generating a wide genetic variation within the offspring. Depending on the utilization, rootstocks tolerant or resistant against phylloxera need to be distinguished from scions with fungal disease resistance with and better berry quality for either table or wine grape.
- The general breeding objectives for cross breeding are listed in Table 24.1. Achievement of the specific breeding goals for table or wine grapes respectively rootstocks requires totally independent breeding programmes and makes use of different kinds of genetic resources.

Table 24.1: General Objective for Grape Breeding Program

Major Trait	*Wine Grapes*	*Table Grapes*	*Rootstocks*
Quality	☆ High wine quality (*e.g.* high sugar, balanced acidity, flavours colour, body of a wine) ☆ Taste ☆ Free of off-flavours	☆ Seedlessness ☆ Taste ☆ Free of off-flavours ☆ Berry texture and colour	
Resistance/ tolerance (biotic)	☆ Phylloxera resistance leaf ☆ Phylloxera resistance root (with perspective for own rooting) ☆ Powdery mildew resistance ☆ Downy mildew resistance ☆ Botrytis rot resistance ☆ Black root resistance	☆ Phylloxera resistance leaf ☆ Powdery mildew resistance ☆ Downy mildew resistance ☆ Botrytis rot resistance ☆ Black root resistance	☆ Phylloxera tolerance or resistance of roots ☆ Nematode resistance
Resistance/ tolerance (abiotic)	☆ Frost resistance ☆ Drought tolerance ☆ Sun burn resistance	☆ Drought tolerance ☆ Sun burn resistance	☆ Lime tolerance ☆ Rooting ability
Others	☆ Climate adaptation ☆ Viticultural properties (*i.e.* upright growth, medium vigour)	☆ Climate change adaptation	☆ Callus formation and affinity for grafting ☆ Growth to support scion

Specific Breeding Objective

Table Grapes

In contrast to wine grape breeders, table grape breeders mainly performed crosses within *V. vinifera* L. subsp. *vinifera*, though recently the entire *Vitis* gene pool in particular breeding strains developed thereof became of increasing relevance in order to extent the genetic basis for the introduction of resistances. Breeding for seedlessness, taste, sweetness, colour, uniformity of colour, crispness, berry size (large but not more than 10 g), symmetric cluster architecture, *Botrytis* resistance, time of ripening (very early to very late for an extended availability on the market), shelf- life (transport stability, no release of berries form the peduncle) are important criteria for table grape breeding.

Subtropical Grapes

- ☆ To develop early maturing, seedless and sweet cultivars.
- ☆ To induce resistance to anthracnose, downy mildew, powdery mildew and chaffer beetle.
- ☆ To develop varieties with medium vigour and productive basal bud.

Tropical Grapes

- ☆ To develop high yielding and high quality varieties with increased fruitfulness of basal buds, less degree of apical dominance, suitability for different purpose such as table, raisin, wine and juice, and resistance to diseases.
- ☆ To develop rootstocks resistant to salinity, nematodes and drought.

Wine Grape

- ☆ High wine quality combined with high disease resistance and good climatic adaptation.
- ☆ Depending on the climatic conditions, cool climate viticulture or hot climate viticulture, the kind of disease resistance required may vary.
- ☆ In recent times environmental concerns of the public are an additional driving force to get improved grapevine cultivars requiring less pesticide applications. Hedrick and Anthony, summarizing work with *Vitis* species in 1915, provided some data for inheritance of self-sterility, sex of the flower, colour of berry peel, berry size, berry shape, berry quality, and berry ripening time. In terms of genetics, the only reliable conclusion which could be drawn was that berry colours black and red are dominant over white and white is homozygous recessive. However, inbreeding depression as a problem in grapevine breeding. The breeding objectives are given in Table 24.2.

Rootstock Breeding Objectives

For rootstock improvement mainly non-vinifera vines from the North American gene pool have been used for interspecific crosses. Despite of phylloxera resistance agronomical performance is the major issue in rootstock breeding since the grafted vine is influenced by many factors as yet poorly understood. Since *V. vinifera* is considered to be rather lime tolerant growing well on calcareous soils in Europe rootstocks need to be equally tolerant. The failure of the first generation rootstocks

Table 24.2: Breeding Objectives for Wine Grapes

Breeding Goals	*Range of Characteristics*		
1. Wine quality			
White	Fruity	Neutral	Muscat/aromatic
Red colour	Dark colour	Moderate colour	
Rich in various components	Tannins, flavonols	Amino acids	Potassium
Augar (hot or cold climate)	Medium	High	
Acidity (hot or cold climate)	High	Medium	
Off-flavour	None	None	None
Other wine taste characters	Well balanced taste	Wine with rich body	Long lasting wine
Aging potential	Medium aging potential	High	
2. Agronomic performance			
Resistances – fungi	*Erysiphe necator* (syn. *Uncinula necator*)	*Plasmopara viticola*	*Botryotinia fuckeliana* (syn. *Botrytis cinerea*)
	Black rot	Anthracnose	*Phomopsis viticola*
Resistance – bacteria	Pierce`s disease	*Agrobacterium*	
Resistances – insects	*Daktulosphaira vitifoliae*	*Xiphinema index* (vector for viruses)	
Resistances – abiotic factors	Frost	Drought	Sunburn
Berry ripening	Early	Middle	late
Wood maturation	Early	Middle	
Fruit characters	Loose cluster	Thickness of the berry peel	
3. Yield traits	< 1 kg/m^2	≤ 1.5 kg/m^2	> 1.5 kg/m^2
Berry size	small	medium	High
Berries per cluster	<200	200-300	>300
Cluster per cane	2	3	4

was mostly due to insufficient adaptation to this kind of soil. Thus, first rootstocks were poor mediators of iron and mineral uptake into the vine. Consequently, rootstock breeding aims at lime tolerance which prevents iron chlorosis on calcareous soils. Similarly, rootstocks should tolerate drought to guarantee high quality berry development even during hot and dry weather periods. A source known for drought tolerance is, *e.g. V. rupestris* Scheele. The quality of the tissue connection between scion and rootstock, so-called "affinity" is another characteristic, which is of crucial importance for the production of grafted vines. Also the ability to establish a good root system is of major importance in order to obtain a well and equally rooted grafted vine that can be established easily in the vineyard. The genetics of these traits still need to be investigated. The grape rootstock breeding objectives are given in Table 24.3.

Table 24.3: Objectives for Rootstock Breeding in Grape

Breeding Goals	*Range of Characteristics*		
1. Pest resistance			
Root phylloxera	Tolerance	Resistance	
Nematodes	Tolerance	Resistance	
Virus vectors	Resistance		
2. Grafting properties			
Affinity to scion	Good callus formation		
Rooting capability	High		
3. Agronomic performance			
Vigour	Low	Medium	High
Adaptation to calcareous soils	High		
Salt tolerance	Medium	High	
Drought tolerance	Medium	High	

Genetics of the Important Traits

Berry and Wine Quality

A first attempt to elucidate berry quality genetically was reported by Hedrick and Anthony (1915). Berry and wine quality is by far the most complex trait in grape breeding. It relies on complex sensory perceptions including taste, smell, and mouthfeel. Selection of good quality genotypes depends on the organoleptic perception of a tasting panel thus being rather subjective. Berry quality is difficult to evaluate for table grapes and even more difficult for wine grapes since must fermentation by yeasts increases the complexity of the trait through metabolic conversions. The amounts of sugars, acids, fermentable nitrogen (amino acids), minerals (*e.g.* potassium), balanced (positive) aroma compounds, and lack of off-flavours in the must are major components to estimate berry quality. In a wine, which is free of sugar after fermentation, any inharmonious taste can easily be recognized and off-flavours quickly emerge. Changes during storage and aging of wine need to be evaluated to uncover sensory deficits which are attributed to the breeding line. Within a breeding programme berry respectively must quality can be recorded only 4 to 5 years after a cross and it is strongly influenced by environmental factors. Up to now the trait "quality" was treated mostly empirically with the help of trained tasting panels and analytical measurements of major must components.

Berry Colour Formation

Berry colour varies in a wide range from green/yellow (considered as white) to many shades of red and purple to black. Several authors found berry colour as a dominant trait (Hedrick and Antony 1915) though the variation in colour expression is influenced by additional factors. Genetic studies during the years could not resolve further details. Genetic maps produced by applying molecular markers (see below) localized the ability to form dark-coloured berries as a single qualitative trait on chromosome 2. Using molecular tools transposon integration in a regulatory *myb* gene (a transcription factor regulating the gene for the last enzymatic conversion in anthocyanin biosynthesis) was identified as causal for the white phenotype. The expression of the *Myb* factor could widely explain the phenotypes qualitatively. The gene was found to co-segregate with the colour locus on chromosome 2.

Mildew Resistances

For a long time resistance breeding was dominated by selecting genotypes resistant to powdery mildew (*Erysiphe necator*, an Ascomycete) and downy mildew (*Plasmopara viticola*, an Oomycete) combined with high wine quality. In the 19th century breeders used resistant genotypes which were available and breeding material carrying some beneficial gene combinations, thus taking advantage of the breeding progress. A systematic approach to take advantage of genetic resources is the introgression of resistance traits from wild Vitis species followed by consecutive pseudo backcrosses with *V. vinifera* L. subsp. *vinifera*. An exceptionally good but also rare example is the introgression of the *run 1* locus of *M. rotundifolia* conferring resistance to powdery mildew. Recurrent pseudo backcrosses, *e.g.* for six generations can be estimated to last about 25 to 30 years and result statistically in less than 1 per cent of genetic material from the wild species remnant in the introgression line. Due to this huge time span it does not surprise that such an endeavour has rarely been taken during the last 200 years.

Classical Breeding of Grapes

A typical breeding programme consists of several consecutive steps decreasing the number of individuals in each selection step. Assuming a current breeding programme for grapes starts with 50,000 seedlings a year, greenhouse testing and screening for mildew resistances results in about 5,000 plants to be planted in a seedling plot (requiring about one hectare). Beyond the seedling stage, all further breeding steps require five to eight years of growth: year one to three to get the vine established and year four to eight for a full crop. Grapes from breeding lines showing good viticultural performance including sufficiently high levels of resistance will be used for table and wine making. This starts already from a single vine yielding frequently no more than one litre of wine. This can be achieved only by the development and application of markers monitoring distinct aspects of quality like sugars, acids, flavours, off-flavours *etc.* or which are correlated to important quality and yield traits like berry size, berry number, cluster size, cluster architecture, ripening time, ripening duration *etc.* At the beginning of the 21st century the tools become available. This marks the beginning of a paradigm shift from empirical to a knowledge-based and much more target-oriented grapevine breeding.

Modern Breeding Tools

Application of molecular markers, genetic mapping and whole genome sequencing combined with high throughput technologies are becoming the modern tools, which are more efficient and precise. One hundred years ago due to non-existence of suitable technologies Hedrick and Anthony (1915) were unable to dissect the genetic base of traits. For roughly a decade now we have started to learn more about where traits are located in the genome, how they are inherited, and how they are molecularly organized. For some traits valuable knowledge is accumulating that is relevant for breeding: Most important is the development of molecular markers.

Marker-Assisted Selection

The rapid development of molecular techniques and genome sequencing capacities will accelerate plant breeding. New tools, particular molecular markers permitting a new endeavour to dissect grapevine genetics. The breakthrough for grapevine came with PCR-based DNA amplification techniques. First genetic mapping studies using RAPD markers (randomly amplified polymorphic DNA) were described by application of a double-pseudo testcross strategy suitable for highly heterozygous plants such as grapevine and the first genetic map of grapevine was published shortly thereafter (Lodhi *et al.*, 1995).

Most successful was the development and application of DNA microsatellite analysis using STMS, sequence tagged microsatellite sites, also called SSR (simple sequence repeats). This type of molecular markers proved to be reliable, comparable, and robust permitting a more detailed analysis of genetically determined traits in grapevine. Microsatellites were used first for genotyping studies to unravel the descent of cultivars and were soon introduced into genetic mapping. Meanwhile several genetic maps have been developed using SSR or other marker types and combinations thereof providing the genetic framework required for QTL (quantitative trait locus) mapping combining genotypic and phenotypic information. This biostatistic analysis permits the dissection of complex traits that are polygenic and governed by several factors as QTL into a genetic map. It provides a rough localisation of the underlying genes and an orientation in the grapevine genome. Single nucleotide polymorphism (SNP) based markers will present the next generation of markers for applications in grapevine breeding. SNPs in grapevine have already been found to be frequent and useful for genetic analysis; their future, however, relies on high throughput analysis. SNP markers proved to be very useful for linkage analysis and could also be transferred within the genus *Vitis*. Their versatility for whole genome association studies, however, is in question since a rapid decay of linkage disequilibrium (LD) was found in grapevine. The LD drops down to background levels at an inter SNP distance of around 10 kb. Even in a small inter SNP distance of 50 bp LD is found to be very low. In this situation SNP analysis with very high marker numbers are necessary to detect any association to neighbouring alleles determining trait expression. It may be more productive to use cost efficient SNP genotyping for genetic mapping of segregating populations followed by QTL analysis rather than expensive high number SNP analyses for whole genome spanning association mapping. An alternative strategy may rely on whole genome sequencing approaches on now emerging second generation and 3rd generation sequencing platforms. Such approaches will become standard once the bioinformatic tools for rapid and correct genome sequence assembly from "sequence generation" sequencing reads become generally available and data management of huge datasets will be quickly possible.

Markers for Resistance

An allele specific marker for powdery mildew resistance was used to monitor inheritance in a segregating population. Eibach *et al.* (2007) gave an example of pyramiding resistance loci, two for resistance against *E. necator* and two for resistance against *P. viticola*. The examples show that for grapevine breeding programmes, which still in our days are operating empirically, marker assisted selection (MAS) is at the onset of utilization. In cultivar 'Regent', one major QTL for powdery mildew (chromosome 15) and two QTLs for downy mildew (on chromosomes 4 and 18) have been identified. Further two loci for powdery mildew resistance are available. A further locus for resistance against powdery mildew, *ren1*, could be identified on chromosome 13 in cv. 'Kishmish vatkana'.

Table 24.4: List of Relevant QTL/Loci for Grape Breeding

Trait	*Symbol*	*Associated Marker*	*Genome Position (Chr/Mb)*	*Source (Origin) Species*
Berry size (berry weight)	*be size*	SCC8	18/25.9	*Vitis vinifera*
Berry size (berry weight)		VMC7f2	18/26.9	*Vitis vinifera*
Monoterpene content		DXS1	5/3.8	*Vitis vinifera*
Linalool content		Cnd41	10/	*Vitis vinifera*
Linalool content		VrZAG64	10/13.4	*Vitis vinifera*
Linalool content		VMC3d7	10/10.8	*Vitis vinifera*
Fleshless berry	*flb*	VMC2A3	10/0.9	Ugni Blanc Mutant
Berry skin colour	*mybA*		2/14.2	*Vitis vinifera*
Pierce's diasese	*Pdr1*	VMCNg3h8	14/25.3	*Vitis arizonica*
Pierce's diasese		VVIn64	14/26.6	
Pierce's diasese		UDV-095	14/26.1	
Daktulosphaira vitifoliae	*rdv1*	Gf13_9	13/21.9	*Vitis cinerea*
Daktulosphaira vitifoliae		VMC8e6	13/22.5	
Plasmopara viticola	*rpv1*	VMC72	12/-	*M. rotundifolia*
Plasmopara viticola		VVIb32	12/10.3	*M. rotundifolia*
Plasmopara viticola	*rpv2*		18	*M. rotundifolia*
Plasmopara viticola	*3pv3*	UDV-112	18/-	'Regent'
Plasmopara viticola		VVIn16(2)	18/23.4	'Bianca'
Plasmopara viticola		UDV-305	18/24.9	
		VMC/F_2	18/26.9	
Plasmopara viticola	*rpv4*	VMC7h3	4/4.7	'Regent'
Plasmopara viticola		VMCNg2e2.1	4/4.7	'Regent'
Plasmopara viticola	*rpv5*	VVIo52b	9/4.0	*Vitis riparia*
Plasmopara viticola	*rpv6*	VMC8G9	12/20.4	*Vitis riparia*
Plasmopara viticola	*rpv7*	UDV-097	7/11.4	*'Bianca'*
Erysiphe necator	*ren1*	UDV-020	13/-	'Kishmish vatkana'
Erysiphe necator		VMC9h4-2	13/18.4	

Trait	Symbol	Associated Marker	Genome Position (Chr/Mb)	Source (Origin) Species
Erysiphe necator		VMCNg4e10.1	13/18.4	
Erysiphe necator	*ren3*	UDV-015b	15/7.1	'Regent'
		VVIv67	15/10.9	'Regent'
Erysiphe (*Uncinula*) *necator*	*run1*	VMC1g3.2	12/10.0	VRH3082-1-42 (*Muscadinia rotundifolia*)
		VMC4f3.1	12/13.1	
Seed development inhibitor	*sdl*	SCC8	18/25.9	
Seedlessness	*sdl*	VMC7f2	18/26.9	'Autumn Seedless'
		VMC6f11	18/23.2	
Sex	*sex*	VVMD34	2/3.7	
		VVS3	2/4.2	
		VVIb23	2/4.9	
	ufgt	SCAR	16/2.3	
Véraison	*ver(4)*	VMC1E11	16/13.7	'Regent'
Xiphinema index	*xir1*	VMC5a10	19/20.9	
Anthocyanin 3,5-diglucosides	*5-gt*	Gf09_01	9/6.5	*Vitis arizonica* 'Regent'

Note: Only few QTLs have been indicated in the table. However, several more QTLs are available in literature.

Markers for Berry Quality Traits

Recently the biosynthesis of tartaric acid contributing to taste, mouthfeel, and aging potential received some interest, since too low acidity in hot climate viticulture is a major quality issue. Tartaric acid in grapevine is a product of vitamin C (ascorbate) catabolism. The ascorbate metabolism regulatory aspects are the important aspects to fix its mechanism. Another important trait is the colour of the berries which is caused by the synthesis of anthocyanins in the berry skin of red and black genotypes in the second ripening phase after véraison. The key biosynthetic enzyme for anthocyanin formation, UDP-glucose:flavonoid 3-O-glucosyltransferase (UFGT), has been mapped on chromosome 16 by using a SCAR marker. More important for colour formation is the transcription factor *MybA* that controls UFGT gene expression. The *mybA* gene is located on chromosome 2. Due to a transposon-based mutation within the promoter of one allele of the *mybA* gene the development of a molecular marker is now possible correlating very tightly with berry skin colour. This transposon insertion is tightly correlated with white berry colour. Colour variants could be explained in 95 per cent of the cases by different alleles of the *mybA*1 gene showing molecular fingerprints of transposon excision. Further modulation of colour can be explained by different expression of genes for anthocyanin modifying enzymes. In terms of genetic unterstanding another modification which has been introgressed into *V. vinifera* L. subsp. *vinifera* has been much easier to be accomplished. Among the anthocyanins two major types exist: anthocyanin 3-glucosides and anthocyanin 3,5-diglucosides (mainly malvin). Anthocyanin 3-glucosides are found in all coloured grapes whereas anthocyanin 3,5-diglucosides occur in most wild *Vitis* species and in derivatives of crosses of *V. vinifera* L. subsp. *vinifera* with wild *Vitis* species. They are absent on very low level in traditional *V. vinifera* L. subsp. *vinifera* cultivars. Anthocyanin 5-glucosyltransferase (5-GT) is the responsible enzyme catalyzing the glycosylation reaction from anthocyanin 3-glucoside to anthocyanin 3,5-diglucoside. Expression of the *5-gt* gene correlates positively with anthocyanin 3,5-diglucoside formation in berry skins of different grape genotypes. Therefore the gene encoding 5-GT was cloned and sequenced from different *Vitis* genotypes. The 5-gt alleles from traditional *V. vinifera* genotypes showed mutations leading to non-functional gene products in contrast to a functional *5-GT* originally descended from a wild *Vitis* species. Based on the sequence differences in the *5-gt* alleles a molecular marker was developed. Using this *5-gt* sequences characterized amplified region (SCAR) marker the *5-gt* gene was mapped on chromosome 9 at the same site where the trait 'malvin' has been previously localized. Since malvin is very intense in colour and quite stable it may be used to develop cultivars with dark coloured berries.

Markers for other Traits

Despite these perspectives current markers have been assigned to traits such as seedlessness or resistances and can be used for selection of particular traits. Seedlessness could be scored easily by markers developed by Striem *et al.* (1992, 1996) or Adam-Blondon *et al.* (2001). Similarly, Doligez *et al.* (2002) developed markers for seedlessness and berry weight. Several publications identified the locus for sex of the flowers on chromosome 2 a trait of interest *e.g.* for developing introgression lines. A major QTL for begin of berry ripening (veraison) was found on chromosome 16 for 'Regent' x 'Lemberger', for 'Italia' x 'Big Perlon', and for Gf.Ga-47-42 x 'Villard blanc'. Markers for resistance against Pierce's disease are available as well as makers for phylloxera resistance. A QTL influencing Magnesium-update was identified on LG 11.

Marker-Assisted Backcrossing

Marker-assisted backcrossing (MABC), offers the possibility at each generation to select for a maximum of *V. vinifera* L. subsp. *vinifera* genome while maintaining the trait of interest. Using the pseudo backcross approach in pBC5 theoretically 1.6 per cent of the non-recurrent (wild ancestor) genome remains in the introgression line. This can be accelerated by background selection identifying those genotypes in a progeny that, due to recombination in meiosis, received a higher proportion of the *V. vinifera* L. subsp. *vinifera* genome. Selecting against the wild ancestor about two generations *i.e.* eight years might be saved calculating with 4 years generation time and cultivation in the vineyard. Based on this calculation introgression requires 16 instead of 24 years. Reducing the generation time due to greenhouse cultivation the goal might be achieved already within 8-10 years. For such an approach five to ten markers per chromosome should be sufficient *i.e.* about 200 markers equally distributed throughout the genome. Preliminary analysis in a MABC population of 300 pBC1 seedlings subjected to background selection revealed three plants showing more than 85 per cent of *V. vinifera* genome when 75 per cent are statistically expected. These three plants were found in the 50 per cent of plants carrying the locus of interest. Thus, as long as a single locus is concerned population sizes of at least 300 plants give a reasonable basis for running a MABC programme to find desirable recombinants. New marker techniques based on SNP analysis will permit the investigation of 300 plants with 200 markers (60,000 data points) within a few days leaving sufficient time to integrate such analyses in breeding programmes and their tight time schedule.

Map-Based Cloning Approaches

To understand the mechanism of how a trait is expressed the responsible gene needs to be isolated. Having genetic maps this can be achieved by map-based cloning approaches. In principle molecular markers are to be identified successively reducing the distance between markers and the trait locus down to a distance permitting cloning of the locus. Two close markers are required flanking the locus of interest. A straight forward approach takes advantage of the reference genome sequence of PN40024 and requires co-linearity between the two genome regions (PN40024 and the locus of interest). Around the desired locus, *e.g.* an SSR based marker can be deduced from PN40024 and placed on the genetic map moving towards the locus of interest. For grapevine an ideal distance would be around 1 cM (statistically ca. 300 kb) or below. Isolation of marker-carrying BACs (bacterial artificial chromosomes) followed by identifying overlapping clones from both sides of a locus will reveal at a certain point in time an overlap of clones and thus a continuous physical map, a BAC contig, spanning the locus. In a final step the BACs will be sequenced and candidate genes for the trait can be identified.

Genome Sequencing

The best marker is a marker identifying the desired allele of the corresponding gene. The new sequencing options in terms of efficiency and low costs open novel possibilities. Since the first grapevine genome sequence was published dozens of cultivars and accessions including *Vitis* species. have been re-sequenced or their re-sequencing is in giving rise to thousands of SNP markers opening a huge potential of applications such as genotyping or high resolution gene mapping. Progress in sequencing technologies and decreasing costs for sequencing will permit within the next few years to sequence any genotype of choice. Thus, a genome sequence of grapevine (500 Mb) will be obtained easily and markers are coming not only for a locus but for the desired allele or haplotype. Further map-based approaches will no longer rely on BAC clones to get the gene. The genome sequence and a fine genetic map will permit identification of the corresponding gene and its alleles.

In vitro Culture and Genetic Engineering

High heterozygosity and inbreeding depression prevent an improvement of existing cultivars by classical cross breeding techniques. Thus, for marketing and from biological point of view improvements of traditional cultivars are exclusively possible by genetic modifications. Only in this case the cultivar name eventually could be maintained and the characteristics of a cultivar like quality traits will be preserved while deficiencies like disease susceptibility can be improved. Thus, primary genetic modifications within a grape breeding programme should be focussed on the improvement of traditional cultivars for tolerance or resistance against biotic (*e.g.* fungus, insect, virus resistance) or abiotic stress factors (*e.g.*, heat, drought, cold tolerance).

Limitations of Grapevine Transformation

Grapevine transformation proved to be very difficult as recalcitrant. Though most grapevine cultivars are a host for *Agrobaterium vitis* infection, highly efficient transformation protocols are restricted to specific cultivars like 'Thompson Seedless' as a table grape, or the wine grapes 'Cabernet Sauvignon', 'Chardonnay', 'Chancellor' or 'Merlot' as well as the rootstock cultivars '41B' or '3309 C'. Thus, generally speaking transformation suffers from insufficient regeneration systems. This particularly includes the crucial differentiation step from a somatic embryo to an entire plant, the so called "conversion" of the germinating embryo to intact rooted plantlets. Currently, no GM-vine of variety has reached the market. Public concern seems to be a more general retarding aspect but, except for a few examples, good genes for traits are the other missing issue. Since virus resistance was not found in *Vitis*, GM-vines could be an interesting solution and have attracted researches since the early 1990s. Rootstocks showing virus resistance have been obtained but a cultivar is not yet available. *Vitis* does not carry resistances against the wood disease eutypa dieback caused by *Eutypa lata*. In a transgenic approach developed for rootstock plants expressing a gene from *Vigna radiata* encoding a NADP- dependent aldehyde reductase (Vr-ERE), an enzyme converting eutypine, a toxin from *Eutypa lata*, into its corresponding non-toxic alcohol. Transgenic plants cultivated *in vitro* showing a high VrERE expression were not affected by relatively high concentrations of eutypine whereas growth and development of untransformed control plants were substantially retarded. Several attempts have been made to improve grapevine for mildew resistance. Field resistance has not yet been observed. A promising approach could be the expression of the *run1* gene. Results of a transgenic approach are pending. Quality aspects for particular purposes have been addressed.

Gene Function Analysis

Genome analysis carried out around the world aims at resolving the molecular basis of important traits and how to elucidate gene function. Within the next few years plenty of candidate genes for interesting traits will become available. Transformation is highly important to elucidate their function but time consuming as it requires about one year to get a transgenic plant for analysis. If berry traits are to be studied it takes even longer. Thus, fast systems for functional studies become more and more important. Recently transient gene expression system based on agroinfiltration in homologous or heterologous systems have been used. The methods need to be refined. However, transient gene expression analysis will provide a shortcut only for some gene function studies. It will not replace stable transformation and field testing.

Grape Breeding in India and Accomplishments

Indian Agricultural Research Institute (IARI) is the founder institution in the country which started grape breeding programme through introduction of genotypes from USA, Former USSR, France, Italy *etc.* The major breeding aim was to develop a variety for subtropical climate with early maturity, acceptable level of TSS/acid ratio, seedlessness (table purpose), coloured (juice type), bold berry, self-thinning, and higher yield. In addition, several hybrids were developed at ICAR-Indian Institute of Horticultural Research, Bengaluru, ICAR-National Research Centre on Grapes, Pune and State Agricultural Universities. Several varieties/ hybrids were developed for table and juice purposes. Salient characteristics of some of them have been given below:

Arkavati: It is a cross between 'Black Champa' and 'Thompson Seedless'. Berries are medium, yellowish-green, ellipsoidal-spherical, sweet, seedless with 22-25 per cent TSS. It is multi purpose variety used for making raisins and wine (Negi and Randhawa, 1980).

Arka Hans: It is a cross between 'Bangalore Blue' and 'Anab-E- Shahi' released in 1992. Berries are yellowish-green, ellipsoidal-spherical, seeded with pleasant flavour. The TSS is 18-21 per cent. Variety is used for wine making (Negi and Randhawa, 1980).

Arka Kanchan: This late maturing variety is a cross between 'Anab-E- Shahi' and 'Queen of the Vineyards' or 'Koenigin der Weingaerten' released in 1996. Berries are golden yellow, ellipsoidal to ovoid, seeded, with pleasant 'Muscat' flavour. The variety has a TSS of 19-22 per cent and suitable for table purpose and wine making. Variety has a high yield potential but with poor keeping quality (Negi and Randhawa, 1980).

Arka Shyam: It is a cross between 'Bangalore Blue' and 'Black Champa'. Berries are medium large, shiny black, spherical to ovoid, seeded, with mild flavour. The variety is resistant to anthracnose. It is suitable for wine making and table purpose (Negi and Randhawa, 1980).

Arka Neel Mani: It is a cross between 'Black Champa' and 'Thompson Seedless' released in 1976 from IIHR, Bangalore. Berries are black seedless with crispy pulp having 20-22 per cent TSS. The variety is tolerant to anthracnose. Average yield is 28 t/ha. It is suitable for wine making and table purpose.

Arka Shweta: It is a cross between 'Anab-E-Shahi' and 'Thompson Seedless' released in 2006. Berries are yellow, ovoid, seedless with 18-19 per cent TSS. Average yield is 31 t/ha. Variety is used for table purpose and has a good export potential.

Arka Majestic: It is a cross between 'Angur Kalan' and 'Black Champa' released in 1983. Berries are deep tan coloured, uniform, round, seeded with 18-20 per cent TSS. Variety is tolerant to anthracnose. Average yield is 38 t/ha. Variety has a good export potential.

Arka Chitra: It is a cross between 'Angur Kalan' and 'Anab-E-Shahi' released in 1989. Berries are golden-yellow with pink blush, slightly elongated having 20-21 per cent TSS. Variety is tolerant to powdery mildew. Average yield is 38 t/ha. The berries are very attractive and thus suitable for table purpose.

Arka Krishna: It is a cross between ''Black Champa' and 'Thompson Seedless' released 1996. Berries are dark coloured, seedless, round to ovoid with 20-21 per cent TSS. Average yield is 33 t/ha. The variety is suitable for juice making.

Arka Soma: It is a cross between 'Anab-E-Shahi' and 'Queen of Vineyards' released in 2006. Berries are greenish-yellow, round to ovoid. Pulp is meaty and has 'Muscat' flavour with 20-21 per cent TSS. The variety is tolerant to anthracnose, downy mildew and powdery mildew. Average yield is 40t/ha. It is good for preparing white desert wine.

Arka Trishna: It is a cross between 'Bangalore Blue' and 'Convent Large Black' released in 2006. Berries are deep tan in colour, round to ovoid with 22-23 per cent TSS. Variety is resistant to anthracnose and tolerant to downy mildew. Average yield is 26 t/ha. It is a good variety for wine making.

Cheema Sahebi: A selection (Sel. 7) made by Dr G. S. Cheema from open-pollinated seedlings of Pandhari Sahebi at Ganeshkhind nursery, Pune. It is a vigorous and a very heavy yielder though very late in ripening.

Kali Sahebi: This variety is grown in the states of Maharashtra and Andhra Pradesh on a small scale. Berries are large, oval cylindrical, reddish-purple and seeded. The TSS is 22 per cent. Variety is susceptible to rust and downy mildew. Average yield is 12-18 t/ha. The variety is suitable for table purpose.

Manjari Naveen: Its is a clonal selection from the Centennial Seedless (Gold x Q25-6 (F_2 Emperor x Pirovana 75) bred by Drs H.P. Olmo and Albert T. Koyoma at the University of California, Davis, California, USA and released by NRC on Grapes, Pune in 2002. The optimum ripe fruit have 16-18°Brix with 0.45-0.55 per cent total acidity. The fruits are ready for harvest between 115-120 days after pruning being an early maturing variety.

Punjab Purple: It is hybrid of Catawbax Beauty (H-516) Seedless breed at ARI, Pune. It ripens in first week of June and is rich in anthocyanins - a source of antioxidants and is suitable for processing into juice, nectar and ready of serve beverage. It is black in colour and juicy. Potential yield 24.52 t/ha under Y system had significantly higher TSS (17.7 per cent).

Pusa Seedless: It is a selection made from Thompson Seedless at IARI, New Delhi. It resembles its parent Thompson Seedless with regard to most of the characters but its berries are more elongated. It was recommended for commercial cultivation in 1976. It is highly responsive to GA_3 application. The berries have high TSS (22-24 °Brix) content. It is suitable for both table purpose and raisin making.

Pusa Navrang: It is a hybrid between 'Madeleine Angevine' and 'Rubired' developed in 1996. It is an early ripening (1st week of June), basal bearer and teinturier variety containing red pigment both in peel and pulp. The bunch is loose, medium in size with round and medium sized berries with TSS ranging from 18-20 per cent. It is ideally suited for coloured juice and wine making and is resistant to anthracnose disease.

Pusa Urvashi: It is a hybrid between 'Hur' and 'Beauty Seedless' developed in 1996. It is early ripening (1st week of June) and a basal bearer. Its bunch is loose and medium in size with seedless greenish-yellow berries. The berry TSS varies from 20 to 22 per cent. It is suitable for table purpose and raisin making.

Pusa Aditi: It is a hybrid between 'Banqui Abyad' and 'Perlette'. It is early maturing (1st week of June), seedless having large round berry (2.7 g), yellowish-green in colour, firm pulp with good TSS (19.3°Brix). Vine is moderately vigorous, and is a spur pruned variety. It is tolerant to anthracnose and powdery mildew. Grape bunches and berries are GA3 responsive. The fruits are good for table purpose and making juice. The average yield is about 12-15 t/ha.

Pusa Trishar: It is a hybrid between {('Hur' x 'Bharat Early') x 'Beauty Seedless'}'. It is early maturing, suitable for sub-tropical conditions, semi-vigorous and spur-pruned. It matures by 10th June. Berries are round (2.15 g), yellowish-green in colour, with good TSS (18.4°Brix) and firm pulp. Grape bunches and berries are GA3 responsive. The fruits are good for table use and juice making purpose. The average yield is about 14-16 t/ha.

Pusa Swarnika: It is a hybrid between 'Hur' x 'Cardinal'. Large berry, round golden-yellow colour with firm pulp. Berries are very sweet having high TSS (20-22 °Brix). It has natural loose bunches with natural bold berries. Average bunch size is 409 g and bunch length 20 cm. The fruits are ready for harvest between 75-80 days after full bloom being early maturing variety. It is suitable for table purpose, juice making and *Munnakka* preparation.

Sharad Seedless: It is a clonal selection from Kishmish Cherney. It is seedless and the berries are seedless, black, crisp and very sweet. The TSS is upto 24°Brix. It responds well to GA and has a god shelf-life. It is grown mainly as table purpose variety. Berries are oblong to elliptical in shape and bluish black in colour with TSS 18-21°Brix and acidity ranging from 0.5-0.7 per cent.

Prominent Grape Varieties Worldwide

Autumn Royal: It a hybrid (Autumn Black x C74-1) and a late-maturing black seedless table grape developed by Dr David Ramming and Ron Tarailo of the USDA-ARS in Fresno, California and released in 1996.

Beauty Seedless: It is a hybrid of Scolokertek Kiralynoje x Queen of Vineyards bred at the University of California bred by Dr H.P. Olmo in 1954. It was evaluated and recommended from IARI, New Delhi for commercial cultivation. It has medium vine vigour, medium bunch weight (250 g), well-filled bunches, berry red to bluish-black, medium size (1.5-1.7 g) berry, good for juice making, It matures by end May to first week of June. Fruit responds well to girdling and gibberellin. Berries are ovoid in shape and the skin is bluish-black in color

Cardinal: Evolved from cross between Tokay x Ribier made by E. Snyder at California. It is medium in vigour, early ripening, red coloured, seeded table grape, loose bunch, medium size berries, sweet and changes to reddish-black on ripening. Bunch may suffer from uneven berry sizes.

Crimson Seedless: It was previously known as selection C102-26, resulted from a cross of Emperor x C33-199 made by Drs David Ramming, Ronald Trailo, Sayed Badr at the Fresno campus of University of California. Its parentage also includes Italia, Calmeria, Muscat of Alexandria and Sultanina. The cultivar produces medium sized (500 g), conical, well filled to slightly compact clusters with a shoulder. Crimson Seedless is the latest ripening seedless table grape currently produced in California and other grape crowing counties. The berries are medium in size (3-4 g) and are cylindrical to oval in shape. The skin is medium in thickness and the flesh is translucent and firm to crisp.

Christmas Rose: It was evolved by Dr. H.P. Olmo and Albert T. Koyama as a complex hybrid {S44-35c x 9117D (complex hybrid derived from Emperor, Hunisa, Piranova 75 and Nocera)}. It was released in 1980 from the University of California Davis. Its berries are large, dark red, and seeded with a tart and sweet flavour. In the store, look for grape bunches that are heavy and well coloured with plump berries well attached to the stems.

Delight: It is sister seedling of Perlette bred by Dr H.P. Olmo and released in 1947. It is a hybrid between Scolokertekhiralynoje 26 x Sultana Marble. Very early ripening and can be grown in North Indian conditions. Mature by first week of June have medium bunches with small, round green berries. Fit for both table purpose and raisin making.

Fiesta: It was evolved as a result of a complex cross {(Calmeria x 43-13N x (Cardinal x Sultanina x 14-13S; parentage of 43-13N and 43-13S Red Malaga x Tifafihi Ahmer x (Muscat of Alexandria x Sultanina)} made in 1965 by Drs John H. Weinberger and F.N. Harmon of the USDA Agricultural Research Service at the U.S. Horticultural Field Station in Fresno. It was released for commercial production in 1973.

Flame Seedless: It was bred by Dr J.H. Weinberger and F.N. Harmon at University of California in 1973 as result of complex cross (Cardinal x Sultanina) x [(Red Malaga x Tifafihi Ahmer) x (Muscat of Alexandria x Sultanina)]. It is a vigorous, heavy-bearing table grape cultivar that keeps well in storage. The Flame Seedless was first introduced in the 1970s. It is the second-most popular grape in the United States

Perlette: Evolved by Dr H.P. Olmo as a cross between Scolokertekhiralynoje 26 x Sultana Marble at the University of California, USA in 1936 and was introduced in India during 1958. It has small round berries, early ripening, seedless and basal bearer. It is very popular in sub-tropical plains. It possess compact berries and sometimes bear irregular berries. The juice is clear and green with 16-18 per cent TSS. Keeping quality is good and is used for table purpose. Average yield is up to 35 t/ha.

Red Globe: It is a table grape variety having very large, red seeded berries with firm flesh. It can be grown outdoors in very warm areas with long growing seasons such as California or Australia, but in most parts of the world it is strictly a greenhouse grape. In Japan it is in great demand. A large chunk of Red Globe production in the USA and Australia are exported to Asian markets.

Thompson Seedless: Originating in Asia Minor, *i.e.* Turkey this variety was grown by William Thompson in Yuba City in California and it was John Onstott who realized the commercial potential of the cultivar and achieved great success with it. Berry is yellowish-green and very responsive to GA_3 application. It is popular variety of tropical region but in northern plains, and is mid-ripening. The juice is straw coloured, sweet with a TSS of 20-22 per cent. Variety has a good keeping quality and is used for table purpose and raisin making. Average yield is 20-25 t/ ha. Several superior clones of it have been made namely Tas-a-Ganesh, Sonaka, Manik Chaman and Monika.

Future Thrusts

Since immemorial time grapes are important fruit crops and used widely as fresh, wine, juice and resins. With the dissemination of the two mildew pathogens, other fungi, and phylloxera around the world the old tradition of viticulture experienced major changes: cultivation of grafted vines and intense chemical plant protection became necessary. Environmental concerns and new threats coming along with climatic change enforce adaptations on the plant itself. On the long term the only solution will be a genetic improvement of the grapevine plant to face the major threats either by growing newly selected cultivars or bringing existing cultivars to perfection. As the history of grapevine breeding taught us, continuous and sustainable efforts of breeders will provide the solutions even if it requires decades. However, there is considerable room for the expectation of reaching solutions much faster than in previous decades: (i) we face about 200 years of progress within the breeding material and (ii) molecular genetic technologies offer unprecedented possibilities for selection. Soon breeding will no longer require 25 to 30 years to get a new cultivar. This time span is expected to be reduced by up to 10 years. In spite of all expectations for acceleration of grapevine breeding there are some biological restrictions: simply the availability and propagation of the plant material will become a limiting factor within the breeding process. Thus, irrespective of the shortening of time, sustainability of improvements will become more relevant. Modern breeding tools make this challenge accomplishable.

- Today it is possible to address single loci by molecular markers (MAS) and to combine (pyramide) several resistance loci acting against a disease in a plant to achieve a better chance of durability of resistance. Several loci for powdery mildew and for downy mildew resistance are known other loci will follow as well as resistance loci directed against other pathogens. Moreover, resistances against various threats can be easily combined by MAS. The next generation of grapevine cultivars will have multiple resistances against several biological stress factors.
- Now it is possible to run marker assisted backcrossing (MABC) programmes to introgress traits of interest into the *V. vinifera* gene pool within a reasonable time frame.
- High throughput techniques are available for genotyping and for genome sequencing upgrading the breeder's toolbox. A description of the

haplotype is possible and desired alleles can be addressed and monitored within a breeding programme. Genes can be isolated and their function and the underlying mechanisms can be elucidated.

- Existing cultivars can be improved though some technical difficulties need to be overcome.
- GM-grapevines are a current possibility. Fungus resistance for environmentally friendly viticulture could be an argument in the public debate for acceptance.
- Phylloxera resistance as breeding goal can be reconsidered directing viticulture on the long term back to own-rooted cultivation of *Vitis vinifera*.
- Several developments are very much advanced and their contribution or their output will become soon visible. However, there are still some missing links which require further research and development and some more time.
- Major missing links are highly efficient phenotyping tools. Today phenotyping possibilities for grapevine are far behind the genotyping options.
- Markers describing quality are required. The marker description of positive and negative characters will surely be developed. Markers can be imagined for sugars, acids, certain aroma compounds, off-flavours, tannins, *etc.* It is an open question how deep a quality description can go. Unknown minor aroma compounds can have a major impact on sensory perception. The bouquet of a wine is influenced by the matrix of the wine. The body of a wine is not described in terms of compounds. Quality is probably the trait most deeply influenced by the environment. In order to reduce it to a genotypic description requires a very deep evaluation.
- Elucidation of the various mechanisms of resistance in orders to pyramide the best suited resistances.
- Genetic resources – more precisely wild species – should be evaluated in an internationally complementary manner. Core collections for a species could be developed based on genetic distance determined by markers to maintain and manage that gene pool efficiently. This would provide the opportunity to make important traits accessible on the long term within a minimal set of individuals and eventually to develop introgression line. Otherwise breeders will select their material at a given time for a particular trait and discard plants valuable from a different perspective material.

Finally, if the appropriate methods are established, a cost-benefit calculation will show what will be accomplishable in the breeder's hands and what will remain a dream. The shift from empirical to a systematic knowledge based breeding is taking place. As a consequence the chances of success in grapevine breeding have become more promising since ever.

REFERENCES

APEDA 2017. APEDA. www.agriexchange.apeda.gov.in. Data retrieved on 7th January, 2019.

FAO 2017. FAOSTAT. Data retrieved on 7th January, 2019.

NHB 2017. NHB. Data retrieved on 7th January, 2019.

Bouquet Alain, Torregrosa Laurent, Iocco Pat and Thomas Mark R. 2008. Grape. In: *Compendium of Transgenic Crop Plants: Transgenic Temperate Fruits and Nuts* (Eds.) Kole Chittaranjan and Hall, Timothy C., Wiley-Blackwell Publishing House, Singapore, pp. 189-220.

Olmo, H. P. 1986. The potential role of (*vinifera* x *rotundifolia*) hybrids in grape variety improvement. *Experientia*, 42: 921-926.

Owens, C.L. 2008. Grapes. In: *Temperate Fruit Crop Breeding: Germplasm to Genomics* (Ed.) James F. Hancock, Springer, East Lansing, USA, pp. 197-234.

Reisch, Bruce J. and Pratt Charlotte 1996. Grapes. In: *Fruit Breeding, Vol-II: Vie and Small Fruits*. John Wiley and Sons, Inc, New York, USA, pp. 297-370.

Riaz, S., Doligez, A., Henry, R.J. and Walker, M.A. 2007. Grape. In: *Genome Mapping and Molecuar Breeding in Plants: Fruits and Nuts* (Ed.) Chittaranjan Kole, Springer-Verlag Berlin Heidelberg, pp. 63-94.

Töpfer Reinhard, Hausmann Ludger, Harst Margit, Maul Erika, Zyprian Eva, Eibach Rudolf. 2011. New Horizons for Grapevine Breeding. Julius Kühn-Institut - Federal Research Centre for Cultivated Plants, Institute for Grapevine Breeding Geilweilerhof, 76833 Siebeldingen, Germany.

Wan, Y., Schwaninger, H.R., Baldo, A.M., Labate, J.A., Zhong, G.Y. and Simon, C.J., 2013. A phylogenetic analysis of the grape genus (*Vitis* L.) reveals broad reticulation and concurrent diversification during neogene and quaternary climate change. *BMC Evol Biol*. 13: 141. doi: 10.1186/1471-2148-13-141.

Breeding of Perennial Horticultural Crops *Pages* **315–331**
Editors: S.K. Singh, V.B. Patel, A.K.Goswami, Jai Prakash and Chavlesh Kumar
Published by: **BIOTECH BOOKS, NEW DELHI**

***Chapter* 25**

Guava Improvement

A.K. Goswami, A. Nagaraja, Madhubala Thakre and Chavlesh Kumar

Division of Fruits and Horticultural Technology,
ICAR-Indian Institute of Agricultural Research, Pusa, New Delhi – 110 012

INTRODUCTION

Guava tree (*Psidium guajava* L.) has its origin in the American tropics and is today distributed throughout the tropical and subtropical areas of the world. It belongs to the Myrtaceae family, comprising a large number of fruit yielding species. Guava is one of the most popular fruits of Indian subcontinent. Guava fruit is important for fresh consumption and for substantial industrial preservation. Owing to high production and wide adoptability, it is considered common man's fruits and is rightly called 'the apple of tropics'. The fruit is rich source of vitamin C and pectin apart from other nutrients. In India guava is producing in almost all the states producing at least two crops in a year. In India, guava rank 5th in area and production. The total area of guava in India is 0.27 million hectare with a production of 3.66 million tonnes (anonymous, 2015). Guava fruit is commercially important in many countries like Australia, Bangladesh, Brunei, Cambodia, Cameroon, China, Costa Rica, Cote d'Ivoire, Cuba, Dominican Republic, Ecuador, Eritrea, Ethiopia, Fiji, Gabon, Gambia, Greece, Guyana, Haiti, India, Indonesia, Israel, Kenya, Laos, Malawi, Malaysia, Myanmar, Nigeria, Pakistan, Panama, Philippines, Puerto Rico, Samoa, Senegal, South Africa, Sri Lanka, Sudan, Tanzania, Thailand, Togo, Uganda, Venezuela, Vietnam and is popular due to its year round availability, rich nutritional and medicinal value, and affordable price, suitability for transportation, handling and consumer preference. Guava is rich in vitamins 'A' and 'C' with seeds that are rich in omega-3, omega-6 polyun- saturated fatty acids and especially dietary fibre and also has good level of the dietary minerals, potassium, magnesium, and generally a broad, low-calorie profile of essential nutrients. Guava contains both carotenoids and polyphenols giving them relatively high dietary antioxidant value among plant foods. Red or orange colour guavas have more potential value as antioxidant sources than yellowish-green ones.

Cytology

Most of the commercial varieties in guava are diploids, the chromosome number being 2n = 22, except the seedless types which are triploids in nature (Kumar and Ranade, 1952). Cytological studies made on structure and behaviour in different varieties of *P. guajava* by several workers indicated that meiosis was normal with formation of 11 bivalents at diakinesis, and normal distribution of chromosomes at later stages (Raman *et al.*, 1969). A natural triploid with somatic chromosome number of 2n = 33 was reported by Kumar and Ranade (1952). The same chromosome number was reported by Raman *et al.* (1971) in a seedless variety of *P. guajava* suggesting that triploidy is the cause of seedlessness in guava. Shafaat Mohammed (1975) studied breeding behaviour of the aneuploids in guava (*Psidium guajava* L.) such as trisomic, tetrasomic and higher aneuploids. He observed that reciprocal crosses between aneuploids and diploids indicated less than 100 per cent crossability. The aneuploids, when used as the male parent, crossed less frequently than as female parents and some aneuploids crossed more readily than others. Differences were observed in fruit size, fruit weight, and seed number in reciprocal crosses. The extra chromosome was found to be transmitted through both the egg cell and the pollen. However, frequency of transmission was greater through the egg cell than the pollen. As high as 26 per cent transmission of extra chromosomes were obtained through the egg cell. There was no clear cut difference between trisomics and higher aneuploids with regard to frequency of transmission of the extra chromosomes. In guava, where large number of seeds is a disadvantage, aneuploidy breeding appears to be beneficial.

Floral Biology

The knowledge of flower bud development, time of anther dehiscence and anthesis, extent of fruit set and degree of cross pollination are a pre-requisite for planned hybridization for crop improvement. In guava, flower buds are borne in leaf axil on current season's growth, either singly or in cymose of two or three (Braganza, 1990). Guava is reported to require about 30 days in Northern India from flower bud differentiation to complete development upto the calyx cracking stage (Singh and Sehgal, 1968). However, under Bangalore conditions, Braganza (1990) reported that the period varied from 45 to 51 days. The flowers consist of a superior calyx with five lobes and the corolla consists of 6 to 10 petals arranged in one or two whorls. The androecium consists of 160 to 400 thin filaments carrying bilobed anthers, closely packed together. The gynoecium consists of an inferior ovary, syncarpous, with axillary placentation and subulate style. The style is smooth and bearded at the summit. Three flowering seasons, *viz.*, '*Ambe bahar*', '*Mrig bahar*' and '*Hasta bahar*' have been reported out of which maximum fruit setting were reportd in *Ambe bahar*. In guava, it has been observed that the flowering season does vary between regions. In guava, peak anthesis was found to be between 6 and 8.00 A.M.

under North Indian conditions. Dehiscence of anthers was observed to take place 15 to 30 minutes after anthesis and continued up-to 2hrs. Pollen fertility has been generally found to be high in guava (78 to 91 per cent) in diploid varieties. Stigmatic receptivity found to be maximum during anthesis. Stigma was found to be receptive two days before dehiscence, extending upto 4 days (Singh and Sehgal, 1968).

Botanical Description

Taxonomy

The term myrtle, a common name for some species in the genus Myrtus is also used as a common name for numerous other plants. These are not to be confused with species from the Myrtaceae. Myrtaceae is subdivided into two main subfamilies – the Leptospermoideae, which is distributed mostly in Asia and Africa, and the Myrtoideae, found in tropical America, Asia, Australia and the Pacific. Myrtoideae subfamily has fleshy fruit and opposite leaves and Leptospermoideae has dried fruit. Most of the genus of edible fruits such as the *Acca, Eugenia, Myrcia, Myrciaria, Myrcenthes, Myrtus, Rhodomyrtus, Psidium, Syzygium* comes under the subfamily Myrtoideae.

The scientific classification or the taxonomical hierarchy of guava is given below.

Domain: Eukaryopta

Kingdom: Plantae

Phylum: Spermatophyta

Subphylum: Angiospermae

Class: Dicotyledonae

Superorder: Myrtanae

Order: Myrtales

Family: Myrtaceae

Genus: *Psidium*

Species: guajava

Genetic Resources

The guava tree belongs to Myrtaceae family, comprising more than 70 genera and 2800 species. In *Psidium* genus only, there are about 150 species of evergreen trees and shrubs of which the important species are, *Psidium guajava* L., *P. friedrichsthalianum* (O. Berg.), *P. cattleianum, P. Guineense* Swartz.), *P. acutangulum* and *P. angulatum*. The wild species of guava are of considerable importance in breeding programmes although most of them are not cross compatible in nature.

P. Cattleianum (Sabine) var. *lucidum* syn: *P. littorale* (Raddi) var. *littorale* (Berg).

It is also called 'lemon guava'. It is a relatively hardy subtropical species. The fruits are small, globose, juicy, acidic and sulphur yellow in colour.

P. *Cattleianum* var. *cattleianum* (Sabine) syn: P. *littorale* (Raddi) var. *longipes* (Berg.)

It is also known as 'Strawberry guava' because of the sweet aroma of strawberry. It is a wild subtropical species closely related to guava. It can adapt to many soil types and is quite cold resistant. It is a small tree or shrub with a smooth bark. Leaves are obovate elliptic and glabrous. Fruits are round, about 2.5 cm in diameter and very fragrant. The skin is thin; pulp is soft with numerous seeds. It has a sweet flavour and good aroma. Since this lacks muskiness of the common guava, it is preferred among certain tribals.

P. *guineense* (Sw). Syn: P. *molle* (Bertol), P. *araca* (Raddi), P. *Schiedeanum* Berg.

It is also called Brazilian or Castilian guava. It is a slow growing shrub, about 1 to 3m long and withstands short periods of drought. The leaves are oblong, scantily hairy on the upper side but coated beneath with pale or rusty hairs and distinctly dotted with glands. The fruits are round with yellow skin, pale yellow pulp surrounding the white central pulp. It contains numerous hard seeds.

P. *friedrichsthalianum* (Niedenzu)

It is a tall tree about 7-10m high. The branches are slender and smooth. Leaves are oval or oblong/oval, smooth, almost glossy above and pubescent below. Fruits are globose, small and sour. The fruits are good for jelly making because of their high acidity. Reported to be wilt and nematode (*M. incognita*) resistant, it is also called Chinese guava or Costa Rican guava.

P. *montanum* (Swartz)

It is generally found in the mountains of Jamaica. The branchlets are four angled, leaves oblong to oval, glabrous, fruits are round, pulp white with more number of seeds. It produces fruits of poor quality.

P. *araucanum* (Soares-Silva and Proenca)

A large tree with membranous leaves, brochidodromous venation, long petioles and peduncles, flowers solitary, axillary or ramiflorous or in short racemes, with two pairs of flowers. Fruits globose or pear shaped, thin pericarp, fruits yellowish green when mature, seeds angular or lenticulate.

P. *acutangulum* DC

The shrub or tree ranges in height from 26 to 40 ft. Its branchlets are quadrangular and winged near the leaf base and the new growth is finely hairy. Leaves are elliptical with very short petioles. Fruits are round to pear shaped, pale yellow to yellowish white acidic pulp but well flavored pulp containing few hard, triangular seeds. The fruits are mixed with honey and eaten or, made into acid drinks or preserves.

***Psidium guajava* L.**

Psidium guajava is a large dicotyledonous shrub, or small evergreen tree, generally 3-10 m high, many branches; stems crooked, bark light to reddish brown, thin, smooth, continuously flaking; root system generally superficial and very extensive, frequently extending well beyond the canopy, there are some deep roots but no distinct taproot. Leaves opposite, simple; stipules absent, petiole short, 3-10 mm long; blade oblong to elliptic, 5-15 × 4-6 cm, apex obtuse to bluntly acuminate, base rounded to subcuneate, margins entire, somewhat thick and leathery, dull grey to yellowgreen above, slightly downy below, veins prominent, gland dotted. Inflorescence, axillary, 1- to 3-flowered, pedicles about 2 cm long, bracts 2, linear. Calyx splitting irregularly into 2-4 lobes, whitish and sparsely hairy within; petals 4-5, white, linear-ovate 2 cm long, delicate; stamens numerous, filaments pale white, about 12 mm long, erect or spreading, anther straw coloured; ovary inferior, ovules numerous, style about 10 cm long, stigma green, capitate. Fruit an ovoid or pear-shaped berry, 4-12 cm long, weighing up to 500 g; skin yellow when ripe, sometimes flushed with red, pulp juicy, creamy-white or creamy-yellow to pink or red; mesocarp thick, edible, the soft pulp enveloping numerous, cream to brown, kidney-shaped or flattened seeds. The exterior of the fruit is fleshy, and the centre consists of a seedy pulp. The pollen is viable for up to 42 hours and the stigmas are receptive for about 2 days. Bees are the principal pollinators. There is some self and cross-incompatibility but several cultivars have set fair crops of seedless or few-seeded fruit. One of the most critical botanical characteristics of guava is that flowers are borne on newly emerging lateral shoots, irrespective of the time of year. The floral structure, which makes emasculation difficult and with a juvenile period of 3-5 years limit conventional breeding. Trees reach full bearing after 5- 8 years, depending on growing conditions and spacing. The guava is not a long-lived tree (about 40 years), but the plants may bear heavily for 15-25 years.

Important Cultivars Worlwide

Several breeding programme throughout the world reported many cultivars but the easiest and most common way of getting varieties is through selection due to great diversity available in seedling populations. There are possible more than 400 cultivars around the world, but only a few dozen are responsible for the majority of plantings (Table 25.1).

Redland, the first named cultivar in Florida, was developed at the University of Florida Agricultural Research and Education Center, Homestead, and described in 1941. Very large, with little odor, white-fleshed and with relatively few seeds, it was at first considered promising but because of its excessively mild flavor, low ascorbic acid content, and susceptibility to algal spotting, it was abandoned in favor of better selections.

Supreme came next, of faint odor, thick, white flesh, relatively few, small seeds, high ascorbic acid content and ability to produce heavy crops over a period of 8 months from late fall to early spring.

Red Indian, of strong odor, medium to large size, round but slightly flattened at the base and apex, yellow skin often with pink blush; with medium thick, red flesh of sweet flavor; numerous but small seeds; agreeable for eating fresh; fairly productive in fall and early winter.

Table 25.1: Commercial Cultivars of Guava in the World

Country	*Cultivars*
Australia	Allahabad Safeda; Beaumont; Lucknow-49; Ka Hua Kula
Bangladesh	Swarupkathi; Mukundapuri; Kanchannagar; Kazi
Brazil	Paluma; Rica; Pedro Sato; Kumagai; Sassaoka; Ogawa; Yamamoto; Século XXI, Cortibel
Colombia	Puerto Rico;; Trujillo, Rojo Africano; Extranjero
Costa Rica	Tai-kuo-bar
Cuba	Enana Roja Cubana; EEA 1-23
Egypt	Bassateen El Sabahia; Bassateen Edfina; Allahabad Safeda
Malaysia	Kampuchea (Vietnam, GU8); Hong Kong Pink; Jambu Kapri Putih; Maha 65; Bentong Seedless (Malaysian S.); Taiwan Pear
Mexico	Media China, Regional de Calvillo, China, la Labor, Acaponeta, Coyame
Puerto Rico	Corozal Mixta; Corriente; Seedling 57-6-79
South Africa	Fan Retief, Frank Malherbe
Taiwan	Tai-kuo-bar
Thailand	Glom Sali; Glom Toon Klau; Khao Boon Soom
Vietnam	Glom Toon Klau; Khao Boon Soom
USA (Hawaii)	Beaumont; Pink Acid; Ka Hua Kula

Source: Pommer and Murakami, 2009.

Ruby, with pungent odor, medium to large size; ovate; with thick, red flesh, sweet flavor, relatively few seeds. An excellent guava for eating fresh and for canning; fairly productive, mainly in fall and early winter.

Blitch - Strong odor, medium size, oval, with light-pink flesh, numerous, small seeds; tart, pleasant flavor; good for jelly.

Patillo - Very mild odor, medium size, ovate to obovate, with pink flesh, moderate number of small seeds; subacid, agreeable flavor; good for general cooking. (As grown in Hawaii it is highly acid and best used for processing).

Miami Red and **Miami White,** large, nearly odorless and thick-fleshed, were released by the University of Miami's Experimental Farm in 1954.

Webber (formerly 'Riverside'), of medium-large size, pale-yellowish flesh, good flavor and 9.5 per cent sugar.

'Rolfs', of medium size with pink flesh; of good quality and containing 9 per cent sugar.

'Hart', fairly large, with pale-yellow flesh, and 8 per cent sugar content.

Important Cultivars in India

Varietal description and nomenclature of different guava varieties grown in India are greatly confusing. Some varieties were named according to shape of the fruit, skin colour and pulp colour, while, several other varieties were named after in the place of origin. Guava is largely a self-pollinated crop, but cross- pollination also does occur. This results in a large variability in the seedling population from which promising genotypes have been selected in different agro-climatic regions of the country.

Allahabad Safeda: It is the most important variety of guava used for table as well as processing purposes. The tree is medium to tall in size, upright growth habit, heavy bearer, dense foliage and has a tendency to produce long shoots. Crown is broad and compact, often dome-shaped and rarely loose. Fruit is of medium size, round in shape with smooth skin and white flesh. The fruits are relatively soft with less seeds. Its keeping quality is good.

Sardar (Lucknow 49): Tree is vigorous, spreading and profuse bearer, heavy branching type

with flat crown. Fruit is large, round to ovate in shape, primrose-yellow skin colour, white flesh and seeds are in plenty and harder than that of Allahabad Safeda.

Apple Colour: The trees are medium in vigour and are moderate yieldes. Fruits are medium sized, with apple coloured skin; cool temperature is required for good colour development. The pulp is white and firm, sweet to taste.

Lalit: Fruits are of saffron yellow colour with red "blush weighing 185-200 g fruif1. Its flesh is firm and pink with a good blend of sugar and acid. It is suitable for both table and processing purposes. The jelly made from this variety has better flavour and attractive appearance.

Swetha: The variety recently developed by Central Institute for Subtropical Horticulture, Lucknow is suitable for commercial cultivation. Tree is semi vigorous with medium height and is a prolific bearer. It is a variety with medium size globose fruits, creamy white skin with red spots or blush and snow-white flesh. Fruits are attractive and have good nutritive value.

Arka Mridula: At the Indian Institute of Horticultural Research, Bangalore, out of 200 open pollinated seedlings of the variety 'Allahabad Safeda' (collected from Uttar Pradesh), one seedling selection, 'Selection-8', was found to be promising and was released as 'Arka Mridula'. The plants are semi-vigorous and spreading in nature. Fruits are round in shape and weigh about 180g. Fruit are yellow in colour with smooth skin. The pulp is white, firm, sweet with few soft seeds. The TSS is 12.0°B, 100 seed weight is 1.6g. Keeping quality is good. Pectin content is 1.04 per cent. The variety is good for jelly making.

Allahabad Surkha: It is an outstanding variety oflarge uniform pink fruit with deep pink flesh. The plants produce up to 120 kg fruits in the sixth year of fruiting. Trees are vigorous, dome-shaped and compact. The fruit is sweet and strongly flavoured with a few seeds. The plants are vigorous, dome shaped and compact. Trees are high yielding, producing upto 120kg per plant. The fruits are round with uniform, pink skin and deep pink pulp, sweet, strongly flavoured and with few seeds.

G. Vilas Pasand: The trees are vigorous, wide spreading with bushy, low growing habit. Fruits are round to ovoid, skin texture is course to smooth, fruit skin pale yellow to golden, colour of flesh is creamy white texture creamy soft, very large (400g to 800g) fruit, less seeds, very productive throughout the year. High content of Vitamin C makes it stand out among guava varieties.

Pant Prabhat: A selection was made as 'Pant Prabhat' for commercial cultivation at GBPUA and T, Pantnagar (Uttarakhand). Plant growth in this line is upright; with broad leaves, the tree is highly productive (100 -125kg). Fruit skin is smooth and light yellow in colour, fruits medium sized with average fruit weight of 150-172g, pulp is white, seeds are small and medium soft, the fruit has a sweet taste with pleasant flavour, ascorbic acid content varies from 125mg (rainy season) to 300mg/100g fruit weight (winter season). TSS varies from 10.5 to 13.5^{o}B.

Dhareedar: A selection from an old, seedling orchard was made as 'Dhareedar' at the Fruit Research Station, Kuthulia, Rewa. The trees are vigorous, medium tall with erect and upright branching and a flat crown. Fruits are medium to large sized, roundish ovate in shape with 5-7 raised lines on the surface of mature fruits, the peel in greenish yellow, the pulp soft and sweet (TSS 11.7^{o} Brix).

Bangalore: Fruits are medium to large in size, pulp is white with good taste and flavour.

Chittidar: This variety is very popular in Western Uttar Pradesh. The fruits are characterized by numerous, red dots on skin. Fruits are sub globose, with white pulp, high TSS and Vitamin C content of 240 mg/100g pulp.

Hafsi: Fruits are spherical in shape with thin skin and medium size. The pulp is red with good taste and flavour. Seeds are comparatively less in number, but hard.

Red Fleshed: Fruits are medium sized with red pulp, round, smooth skinned, seeds are plenty and medium soft. Fruits possess sweet flavour, are rich in Vitamin C (386 mg/100g pulp)

Smooth Green: Fruits are round and medium sized. Skin is glossy and greenish yellow when mature. Pulp is white, good in taste and flavour

Nasik: Fruits are medium sized, round, with white pulp, sweet with good flavour

Banarsi Surkha: Trees are medium sized (5.1m) with a broad crown, fruit shape is round and surface smooth, skin colour golden yellow, pulp colour pink, seed number very high, seed texture very hard.

Seedless: The plants are very vigorous. There are different varieties, like, Saharanpur Seedless, Nagpur Seedless, Sringeri Seedless, *etc.*, which are nearly identical. Two types of fruits, *viz.*, long, big sized with warty surface, and yellow, thin skin with swollen calyx end and round; small, with very few seeds. The pulp is white, good to taste and has aroma, contains moderate to high levels of Vitamin C (240 mg/100g pulp).

Anakapalli: Fruits are medium sized with red pulp. Seeds are soft and plenty. Fruits are slightly oval.

Apart from these prominent commercial cultivars, other cultivars grown in localized areas are: Pear Shaped, Apple Colour, Banarasi Surkha, Sangam, Dholka,

Sindh, Karela, Mirzapuri Seedling, Superior, Pourtgal, Spear Acid, Superior Sour Licidium, White Fleshed, Behat Coconut, Smooth white, Amsophri, Madhuriam, Bariampur, Harijha,Dharwar, Safri or Payera, Soh-Pryiam, Am-Sophri, Kaffree, Supreme, White Supreme, Bangalore, Bhavnagar, Gwalior- 27, Kafri (Pear shape), Kafri (Round shape), Rewa-72, Hazi sahib, Kohir, *etc.*

Apart from this, promising cultivars of different Indian states are given Table 25.2.

Table 25.2: State-wise Distribution of Important Cultivars of Guava in India

State	*Cultivars*
Andhra Pradesh	Allahabad Safeda, Anakapalli, Banarasi, Chittidar, Hafsi, Sardar, Smooth Green and Smooth White
Assam, Bihar	Amsophri, Madhuriam, Safrior Payele
Gujarat Karnataka Maharashtra Tamil Nadu	Allahabad Safeda, Chittidar, Hafsi (Red Fleshed), Harijha, Seedless
Uttar Pradesh	Nasik, Seedless, Sindh
West Bengal	Allahabad Safeda, Arka Mridula, Sardar, Navalur
Andhra Pradesh	Dharward, Dholka, Kothrud, Lucknow-24, Sardar
Assam Bihar	Anakapalli, Banarasi, Bangalore, Chittidar, Hafsi, Nagpur Seedless and Allahabad Safeda
Gujarat Karnataka Maharashtra Tamil Nadu	Allahabad Safeda, Apple Colour, Chittidar, Red Fleshed, Banarasi Surkha, Sardar, Mirzapur Seedless
Madhya Pradesh	Allahabad Safeda, Sardar, Rewa 27, Gwalior 72, Lalit, Shweta

Breeding Objectives, Methodologies and Achievements

Breeding Objective

The majority of problems with guava crop deal with pests and diseases, seed content, low yield, poor quality of fruits and post-harvest. In India, **Negi and Rajan (2007)** explained the breeding objectives:

1. **Scion Improvement**: Wilt tolerant, high yielding, high quality dwarf cultivars, with fruits of uniform shape, good size, attractive skin and pulp colour, fewer and soft seeds, long storage life and suitable for table and processing purposes, resistant to insect pest like fruit fly.
2. **Rootstocks Improvement**: Wilt resistant and dwarf, tolerant to various biotic and abiotic stress, drought tolerant.

Methodologies and Achievements

Plant Introduction

Most of the guava varieties have evolved as selections from seedling variants. Variability has come about because of open pollination from highly heterozygous parents. Several introductions of promising genotypes have been made in guava growing countries. In India, many introductions made from Hawaii, Brazil, Thailand, *etc.* are being cultivated and used in breeding programmes. Similarly, introductions of Indian cultivars like Allahabad Safeda and Sardar have given excellent results in other parts of the world (Gonzaga *et al.*, 1999). Introduction is a potential tool in any crop improvement programme as it considerably saves time, there is also the danger of new diseases getting introduced.

Selection

In guava majority of successful varieties are released through selection. Important varieties released through selection are:

Lucknow- 49: In 1907, guava improvement work was initiated at Ganeshkhind Fruit Experimental Station, Pune, and one strain from open pollinated seedlings of 'Allahabad Safeda' collected from Lucknow was selected and released as 'Lucknow-49' which has now been renamed as 'Sardar'.

CISH-G-1: It is a selection at CISH, Lucknow, from local red fleshed type, with attractive fruits having deep red skin, firm pulp with high TSS, soft seeds.

CISH-G-2: It is a selection at CISH, Lucknow from local red fleshed type, crimson colored attractive fruit, stripes in groove, seeds soft.

CISH-G-3 (Lalit) and CISH-G-4 (Swetha): Selection at CISH, Lucknow.

Arka Mridula: At the Indian Institute of Horticultural Research, Bangalore, out of 200 open pollinated seedlings of the variety 'Allahabad Safeda' (collected from Uttar Pradesh), one seedling selection, 'Selection-8', was found to be promising and was released as 'Arka Mridula'.

Allahabad Surkha: At Allahabad Agricultural Institute, Allahabad, a selection from the local red pulp type has been released as 'Allahabad Surkha'.

Pant Prabhat: A selection was made as 'Pant Prabhat' for commercial cultivation at GBPUA and T, Pantnagar (Uttarakhand).

Dhareedar: A selection from an old, seedling orchard was made as 'Dhareedar' at the Fruit Research Station, Kuthulia, Rewa.

Hybridization

In guava, flowers that are chosen for crossing are emasculated when at the 'calyx break stage', a day before opening. Pollen from the pollen parent is brought from an unopened flower, preferably at calyx break stage. The stigmatic surface is gently smeared with the pollen and flowers are bagged and pollination carried out during the morning hours between 08 AM to 12.00 noon has given better results.

Inter-Varietal Hybridization

In general, inter-varietal crosses in guava are successful, having no crossability barriers. However, varietal cross incompatibility in guava is reported in crosses made between 'Behat Coconut' and 'Sardar', 'Behat Coconut' and 'Apple Colour'. The majority of existent breeding programs are under development in the countries where guava represents an important economic crop, for instance Brazil, India, Mexico, Pakistan, Cuba, Thailand and others. In India, breeding work for guava improvement has been going on at several research institutions and very few guava hybrids were released over the years and important one are listed below:

Safed Jam: This is a hybrid between 'Allahabad Safeda' × 'Kohir' developed at FRS, Sangareddy. The tree is medium in height and a heavy yielder. Fruits are round in shape, large in size with a thin peel, good taste and few, soft seeds.

Kohir Safeda: it is a cross between a selected, heavy yielding line of 'Kohir' × 'Allahabad Safeda'. The tree is vigorous, fairly large in size and dome shaped. Fruits are large, with few, soft seeds and white pulp.

Arka Amulya: This is a hybrid between 'Allahabad Safeda' × 'Triploid' developed at the Indian Institute of Horticultural Research, Bangalore. Plants are semi-vigorous and spreading type. Fruits are medium sized (180-200 g), weight of 100 seed is 1.8 g, pulp is white, with high TSS (12.5°B), fruits have good keeping quality.

Arka Kiran: It is from the hybrid between 'Kamsari' × 'Purple Local' developed at the Indian Institute of Horticultural Research, Bangalore. Plants are semi vigorous, amenable to high density planting. The fruits are sub globose, weighing about 200-230 g The pulp is deep pink, thick and has good flavour. The seeds are medium soft (9.0 kg cm-2), with high lycopene content (7.45 mg/100 g) and TSS of 12.0 to 12.5°B.

Hisar Safeda: It is from the cross 'Allahabad Safeda' × 'Seedless', developed at CCSHAU, Hisar. Plants are upright, trees have a compact crown. Fruits are round, with a smooth urface, creamy yellow skin; average fruit weight is 92 g, creamy white pulp, few soft seeds high TSS (13.4 per cent).

Hisar Surkha: it is from the cross 'Apple Colour' × 'Banarasi Surkha'. The tree crown is broad to compact. Fruits are round, skin yellow with red dots, average fruit weight 86 g, pulp pink, seed count medium, TSS high (13.6 per cent).

Polyploidy

The basic chromosome number (haploids) of *P. guajava* is X = 11. Mainly the seeded types are diploids while the 'highly-seedless' types are triploids. Seedless type or less and soft seeded type is an important commercial attribute. However, those commercially recognized as 'seedless' are not completely seedless and types ranging from 'seedless' to 'less-seeded' can be identified. The seedless types are nevertheless found to be pollen fertile. Cytological examination of different 'seeded', 'less-seeded' or 'seedless' types has shown that differences in chromosome number as well as gene dissimilarities contribute to this feature. Several parts of our country reported seedless varieties. Among polyploidy, triploids are of utmost importance as they produce seedless fruits. Kumar and Ranade (1952) reported a auto-triploid guava variety with 33 chromosomes at Poona. At IARI and Saharanpur, Majumder and Singh (1964) also suggested the same chromosome status of seedless varieties and were found to be auto-triploids. Naithani and Srivastava (1966) also reported natural auto-tetraploid in *guava*. Tetraploidy in guava can also be induced with colchicine treatment. Chromosome set of 'Indonesian Seedless' and 'Bangkok

Apple' is triploid (2n = 33). Sharma (1992) identified a promising tetrasomic dwarfing rootstock (Aneuploid No. 82 renamed as Pusa Srijan), through selection, out of 48 aneuploid seedlings at IARI, New Delhi. Studies conducted on the effect of Pusa Srijan rootstock on growth and yield of 'Allahabad Safeda' showed that the aneuploid induced substantial dwarfing in Allahabad Safeda in terms of plant height, plant spread and tree volume. Overall yield/unit volume of the plant was highest in Pusa Srijan, which indicated its strong potential for use as dwarfing rootstock on a commercial scale, for increasing production and profitability of guava orchards.

Naturally occurring mutations are not very rare in guava. Brar and Bal (2003) investigated the effect of gamma rays on buds of guava cv. Sardar and observed maximum variability in growth parameter with 2 kR treatment. In vitro mutagenesis was carried out by Zamir *et al.* (2003) and reported that Shoot tips irradiated with gamma rays at 15-90 Gy and cultured in MS medium containing 3 per cent sucrose, 6- benzylaminopurine (benzyladenine) (BAP) and L-glutamine. Optimum shoot proliferation was recorded in MS medium supplemented with 1.0mg BAP and 250mg L-glutamine/litre. Rooting of cultured shoots was observed in half-strength MS medium supplemented with IAA and IBA. LD50 was observed at 45 Gy. Rates of more than 75 Gy were lethal to explants.

Biotechnology in Guava Improvement

Biotechnological techniques can be useful chiefly in breeding for disease resistance and in germplasm storage using tissue culture techniques. Use of tissue culture and micro-propagation of superior guava cultivars has been discussed by several researchers. Jaiswal and Amin (1992) reported that somatic cell genetics could be useful in guava breeding with specific objectives. A technique for successful in vitro propagation using shoot-tip explants from mature trees was reported by Jaiswal and Amin (1987). These workers also demonstrated that adding activated charcoal to the medium enhanced rooting of explants and vegetative growth of established plantlets. Risterucci *et al.* (2005) constructed a library of microsatellite-enriched (GA)n and (GT)n and, 23 nuclear simple sequence repeat (SSR) loci were characterized in the guava. All the SSR loci were found to be polymorphic after screening for diversity in different cultivars, and across- taxa amplification tests showed potential transferability of most SSR markers in three other *Psidium* species. First to be published for *P. guajava*, this new SSR resource will be a powerful tool for genetic studies in guava, including cultivar identification and linkage mapping, as well as potentially, for, interspecific genetic studies within the genus *Psidium*.

Somaclonal variation could be useful technique for selecting guava plants resistant to the wilt disease. An efficient protocol for plant regeneration from callus culture would also be helpful in selecting plants resistant to disease or environmental stresses. Recovery of plants of haploid origin from anther/pollen culture of guava could offer advantages in breeding (Jaiswal and Amin, 1992).

Usefulness of Wild Species for Crop Improvement

The contribution of wild species to crop improvement and management programmes mainly involves their utilization as rootstocks for regulation of vigour, yield, fruit quality and, disease and pest resistance. In guava, utilization of wild species for crop improvement has been very useful to introduce certain gene(s) of interest for specific purposes like soft seeded characteristics, disease and pest resistance, *etc.* Exploitation of wild species requires extensive knowledge of taxonomy, reproductive biology and cytogenetics, genetics, crossability barriers and fertility of the hybrids (Dinesh andVasugi, 2010). Although success obtained in fruit crops is very low, many desirable traits for crop improvement are found in the wild species. Sometimes, progeny of inter-specific crosses in many fruit crops are partially sterile due to ploidy level differences, genomic incompatibilities and cytoplasmic imbalances. Work in many institute have already been initiated to develop a rootstock tolerant/resistant to guava wilt. Two species, *P. molle* and *P. guineense* were found to be morphologically similar with minute differences in their chromatographic pattern (Das and Prakash, 1981). These workers also found a close affinity between *P. guineense, P. pumilum* and *P. chinensis*. It was observed that P. guajava and P. chinensis were cross compatible in nature. However, P. guajava and P. molle are cross incompatible when P. guajava is used as the female parent (Subramanyam and Iyer, 1982). Leslie *et al.* (1995) reported that *Psidium friedrichsthalianum* Niedenzu to be resistant to guava wilt. The other species reported to be tolerant to the wilt are *P. cumuni, P. cattleianum var. lucidum, P. molle*, and *P. guineense.* Hence, intensive work is needed to develop useful inter-specific hybrids resistant to wilt using the resistant species in breeding programmes. Pathak and Ojha (1993) enumerated their potential uses: *P. cujavalis, P. molle, P. cattleianum* and *P. guineense* can be used as rootstock. Chinese guava (*P. friedrichsthalianum*) and Philippine guava are compatible rootstocks and have been reported to be resistant to the wilt disease. Allahabad Safeda trees grafted on to *P. pumilum* had a dwarfing influence. At the Central Institute for Subtropical Horticulture, Lucknow, Uttar Pradesh, interspecific hybridization was carried out between *P. molle* and *P. guajava.* The interspecific hybrids have been found resistant to guava wilt and are graft compatible with commercial varieties of P. guajava (Anon., 2003-04).

Inheritance Studies

Genetic studies conducted in guava at IIHR, Bangalore, indicated that red pulp colour was dominant over white and that this character was governed monogenically. Many cultivated red fleshed varieties were found to be heterozygous for this character. Bold seeds were found to be dominant over soft seeds and this was also found to be determined monogenically. Linkage was also found between flesh colour and seed size, *i.e.*, red flesh with bold seeds (Subramanyam and Iyer, 1982). Mitra and Bose (1985) have reported heterosis in guava. Studies conducted at Coimbatore by Raman *et al.* (1969 and 1971), have shown that triploidy and genetic factor(s) are responsible for female sterility and, that; variation among triploids is due to their independent origin from a distinct diploid variety. Iyer and Subramanyam (1971) opined that seedless varieties of guava were triploids, grew vigorously and bore fruits that were irregular in shape with ridges, because of irregular distribution of seeds of various sizes. Dinesh and Yadav (1998) carried out half-sib analysis in progenies of the variety 'Apple Colour'. They observed that genotypic variance was lower than the phenotypic variance, and heritability was moderately high for all the characters implying, that, selection may be practiced for improvement of fruit characters. Hence, hybridization among less seeded diploids can be adopted in an improvement programme. Inheritance pattern using 'Purple guava' and 'Arka Mridula' as parents that hybrids segregate in a ratio of 3:1 for green leaf types and purple leaf types (Dinesh andVasugi, 2010).

Future Needs

Priority needs to be given to develop a wilt tolerant having high yield and good fruit quality include high TSS, good sugar-acid blend, good aroma, attractive skin pulp colour and soft seeds including processing traits like juice colour, high Vitamin C content, higher lycopene content, good pectin content and good flavour. Shelf life may be accorded adequate importance. In the recent past, efforts have been intensified to develop attractive red coloured cultivars for table and processing purpose. However, deep red colour has been found to be a very unstable character, hence, 'stable' types need to be identified and great care should be taken while selecting hybrids from a large population. Aneuploidy breeding should be intensified to develop high yielding, high quality varieties with fewer and soft seeds, and to develop dwarfing rootstocks. Auto-tetraploids of less-seeded, superior diploid varieties may be developed. Induction of mutation by physical and chemical mutagens may be attempted where improvement in a specific character is required in an otherwise acceptable variety. Work on interspecific hybridization to develop wilt resistant rootstocks and scion varieties should be intensified. Varietal introduction, though, becomes an essential part of any crop improvement programme and needs to be made with great concern. Molecular characterization of germplasm needs to be accelerated with a view to work out genetical distance so that good recombinants can be arrived at by crossing suitable parents. Biotechnological tools need to be employed extensively. Attempts to spot genetic markers for wilt resistance may be made to improve efficiency for developing wilt resistant clones and rootstocks.

REFERENCES

Amin, M.N. and Jaiswal,V.S. 1988. Micropropagation as an aid to rapid cloning of a guava cultivar. *Scientia Hort.*, 36: 89-95.

Anonymous. 1996. Research Programmes and Progress, Indian Institute of Horticultural Research, Bangalore, pp. 10-11

Anonymous. 2003-04. Annual Report, Central Institute for Subtropical Horticulture, Lucknow, pp. 10-11.

Aranguren, Y., Valecillos, C. and Fermin, G. 2010. Variability of Venezuelan guava geographic landraces employing phenotypic markers. *Acta Hort.*, 849: 87-94.

Aswath, C., Padmakar, B., Kanupriya, Vasugi, C. and Dinesh, M.R. 2011. Genotyping of parental lines in *Psidium guajava* using microsatellite markers. 7 International symposium on In vitro culture and horticultural breeding, Belgium.

Bailey, L.H. 1919. Standard encyclopaedia of Horticulture. Macmillan, New York, USA pp. 2847-2849

Balasubramanyam, V.R. 1959. Studies on blossom biology of guava (*Psidium guajava* L.). *Ind. J. Hort.*, 16: 69-75.

Braganza, M.A. 1990. Floral Biology studies and varietal evaluation in genus Psidium. M.Sc. (Ag). Thesis submitted to University of Agricultural Sciences, Bangalore.

Brar, H.S. and Bal, J.S. 2003. Studies on the use of gamma rays on the performance of guava budlings. *Ann. AgriBio Res.*, 8: 213-217.

Campbell, C.D. 1989. 'Homestead', a superior guava for fresh market and for processing. *Proc. Fla. State Hort. Soc.*, 102: 202-204.

Chadha, K.L. 1998. Improvement in tree fruit and plantation crops. *Ind. J. Hort.*, 55: 265-296.

Chadha, K.L., Harmail, S. and Tandon, D.K. 1981. A varietal trial of guava. National Symposium on Tropical and Sub-tropical Fruit Crops, Bangalore, p. 17.

Chandra., R.A., Bajpai, Soni Gupta and Tiwari, R. K. 2004. Embryogenesis and plant regeneration from mesocarp of *Psidium guajava* L. (guava). *Ind. J. Biotech.*, 3: 246-248.

Cheema, G.S. and Deshmukh, G.B. 1927. Culture of guava and its improvement by selection in Western India. Bull. Dept. Agri., Bombay, No. 148.

Cheema, G.S., Bhat, S.S. and Naik, K.C. 1954. Commercial fruits of India. MacMillan and Co., New York, USA. Dass, H.C. and Prakash, D. 1981. Phylogenetic affinities in *Psidium* spp. as studied by flavonoid patterns. National Symposium on Tropical and Sub-tropical Fruit Crops, Bangalore, p. 15.

Dass, H.C. and Prakash, D. 1981. *Phylogenetic affinities in Psidium spp. as studied by flavonoid patterns.* National Symposium on Tropical and Sub-tropical Fruit. Crops, Bangalore, p. 15.

De Candolle, A.P. 1904. Origin of cultivated plants. Kegal Paul, London.

Dinesh, M.R. and Reddy, B.M.C. 2001. Evaluation of *Psidium guajava* accessions and some other Psidium species for fruit characters. *J. Appl. Hort.*, 3: 41-43.

Dinesh, M.R. and Vasugi, C. 2010. Guava improvement in India and future needs. *J. Hortl. Sci.*, 5: 94-108.

Dinesh, M.R. and Yadav. I.S. 1998. Half-sib analysis in guava (*Psidium guajava*). *Ind. J. Hort.*, **55**: 20-22

Edward, J.C. and Shankar, G.1964. Rootstock trial for guava (Psidium guajava L.). Allahabad Farmer, 38: 249-50 Evans, D.A. and Sharp, W.R. 1988. Somaclonal variation and its application in plant breeding. Feature article. IAPTC Newslett. 54: 2-10

Golberg, L. and Levy, L. 1941. The vitamin C content offresh, canned and dried guava. Nature, 148: 286. (citedfrom S.K.Mitra and T.K.Bose,1990. Nature, Guava. In: Fruit: Tropical and sub tropical. T.K.Bose and S.K.Mitra (Eds.). Naya Prakash, Calcutta-6, pp: 280-303)

Gonzaga, N. L., Bezerra, J.E.F and Montano, J.C. 1999. Introduction and evaluation of Indian varieties of guava in the region of Submedio San Francisco. Pesquisa em Andamento da Embrapa Semi Arido, 95: 3.

Hayes, W.B. 1953. Fruit Growing in India. Kitabistan, Allahabad.

Iyer, C.P.A. and Subramanyam, M.D. 1971. Problems with triploidy in guava. *SABRAO Newslett.*, **3**(1): 31-33.

Jaiswal, V.S. and Amin, M.N. 1987. In vitro propagation of guava from shoot culture of mature trees. *J. Pl. Physiol.*, 130: 7-12.

Jaiswal,V.S and Amin, M.N. 1992. Guava and jackfruit. Biotechnology of perennial fruit crops. *Mol.Biol.*, 39: 413-437.

Kumar, L.S.S. and Ranade, S.G. 1952. Autotriploid in guava (*Psidium guajava* L.). *Curr. Sci.*, **21**: 75-76.

Leslie, R.W. Landrum, Dennis Clark, P. William and Jeff Brendecke. 1995. Hybridization between *Psidium guajava* and *P. guineense* (Myrtaceae), *Econ. Bot.*, **49:** 153-161.

Mitra, S.K. and Bose, T.K., 1985, Guava. *Fruits of India- Tropical and Sub-tropical* ed. by Bose Naya Prokash, Calcutta.

Naithani, S.P. and Srivastava, H.C. 1966. Autotetraploidy in *Psidium guajava* L. *Naturwissenchaft.*, **8**: 205-206.

Pathak, R.K. and Ojha, C.M. 1993. Genetic resources of guava. In: Advances in Horticulture (Vol I). Chadha, K.L and Pareek, O.P. (Eds.) Malhotra Public House, New Delhi.

Raman, V.S., Rangaswamy, S and Manimekalai, F. 1971. Triploidy and Seedlessness in guava (*Psidium guajava* L). *Cytologia*, **36:** 392-399.

Raman, W.M., Manimekalai,G and Ramalingam, R.S. 1969. Observation on seedlessness, fruit development and cytology of varieties of guava. *Madras Agri. J.*, **56**: 255-61.

Sehgal, O.P. and Singh, R. 1967. Studies on blossom biology of guava (Psidium guajava L.) I. Flowering season, flowering habit, floral bud development, anthesis and dehiscence. *Ind J. Hort.*, 24: 118-112.

Shafaat Mohammed.1975. Investigations on the breeding behaviour of aneuploids of guava. Thesis submitted to the Division of Fruits and Horticultural Technology, IARI, New Delhi

Shanmugavelu, K.G. Selvaraj, M. and Thamburaj, S. 1987. Review of research on fruit crops in Tamil Nadu. *South Ind. Hort.* 35: 1-3.

Sharma, Y.K. 1982. Rootstock investigation in guava (*Psidium guajava* L.). Thesis submitted for the award of Ph.D. degree to Meerut University, Meerut.

Sharma, Y.K., Goswami, A.M. and Sharma, R.R. 1992 Effect of dwarfing aneuploid guava rootstock in high density orcharding. *Ind. J.Hort.*, 49: 31-36.

Singh, R. and Sehgal, O.P. 1968, Studies on blossom biology of *Psidium guajava* L. (guava). II. Pollen studies, stigma receptivity, pollen and fruit set. *Ind. J. Hort.*, **25**: 52-59.

Singh, V.R., Dhar, L. and Singh. G., 1977. Note on the performance of guava cultivars and *Psidium* species against wilt disease under natural field conditions. *Haryana J. Hort. Sci.*, 6(3-4): 149-50.

Subramanyam, M.D. and Iyer, C.P.A. 1982. Improvement of guava by breeding. Report, Fruit Workshop, Nagpur. pp. 117-118.

Breeding of Perennial Horticultural Crops *Pages 333–356*
Editors: S.K. Singh, V.B. Patel, A.K.Goswami, Jai Prakash and Chavlesh Kumar
Published by: **BIOTECH BOOKS, NEW DELHI**

Chapter **26**

Litchi Improvement

Vishal Nath, Alok Kumar Gupta, Manvendra Singh, Awtar Singh, Swapnil Pandey, Evening Stone Marboh, Devesh Pandey and Ankit Kumar Pandey

ICAR-National Research Centre on Litchi, Muzaffarpur – 842 002, Bihar

INTRODUCTION

Litchi (*Litchi chinensis* Sonn.) is a subtropical evergreen fruit crop of the Sapindaceae family (Leenhouts 1978). Its climatic requirements are highly specic, and thus its commercial cultivation is restricted to only a few subtropical countries in the world. The major litchi-producing countries are China, India, Taiwan, Thailand and Vietnam. A relatively small amount of litchi is produced in the United States, Mexico and Central and South America (Menzel and Waite 2005). World production of litchi is estimated to be around 2.11 million tons, with more than 95 per cent of the area and production shared by Asia. India and China account for 91 per cent of the world production, but it is mainly marketed locally. Over the years, signicant growths in production and productivity of litchi have been recorded coupled with a fast-expanding market at both the national and international level (Singh *et al.*, 2012).

There has been a substantial increase in the area and production of litchi in India during the past 50 years. The area has increased from 9400 ha (1949–1950) to 84,950 ha (2014–2015). The contribution of litchi to the total area under fruit cultivation has increased from 0.75 per cent to 1.70 per cent. Increase in area during 1991–1992 to 2014–2015 (22 years) has been more than 80 per cent, whereas production increase during the same period is to the tune of more than 150 per cent (NHB 2015). Productivity also recorded an increase of about 50 per cent during the same period. Evidently, production and productivity of litchi are constantly increasing in the country. The fresh fruit market dominates the trade, followed by dried and canned fruit. The main importers are the European Union, the United States, Hong Kong, Singapore, Japan and Canada, and China, Taiwan, Thailand, Madagascar, South Africa, Australia and Mexico are the main exporting countries.

Globally, the litchi industry is beset with myriads of problems. Low success rate of establishment, lesser fruiting span, low and irregular yields because of poor owering and fruit set, fruit cracking, browning and rotting of fruits, fruit borers and mites, poor shelf life, and lack of suitable varieties with early and late maturity and good-quality fruits are some of the factors encountered.

Origin and Distribution

The Litchi genus contains three subspecies: *Litchi chinensis* ssp. *chinensis*, L. chinensis ssp. philippinensis, and *L. chinensis* ssp. *javensis* (Leenhouts 1978). The subspecies philippinensis is found growing wild in the Philippines from sea level to 500 m altitude (Sotto 2002). The fruits are distinguished from cultivated litchi by their long and oval shape, with long thorny protuberances and inedible esh that partially covers the seed (Menzel 1991). Fruits of the subspecies javensis from the Malay Peninsula and Indonesia produce an aril thinner than that of the cultivated litchi. Neither of these two subspecies is grown commercially. The subspecies chinensis, a litchi of commerce, originated in Southern China and Northern Vietnam from wild populations of these regions (Wu 1998b; Hai and Dung 2002). In China, enormous diversity has been observed in the moist forests of Hainan from low elevations up to 1000 m, below 500 m in hilly areas of western Guangdong and eastern Guangxi, and in valley or hilly regions of southern Yunnan below 1000 m (Wu 1998a). Litchi dominates most of these forest compositions and often grows in mixed stands with Vatica astrotricha (green plum), Hopea hainanensis, Heritiera parvifolia of the Chinese parasol family, Coelodepas hainanensis, Polyathia laui, and Diospyros hainanensis, which belongs to the Ebenaceae family. The small area of wild litchis have also been found growing in the natural rainforest of Vietnam at low elevation in the Ba Vý ' mountains and in the forests of Tam Dao (Vinh Phuc Province) and Tuye ˆn Ho ´a of Quang Binh Province (Hai and Dung 2002). Generally, in Hainan the wild trees look similar to cultivated litchi and produce edible fruit, but the esh is relatively thin and sour. Fruits are variable and can be characterized into three groups based on their shape and arrangements of skin segments and protuberances: group 1 possesses sharp and pointed protuberances, group 2 has protruded but obtuse skin segments, and group 3 has at skin segments. The wild types evolved in two directions, with skin segments becoming protruded and long, as in 'Dazao' ('Tai So') and 'Guiwei' ('Kwai May'), or attened, as in 'Sanyuehong' ('Sum Yee Hong'), 'Shuidong' ('Souey Tung'), 'Nuomici' ('No Mai Chee'), and 'Huaizhi' ('Wai Chee') (Wu 1998a). In Yunnan, a population of wild litchis was identied in which the trees required less cool weather to initiate owering. These trees mature earlier and produce a crop in warmer climates than can the traditional subtropical ecotypes. Flowers from these wild specimens have sepals with brownish pubescence, which gave the species the name 'brown-hair litchi' or 'Hemaoli.'

Litchi, a luscious fruit crop, originated from the Kwantung and Fukien Provinces of Southern China (Tao 1955). According to de Candolle (1909), 'Chinese authors living at Peking knew about litchi only late in the third century of our era.' Although controversial, the rst citation for this fruit in the literature probably can be traced back to as early as 1766 BC. However, a clear reference has been mentioned in the literature of the Han dynasty (140–86 BC). Possibly the rst complete book on litchi in English (Groff 1921) was a monograph published by Ts'a' Hsiang (1059 AD).

Until the tenth century, litchi was propagated through seedlings. Later, vegetative propagation through air layering or marcotting became widely accepted among the growers, being rst practised in the fourth century AD. By the sixteenth century, grafting was also recorded (Anonymous 1978). However, their detailed application was rst documented in the Registers by Xu Bo in 1579 and by Deng Qingcai in 1628, respectively. Until the late seventeenth century, litchi cultivation was restricted to Southern China and Northern Vietnam (Tindall 1994; Hai and Dung 2002). It then spread to other regions through the route suggested by Gala ´n Sau ´co and Menini (1989). Apparently, it reached Burma and Eastern India in the late seventeenth century (Hayes 1957). Subsequently, by the end of the eighteenth century, it made its way to Bengal, whence it diversied to other parts of India on a commercial scale and later to Nepal (Budathoki 2002) and Bangladesh (Abu Baker Siddiqui 2002). The litchi industry has grown signicantly in these countries. Litchi was rst introduced into Thailand from China 300 years ago by merchants who carried the fruit with them. Some plants adapted successfully to the tropical conditions of the central region of the country. Promising seedlings were selected and accordingly named by local growers as lowland litchi or tropical litchi, as the trees do not require a long period of cold to initiate owers (Subhadrabandhu 1990). In Chiang Mai, litchi was planted around 1890, through air layers brought by emigrants who migrated from Yunnan through Laos or Myanmar (Boonrat 1984). These are truly subtropical types, requiring a longer period of low temperature for owering. Many of them still retain their Chinese names, such as 'O-Hia,' 'Hong Huay,' and 'Kim Cheng'; however, the Thai spellings and pronunciations are different (Subhadrabandhu and Yapwattanaphun 2001a). Litchi reached the Philippines from China before 1916 but failed to ower and set fruit at low altitudes until 1931 (Sotto 2002). Later, in 1931, trees were introduced from other sources that bear fruit and gave hope for litchi growing in the more elevated areas. The crop was introduced in Australia through Chinese migrants attracted by gold rushes around 1854 (Menzel *et al.*, 1988) and reached Southern Africa about 50 years earlier.

In Madagascar the arrival of litchi dates back to 1802 and the 1940s, with many old and relatively few new plantations still in existence at a commercial level (Jahiel and Abraham 2001). Evidence revealed that the rst litchi trees were imported into South Africa from Mauritius in 1876. However, trees had already been observed in Natal in 1875, suggesting there must have been earlier imports (Oosthuizen 1991). From Natal, trees traveled to the Transvaal Lowveld as well as to other suitable frost-free areas. The species arrived in Florida in the 1880s, but commercial cultivation started in the 1940s. 'Brewster' ('Chenzi') is the main cultivar of the industry in the region, but currently it is replaced by 'Mauritius' ('Dazao,' 'Tai So') (Knight

2001). The rst trees, known as 'Afong,' which were transported to Hawaii in 1873 by Chinese merchants, later were identied as being similar to 'Dazao' (Nakasone and Paull 1998a). In Israel litchi was introduced in the 1930s from different parts of South Africa ('Mauritius'), California ('Floridian'), and India ('Bengal'); however, a commercial industry developed only after the 1980s (Goren *et al.*, 2001).

At a later event, litchi was introduced to several other countries. In Hawaii, it was probably successfully introduced in 1873. Its cultivation in the West Indies and Natal, South Africa, dates back to 1775 and 1869 (Marloth 1947), respectively. Commercial cultivation of litchi in Queensland (Australia) was of late occurrence (Batten and Lahav 1994), although it was introduced as early as 1854. In the United States, it reached from Saharanpur (India) to Florida in 1883, to California in 1897, and subsequently again to Florida from Fukien Province of China in 1906 (Singh and Singh 1954), where it was named 'Brewster' litchi. Early in the nineteenth century, it also reached England and France but was unsuccessful (Pandey and Sharma 1989).

The Litchi Fruit

Botanically, litchi fruit is pendant in a loose cluster (up to 0.6 m long) of several dozen fruits. Fruits are round, ovoid, heart shaped or even kidney shaped and vary in size depending on the cultivar but can measure up to 5 cm in length and up to 4 cm in diameter. The fruit possesses shoulder, suture, apex and skin segments as observed externally (Deng *et al.*, 1999; Huang *et al.*, 2005a). Different parts of the fruits are briey described in following paragraphs (Menzel and Waite 2005; Singh *et al.*, 2012).

Epicarp

He rind of the fruit (pericarp) is thin, tough, hard, brittle and frangible. At maturity, its colour changes from green to bright red. The intensity of redness varies among cultivars, and fruits may even be yellow or green depending upon cultivars. The skin possesses prominent protuberances but is less marked than in related species such as rambutan and pulusan. When overmature, the skin assumes a dirty brown colour and loses its extensibility (Huang *et al.*, 2005b); this occurs rapidly, even when the fruit is still perfectly edible (Menzel and Waite 2005; Singh *et al.*, 2012).

Mesocarp

The mesocarp tissue has a loose consistency without clear demarcation between the epicarp and the mesocarp. In the initial stages, this tissue is composed of ordinary parenchymatous cells which become separated with time as large intercellular spaces develop. Botanically, the edible part of the fruit is known as the aril. In a developing fruit, the aril grows continuously as an outgrowth of the outer cells of the seed coat (outer integument); the rate of outgrowth varies greatly among cultivars (Huang 2005). The aril is translucent, white, slightly acid, juicy and sweet, with a faint and pleasant aroma. Many cultivars of litchi can be distinguished by their avour and aroma (Menzel and Waite 2005; Singh *et al.*, 2012).

Endocarp

Two distinct tissues have been noted in the endocarp. Lying next to the aril is a dual layer of rectangular cells which are considerably thickened and lignied; just above it is the tissue composed of thick, elongated parenchymatous cells with tapering walls (Huang 2005; Menzel and Waite 2005; Singh *et al.*, 2012).

Cytogenetics

There is a dearth of information on the aspects related to the cytogenetics of litchi (Liu 1954; Chapman 1984) and this has not drawn much attention among litchi researchers. The species is probably a natural hybrid involving more than one wild progenitor. Variable haploid chromosome numbers of 14, 15, 16, and rarely 17, have been reported in litchi, thus suggesting its polyphyletic origin (Chapman 1984; Sarin *et al.*, 2009).

Germplasm Collection and Conservation

Historically, litchi cultivar breeding relied primarily on seedling selection. To date, more than 300 cultivars have been planted in the National Litchi Germplasm repository located in Guangzhou, China, established in 1998, which is the largest litchi germplasm gene bank in the world (Wu *et al.*, 2007). Globally, standardized protocols are being followed for cryopreserved tissues and excised embryos, recalcitrant seeds, and vegetative propagated plants of litchi for long-term storage at -196 °C. Besides seed banks, in vitro gene banks have also been established for germplasm conservation.

Selection of improved cultivars in litchi has been practised for thousands of years in Asia. Fay Zee Siu, Bah Lup, Lanzhu, Baitang-ying, Haak Yip, Kwai May (Red), No May Chee and Wai Chee, the main cultivars evolved in China, form the basis of the litchi industry in many other countries; for example, Tai So and Wai Chee in Thailand; Tai So, Kwai May Pink and Wai Chee in Australia; and local seedling selections of Chinese origin in Vietnam, India, Nepal, Bangladesh and Southern Thailand. Breeding effort in the past 50 to 60 years led to the development of popular seedless or small-seeded Dongguan Seedless and Hexiachuan and late maturity Maguili in Guangdong (China), SahKeng in Taiwan (China), Kom and Chacapat in Thailand, UPLB Red in the Philippines and Salathiel in Australia. There is a great potential for improving productivity through breeding new selections, with emphasis on traditional breeding rather than on biotechnology Bose (2001). According to the Chinese, litchi has more cultivars than any other fruit Bose (2001). Bah Lup, Baitang-ying, Haak Yip, Fay Zee Siu, Kwai May, No Mai Chee and Wai Chee are the important cultivars in Guangdong with Wai Chee accounting for more than 80 per cent of the area in Guangxi. However, in Fujian, Lanzhu is the predominant cultivar. No Mai Chee and Kwai May with high proportion of chicken-tongued or aborted seeds and Fay Zee Siu having good fruit size (24–32 g) with excellent eating quality Bose (2001). Haak Yip dominates in the Taiwan Province of China and accounts for more than 50 per cent of the area. Other important cultivars include Sum Yee Hong, Chong Yun Hong, No Mai Chee and SahKeng Bose (2001).

In Vietnam, 80 per cent of the area is occupied by Vaithieu Bose (2001). Tai So (Hong Huay) is the main cultivar in Northern Thailand, besides Wai Chee, O-Hia (Baidum) and Chacapat (Chakrapad). Kom, Luk Lai, SampaoKaow, KalakeBaiYaow and Red China are ecotypes developed for Bangkok regions but of low fruit quality. In India, most cultivars are of seedlings origin introduced from China. Selections of more than 30 cultivars are grown in this region, but only Shahi (Muzaffarpur), China, Calcuttia, Bedana, Late Bedana and Longia, possessing large fruits of excellent quality, are commercially important. In West Bengal, Bombai, Shahi and Rose Scented yield up to 40 kg/tree compared to 15–25 kg/tree in many other cultivars Bose (2001). In Nepal, Majfpuri, Raja Saheb, Dehraduni, China and Calcuttia are the established cultivars probably introduced from India, whereas commercial production in the hilly tract is primarily based on seedling populations. Bombai, Muzaffarpuri, Bedana and China Number Three are the popular cultivars of Bangladesh. In the hilly areas of the Philippines, Mauritius and a local selection from China, Sinco, dominates production, and an introduction from Thailand, UPLB Red, is planted in the lowlands. In Australia, Kwai May Pink accounts for more than 50 per cent of the area, besides Tai So, Souey Tung, Fay Zee Siu, Salathiel and Wai Chee Bose (2001).

Litchi has a narrow genetic base. However, ecological variations do exist in litchi, bringing the total of cultivars to more than 60, although commercially only a few varieties such as Shahi, China, and Bedana are popular. The National Active Germplasm Site at NRCL maintains more than 50 such variants, 20 clonal variants, and a large number of seedling populations arising from various parental combinations. The crop improvement programme in litchi started back in the 1950s at Sabour and Ranchi, resulting in promising cultivars such as Swarna Roopa, Sabour Bedana, and Sabour Priya. The main target of crop improvement in litchi is the integration of such characters as larger fruit size, smaller seed, high pulp ratio and acceptable pulp quality. In clonally propagated species, genetic characterization of clones held in gene banks is important for proper documentation and assessment of variability in the collection, thereby enhancing the effectiveness of gene conservation efforts. The pulp ratio of the existing varieties is moderate with associated problems related to irregularity in bearing, incidence of fruit cracking, and susceptibility to pest and disorder, thus affecting the overall yield potential. Development of improved cultivars coupled with efcient cultural practices is vital for improving crop production in a region. Improved varieties of litchi have been identied and evaluated for their potential from ICAR-NRC on Litchi, Muzaffarpur. The striking features of these varieties have been described.

Shahi: Shahi is the most popular cultivar, grown in North Bihar, Jharkhand, Uttarakhand and Uttar Pradesh. Shahi fruits are known for excellent aroma and quality with 65–70 per cent pulp recovery. Fruits are globose, heart or obtuse in shape; pulp is greyish white, soft, moderately juicy and sweet [19–21 B total soluble solids (TSS)]; seeds are large but 6–8 per cent small shrunken seeds are produced. Besides high-quality fruit, Shahi possesses a rosy avour and heavy fruit weight (22–26 g), earliest in maturity (20–30 May); this is the rst variety to reach the market. Tree size is large with high-yield potential (140–150 kg^{-1} tree) and fruit is moderately prone to cracking (Figure 26.1a).

Figure 26.1: Important Litchi Varieties at ICAR-NRC on Litchi, Muzaffarpur. a: Shahi, b: China, c: Gandaki Lalima, d: Gandaki Yogita, e: Gandaki Sampda; f: Seedlessness in litchi due to embryo abortion.

China: This variety has a very high yield potential and is tolerant to heat waves and uctuations in soil moisture. It is medium to late in maturity (30 May–10 June), with comparatively compact trees and high-yield potential (150–160 kg^{-1} tree), but prone to alternate bearing. Fruits of China are large (22–25 g), oblong in shape and tyrant rose in colour with dark tubercles at maturity. The aril is creamy white, soft, juicy and sweet (18–19 B); seeds are large and less prone to cracking with good esh recovery (60–67 per cent). The maturity period of this variety coincides with pre-monsoon showers (Figure 26.1b).

Swarna Roopa: A late-maturing, cracking-resistant cultivar selected at Ranchi has attractive red-coloured fruits with small seeds and high (65–70 per cent) aril content; fruits are medium in size weighing 15–17 g and have a high pulp content. The pulp contains high TSS and low acidity. The cultivar is suitable for extended harvest as it matures after China and is prized for its attractive fruit colour.

Purbi: Purbi is mostly grown for table purposes in the eastern part of Bihar. Fruits are medium large, oblong conical in shape, and ripen at the end of May or rst week of June. At maturity red tubercles appear on the pinkish-brown background. The average yield is 100–120 kg/tree.

Gandaki Sampada: A late-maturing strain ripens during mid-June. Fruits are large in size, conical in shape and vermilion to carmine in colour at maturity with dark blackish-brown tubercles (Figure 26.1e). Pulp is creamy white, soft and juicy, with large fruit size (35–42 g), cracking resistance, pleasant aroma and good yield potential (120–140 kg^{-1} tree). It has a very high percentage of shrivelled and small-seeded fruits, resulting in 80–86 per cent pulp recovery. This variety has potential for export purpose.

Gandaki Yogita: This is a dwarf plant, comparatively compact, and tolerant of heat waves and uctuations in soil moisture. The fruits are free from fruit borers and are of very late maturity (5–15 June). Fruits are round in shape, tyrant rose in colour with dark tubercles and exible seeds at maturity. The aril is creamy white, soft, juicy and sweet; fruit size ranges from 15 to 19 g with small seeds (Figure 26.1d). Pulp recovery is 70–75 per cent, possessing melting texture with pleasant aroma and a good blend of sugar and acid. This variety has good yield potential (70–80 kg^{-1} tree) and can be recommended for high-density planting as a speciality variety.

Gandaki Lalima: Trees are moderately vigorous, attaining an average height of 10–12 m and almost uniform spread. It is the most nutrient-efcient strain having dark green leaves and the capacity to withstand climatic aberrations. Fruits are conical, bright marigold-orange red in colour with sharp tubercles. A late-maturing cultivar ripens in the second week of June; a heavy yielder has an average yield of 130–140 kg per tree and the fruit weighs between 28 and 32 g. The fruit pulp is creamy white in colour, sweet (18–19 B TSS), soft and juicy with agreeable avour. Seeds are moderate in size and pulp recovery is above 60 per cent (Figure 26.1c).

These varieties can offer a better alternative for commercial exploitation for export of litchi from India.

Breeding for Litchi Improvement

Breeding Objectives

Litchi, being a cross-pollinated crop, has a high degree of heterozygosity which does not produce true-to-type plants through sexual means. Most commercial cultivars clonally selected under Chinese or Indian conditions have been adapted to limited climatic conditions (Pandey and Sharma 1989). Clonal selections as the basis of cultivar selection rely on limited characters such as fruit size, quality, and period of maturity. Creation of variability within the litchi gene pool is of paramount importance for desirable characters such as precocity, dwarfness, regularity of bearing, wider adaptability in tree characteristics and resistance to physiological disorders. Cross pollination results gametic variations which serve as the baseline for new selections through genetic manipulation; this stresses the need for initiating a planned breeding programme and raising plants from seeds (Sarin *et al.*, 2009; Singh *et al.*, 2012). The literature reveals that less attention has been paid to breeding work for raising new varieties; although programmes were initiated and conducted sporadically at Hawaii (Storey *et al.*, 1953) and Florida, USA (Knight 2001), Queensland, Australia (Cull 1977), and Saharanpur, India (Lal and Nirwan 1980). As a result, important cultivars such as 'Groff' and 'Brewster' were developed. A strategic effort for a breeding programme on a large scale is needed with consideration of a comprehensive survey of various genotypes and trait inheritance patterns because of the obvious difculties in litchi breeding (Sarin *et al.*, 2009; Menzel and Waite 2005; Singh *et al.*, 2012).

Inheritance Pattern or Linkage

Studies on the trait inheritance pattern in litchi have been very limited, complicating our understanding of the segregation pattern of genetic traits (Galen Sauco 1989). Differentiating cultivars through genetic markers revealed that most cultivars shared similar parentage. Variability among litchi cultivars is still unknown because breeding for new cultivars is done by growers and is based on a small number of parents (Kumar *et al.*, 2006). However, high heritability and high genetic advances were recorded for fresh seed weight, fruit weight, fruit volume and fresh and dry pulp weight. Fruit and seed weight had strong positive correlations with total sugars, ascorbic acid, protein and tryptophan content but a signicant negative correlation with acidity and phenol content (Singh *et al.*, 1987). A negative partial correlation between embryo and aril and a direct repressive effect of the former on the latter were conrmed. These results may be taken into account when breeding varieties resistant to cracking. In nature, the extent of outcrossing in litchi varies from 65 per cent to 87 per cent depending on the proximity of the pollen source. Inbreeding depression occurs on selng with respect to fruit and seed weight (Stern *et al.*, 1993). Understanding the genetic diversity within a collection is of prime importance in facilitating germplasm utilization (Strauss *et al.*, 1998). To identify genetic materials containing desirable traits, a systematic evaluation of genetic diversity is required to understand the genetic structure and relationships among accessions and their corresponding collecting site environment (Steiner and Greene 1996). Comparison of parents using different DNA markers is one of the approaches by which breeders can increase the probability of selecting parents with different gene sets. Cultivars with high percentages of aborted seeds (chicken tongue) are highly sought after in litchi. Even within the same cultivar, some trees consistently produce higher percentages of chicken-tongued seeds, and this stresses the needs to determine whether this phenomenon is attributable to environmental conditions or to genetic variation (Madhou *et al.*, 2010).

Breeding Strategies

Up to the present, the majority of litchi breeding work has been attempted through conventional methods with minor attention paid to modern biotechnological tools. Reducing the breeding period in evolving new improved cultivars is very desirable from a breeder's view. Breeding methods combined with novel technologies including genetic engineering, in vitro mutagenesis and molecular assisted breeding would assist litchi breeders to develop new cultivars in a cost-effective manner.

Conventional Breeding

Seeds are the main source of variability. However, the lengthy juvenile period associated with seedling populations hinders the breeding progress. Breeding programmes have been undertaken in several countries within the past two decades. One was initiated during 1992–1993 at the Institute for Tropical and Subtropical Plants (ITSP), South Africa, for the purpose of cultivar selection for South African conditions (Froneman and Oosthuizen 1995). In China, breeding programmes and selection of cultivars with longer duration of bearing period and good-quality fruits have been initiated (Huang *et al.*, 2005a). A seedless litchi variety was produced by conventional breeding in Hainan, China (Sarin *et al.*, 2009; Menzel and Waite 2005; Singh *et al.*, 2012).

Reciprocal controlled crosses among several existing cultivars have been reported in Australia (Dixon *et al.*, 2005). Most seedlings bear fruit within 3 years under subtropical and tropical conditions, and one promising genotype producing 32 kg of fruits with an individual fruit weight of 35–45 g was identied. According to Miao *et al.* (1998), four new litchi selections were recommended for cultivation in Hainan Province of China. Of these, Aili is a dwarf selection producing an average fruit weight of 24.8 g, and Ziangxi has large fruits (39.1–59.5 g) with acceptable eating quality. Apart from yield and quality characters, there is still a great scope for evolving cultivars that are less prone to cracking and browning as these are directly related to fruit quality. Extension of bearing period, irregular bearing, poor shelf life and disease resistance are the other signicant areas to be addressed in future programmes on litchi breeding (Sarin *et al.*, 2009; Menzel and Waite 2005; Singh *et al.*, 2012).

Crossing Techniques

Litchi requires cross-pollination, carried out primarily by bees. Although selfsterility has been reported in litchi, bagging of individual panicles is necessary to avoid outcrossing. Male ower development of litchi was dened in three stages. In the rst stage, the anther extended just beyond the receptacle, and the laments were not visible. In the second stage, the laments extended by half. In the third stage, the laments extended further before the anthers dehisced. The maximum number of mature pollen grains was found in the third stage before anther cracking (Wang *et al.*, 2015); however, exclusive collection was inefcient because anther development in the same male ower was asynchronous. Therefore, anthers between stage 2 and stage 3 can be collected to ensure sufcient pollen for a crossing programme. For effective crossing, owers should be collected at the second stage (half-extended laments light yellow in colour), and at the third stage, bags are removed and fresh male owers are collected when the anthers turn light yellow in colour. The pollen can be collected either directly from freshly dehisced anthers, if used immediately, or from indehiscent mature anthers when it will be stored for a day or two. Anthers are rst separated from owers by passing them over a screen, which allow them dehisce overnight in petri plates at room temperature. Moist cotton is placed inside the petri plates for obtaining fresh, quality and abundant pollens used for crossing. Pollen obtained in this way can be used directly in crosses and remains viable for 3–5 days at room temperature. It can remain for several weeks if refrigerated under low relative humidity. For longterm storage, litchi pollen can be stored at -86°C for 2 years. Litchi pollen stored at 4C remained viable after 52 days, which may meet the demand for articial supplementary pollination in production, but the demand for articial crossbreeding is not completely met (Wang *et al.*, 2015).

Emasculation and Pollination

To emasculate owers, scissors are used to remove excess owers; only 10–15 owers per panicle are retained for effective crossing, and owers can be pollinated on the same or the next day. By dipping a small brush into vials or petri plates, pollens are placed on the stigma. Holding the dehisced anthers by their laments using forceps can also be adopted to transfer pollen. After pollination, owers are bagged. Successful fertilization occurs after a period of several days.

Technique/Procedure of Artificial Hybridization/Artificial Crossing

M2 pollen has a higher germination rate over a wide range of temperatures. Hand pollination with M2 pollen produces a satisfactory fruit set compared to M1 pollen. Hot (32/27 C) and warm (27/22 C) temperature regimes during ower development had a pronounced effect on pollen viability compared with cool regimes (22/17 C) (Stern and Gazit 1998). The pollen germination percentage varied greatly among cultivars and was also signicantly affected by weather conditions before the opening of staminate owers, especially during the anther development stage (Xian *et al.*, 1994). Fruit and leaf characters are widely used to identify litchi hybrids, although it is known that phenotypic traits can be inuenced by environmental conditions and are not reliable indicators (Anuntalabhochai *et al.*, 2002). Hence, along with morphological characterization, molecular studies are essential to conrm the identity of litchi cultivars for optimum germplasm management and establishment of appropriate breeding programmes.

Germplasm Evaluation and Clonal Selection

Because many of the existing cultivars originated from a relatively limited ancestral stock, the introduction of new germplasm from wild forms and varieties into the genetic composition of existing cultivars is of paramount importance for achieving the breeding objectives. The important genetic resources of litchi identied for different characters are listed in Table 26.1. At present, most cultivars grown are of Chinese origin, and the genetic base of the commercial cultivars is relatively small. the majority of the cultivars arose through clonal propagation of high-performing parents. Wild forms or types of the three known litchi subspecies have been widely collected but have been minimally used for integration in breeding programmes, and genetic enhancements are largely based on collections of commercial cultivars selected from open-pollinated Chinese seedlings.

Table 26.1: Special Characteristics of some of the Litchi Cultivars

Special Traits	*Source of Traits*
Small Seed	Bedana. Nuomici, LingshanXiangli, Hainan Xiaodingxiang, Guangxi Zhangluoli.
Crisp and sweet flavour with low tannin	LingshanXiangxi
Early Maturity	Shahi, Shanyuedong Red, Early Bedana, Dehra Rose
Big fruit	Edanli (60-70g), Vnsdasu (40-45g)
High yields	Heiye (Black Leaf), Baitangying (White Super Poppy)
High temperature tolerance	Shanyehong, better flower bud differentiation at 20°C as compared to 12°C to common varieties
Good Canning purpose	Heiye, Xuangxi
Drought tolerance	Tianyan (Sweet Stone)
Good on tree storage	Huaizhi, fully ripe fruits can be left on trees for fresh picking
Late maturity	Xuehuazi, ripening during early/mid July, with high yielding characteristics, Fijian Xiafanli (Fujian Xiafan litchi) maturing during the late July or early August, enabling extended supply period in combination with early and intermediate maturity types, Longia, Kaselia, Yogda

Hybridization

Selection for high-yielding and good-quality litchi has been attempted over a long period, but the breeding of new hybrids has not yet attained a signicant impact. Recent breeding efforts at Sabour (India) led to the development of two hybrids, namely, Sabour Madhu, Sabour Priya and Sabour Litchi-1 (Table 26.2 and Figure 26.2). The growth of the hybrid seedlings is slow, and only 4 per cent of the total population reached maiden owering at the age of 14 years (Thakur 1992). Thus, in addition to a short period of seed viability, the late-bearing habit of the seedlings poses serious problems for hybridization work. Besides, these hybrids showed erratic owering which made it difcult to obtain pollen for breeding purpose.

Distant Hybridization

Hybridization was also attempted between litchi and longan, and seedling progenies were variable with small fruit size, which appeared to be a dominant character. Attempts on diallele crosses between Nephelium lappaceum, N.

rambutanakee, Dimocarpus longan, and Litchi chinensis were made, and only intergeneric crosses between longan and litchi were successful. The pollen of both litchi and longan germinated on stigma of rambutan but was arrested in the embryo sac. However, there appears to be no breeding barrier between cultivars or species within a genus except when seedless fruits are commonly produced (Figure 26.1f). McConchie *et al.* (1994) attempted reciprocal crosses between commercial cultivars of litchi (Bengal and Kwai May Pink) and longan (McLeans Ridges and Duan Yu) and found that hybrid progeny developed only when litchi was used as the female parent. Morphologically the hybrid plants were similar to litchi, but the leaves were smaller. Three types of seeds developed in litchi following pollination with longan pollen: (1) normal seeds with well-developed testa and embryo, (2) seeds with aborted embryo but normal testa development, and (3) seedless, wherein the ovule development is arrested.

Table 26.2: Litchi Cultivars Developed through Hybridization

Hybrids	*Discription*
Sabour Madhu	This hybrid resulted from Purbi x Bedana. It has higher number of fruits (24) per panicle and ripens 8 days later than another late maturing cultivar, Kasba. It has higher TSS and aril percentage than Purbi. Fruit shape resembles Purbi.
Sabour Priya	This is a product of Purbi x Bedana. It has better fruit quality than Purbi in terms of higher aril percentage and TSS content. The fruit shape is intermediate between both the parents. The fruit weight is higher than the better parent (Purbi).
Sabour Litchi-1	A hybrid between Purbi x Bedana, matures in first week of June. Attractive coloured fruits, large, round, Fruit weight (22.90g) having 70.75 per cent aril, Small seeded (3.18g per seed). Light pink juicy arils, pleasant aroma, TSS 19.4 per cent, very good sugar acid blend (Figure 26.2)

Figure 26.2: Sabour Litchi-1, A Hybrid of Purbi x Bedana, Developed by BAU Sabour.

Characterization of Litchi Based on Morphological Traits

Litchi cultivars were classied into four categories according to fruit shape: long oval, heart shaped, short heart shaped and round (Li and Fang 1956). Guangdong litchi cultivars were grouped into seven types in 'The Flora of Guangdong Litchi' (Guangdong Academy of Agricultural Sciences 1978). These types include Guiwei, Xiaozhi, Jinfeng, Sanyuehong, Heiye, Nuomici and Huaizhi, based on the shape of tortoise shell-like cracking segments and pericarp with or without sharp protuberances as well as the shape of leaf, inorescence and fruit, maturity time and quality of fruit. Later, in 1986, litchi cultivars were separated into three types and seven groups according to a four-rank classication criterion considering pericarp, fruit shape and other morphological characteristics. These classication principles were adopted in the subsequent 'The Flora of Chinese Fruit Tree, the Volume of Litchi' (Wu 1998b); Wu *et al.* (2016) classied litchi cultivars into three types based on the smoothness or roughness of the pericarp. Morphological characteristics such as tree height, canopy spread, tree shape, foliage texture and colour, leaf length, width, shape and orientation, intermodal distance, number of leaets per leaf, number of leaves per ush, ush colour, panicle length, number of anthers and carpels per ower, lament and style size and fruit colour and size have been used to quantify genetic diversity in litchi (Khurshid *et al.*, 2004).

Characterization of Litchi Based on Isozyme Analysis

Isozyme markers were effective in differentiating litchi accessions. Ambiguities resulting from synonyms and homonyms and sexual propagation of accessions have been resolved to some extent by comparing isozyme ngerprints of different accessions. However, an authentic collection and evaluation of native Chinese cultivars would help in establishing the identity of the present litchi collection. Liu *et al.* (1989) and Zhou *et al.* (2001) analysed the PER isozyme of 24 Guangxi litchi cultivars and 35 Guangdong litchi cultivars and produced almost identical results (Wu *et al.*, 2007). Liu *et al.* (1989) also reported the stability of the litchi PER isozyme pattern as unaffected by the age of leaves. Aradhya *et al.* (1995) reported genetic diversity among 49 litchi (Litchi chinensis Sonn.) accessions using eight enzyme systems encoding 12 loci and revealed moderate to high levels of genetic variability. Comparison of isozyme ngerprints revealed different isozyme patterns in 'Nomai tsz', 'Kwai mi' and 'Hak ip,' believed to be synonyms, whereas some others with different names displayed identical patterns. Wu *et al.* (2016) reported the usefulness of morphological traits of leaf and branching to discriminate 146 Chinese litchi germplasms.

Characterization of Litchi Based on Molecular Markers

Molecular studies on genetic diversity of litchi are limited. The presence of variable litchi cultivars in China and India provides a good basis for development of new cultivars (Groff 1921). Recent development in molecular tools further enhances our efciency in understanding the genomic variability and the diversity between and within different species of litchi.

Initially, studies on genetic diversity in litchi have been largely based on morphological characters. Development of molecular markers in the past few decades yielded accurate and precise information on the extent of genetic diversity in litchi. The most commonly used DNA marker is random amplication of polymorphic DNA (RAPD) (Ding *et al.*, 2000; Chen *et al.*, 2004, 2005; Liu and Mei 2005; Wang *et al.*, 2006; Bajpai *et al.*, 2016), followed by amplied fragment length polymorphism (AFLP) (Yi *et al.*, 2003; Peng *et al.*, 2006) and simple sequence repeat (SSR) (Li and Zheng 2004). However, in most of the tested litchi cultivars, contradictory results were obtained from different markers, and even those obtained from the same DNA marker are not totally identical. Results of RAPD and AFLP analysis indicated a low level of genetic diversity within litchi collections; several workers characterized litchi cultivars using molecular markers and reported the occurrence of synonyms (Ding *et al.*, 2000; Yi *et al.*, 2003; Chen *et al.*, 2004) such as 'Nongmei No. 9' and 'Qiongshan No. 27' (Chen *et al.*, 2004), 'Dazao' and 'Zaohong,' 'Baiye' and 'Guahong,' 'Feizixiao' and 'Zhimali,' 'Ziniangxi' and 'Zengchengdaguoli' (Liu and Mei 2005) and 'Fengshuang' and 'Tunchangfengshuang' (Wang *et al.*, 2006). However, Ding *et al.* (2000) reported that 'Dazao' and 'Zaohong' are two entirely different cultivars. Viruel and Hormaza (2004) obtained 12 microsatellites enriched in CT repeats from a genomic library of the litchi cultivar 'Mauritius.' Assessment of genetic diversity of litchi at Kalyani, West Bengal (India), revealed the segregation of cultivars into two clusters which can give heterotic hybrids when intercluster cultivars are crossed (Dwivedi and Mitra 1995, 1996). Recently, Liu and Mei (2005) reported an average of 15.8 per cent polymorphic and 0.10 per cent monomorphic RAPD markers from a population comprising 60 litchi cultivars, one longan cultivar, and one tentative intergeneric hybrid of litchi and longan. Two accessions (LH80 and LH109) were genetically distinct as revealed by RAPD and AFLP markers. In Thai litchi cultivars, the percentages of polymorphic markers for RAPD and AFLP were 34.6 per cent and 36.3 per cent, respectively (Tongpamnak *et al.*, 2002), and each marker system was able to differentiate all accessions. RAPD markers have been widely used to study the genetic relatedness among litchi cultivars (Kumar *et al.*, 2006); Ding *et al.* (2000) investigated the segregation patterns of RAPD markers in an F_1 population of litchi from a cross between 'Wuye' and 'Luhebao.' Bajpai *et al.* (2016) reported genetic diversity of 20 litchi cultivars from the Indian peninsula, and phylogenetic analysis based on RAPD and microsatellites revealed clustering of the cultivars into four major groups, although within a very narrow range (0.63–0.90) of similarity, *viz.* Seedless (*i.e.*, Bedana), Mandarji, Shahi and China groups. Simple sequence repeats (SSRs) markers previously developed for litchi have also been evaluated for polymorphism in different populations (Ekuae *et al.*, 2009). The Power Core programme based on 30 EST-SSRs and a combined dataset of the EST-SSRs and 16 phenotypic traits has been used to construct two core collections from 96 accessions (Sun *et al.*, 2012). Madhou *et al.* (2013) conducted molecular characterization study of litchi accessions from Mauritius and Reunion and compared them with Spanish litchi cultivars. Liu *et al.* (2015) obtained the potential of single nucleotide polymorphisms (SNP) for the identication of 96 representative litchi accessions, and their genetic relationships in China were evaluated using 155 SNPs that were evenly spaced across the litchi genome. Ninety SNPs with minor allele frequencies above 0.05 and a good genotyping success rate were used for further analysis. A relatively high level of genetic variation was observed among litchi accessions, as quantied by the expected heterozygosity (He ¼0.305).

Mutation Breeding

So far, induced mutations have not been in vogue in litchi improvement. It was reported that seedlessness in the litchi cultivars Lanzhu and Luhebao might have resulted from mutation, gene interaction and long-term articial selection and is not the result of variation in chromosome number or structure (Lu *et al.*, 1987).

Qinzhou red litchi, a new variety of litchi, was derived from spontaneous mutation of the cultivar Black Leaf (Peng *et al.*, 2001).

Transgenic Approach

Green uorescent protein (GFP) gene expression in leaf tissues of litchi after transformation is using Agrobacterium (Puchooa 2004a). In vitro, grown leaf tissues were used for transformation. After 4 weeks in culture, expression of GFP was apparent when the regenerated callus and the leaves were observed under a uorescence microscope tted with a blue exciter lter, a blue dichroic mirror and a barrier lter. Although no transformed litchi plantlets were regenerated, screening for GFP gene expression may prove useful to improve transformation efciency and to facilitate the detection of transformed litchi plants. Ouyang *et al.* (1985) reported T-DNA transfer and tumour formation induced by Agrobacterium tumefaciens on litchi. Therefore, genetic transformation of litchi using Agrobacterium could be exploited in the future.

Constraints in Litchi Breeding

The development of improved cultivars in litchi is a lengthy and cumbersome process, and trees when Fruited only produce less than 1 per cent of the seedlings worthy of selection. Breeding for a new hybrid cultivar through conventional methods would require about 40 years in fruit trees such as litchi (Zheng *et al.*, 2001). In addition to poor seed viability and wide variability in seedling progenies, erratic owering poses a hindrance for pollen collection during the breeding process. Although crosses between litchi and longan are successful, the use of longan as parents in a hybridization programme is limited by its biennial bearing tendency. In crosses involving plants that have a tendency to produce chicken-tongued seeded fruits as the female parent, many of the most valuable progeny are lost before the harvest (Puchooa 2004b). Only a few cultivars have been bred through conventional methods because of the long juvenile period, the apparent lack of genetic variability in the existing germplasm, and the great expenditure that is required in terms of land, time and money (Litz *et al.*, 2005). Thus, the improvement of litchi appears to be conûned mainly to selections of improved chance seedlings or genotypes. Future efforts in plant breeding need to emphasize identication of potential parents followed by their reciprocal crossing.

Future Prospectives

Despite constraints and the long time frame needed for conventional breeding, opportunities do exist to develop new cultivars with improved traits that can improve productivity and fruit quality against the endogenous and exogenous attack of pathogens, mites and insects. New ideotypes through appropriate blending of traditional breeding and biotechnological tools can be used in the future for disease management in the litchi system. Recent and future development in highthroughput MAS can substantially improve the efûciency of the conventional litchi breeding to attain the resistance in susceptible litchi crops. Application of molecular markers in the creation of a genetic map and other pre-selection techniques has the potential to substantially reduce selection time frame in litchi breeding programmes. Identication and cloning of genes of horticultural interest from litchi and its wild relatives will enhance the rate of conservation of plant survival from germinated transgenic somatic embryos and would certainly revolutionize the genetic improvement of litchi cultivars. The advantage of mutation induction in obtaining the hidden genetic variation and improvement of vegetatively propagated plants when one or few characters of an outstanding cultivar are to be modied can also be exploited to a great extent. Biotechnological developments and statistical analysis of current breeding populations would immensely improve our understanding of the genetic traits and their inheritance in litchi which, in turn, make it easier for the breeders to select parents and design a systematic breeding programme with more specic breeding goals than has been pursued in the past. However, there is still signicant constraint in controlled hybridization that needs to be addressed.

REFERENCES

Abu Baker Siddiqui S.B.M. 2002. Lychee production in Bangladesh. In: Papademetriou, M.K. and Dent, F.J. (eds) Lychee Production in the Asia-Pacific Region. Food and Agricultural Organization of the United Nations, Bangkok, Thailand, pp. 28–40.

Anonymous.1978. Studies on tropical fruit trees. *Fruits* 33: 647–665.

Anuntalabhochai S., Chundet R., Chiangda J., Apavatjrut P. 2002. Genetic Diversity within Lychee (*Litchi chinensis* Sonn.) based on RAPD Analysis. *Acta horticulturae*, 575: 253-260.

Aradhya M.K., Zee F.T., Manshardt R. 1995. Isozyme variation in lychee (*Litchi chinensis* Sonn.). *Scientia Horticulturae*, 63: 21-35.

Bailey L.H. (1949) Manual of Cultivated Plants. Macmillan, New York, 1116 pp.

Bajpai A., Muthukumar M., Singh A., Nath V., Ravishankar H. 2016. Narrow genetic base of Indian litchi (*Litchi chinensis*) cultivars based on molecular markers. *Indian Journal of Agricultural Sciences*, 86: 448-455.

Batten D.J. and Lahav E. 1994. Base temperatures for growth processes of lychee, a recurrently flushing tree, are similar but optima differ. *Australian Journal of Plant Physiology*, 21(5): 589–602.

Batten D.J. and McConchie C.A. 1995. Floral induction in growing buds of lychee (*Litchi chinensis*) and mango (*Mangifera indica*). *Australian Journal of Plant Physiology*, 22: 783–791.

Boonrat S. 1984. Lychee Cultivation. Horticultural Research Institute, Department of Agriculture, Bangkok, Thailand, 86 p.

Bose T.K. 2001. *Fruit: Tropical and Subtropical*. Naya Yug,.721p.

Budathoki K. 2002. Lychee production in Nepal. In: Papademetriou, M.K. and Dent, F.J. (eds) Lychee Production in the Asia-Pacific Region. Food and Agricultural Organization of the United Nations, Bangkok, Thailand, pp. 81–93.

Butcher F.G. 1957. Pollinating insects on lychee blossoms. Proceedings of the Florida State Horticultural Society, 70: 326–328.

Chadha K.L. and Rajpoot M.S. 1969. Studies on floral biology, fruit set, and its retention and quality of some litchi varieties. *Indian Journal of Horticulture*, 26: 124–129.

Chapman K.R. 1984. Sapindaceae. In: P.E. Page (ed.), Tropical Tree Fruits for Australia. Queensland Department of Primary Industry, Queensland, Australia, pp. 179-191.

Chaturvedi R.B. 1965. Preliminary studies in the sex distribution, pollination and fruit development in litchi (*Litchi chinensis* Sonn.). *Allahabad Farmer*, 39: 49–51.

Chen H., Huang H. 2001. China litchi industry: development, achievements and problems. *Acta Horticulturae*, 558: 31–39.

Chen Y.T., Lai Z.X., Chen J.Y., Guo Y.Q., Cai Y.Q., Wu J.S. 2005. RAPD analysis of genetic relationship among some ancient litchi genetic resources. *J. Fujian Agric. Forest. Univ.*, 34 (4): 458–463.

Chen Y.Y., Deng S.S., Zhang X., Wei S.X., Gao A.P., Wang J.B., Zhou B. (2004) RAPD analysis of genetic relationship among partial litchi germplasms in Hainan island. *Acta Hortic. Sin.*, 31 (2): 224–226.

Choo W.C. and Ketsa S. 1991. *Dimocarpus longan* Lour. In: Verheij, E.W.M. and Coronel, R.E. (eds) Plant Resources of South-East Asia No. 2: Edible Fruits and Nuts. Pudoc, Wageningen, The Netherlands, pp. 146–151.

Costés E. 1988. Analyze architectural et modelisation du litchi (*Litchi chinensis* Sonn.). PhD. dissertation, University of Montpellier, France.

Cull B.W.1977. Report on Overseas Study Tour" Queensland Department of Primary Industries, pp 1-92.

Das C.S. and Choudhury K.R. 1958 Floral biology of litchi (*Litchi chinensis* Sonn.). *South Indian Horticulture*, 6: 17–22.

Davenport T. L. 1990. Citrus flowering. *Horticultural Reviews*, 12: 349–408.

De Candolle. 1909. "Origin of cultivated plants ", 2nd. Ed. International scientific Series, Kegan Paul, Trench,Tribner and Co.Ltd., London Dep. Primary Ind. p-123.

Dhaliwal H.S., Srivastava S., Adlakha R.L. 1977. Insect pollination of lychee, *Litchi chinensis* Sonn. in the valley areas of the Indian Himalayas. In: Proceedings of the Twenty-Sixth International Apiculture Congress. Apimondia Publishing House, Bucharest, Hungary, p. 396.

Ding X.D., Xin L.L., Jing C.X., Xiong G., Ding L., Lu X.D., Chen X., Guan X. 2000. Identifying litchi cultivars and evaluating their genetic relationships by RAPD markers. J.*Trop Subtrop.Bot.*, 8: 49-54.

Dixon T., Batten D. and Mcconchie C. 2005. Genetic improvement of lychee in Australia. Acta Horticulture 665: 107-110.

Du Toit A.P. 1994. Pollination of avocados, mangoes and litchis. Plant Protection Research Institute, South Africa Plant Protection News Bulletin 35, pp. 4–5.

Dwivedi A.K., Mitra S.K. 1996. Divergence analysis of litchi (*Litchi chinensis* Sonn.) cultivars grown inWest Bengal. *Indian J. Genet. Plant Breed.*, 56 (4): 486–489.

Dwivedi A.K., Mitra S.K. 1995. Genetic diversity of fruit quality traits in litchi (*Litchi chinensis* Sonn.). *Hortic. J.*, 8 (2): 113–118.

Ekua/e M.R.M., Gailing O., Finkeldey R. 2009. Transferability of Simple Sequence Repeat (SSR) Markers Developed in *Litchi chinensis* to Blighia sapida (Sapindaceae). *Plant Molecular Biology Reporter*, 27: 570-574.

Froneman J.J., Oosthuizen J.H. 1995. ETSG Litchi breeding programme. *Yearbook-South African Litchi Growers Association*, 7: 35-37.

Galan Sauco V. 1989. Litchi Cultivation. Food and Agriculture Organization of the United Nations, Rome, 136 p.

Galán Saúco V., Menini U.G. 1989. Lychee Cultivation. FAO Plant Production and Protection Paper No. 83. Food and Agricultural Organization of the United Nations, Rome, 205 pp.

Ghosh S.P. 2001. World trade in litchi: past, present and future. *Acta Horticulturae*, 558: 23–30.

Goren M., Degani O., Gazit S. 1998. Flowering phenology of 10 lychee cultivars in the Volcani Center, Israel. *Alon Hanotea*, 52: 59–63.

Goren M., Tomer E., Gazit S. 2001. Litchi cultivars in Israel. *Acta Horticulturae*, 558: 129–133.

Groff G.W. 1921. The Lychee and Longan" Orange Judd company and Canton Christian College. *China Science and culture*, 38: 39-40.

Guangdong Academy of Agricultural Sciences (1978) The Flora of Guangdong Litchi. Guangdong Scientific and Technical Press, Guangzhou, pp. 25–36.

Hai V.M.and Dung N.V. 2002. Lychee production in Viet Nam. In: Papademetriou, M.K. and Dent, F.J. (eds) Lychee Production in the Asia-Pacific Region. Food and Agricultural Organization of the United Nations, Bangkok, Thailand, pp.114–119.

Hayes W.B. 1957. Fruit Growing in India. Kitabistan, Allahabad, India, 180 pp.

Huang X., Zeng L. and Huang H.B. 2005. Lychee and Longan production in China. Acta Hort 665: 27-36.

Jahiel M. and Abraham D. 2001.The growth of the Malagasy litchi field face to the hardening of European control in the matters of sulphured residue. *Acta Horticulturae*, 558: 63–71.

Joubert A.J. 1985. *Litchi chinensis*. In: Halevy, A.H. (ed.) Handbook of Flowering, Vol. 5. CRC Press, Boca Raton, Florida, pp. 204–210.

Khurshid S., Ahmad I., Anjum M.A. 2004 Genetic diversity in different morphological characteristics of litchi (*Litchi chinensis* Sonn.). *Int. J. Agric. Biol.*, 6 (6): 1062–1065.

King J., Exley E.M., Vithanage V. 1989. Insect pollination for yield increases in lychee. In: Batten, D.J. (ed.) Proceedings of the Fourth Australian Conference on Tree and Nut Crops. Exotic Fruit Growers' Association, Lismore, Australia, pp. 142–145.

Knight R.J. Jr. 2001. The lychee's history and current status in Florida. *Acta Horticulturae*, 558: 41–44.

Kumar M., Gupta, M., Shrivastava D, Prasad M and Prasad U.S. 2006. Genetic Relatedness among Indian Litchi Accessions (*Litchi chinensis* Sonn.) By RAPD Markers. *International Journal of Agricultural Research.* 1: 390-400.

Lal B.S. and Nirwan R.S. 1980. Saharanpur Selection a new litchi. *Prog. Hort.*, 12: 75-77.

Leenhouts, P.W. 1978. Systematic notes on the Sapindaceae-Nephelieae. *Blumea,* 24: 395–403.

Li L.R. and Fang Y. 1956. The Research of Litchi and Longan. Scientific Press, Beijing, pp. 1–10.

Li M.F. and Zheng X.Q. 2004. Development of SSR markers in litchi (*Litchi chinensis*). *Hereditas* 26 (6): 911–916.

Litz R.E., Raherjo S., Matsumoto K. and Witjaksono 2005. Biotechnology. In: C.M. Menzel and G.K. Waite (eds.), Litchi and Longan: Botany, Production and Uses, CAB International, pp. 49-58.

Liu H.B., Liu Y.Q. and Liu R.G. 1989. The peroxidase isozymes analysis of litchi cultivar. *J. Guangxi Agric. Sci.* 5: 27–29.

Liu S.Y. 1954. Studies of *Litchi chinensis* Sonn. PhD thesis, University of Michigan, Ann Arbor, Michigan. Costés, E. (1988) Analyze architectural et modelisation du litchi (*Litchi chinensis* Sonn.). PhD dissertation, University of Montpellier, France.

Liu W., Xiao Z., Bao X., Yang X., Fang J. and Xiang X. 2015. Identifying Litchi (*Litchi chinensis* Sonn.) Cultivars and Their Genetic Relationships Using Single Nucleotide Polymorphism (SNP) Markers. *PloS One, 10(8): p.e0135390.*

Liu, C. and Mei M. 2005. Classification of lychee cultivars with RAPD analysis. *Acta Horticulturae,* 665: 149–159.

Lu L.X., Chen J.L.and Chen X.J. 1987. A study on chromosome number and meiosis in pollen mother cells in *Litchi chinensis* Sonn. *Journal of Fujian Agricultural College,* 16 (3): 224-228.

Madhou M., Bahorun J., Hormaza J.I. 2010. Phenotypic and molecular diversity of litchi cultivars in Mauritius. *Fruits,* 65: 141-152.

Madhou M., Normand F., Bahorun T.and Hormaza J.I.2013. Fingerprinting and analysis of genetic diversity of litchi (*Litchi chinensis* Sonn.) accessions from different germplasm collections using microsatellite markers. *Tree Genetics and Genomes,* 9: 387–396.

Marloth R.H. 1947. The Litchi in South Africa I,.Varieties and Propagation II Cultural Practices and Marketing. *Farming South Africa,* 22: 824-30.

McConchie C.A., Vithanage V. and Batten D.J. 1994. Intergeneric hybridization between litchi (*Litchi chinensis* Sonn.) and longan (*Dimocarpus longan* Lour.). *Annals of Botany*, 74 (2): 111-118.

Menzel C.M. 1991. *Litchi chinensis* Sonn. In: Verheij, E.W. and Coronel, R.E. (eds) Plant Resources of South East Asia No.2: Edible.

Menzel C.M., Olesen T., McConchie C., Wiltshire N., Diczbalis Y. and Wicks C.2000. Lychee, Longan and Rambutan: Optimizing Canopy Management. Rural Industries Research and Development Corporation No. 00/29, Canberra, Australia.

Menzel C.M. and Simpson D.R. 1990. Performance and improvement of litchi cultivars: a review. *Fruit Varieties Journal*, 44: 197–215.

Menzel C.M., Simpson D.R. and Watson B.J. 1993. Fruits of tropical climates: fruits of the Sapindaceae. In: Caballero, B., Trugo, L. and Finglas, P (eds) Encyclopedia of Food Science, Food Technology and Nutrition. Academic Press, London, pp. 2114–2120.

Menzel C.M. and Waite G. 2005. Litchi and longan: botany, production, and uses. Cambridge, MA, CABI Pub. http: //www.SLQ.eblib.com.au/patron/FullRecord.aspx?p=297524.

Menzel C.M., Watson B.J. and Simpson D.R. 1988. The lychee in Australia. Queensland Agricultural Journal 114, 19–27.

Milne D.L. 1999. Lychee production and research in Southern Africa. In: Menzel, C.M, Waite, G. and Bronson, R. (eds) Proceedings of the Fifth National Lychee Conference. Australian Lychee Growers' Association, Twin Waters, Queensland, pp. 25–31.

Mustard M.J., Liu S. and Nelson R.O. 1953. Observations of floral biology and fruit setting in lychee varieties. *Proceedings of the Florida State Horticultural Society*, 66: 212–220.

Nakasone H.Y. and Paull R.E. 1998 Litchi, longan and rambutan. In: Tropical Fruits. CAB International, Wallingford, UK, pp. 173–207.

Nakasone H.Y. and Paull R.E. 1998. Tropical Fruits. CAB International, Wallingford, UK, 445 pp.

Naphrom D., Potjanapimon C., Subhadrabandhu S. 2001. Changes in cytokinin and gibberellin-like substances in stem apex of lychee cv. Hong Huay prior to flowering. *Acta Horticulturae*, 558: 199–203.

Oosthuizen J.H. 1991. Lychee cultivation in South Africa. Yearbook of the Australian Lychee Growers' Association 1, 51–55.

Padilla G., et al., 2013. Agrobacterium tumefaciens-mediated transformation of 'Brewster' ('Chen Tze') litchi (*Litchi chinensis* Sonn.) with the PISTILLATA cDNA in antisense. In Vitro Cellular and Developmental Biology - Plant. 49, 510-519.

Pandey R.M. and Sharma H.C. 1989. The Litchi. ICAR, New Delhi, India, 80 p.

Pandey R.S. and Yadava R.P.S. 1970. Pollination of litchi (*Litchi chinensis*) by insects, with special reference to honeybees. *Journal of Apiculture Research*, 9: 103–105.

Peng H.X., Cao H.Q., Zhu J.H., Liao H.H., Huang F.Z., Li J.Z. and Liang W. 2006. Studies on the application of AFLP molecular markers on genetic diversity and classification of good and rare litchi resources in Guangxi. Southwest. *China J. Agric. Sci.* 19 (1): 108–111.

Peng H.X., Li Y.C., Huang D.J., Su W.Q. and Liu R.G. 2001. 'Qinzhou red litchi' - a new big fruit form variety of litchi. *Acta Horticulturae Sinica*, 28 (3): 276.

Puchooa D. 2004. Expression of green fluorescent protein gene in litchi (*Litchi chinensis* Sonn.) tissues. *J. Appl. Hort.* 6(1): 11-15.

Puchooa, D. 2004 *In vitro* regeneration of litchi (*Litchi chinensis* Sonn.). *African Journal of Biotechnology*, 3 (11): 576-584.

Robbertse H., Fivaz J., Menzel C.M. 1995. A reevaluation of tree model, inflorescence morphology, and sex ratio in lychee (*Litchi chinensis* Sonn.). *Journal of the American Society for Horticultural Science* 120: 914–920.

Sarin N., Prasad U., Kumar M. and Jain S.M. 2009. Litchi Breeding for Genetic Improvement Litchi breeding for genetic improvement. In Breeding Plantation Tree Crops: Tropical Species (pp. 217-245). Springer, New York, NY.

Scholefield P.B. 1982. A scanning electron microscope study of flowers of avocado, litchi, macadamia and mango. *Scientia Horticulturae*, 16: 263–272.

Shukla R.K. and Bajpai P.N. 1974. Blossom bud differentiation and ontogeny in litchi (*Litchi chinensis* Sonn.). *Indian Journal of Horticulture*, 31: 224–228.

Singh A., Abidi A.B., Srivastava S. and Singh I.S. 1987. Correlations among physical and biochemical traits in litchi (*Litchi chinensis* Sonn.). *Narendra Deva Journal of Agricultural Research*, 2 (2): 160-163.

Singh G., Nath V., Pandey S.N., Ray P.K. and Singh H. S.2012. Litchi. New Delhi, FAO.

Singh L.B. and Singh U.P. 1954. The Litchi. Lucknow Superintendent of Printing and Stationary, Uttar Pradesh, India, 87 pp.

Singh L.B. and Singh, U.P. 1954. The Litchi, Supdt. Printing and stationary, U.P. Allahabad.

Sotto R.C. 2002. Lychee production in the Philippines. In: Papademetriou, M.K. and Dent, F.J. (eds) Lychee Production in the Asia- Pacific Region. Food and Agricultural Organization of the United Nations, Bangkok, Thailand, pp. 94–105.

Sotto R.C. 2002. Lychee production in the Philippines. In: Papademetriou, M.K. and Dent, F.J. (eds) Lychee Production in the Asia- Pacific Region. Food and Agricultural Organization of the United Nations, Bangkok, Thailand, pp. 94–105.

Steiner J.J. and Greene S.L. 1996. Proposed use of ecological descriptors and their utility for plant germplasm collections. Cr*op Science*, 36: 439-451.

Stern R.A. and Gazit S.1996. Lychee pollination by the honeybee. *Journal of the American Society for Horticultural Science,* 121: 152–157.

Stern R.A. and Gazit S. 1998. Pollen viability in lychee. *Journal of the American Society for Horticultural Science* 123: 41–46.

Stern R.A. and Gazit S. 2003. The reproductive biology of the lychee. *Horticultural Reviews,* 28: 393–453.

Stern R.A., Gazit S., El-Batsri R. and Degani C. 1993. Pollen parent effect on outcrossing rate, yield, and fruit characteristics of 'Floridian' and 'Mauritius' lychee. *Journal of the American Society for Horticultural Science,* 118 (1): 109-114.

Storey W.B., Hamilton R.A., Nakasone H.Y. 1953). Groff – a new variety of lychee. *Circular Hawaii Agric.Exp. St.* 39: 1-8.

Strauss M.L., Pino J.A. and Cohen J.I. 1998. Quantification of diversity in ex situ plant collections. *Diversity,* 16: 30-32.

Subhadrabandhu S. 1990. Lychee and Longan Cultivation in Thailand. Rumthai Publishers, Bangkok, Thailand, 40 pp.

Subhadrabandhu S. and Yapwattanaphun C. 2001. Lychee and longan production in Thailand. *Acta Horticulturae,* 558: 49–57.

Sun Q., Bai L., Ke L. Xiang X., Zhao J., Ou L. 2012. Developing a core collection of litchi (*Litchi chinensis* Sonn.) based on EST-SSR genotype data and agronomic traits. *Scientia Horticulturae,* 146: 29-38.

Tao R. 1955. The Superior lychee" Proc. Florida Lychee Growers' Association 2: 73-74 Technoques. Ph.D. Thesis, Bhagalpur University, Bhagalpur.

Thakur S.1992. Hybridization in litchi. Souvenir, National Seminar on Research Developments in Litchi Production, RAU, Pusa, p. 13.

Tindall H.D. 1994. Sapindaceous fruits: botany and horticulture. *Horticultural Reviews,* 16: 143–196.

Tongpamnak P.A., Patanatra P.S. and Babprasert C. 2002. Determination of genetic diversity and relationship among Thai Litchi asccessions by RAPD and AFLP markers. *Kasetsart J (Nat Sci)* 36: 370–80.

Viruel M.A. and Hormaza J.I. 2004. Development, characterization and variability analysis of microsatellites in lychee (*Litchi chinensis* Sonn., Sapindaceae). *Theor. Appl. Genet.,* 108 (5): 896–902.

Wang B and Qiu Y.1997. Growth and fruiting. In: Zhang ZW, Yuan PY, Wang BQ and Qui YP. (eds) Litchi Pictorial Narration of Cultivation. Pomology Research Institute, Guangdong Academy of Agricultural Sciences, Guangzhou, China, pp. 66–88.

Wang J.B., Deng S.S., Liu Z.Y., Liu L.Z., Du Z.J., Xu B.Y. and Chen Y.Y. 2006. RAPD analysis on main cultivars of litchi (*Litchi chinensis* Sonn.) in Hainan. *J. Agric. Biotechnol.* 14 (3): 391–396.

Wu D., Lin X., Ye Q. and Wang W. 2001. Improvement of fruit set in secondary panicles of 'Feizixiao' litchi by removal of the primary panicles. In: Huang, H.B and Huang, X.M. (eds) First International Symposium on Litchi and Longan. Programme and Abstracts. South China Agricultural University, Guangzhou, China, p. 42.

Wu J., Zhanga C., Chena J., Caia C., Wanga L., Fua D. and Oua L. 2016. Morphological diversity within litchi (*Litchi chinensis* Sonn.) based onleaf and branch traits. *Scientia Horticulturae,* 207: 21–27.

Wu S.X. 1998. Encyclopedia of China Fruits: Litchi. China Forestry Press, Beijing, China, 221 pp.

Wu S.X. 1998. The Flora of China Fruit Tree, the Volume of Litchi. China Forestry Press, Beijing, pp. 94–198.

Wu Y., Yi G., Zhou B., Zeng J. and Huang Y. 2007. The advancement of research on litchi and longan germplasm resources in China. *Scientia Horticulturae,* 114: 143-150.

Xian X., Zhang Z.W., Wang B.Q., Yuan P.Y. and Qiu Y.P. 1994. Fertility and storability of lychee pollens. *Guangdong Agricultural Sciences* 4: 25-27.

Yi G.J., Huo H.Q., Chen D.C., Huang Z.R., Cai C.H., Qiu Y.P. 2003. Studies on genetic relationship among litchi cultivars by using AFLP. *Acta Hortic. Sin.* 30 (4): 399–403.

Zhang Z. 1997. Local cultivars and selections with potential. In: Zhang ZW, Yuan PY, Wang BQ and Qui YP. (eds) Litchi Pictorial Narration of Cultivation. Pomology Research Institute, Guangdong Academy of Agricultural Sciences, Guangzhou, China, pp. 48–57.

Zheng Q., Davenport T.L. and Li Y.C. 2001. Stem age, winter temperature and flowering of lychee in south Florida. *Acta Horticulturae,* 558: 237–240.

Zheng Q., Hu G., Chen D. 2001. Biotechnology in longan and litchi breeding: applications andprospects. *Acta Horticulture* 558: 143–148.

Zhou J.H., Luo H.N., Huang J.B. and Chen C.M. 2001. The peroxidase isozymes analysis of 35 cultivars of lychee in Guangdong. J. *Zhongkai Agric. Coll.* 14 (2): 20–25.

Breeding of Perennial Horticultural Crops *Pages* 357–363
Editors: S.K. Singh, V.B. Patel, A.K.Goswami, Jai Prakash and Chavlesh Kumar
Published by: **BIOTECH BOOKS, NEW DELHI**

***Chapter* 27**

Mango Improvement

Manish Srivastav, S.K. Singh and A. Nagaraja

Division of Fruits and Horticultural Technology,
ICAR-Indian Institute of Agricultural Research, Pusa, New Delhi – 110 012

INTRODUCTION

Mango is considered to be a difficult plant species to handle in breeding programme because of certain inherent characteristics including long juvenile phase, high level of heterozygosity, heavy fruit drop and the large progeny population and acreage required for a meaningful assessment of hybrids. Despite its huge economic significance genomic resources for mango are limited and genetics of useful horticultural traits are poorly understood. In India, mango breeding programme was initiated in the year 1911 at Poona (Burns and Prayag, 1921) but not carried far enough to yield useful information. The first report on systematic mango improvement work was from the Indian Agricultural Research Institute, New Delhi. In the recent past mango breeding programmes has shown a significant pick up, with several hybrids has been released from various research centres. The breeding programme has also been focussing on specific/well defined objectives in many research stations based on their priority.

In the present decade, rapid advancement in DNA sequencing and molecular genetic techniques have generated a wealth of information plant genomics, genetics and breeding. Apart from conventional breeding approaches, efforts have been made in recent past to utilise the biotechnological tools for marker aided breeding and deciphering genome of this 'King of fruits'.

Breeding of Mango for Red Peel Colour at IARI, New Delhi

At IARI, New Delhi, the systematic hybridization work was started in 1961. The objective of mango breeding programme was to develop regular bearing variety(ies) for north Indian conditions. Commercial north Indian mango varieties namely, Dashehari, Langra, Chausa and Bombay Green were having the tendency of alternate bearing and Neelum and Banglora varieteis of south Indian have been utilized for imparting regular bearing tendency in north Indian varieteis. The extensive hybridization and selection among segregating F_1 progenies resulted in the development of two full-sib regular bearing hybrids released in the name of 'Mallika' ('Neelum' x 'Dashehari') and 'Amrapalli' ('Dashehari' x 'Neelum') (Singh *et al.*, 1972). However, lately the emphasis shifting toward exporting the mango fruits to western countries and red peel colour was given due importance in mango breeding programmes. This has resulted in the development and release of 'Pusa Arunima' in 2002, and 'Pusa Pratibha', 'Pusa Shresth' and 'Pusa Lalima' hybrid varieties in 2011 by Delhi Seed Sub Committee for commercial cultivation in the NCR region. These hybrids have been placed under AICRP STF for multi-location testing. Sincere efforts have been made for rapid dissemination of Pusa Arunima among growers from all corners of the country. In this regard, grafted plants/ scion-sticks of Pusa Arunima have been supplied to the all AICRP mango centres, SAUs, Nurseries and orchardists. In last one-and-half decade, Pusa Arunima has gained popularity among mango growers. There is a great demand of this hybrid from the fruit growers and the nurserymen. The hybrid has been displayed and appreciated at the Annual Mango Festivals organized by the Department of Tourism, Government of Delhi. The brief description of these mango hybrids having red peel colour are given hereunder:

Pusa Arunima (Amrapali x Sensation)

Pusa Arunima is a hybrid, derived from the cross between Amrapali and Sensation. The plants are semivgoruous and are suitable for closer planting (6 m x 6m). It is a regular bearer and start bearing after 4 years of planting. It ripens late in the season and its fruits are ready for harvesting by 1st week of August. The fruit size is medium (250 g) having attractive red peel colour. It has medium total soluble solids (19.5 °Brix), rich in vitamin 'C' (43.6 mg/100 g pulp) and β carotene content. It has good flavour with very good shelf-life (10-12 days) at room temperature after ripening. Pusa Arunima is suitable for domestic and overseas markets, because of its attractive red coloured peel, fiberless pulp, mild flavour, medium TSS, excellent sugar: acid blend and very good-shelf life.

Pusa Pratibha (Amrapali x Sensation)

Pusa Pratibha, a hybrid between Amrapali x Sensation having regularity in bearing, attractive fruit shape, bright red peel and orange pulp. The red peel colour on golden yellow background makes it very appealing to buyers. It has

oblong shape, uniform sized fruits. It meets all the requirements for the domestic as well as international markets. It has 7 to 8 days shelf life at room temperature after ripening, which is twice as much commercial mango varieties grown in the north India *i.e.* Dushehari. The plants are semi-vigorous and about 278 plants of this hybrid may be accommodated in a hectare (6 m x 6 m) as against 100 plants of Dushehari (10 m x 10 m). Therefore, it will give higher productivity per unit area. On per plant basis, it yields about 3.0 times higher than Dushehari, which includes 'On' and 'Off' year fruiting. It has good sugar: acid blend and above all uniformity in fruit size, which is lacking in Amrapali. Main characteristics of Pusa Pratibha (H-1-1) are; medium statured plant with lanceolate leaves; semi-vigorous growth habit; suitable for planting at a distance of 6 m x 6 m; average fruit weight 181 g; elongated oblong in shape and good TSS and acid blend; mature after 140 days of flowering; TSS 19.6 per cent; shelf life 7 to 8 days at room temperature; high pulp content (71.1 per cent) and ascorbic acid content (34.89 mg/100 g pulp).

Pusa Shreshth (Amrapali x Sensation)

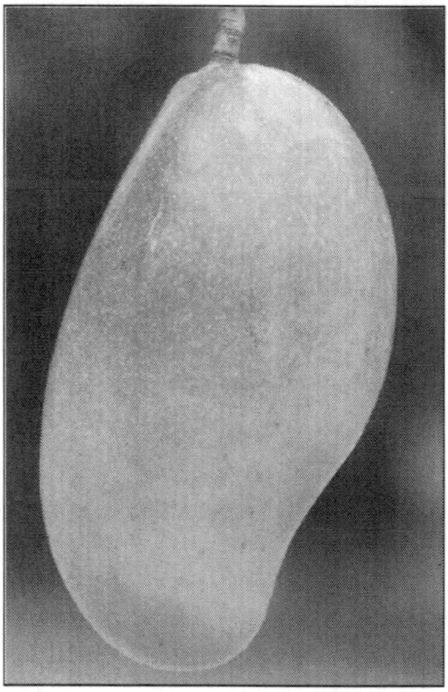

Hybrid H-1-6 (Amrapali x Sensation) has been identified at the Division of Fruits and Horticultural Technology, IARI, New Delhi for red peel colour, semi dwarf stature and moderate tolerance to major pests. Hybrid H-1-6 is a unique hybrid having regularity in bearing, attractive elongated shape, red peel and orange pulp. Due to its elongated shape, it is quite suitable for uniform packaging. It meets all the requirements of the domestic as well as international markets. It has 7 to 8 days shelf life at room temperature after ripening, which is twice as much of commercial mango varieties grown in the north India *i.e.* Dushehari. Furthermore, being semi-vigorous in stature around 278 plants of this hybrid may be accommodated in a hectare (6 m x 6 m) as against 100 plants of Dushehari (10 m x 10 m). Therefore, it gives higher productivity per unit area. On per plant basis, it yields higher than Dushehari, which includes 'On' and 'Off' year fruiting. Being a hybrid evolved using Amrapali as female parent and Sensation as pollen parent; it has all the desirable characters like red colour peel, moderate sugar: acid blend and above all uniformity in fruit size. This hybrid is moderately tolerant to major pests of mango *i.e.* hoppers and mealy bug. The characteristics features of H-1-6 mango hybrid include late maturity (IIIrd week of July); yield potential (20.95 kg/ tree, 10 year plant); fruit weight (211-241g), pulp content (70-72.5 per cent ????g), pulp: stone ratio (3.65-4.59), TSS (18.5-21.5 °brix), acidity (0.15-0.25 per cent), ascorbic acid content (38.8-40.5 mg/100g pulp).

Pusa Lalima (Dashehari x Sensation)

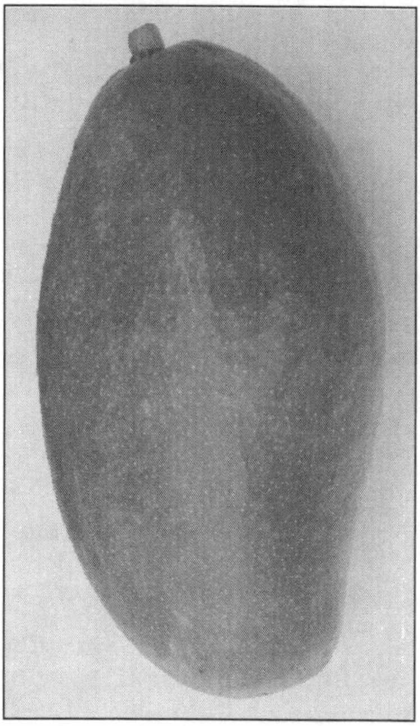

It is a hybrid between Dushehari x Sensation having regularity in bearing, attractive fruit shape, bright red peel and orange pulp. The red peel colour on yellowish green background makes it very appealing to the buyers. It has oblong shape and uniform sized fruits. It meets all the requirements for the domestic as well as international markets. It has 5 to 6 days shelf life at room temperature after ripening. The plants are semi-vigorous and about 278 plants of this hybrid may be accommodated in a hectare (6 m x 6 m) as against 100 plants of Dushehari (10 m x 10 m). Therefore, it will give higher productivity per unit area. Main characteristics of Pusa Peetamber are; medium statured plant with lanceolate leaves, semi-vigorous growth habit, suitable for planting at a distance of 6 m x 6 m; mature 125 days after flowering; average fruit weight 209.0 g; TSS 19.7 per cent, shelf life 5 to 6 days, high pulp content (70.1 per cent) and ascorbic acid content (34.73 mg/100 g pulp). It yields four times higher than Dushehari, the commercial variety of north India.

Mango Biotechnology

Molecular diversity analysis and fingerprinting of mango cultivars has been carried out using different types of DNA markers, including random amplified polymorphic DNA (RAPD) by Schnell *et al.* (1995); Ravishankar *et al.* (2000); Karihaloo *et al.* (2003); inter simple sequence repeats (ISSR) by Eiadthong *et al.*, 1999 amplified fragment length polymorphism (AFLP) by Eiadthong *et al.* (2000); Kashkush *et al.* (2001) and simple sequence repeats (SSR) by Schnell *et al.* (2006); Ravishankar *et al.* (2011). However, most of these studies were carried it on small sets of genotypes. Recently, Ravishankar *et al.* (2015) using 14 carefully selected SSR markers on a comprehensive set of 367 Indian mango cultivars identified two main sub-populations representing mango cultivars from the North-East and South- West regions of India.

Significant progress has been made in the area of whole genome sequencing of mango in recent past. A group led by Professor N.K. Singh identified 1.67 million high quality SNPs by sequencing double digested restriction site associated DNA (ddRAD) from 84 diverse mango cultivars from different regions of India and abroad and a database of SNPs was created. They have developed large number of genic simple sequence repeat (SSR) and single nucleotide polymorphism (SNP) markers from RNA sequence data and generated 60 and 58 million short sequence reads of pooled cDNA libraries of RNA from the leaves of mango varieties 'Neelam' and 'Dashehari' using SOLiD sequencing technology and assembled these into 27,528 and 20,771 transcriptome shotgun assembly (TSA) contigs, respectively (Singh *et al.*, 2016).

Mango has a relatively small genome size of 439 Mb that should have made it relatively easy to sequence and assemble the genome using high throughout second generation sequencing technologies. However, high heterozygosity is a real challenge in achieving a high quality reference genome assembly of mango. The first draft genome assembly of mango cultivar 'Amrapali' using Illumina MiSeq overlapping paired-end reads at San Diego PAG meeting (Singh *et al.*, 2014), but the assembled genome size of 492 Mbp in 211,141 contigs was unexpectedly higher than the actual genome size, indicating redundancy in the contig assembly due to high heterozygosity.

However, as mango is highly heterozygous perennial with a long juvenile period, it is not easy for the breeders to directly determine genes controlling target traits. Thus, revealing the location of trait-control genes on linkage groups and obtaining applicable markers for marker-assisted selection by the construction of genetic linkage maps and agronomic trait mapping is of significant value. The first genetic linkage map in mango was reported by Kashkush *et al.* (2001) utilizing AFLP markers and 29 progeny from a cross of 'Tommy Atkins' × 'Keitt' in Israel. They were able to map 34 AFLP loci and produced a crude linkage map that identified 13 of the 20 linkage groups covering 160 cM. A second map has been produced using 60 progeny from a cross of 'Keitt' × 'Tommy Atkins' in China using AFLP markers. Eighty-one markers with the correct segregation ratios were identified, and 39 of these were used to identify 15 linkage groups. The average distance between two adjacent markers was 14.74 cM. The first population is an F_2 population derived from self pollination of 'Tommy-Atkins' consisting of 168 seedlings. The second population is an F_2 population derived from self-pollination of 'Haden'. The development of a saturated linkage map and the identification of QTL for important traits are objectives for the USDA-ARS program at SHRS in Miami for the next 5 years. Similarly, in Australia, three mapping populations have been developed using controlled pollinations. The largest populations are 'Irwin' × 'Kensington Pride', 'Tommy-Atkins' × 'Kensington Pride', and 'Creeper' × 'Kensington Pride'. A consensus linkage map has been produced using 700 AFLP and 10 SSR markers. Genetic maps are particularly important and valuable tools for quantitative trait locus (QTL) mapping and marker assisted selection (MAS) of plant with desirable traits. 173 F_1plants from a cross between *Mangifera indica* L. "Jin-Hwang" and *M. indica* L. "Irwin" and their parent plants were subjected to high-throughput sequencing and specific-locus amplified fragment (SLAF) library construction. After preprocessing, 66.02 Gb of raw data containing 330.64 M reads were obtained. A total of 318,414 SLAFs were detected, of which 156,368 were polymorphic. Finally, 6594 SLAFs were organized into a linkage map consisting of 20 linkage groups (LGs). The total length of the map was 3148.28 cM and the average distance between adjacent markers was 0.48 cM. This map could be considered, to our knowledge, the first high-density genetic map of mango, and might form the basis for fine QTL mapping and MAS of mango (Luo *et al.*, 2016).

A generation ago, the green revolution culminated, creating surplus of staple crops in many developing countries as a result of nearly a century of efforts of applying Mendelian genetics to crop improvement (conventional breeding) together with the optimization of agronomic practices and the management of insect and pests. However, the improvement of fruit tree crops has lagged behind field crops for several inherent reasons. In the present decade, rapid advancement in DNA sequencing and molecular genetic techniques have generated a wealth of information on mango genome which would provide deciphering the genetics of important horticultural traits, development of DNA markers and marker-assisted mango breeding efforts in coming years.

REFERENCES

Burns, W. and Prayag, S.H. 1921. The book of the mango. Department of Agriculture, Bombay, Bulletin No. 103.

Karihaloo, J.L., Dwivedi, Y.K., Archak, S. and Baldev Gaikwad, A. 2003. Analysis of Indian mango cultivars using RAPD markers, *The Journal of Horticultural Science and Biotechnology*, **78**: 285-289.

Kashkush, K., Jinggui, F., Tomer, E., Hillel, J. and Lavi, U. 2001. Cultivar identification and genetic map of mango (*Mangifera indica*), *Euphytica*, **122**: 129-136.

Kostermans, A. J.G.H. and Bompard. J. M. 1993. *The Mangoes. Botany, Nomenclature, Horticulture and Utilization*. Pub. For the Inter. Board for Plant. Genetic Resources and the Linnean Society of London, Academic Press, London. pp. 332-338.

Mehrotra, R.C.; Dilcher, D.L.; Awasthi, N. 1998. A Paleocene Mangifera - Like leaf fossil from India, *Phytomorphology*, **48**: 91-100.

Oppenheimer. 1956. Study tour report on sub-tropical fruit growing and research in India and Ceylon. Special Bull. No. 3, State of Israel Min. Agric, Agric. Res. Stat., Rehovat.Oppenheimer, C., 1967. Nimrod, a new mango variety selected in Israel. Proc. Fla St. Hort. Soc. **80**: 358–359.

Pandey, S.N. and Dinesh, M.R. 2010. Mango. Indian Council of Agricultural Research, New Delhi. Pp. 153. Pp. 7.

Ravishankar, K.V.; Mani, B.H.; Anand, L.; Dinesh, M.R. 2011. Development of new microsatellite markers from Mango (*Mangifera indica*) and cross-species amplification, *American Journal of Botany*, **98**: 96-99.

Schnell, R.J., Brown, J. S., Olano, C.T., Meerow, A.W., Campbell, R.J. and Kuhn, D.N. 2006. Mango genetic diversity analysis and pedigree inference for Florida cultivars using microsatellite markers, *Journal of the American Society for Horticultural Science*, **131**: 214-224.

Sen, P.K., Mallik, P.C. and Ganguly, B.O. 1946. Hybridisation of mango. Indian J. Hort. **4**: 4.

Sharma, D.K., Majumdar, P.K. and Singh, R.N. 1972. Inheritance pattern in mango. Proc. Symp. on Recent Advances in Horticulture. U.P. Inst. of Aqric. Sci. Kanpur (India). pp: 66–68.

Sharma, D.K. and Singh, R.N. 1970. Self-incompatibility in mango (Mangifera indica L.) Hort Research **10**: 108-118.

Singh, N.K...... *et al.*, 2014. *A Draft Genome of the King of Fruit, Mango (Mangifera indica* L.). PAGXIII 2015, PAG 2014.

Singh, N.K...... *et al.*, 2016.Origin, Diversity and Genome Sequence of Mango (*Mangifera indica* L.). *Indian Journal of History of Science,* **51.2.2**: 355-368.

Viruel, M.A., Escribano, P., Barbieri, M., Ferri, M. andHormaza J.I. 2005. Fingerprinting, embryo type andgeographic differentiation in mango (*Mangifera indica*L., Anacardiaceae) with microsatellites, *Molecular Breeding,* **15**: 383-393.

Wagle, P.V. 1929. A preliminary study of the pollination of the Alphonso Mango. *Agr. J. India,***24**: 259–63.

Whiley, A.W., Mayers, P.E., Saranah, J.B. and Bartley, J.P. 1993. Breeding mangoes for Australian conditions. *Acta Horticulturae,* **341**: 136-145.

Breeding of Perennial Horticultural Crops *Pages* **365–375**
Editors: S.K. Singh, V.B. Patel, A.K.Goswami, Jai Prakash and Chavlesh Kumar
Published by: **BIOTECH BOOKS, NEW DELHI**

Chapter **28**

Papaya Improvement

Jai Prakash, Kanhaiya Singh, A.K. Goswami and Preeti Singh

Division of Fruits and Horticultural Technology,
ICAR-Indian Institute of Agricultural Research, Pusa, New Delhi – 110 012

INTRODUCTION

Papaya (*Carrica papaya*) is a native of tropical America and scientifically known as *Carica papaya* Linn. belongs to the family Caricaceae. Large scale consumer demand in country arises due health awareness. The farmers are also showing keen interested due to early and sizable economic return. Following problems are limiting expansion and sustainability in production needs to mitigate through hardcore breeding.

Papaya ring spot virus, Leaf curl virus, Frost in North India (no. one problem in North India), Shelf life of fruit, Polygamous nature of crop and susceptibility to *Pythium* and *Phytophtora* diseases.

Cytogenetics and Breeding

Carica papaya is diploid (2n=2x=18) with basic chromosome number (x=n=9). Earlier 31 species were reported in *Carica* genus but recently another taxonomic revision was proposed and supported by molecular evidence that genetic distance were found between papaya and other related species. *C. papaya* L., belongs to the Caricaceae family which consists of 6 genera and 35 species. The genera belonging to this family are Carica (1 species), Jarilla (3 species), Horovitzia (1 specie), Jacaratia (7 species), Vasconcellea (21 species), and Cylicomorpha (2 species) (Badillo, 1993, 2000).

Vasconcellea is the largest genus of the family Caricaceae consisting of 21 species. Because of its strong similarity in morphology and uses with common papaya, these highland papayas are considered as close relatives. Most were formerly treated in the *Carica* genus, but have been split out on genetic basis. One of the most striking distinction between *Carica* and *Vasconcellea* is that the former consists of a one-celled ovary and the latter have a five-celled ovary. Unlike papaya, most of the *Vasconcellea* species are strictly dioecious, only *V. monoica* is monoecious. Many have edible fruits similar to papaya, and are widely cultivated in South America. The species of *Vasconcellea* genus are basically evergreen shrubs or small trees growing to 5m tall and are native to tropical South America. Southern Ecuador is considered as the centre of its diversity as it holds 16 of the 21 species. Colombia and Peru are the other countries with high species diversity. Most of the *Vasconcellea* species are adapted to mild temperatures with preference to higher altitudes. exceptions are *V. chilensis*, found in desert conditions, *V. pulchra* which are adapted to extremely wet conditions, *V. parviflora* which occurs on the Ecuadorian and Peruvian coasts in warm and relatively dry climates, and *V. cauliflora*, a tropical lowland species. FAO (1992) reports the use of fruit to treat arterial sclerosis and the use of latex to cure skin mycosis and verruca plana. Clearly, *Vasconcellea* species have as many or more potential medicinal uses than *C. papaya*. For introgression of gene of interest breeders are largely depend on *Vasconcella* species as they contain several traits of interest which are lacking in common papaya, particularly for resitance breeding against virus, frost and fungus.

Species

- ☆ ***Vasconcellea cundinamarcensis* (papayuela or mountain papaya) (*V. pubescens*):** It is mainly grown in the Colombia, on a smaller scale in home gardens. It is a polygamous species, with rare cases of andromonoecy. The proteolytic activity of freeze-dried latex of *V. cundinamarcensis* was found to be five and eight fold higher than that of *C. papaya* latex. In addition, the latex of this species also have medicinal properties which can be used to heal gastric ulcers and other wounds. Resistant against cold, PRSV-P, papaya mosaic virus and black spot (Asperisporium caricae) was also reported in this species.
- ☆ ***Vasconcellea quercifolia*:** This species was found to genetically less distant from *C. papaya* and hence considered as the best option for intergeneric hybridizations, high sugar content. This species of *Vasconcellea* has wide adaptation, being distributed from highlands to lowlands in tropical and subtropical regions of South America. It has been reported to be resistant to PRSV-P in Hawaii and Australia but susceptible to PRSV-P in Venezuela. Other than resistance to PRSV-P and wide adaptation, high sugar content of fruits is also an interesting trait that can be transferred to common papaya.
- ☆ ***Vasconcellea stipulata*:** *V. stipulate* was found to be resistant to PRSV-P and were reported to bear highly aromatic fruits. Its smaller size and aromatic fruits makes it more attractive than the bigger and less aromatic babaco. In southern Ecuador, local consumers prefer this species over the more commercial ones due to its attractive colour and flavour. Therefore it should be a priority species for further domestication.

- ☆ ***Vasconcellea cauliflora*** (Jacq.): It is a tropical lowland species and can be used as a rootstock for maintenance of highland species under tropical lowland conditions. It has climatic requirements similar to that of papaya. It was found to be resistant to PRSV-P and papaya mosaic virus.
- ☆ ***Vasconcellea monoica***: *V. monoica* was reported to be resistant to root fungus (*Fusarium oxysporum*), root knot nematode (*Meloidogyne incognita*) and papaya bunchy top. Other traits of potential interest are monoecy, vigor, and the ability to retain its fruits on the trees for long periods of time, remaining firm for up to 12 months.
- ☆ ***Vasconcellea chilensis***: *V. chilensis* was found to be adapted to desert conditions and thus can be a source for drought tolerance.
- ☆ ***Vasconcellea parviflora***: It basically occurs on the Ecuadorian and Peruvian coasts in warm and relatively dry climates and has climatic requirements similar to *C. papaya*. It was also found resistant to papaya die back (a phytoplasma disease).
- ☆ ***Vasconcellea candicans***: It was found in drier regions of southern Ecuador and coastal Peru and has resistance against PRSV-P.
- ☆ ***Vasconcellea* goudotiana:** *V. gouditiana* has resistance *against Phytophthora* but is susceptible to PRSV.
- ☆ ***Vasconcellea weberbaueri*** (Harms) Badillo: Tolerance to root fungus (*Fusarium oxysporum*) and root knot nematode (*Meloidogyne incognita*) was reported in *V. weberbaueri*.
- ☆ ***Vasconcellea* x *heilbornii***: *V. × heilbornii* has resulted from the hybridization of *V. stipulata* and *V. cundinamarcensis*. It includes the "babaco", "higachos" or "baby babacos". *V. × heilbornii* is the only cultigen in the family and is propagated by vegetative methods. Parthenocarpy is a pronounced trait in babaco, whose fruits are formed independently from pollination and seeds are rarely formed in this cultigen.
- ☆ Other species of *Vasconcellea* are *V. microcarpa. V. glandulosa, V. crassipetala, V. horovitziana, V. longifolia, V. omnilingua, V. pulchra, V. sprucei, V. sphaerocarpa V. palandensis* and *V. weberbaueri*

Bottleneck in Papaya Improvement

- ☆ Cross incompatibility in related species and poor response
- ☆ Ambisexual andromonoceious nature of dome commercial varieties.
- ☆ Unpredictable sex expression and identification.
- ☆ Cross incompatibility in related species and poor response.
- ☆ Lack of established asexual multiplication techniques to produce true to type of planting material.
- ☆ Maintenance of purity is cumbersome and tedious.

Floral Biology

Papaya is a polygamous plant and has many sex forms. There are three basic types *viz.* male or staminate, hermaphrodite (bisexual) and female or pistillate. Of these, only female is stable, whereas flowers of hermaphrodite and male vary in sex expression under different environmental conditions.

Staminate or Male Flowers

Male flowers serve as pollinators. These plants show 1 - 1.5 m long flower stalks hanging out from trunk. The individual flower is small, tubular and contains stamens only.

Female or Pistillate Flowers

The female flowers are large, yellow, borne singly or in – group of three in the leaf axils close to the trunk. The flowers have fine large twisted and fleshy petals that surround an ovary, which swells and develops into papaya fruit. The fruits developed from female flower are spherical to oblong shaped having thick, yellow to orange coloured flesh in different cultivars with large to small cavity in which numerous round wrinkled black seeds are attached.

Hermaphrodite or Bisexual Flowers

It has both male and female organs. Individual flowers are 3.5 to 4.5 cm long with tubular base that widens into goblet shape and then spreads out into 5 thick yellow coloured recurred petals. In between these petals, male organs *i.e.* stamens are present and female organs containing oblong ovary which develops into cylindrical fruits.

The dehiscence of anther takes place 24 hours before anthesis in all the species of *Carica,* the dehiscence in staminate and hermaphrodite flowers takes place between 10 to 12 hours before anthesis. High temperature and low humidity hastens the time of anther dehiscence. In papaya, high per cent pollen germination can be achieved with 5 per cent sucrose solution. Papaya pollen can be stored up to 5 years without loosing the viability if proper conditions are created for storing.

Breeding Approaches for Improvement

Most of the conventional methods are useful and applied in self-pollinating species were carried out by various workers and different research station.

Introduction

Introduction is the easiest way which can be employed to enhance the production and productivity of a crop. Exotic varieties such as Sunrise Solo, Red Lady and Sinta has been introduced in India. Sunrise Solo is a precocious, low-bearing and orange-red flesh variety maturing its first fruit about nine months after transplanting at a height of approximately 3 feet. The cultivar Red lady is a vigorous and tolerant to papaya ring spot virus. Plants start bearing fruits when it reaches the height of 60 - 80 cm and contain 30 fruits per plant in season. Sinta is gynodioecious, semi-dwarf, and tolerant to PRSV-P. The plants bear higher quality fruits with firm flesh and also produces good quantity of papain.

Selection

Pusa delicious, Pusa Majesty and Coorg Honey are some of the varieties that has been selected with traits like gynodioecious, whereas other selections like Pusa Giant, Pusa Dwarf, CO6 and Punjab Sweet are dioecious in nature. Other traits which were kept in mind for selection of superior papaya varieties are high productivity, good quality, high papain production, dwarf stature, good keeping quality and resistance to biotic and abiotic stresses.

Hybridization

Hybridization aims to combine trait of interest by sexual crossing. Inter-varietal and inter-specific hybridization has been carried out in papaya to enhance the yield, quality and resistance to biotic and abiotic stresses. Many inter-varietal hybridization program has been carried out in world and in India for fruit and papain production. CO3, CO4, CO7, HPSC-3 and Arka Surya are the varieties released as a result of inter-varietal hybridization. Most of the inter-specific hybridization programs have focused on the introgression of resistance to the PRSV-P. The first systematic inter-specific hybridization program between *Carica* and *Vasconcellea* was carried out by Horovitz and Jiménez (1967). Till date many researchers have carried out inter-specific hybridization involving different *Vasconcellea* species to incorporate PRSV-P resistance and cold tolerance.

Mutation

The work carried out at Pusa, Bihar has resulted in the release of one dwarf mutant 'Pusa Nanha'. Ram and Srivastava (1984) exposed the dry seed samples of Pusa 1-15 to 10, 15, 20, 25, 30, 40, 50, 70 Kr of gamma rays. They observed that the LD 50 seems to lie between 20-25 Kr of gamma rays. In the M6 generation uniformity was obtained for plant stature and fruit size and shape, which was released later as a variety (Ram, 1983). 'Pusa Nanha' is a dwarf mutant of 106 cm height. It has been evolved by treating the seeds with gamma irradiation. It is dioecious in nature and bears fruit at a height of 30 cm above the ground.

Ploidy Manipulations

Endosperm is a unique tissue in its origin, ploidy level and nature of growth. It is mostly formed by the fusion product of three haploid nuclei, one from the male gametophyte and two from the female gametophyte and is, therefore, triploid (Thomas *et al.*, 2000). Among polyploid cultivars, triploidy, a genomic condition that is favorable for vigor and vegetative productivity, has proved the best for agricultural use (Habashy *et al.*, 2004).

Triploid plants are usually seed sterile and is undesirable where seeds are of commercial value. But in cases where the seedlessness is employed to improve the quality of fruits, such as in banana, apple, citrus, grapes, papaya, *etc.*, induction of triploid plants would be of immense use.

Seedlessness continues to be a key commercial objective in papaya improvement programs, being driven by consumer's preference for seedless fruit. Seedless triploid papaya fruits over their diploid counterparts is that they also have a smaller seed cavity. Thus, they are expected to be more attractive to consumers for direct consumption and processing (Rimberia *et al.*, 2007).

Transgenics

For development of resistance to viral disease may be derived through several means such as genes derived from viral sequences or genes from various other sources that can interfere with target virus. The concept of pathogen derived resistance (PDR), which is the gene derived from viral sequences has been used for PRSV management. Coat protein (CP), RNA silencing, and replicase gene technology has been widely used in papaya to develop papaya varieties resistant to PRSV. SunUp and Rainbow are the examples of PRSV-resistant (Hawaiian) transgenic papaya that was developed through transformation of somatic embryos with the CP gene of the Hawaiian PRSV strain (Fitch *et al.*, 1992). The other countries like Thailand, South Africa, Taiwan, and Philippines have also developed transgenic for their respective problem. Not only for resistance breeding, transgenics approach has also been used for enhancing the shelf life of papaya. Co-suppression technique has been used in papaya to dealay its ripening and ultimately extends its shelf life. Co-suppression refers to the introduction of truncated gene constructs in sense orientation that can result in suppression of homologous host genes. In case of papaya, truncated ACC oxidase gene was introduced in the sense orientation to block ethylene production and delay the ripening rate (López-Gómez *et al.*, 2009).

Papaya breeding work in India could be classified in three phases:

1. High yield and quality.
2. Development gynodioecious varieties having medium sized Fruits, pink pulp colour and high TSS.
3. Development of gynodioecious varieties resistance to 'Papaya Ring spot Virus' (PRSV) with red pulp colour.

History and Achievement

The earliest papaya breeding was started at TNAU in 1954 followed by PUSA Bihar in 1966. The variety CO 1 was released in1958. A Promising selection from 'Honey Dew' 'Coorg Honey Dew' 1959 (C.H.E.S-Chettalli) a centre of IIHR banglore. The variety CO. 2 was released during 1979. During the early eighties the focus changed to develop varieties having medium sized fruits with good quality and high TSS. Latter CO 3 and CO 4 were released during 1983 from TNAU.

Breeding for High Papain and Dwarfness

During 1985 and 1986 CO 5 for papain and CO 6 having dwarf stature were released from Tamil Nadu Agricultural University.Later the variety CO 7 was bred with pink pulp colour and gynodioecious nature.

Regional Fruit Research Station, Pusa - early sixties

Sibmating and selection 1966 to 1982. Four promising strains were observed and latter developed as variety.

Pusa Majesty: Gynodioecious developed from the var. Ranchi

Pusa Giant: Dioecious selection from the variety Ranchi.

Pusa Delicious: Gynodioecious variety developed from the var Ranchi

Pusa Dwarf: Selection from the var. Ranchi

Plants are dwarf in stature, dioecious

Fruits are medium sized— ½ to 1 kg

Pulp is yellow in colour

TSS is between 6.5 to 8°Brix.

At Pantnagar, Improvement Work 1972

Pannagar University had also developed 3 superior types during 1984 through Selection and sibmating.

Pant papaya-1: Dwarf plant type, Fruits medium sized (1 to 1.5 kg).

Pant papaya-2: Plants are vigorous, Fruit (1 to 2 kg) of good quality.

Pant Papaya-3: Vigorous Medium sized fruits (0.5 to 0.9 kg) of good quality.

Papaya Improvement Work at PAU, Ludhiana

Punjab Sweet: Sibmating the lines from Kenya, Dioeciously, Fruits weight > 1 kg, Pulp is deep yellow; TSS ranges from 9-10.5° Brix. Fruit cavity is medium, moderately tolerant to cold.

RCTP-1 was developed from ICAR Research Complex, Tripura Centre, Lembucherra (Prakash *et al.*, 2014) through transgresive selection. It is dual type of gynodioecious variety, high yielding capacity with fair post-harvest life of the fruit.

Genetical Studies in Papaya

- Sunrise Solo, Pink Flesh Sweet good combiners for Waimanalo TSS and fruit cavity index.
- Coorg Honey Dew, Pink Flesh Sweet and Washington good combiners for yield.
- Heterosis obtained for hybrids-yield controlled by non additive
- Several attempts to breed varietiesbiotic and abiotic stress have been done and still there is no significant result observed globally at field level.
- Red pulp colour in papaya is governed by recessesive gene, while yellow pulp colour is dominant.
- Metaxenia effect was also noticed in 1998.
- Plant sex is controlled by a single gene having at least three alleles: M_1. M_2 and m controlling all three sex types.

- ☆ Yellow flesh colour is dominant over orange or red.
- ☆ Green colour of the fruit is dominant over yellow.
- ☆ The genomic size of papaya is 372mbp.

Resistance Breeding in Papaya

PRSV

Among biotic stresses, *Papaya ringspot virus* (PRSV) is a major limiting factor in improving papaya production. The virus was first recorded from western India in 1958 (Capoor and Varma, 1958) and since then it has spread throughout the country including Tamil Nadu. The virus isolates have been bio-typed to papaya infecting (Type P) and non-papaya infecting (Type W) types. Once introduced, PRSV has never been successfully eradicated from any region. In India breeding for resistance to *Papaya ringspot virus* was started at Pune. The varieties screened were found susceptible to PRSV. The wild species *Vasconcellea microcarpa, Vasconcellea cundinamarcensis, Vasconcellea stipulata,* and *Vasconcellea cauliflora* were found to be resistant. However, hybridization was not successful due to embryo abortion at 70-90 days and through *in-vitro* culture hybrid plants were raised. In the F_1 generation progenies, some degree of resistant to PRSV was recorded (Phadnis *et al.*, 1970). The most effective and economical way to control PRSV is to incorporate resistance genes into papaya cultivars, thus avoiding the labour and expenses associated with preventive measures. The species *Vasconcellea cauliflora* is not crossable with *C. papaya*. Dinesh *et al.* (2007) were able to generate intergeneric hybrids by using 5 per cent sucrose, the hybrids generated were confirmed by ISSR markers. The most effective way to minimize the losses due to PRSV infection is by combining host resistance and transgenic resistance developed by using viral coat protein (CP) gene. The virus isolates from India exhibited considerable heterogeneity in CP sequences. CP coding region varied in size from 845-858 nucleotides, encoding protein of 280-286 amino acids. PRSV isolates were divergent up to 11 per cent. Though sequence variation could not be correlated with geographical origin of the isolates, yet the isolates from Southern India formed one cluster (Krishnareddy, 2006). Available information only provides a partial profile of PRSV population from India. A number of aphids *viz. Aphis gossypii, A. malvae, A. medicaginis, Myzus persicae, Macrosiphum sonchi, Aphis citricola, A. craccivora, A. nerii, Acyrthosiphon pisum, Rhopalosiphum maidis, Lipaphis psuedobrassicae, Urolecon sanchi, Toxoptera citricida* either breed or colonize on papaya (Capoor and Verma, 1958; Bhaskar, 1983; Cheema and Reddy, 1985;

Prasad and Sarkar, 1989; Ram and Summanwar, 1989). Further studies on virus vector relations revealed that the aphid vectors transmit the virus in a non-persistent manner (Bhargava and Khurana, 1970; Sharma *et al.*, 1988). Studies conducted to document the dispersing aphid species in a papaya ecosystem and temporal fluctuations in aphid population as influenced by weather factors showed that *Aphis gossypii, A. craccivora* and *M. persicae* were the three major species of aphids caught in the trap, although *Aphis gossypii* was the major one and the peak catches of aphids were recorded during March- April and corresponding higher incidence of the disease during May (Kalleshwaraswamy *et al.*, 2007). At the Indian Institute of

Horticultural Research, the intergeneric hybrids developed have shown resistance to 'PRSV'. Efforts are underway to improve the fruit quality of these hybrids.

Studies at IIHR and TNAU and CISH through Biotechnological tools has shown that F_1 with some degree of resistance to varieties Papaya Ring spot Virus.

Cold Tolerance

Muthulakshmi *et al.* (2007) with the objective of introducing cold tolerance gene into cultivated type crossed dioecious variety 'CO 2' with its wild relative *Vasconcellea cundinamarcensis* and they were able to isolate three genotypes that could be forwarded to F_3 generation for further evaluation and development of a cold tolerant papaya line

Arka Surya

It is the offspring of Sunrise Solo x Pink Flesh Sweet from IIHR Banglore. It was selected from F_{14} generation. Hence, seeds can be produced by bagging the hermaphrodite flowers or by crossing the female flowers with hermaphrodite flowers. The plant is gynodioecious in nature with no male plants. Fruits resemble Sunrise Solo in shape. The plants are shorter compared to Solo. Skin is smooth, becomes uniformly yellow in colour on ripening. Fruits are medium in size of about 600 – 800 g with a small fruit cavity. Pulp is about 3 – 3.5 cm thick, deep red in colour and sweet with a T.S.S. of 13.5 – 15°brix.It does not have the odd flavour. Keeping quality of fruits is good. Yield per plant is approximately 55 – 65 kg (60 – 65t/acre).

Characters	*Surya*	*Sunrise Solo*	*Pink Flesh Sweet*
Fruit weight (g)	723.00	533.00	1562.00
TSS (° Brix)	13.80	12.90	11.20
Fruit cavity index (per cent)	20.00	26.50	33.50
Pulp thickness (cm)	2.80	1.40	2.50
Pulp Colour	Red	Pink	Dull pink

International Breeding Achievements in Papaya

Red Lady

Early, vigorous, productive, and tolerant to papaya ring spot virus. Plants begin to bear fruit at 60-80 cm height and have over 30 fruits per plant in each fruit-setting season. Fruits are short oblong on female plants and rather long-shaped on bisexual plants, weighing about 1.5-2 kg. pulp is thick, red, with 13 per cent sugar content, and aromatic. Red lady is one of the most popular hybrids cultivated in India and developed by Known you company from Taiwan.

Sinta

It is a hybrid of the Py-5xPy-3, gynodioecious, dwarf and precocious. Fruits weight ranges from 1.0 to 1.5 kg, pulp is deep yellow, TSS ranges from 11.5° Brix, fruit cavity is medium and is moderately tolerant ring spot virus. It is popular hybrid developed from Institute of Plant breeding, Philippines.

Future Thrust in Variety Improvement

- Exploration, conservation and characterization of genetic variability.
- Development of stable gynodioecious genotypes for papain production.
- Production of genetically pure seeds by controlled pollination.
- Development of early flowering and fruit maturing and stable gynodioecious lines.
- Breeding varieties resistant to biotic and abiotic stresses. Varieties resistant to biotic stresses like 'PRSV' and which can tolerate high temperature as well as tolerant to frost need to be developed so that cultivation can be taken up in other regions as well.
- Introgression of genes from *Jarilla* and *Horovitza* in the light of new facts under molecular studies.

REFERENCES

Bhaskar, R.B.L. 1983. Confirmation of the etiology of papaya mosaic virus. *Indian J. of Afric. Sci.*, p. 479-481.

Capoor, S.P. and Varma, P.M. 1958. A mosaic disease of papaya in Bombay. *Ind. J. Agri. Sci.;* 28: 225-233.

Cheema, S.S. and Reddy, R.S. 1985. Studies on the transmission of papaya mosaic virus by *Rhopalosiphum maidis* (Fitch.). *Ind. J. Viro.,* 1: 49-53.

Cynthia, B., Kumar, N. and Soorianathasundaram, K. 2003. Breeding for red pulped fruits in dioecious lines of papaya (*Carica papaya* L.). *Ind. J. Hort.,* 60(1): 10-15.

Desikan, K.R. 1972. Studies on first generation hybrids in papaya (*Carica papaya* Linn). M.Sc. Thesis submitted to the Tamil Nadu Agricultural University, Coimbatore.

Dinesh, M.R., Rekha, A., Ravishankar, K.V., Praveen, K.S. and Santosh, L.C. 2007. Breaking the intergeneric crossing barrier in papaya. *Scientia Horticulturae,* 114: 33-36.

Dinesh. M.R., Iyer, C.P.A. and Subramanyam, M.D. 1991. Combining ability studies in Carica papaya L. with respect to yield and quality characters. *Gartenbauwissenchaft,* 56(3): 81-83.

Fitch, M. M. M., Manshardt, R.M., Gonsalves, D., Slightom, J.L. and Sanford, J.C. 1992. Virus resistant papaya plants derived from tissues bombarded with the coat protein gene of papaya ringspot virus. *Nature Biotechnology,* 10(11): 1466–1472.

Ghanta, P.K. and Mondal, S.K. 1992. Genotypic variability and correlation coefficients relating to fruit yield and a few other quantitative characters in papaya (*Carica papaya* L.) cultivars. *South Indian Horticulture,* 40(5): 242-247.

Kalleshwaraswamy, C.M., Verghese, A., Ranganath, H.R., Krishnakumar, N.K., Dinesh, M.R. and Venugopalan, R. 2007. Role of transient aphid vectors on the temporal spread of Papaya ringspot virus in South India. *Acta Hort.,* 740: 251-257.

Khadi, B.M. and Singh, I.D. 1980. Estimates of variability, heritability and genetic advance in papaya (*Carica papaya* L). *Pantnagar J. Res.*, 5(2).

López-Gómez, R., Cabrera-Ponce, J.L., Saucedo-Arias, L.J. *et al.* (2009). *Transgenic Res.*, 18: 89. https: / /doi.org/10.1007/s11248-008-9197-0

Muthulakshmi, S., Balamohan, T.N., Amutha, R., Baby Rani, W., Mareeswari, P. and Indira, K. 2007. Breeding for cold tolerance in papaya (*Carica papaya* L.) through Hybridization. *Research Journal of Agriculture and Biological Sciences*, 3(4): 264-267.

Phadnis, N.A., Budrakkar, N.D. and Kaulgud, S.N. 1970. Embryo culture technique in papaya (*Carica papaya* L). *Poona Agric. Coll. Mag.*, 60: 101-04.

Prakash, Jai, Singh, N.P., Sankaran, M. and Das, S.P. 2009. RCTP-1: A New High Yielding Selection of Papaya For Tripura. *Acta Horticulture*, **851**: 99-101.

Prasad, S.M. and Sarkar, D.P. 1989. Some ecological studies on Papaya ring spot virus in Ranchi. *Ind. J. Viro.*, 5: 118-122.

Ram, M. 1983. Papaya Pusa Nanha. Int. Agri. 21: 18-19. Ram, M. 1984. Promising varieties of papaya. p. 37-39. In: *Proceedings of National Seminar on Papaya and Papain Production.*

Ram, M. and Majumder, P.K. 1984. Character association in papaya. In: Proc. of National Seminar on Papaya and Papain Production, March 26-27, TNAU, Coimbatore. Ram, M. and Srivastava, S. 1984. Mutagenesis in papaya. p.26-27. In: National Seminar on Papaya and Papain Production.

Breeding of Perennial Horticultural Crops *Pages 377–398*
Editors: S.K. Singh, V.B. Patel, A.K.Goswami, Jai Prakash and Chavlesh Kumar
Published by: **BIOTECH BOOKS, NEW DELHI**

Chapter **29**

Apple and Walnut Improvement

R.M. Sharma and A.K. Dubey

Division of Fruits and Horticultural Technology,
ICAR-Indian Institute of Agricultural Research, Pusa, New Delhi – 110 012

INTRODUCTION

Apple and walnut are the principal temperate fruit crops of India with immense nutritional and economic importance (Table 29.1). Breeding is a major innovation factor in fruit production worldwide, as consumer requires products with exciting, new and up to date features. However, the breeding process is long term, and continuous with changing environments. It is therefore a great challenge for every breeder to head for new cultivars that might be on the market only 30-40 years after designing the crosses. Compared with annual crops, breeding perennial crops is complex, long lasting and time consuming due to their long juvenile phase and long economic lifespan (Antofie *et al.*, 2007). During the last 20 years the developments in molecular genetics lead to new and more precise breeding approaches. A major step forward in breeding efficiency is being achieved by combining classical phenotypic selection methods with new molecular techniques (Peil *et al.*, 2010). Despite of new tools, fruit breeding will remain a multitasking and interdisciplinary approach requiring highly sophisticated long term management. In this chapter, we have attempted to review the recent advances made in the field of apple and walnut improvement.

Apple

The specific breeding objectives of apple for scion and rootstock, and some identified gene sources are given in Tables 29.2–29.3. In last two decades, several trait specific genes have been identified, and used to develop novel cultivars (Table 29.4). The genome sequence of apple (*Malus* x *domestica* Borkh.) was published more than a year ago, which helped develop an 8K SNP chip to assist

Table 27.1: Nutritional and Economic Value of Apple and Walnut

Fruit	*Area (000 ha)*	*Production (000 MT)*	*Export (MT)*	*Amount (Rs in Crores)*	*Exporting countries*	*Nutritional importance*
Apple	277	2242	21,085.23	57.75	Bangladesh, Nepal, Saudi Arabia *etc.*,	Rich source of phytochemicals (quercetin, catechin, phloridzin and chlorogenic acid), effective in reducing risk of cancers, cardiovascular disease, asthma, and diabetes.
Walnut	92	228	3595.69*	127.21*	France, UK, USA, Germany *etc.*	Besides essential fatty acids, walnuts contain vitamin E and polyphenols. Walnuts has antioxidant and antiinflammatory bioactivity, having potential role of walnuts against disease initiation and progression, including cancer, cardiovascular and neurodegenerative diseases.

*2017-18 (Anon, 2017).

Table 27.2: Breeding Objectives (Apple)

Scion	*Rootstock*
Tree characters Late leafing, columnar/spur type growth habit, regular bearing, self fruitful	Provide anchorage, explore the soil profile, absorb and nutrient transfer to scion, adapt to pedo-climatic conditions, tolerate extreme weather events, resist or cope with the pathogens, easy to propagate and impart positive architectural properties to the scion (vigour control and precocity)
Fruit quality High quality, low chilling, early and late maturing, long storability, red skin and flesh colour	
Processing use Juice production	
Resistance to biotic stress Resistance to scab, powdery mildew, canker, scab apple aphids, alternaria blotch, apple chlorotic leaf virus, apple stem pitting virus	
Resistance to abiotic resistance Cold hardy, hot and sunny climate, late blooming, russet resistance	

Table 27.3: Important Gene Source

Trait	*Resistance Source*
Scab	*M. floribunda*, Russian Seedling R12740-7A (*M. sieversii*), *M. atrosanguinea* 804, *M. micromalus*, Hassan's baccata# 2 and *M baccata* jackii
Powdery mildew	*M.* x *robusta, M. zumi* and D12 (OP Crabapple seed)
Fibre blight	*M.* x *robusta* 5
Red pigmentation	*M. pumila* var *niedzwetzkyana*

Table 27.4: Trait Specific Genes Identified in Apple

Gene	*Trait*
ACS, ACO, ACC	Ethylene biosynthesis
E1 and *E2*	Early ethylene production
pUFGluT	Anthocyanin expression in apple skin
Rt	Anthocyanin in all tissue
Ru	Russetted skin
Ar	Aromatic flavour
Bp-1, Bp-2	Bitter pit resistance
Ebb	Early bud break
Co	Columnar growth habit
Sp-1, Sp-2, Sp-3	Spur type habit
Sd-1, Sd-2, Sd-3	Aphid resistance
Dw-2	Compact habit
Tb	Terminal bearer
St-1, st-2	Sturdy dwarf
Dw-1, dw-2, dw-3	Early dwarf
S-allels	Pollen incompatibility
Er-1	Wooly apple aphid
Pc	Phytophthora cactorum resistance
Pl-1, Pl-2, Pl-w, Pl-d	Podosphaera leucotricha resistance
Vf, Vh, Vm, Va, Vb, Vd, Vbj, Vg	Scab resistance

in implementing genomic selection (GS). In apple breeding programmes, GS can be used to obtain genomic breeding values (GEBV) for choosing next-generation parents or selections for further testing as potential commercial cultivars at a very early stage. Thus GS has the potential to accelerate breeding efficiency significantly because of decreased generation interval or increased selection intensity (Kumar *et al.,* 2012). The apple genome sequence provided insight into the evolution of this globally important fruit crop, and is now being used to speed up the development of new varieties. Availability of genome sequence information along with high throughput genotyping platforms is changing the nature of research experiments to understand evolution of organisms, as well as transforming the strategies for genetic improvement.

Scion Breeding

The juice acidity, affecting juice flavour, colour, clarity and the overall sensory scores, is an important breeding objective for FS specific apple cultivars. To better understand the genetics of traits related to the quality of freshly squeezed apple juice (FS) in apple cultivars, Huang *et al.* (2018) phenotyped hybrids derived from five biparental crosses of *Malus asiatica* 'Zisai Pearl', and *M.*x *domestica* 'Red Fuji', 'Golden Delicious' and 'Jonathan' and studied the inheritance of quality traits with a quantitative trait loci (QTLs) analysis. Fruit juice yield,TSS, pH value, light transmittance (TRA), and colour traits (CIE coordinates, L*, a*, and b*) segregated quantitatively. They detected thirteen QTLs for juice yield, pH, TRA and b*, while QTLs for pH value co-localized with TRA and b* on linkage group 16 of both 'Zisai Pearl' and 'Red Fuji'.

In the integration study of molecular, genomic and functional approaches to identify novel genes involved in the control of fruit AsA (ascorbic acid) concentrations in apple, *MdGGP* paralogues in mature apple fruit have been found the key regulators of AsA and totAsA concentrations, and that allelic variations in these orthologues is closely associated with fruit totAsA concentrations within the mapping population. These SNP-based markers have been suggested to be tested to screen a wider selection of apple germplasm in order to validate associations between allelic variants and high AsA levels. *MdDHAR* has a key role in regulating the redox state of the AsA pool in apple fruits, and that flesh DHA (dehydroascorbate) concentration is linked to susceptibility to flesh browning (Mellidou *et al.*, 2014). This information is useful to target and focus future apple breeding programmes directed towards the development of new nutritionally enhanced vitamin C-rich cultivars through the application of robust marker technology.

Fruit trees have a long juvenile phase. For example, the juvenile phase of apple (*Malus* x *domestica*) generally lasts for 5–12 years and is a serious constraint for genetic analysis and for creating new apple cultivars through cross-breeding. If modification of the genes involved in the transition from the juvenile phase to the adult phase can enable apple to complete its life cycle within 1 year, as seen in herbaceous plants, a significant enhancement in apple breeding can be realized. There is a novel technology that simultaneously promotes expression of Arabidopsis *FLOWERING LOCUS T* gene (*AtFT*) and silencing of apple *TERMINAL FLOWER 1* gene (*MdTFL1-1*) using an *Apple latent spherical virus* (ALSV) vector (ALSV-AtFT/MdTFL1) to accelerate flowering time, and life cycle in apple seedlings. When apple cotyledons were inoculated with ALSV-AtFT/MdTFL1 immediately after germination, more than 90 per cent of infected seedlings started flowering within 1.5–3 months, and almost all early-flowering seedlings continuously produced flower buds on the lateral and axillary shoots. Cross-pollination between early-flowering apple plants produced fruits with seeds, indicating that ALSV-AtFT/MdTFL1 inoculation successfully reduced the time required for completion of the apple life cycle to 1 year or less. *Apple latent spherical virus* was not transmitted via seeds to successive progenies in most cases, and thus, this method will serve as a new breeding technique that does not pass genetic modification to the next generation (Yamagishi *et al.*, 2014).

Most apple cultivars are diploids ($2n = 2x = 34$). Nevertheless, triploid cultivars ($2n = 3x = 51$) do exist; their use dates back at least to the 17th century when natural triploids were selected for their high yield and large fruit size (*e.g.* cv. 'Gravenstein'). The beneficial characteristics of triploid fruit are also supported by their relatively high occurrence. Tetraploids are more rare in apple, howev-er; they are mainly artificially created and used by breeders as a means to develop triploid genotypes (Sedysheva and Gorbacheva, 2013; Sedov *et al.*, 2014). Two ap-ple genotypes were polyploidized: the commercial cv. 'Gala' and the genotype coded G40 from the breeding germplasm of Better3fruit (Belgium). Tetraploid plants were obtained after chromo-some doubling in tissue culture by stirring axillary buds (3–4 mm) during 2 days in a MS multiplica-tion medium, supplemented with colchicine (10 mM). Tetraploid leaves also exhibited a darker leaf colour, which could be correlated to a higher pigment concentration. Furthermore, the results also show a decreased elongation rate and leaf size in tetraploids, which is suggested to be due to the observed lower cell density in the polyploid apple plants (Hias *et al.*, 2017). Autotetraploid apple possesses better salt tolerance than diploid, which may be associated with the higher levels of expression of aquaporin genes (*MdPIP1;1*, and *MdTIP1;1*) in response to salt stress (Xue *et al.*, 2015). These results will make a practical and theoretical contribution to the breeding of polyploid apple cultivars

Red coloration of apple skin is a key factor for consumer acceptance of fruit within the marketplace and is used as a grading standard in the main producing countries. Red colour intensity results from the accumulation of anthocyanin pigments, which is dependent on the cultivar, fruit maturity and environmental conditions, in particular light and temperature. Many red skin apple cultivars grown under warm summer conditions do not redden as much when they are grown in locations with cool summer; however, some strains of 'Gala' (*e.g.*, 'Brookfield Gala™') and other red skin cultivars (*e.g.* 'Modi™'), are capable of reddening under warm conditions, which indicates genetic variability exists for response to temperature. It has been prposed that temperature-induced down-regulation of fruit anthocyanin biosynthesis is primarily due to down-regulation of the anthocyanin regulatory complex, and that the genes encoding the complex have useful loci for the development of marker assisted breeding of new apple cultivars tolerant to warm conditions (Iglesias *et al.*, 2016).

The genetic improvement of apple using wild red-fleshed apple resources (*Malus sieversii* f. *niedzwetzkyana*) have been focused. It has been found found that the MYB transcription factors (TFs) involved in the synthesis of proanthocyanidins can be classified into TT2 and PA1 types. We characterized a PA1-type MYB transcription factor, *MdMYBPA1*, from red-fleshed apple and identified its role in flavonoid biosynthesis using overexpression and knockdown-expression transgenes in apple calli. We explored the relationship between TT2- and PA1-type MYB TFs, and found that *MdMYB9/11/12* bind the *MdMYBPA1* promoter (Wang *et al.*, 2018).

Markers in apple breeding have been linked to various traits and a list of newly developed cultivars in apple is given in Table 29.5a.

Table 29.5a: List of Newly Developed Apple Cultivars

Variety	*Parentage*	*Origin*	*Characters*
AB17	Kidds D-8 x O.P.	Nevis Fruit Co. Ltd., Central Otago, New Zealand	Bicolor apple similar to Gala but larger and less acidic with purple petals; Fruit: round; conical; mottled red blush overcolor and green-yellow ground color; ripens early midseason; annual bearing
ANABP 01 (Bravo™).	Cripps-Two (Cripps Red) x Tenroy Gala (Royal Gala)	D.A.F.W.A. Bentley, Western Australia	Purple-red apple with excellent flavor and Texture, late-season, ripens;27 d after Golden Delicious, 14 d after Cripps and chilling requirement 400-500 h.
Beller	Chance seedling selected	Malott, WA	Attractively colored, late-season apple, Fruit: round, symmetrical with no ribbing or lobes; bicolor type with 50-80 per cent dark purple red stripes and blush over yellow ground color; not prone to russet or cracking; flesh light cream, medium firmness, fine texture; sweet with little aroma; ripens late-season; stores 3-4 months.
Burnett cultivar. having	Limb mutation of Obrogala	Quincy, WA	Nearly full blush overcolor and maturing two weeks earlier than standard Gala strains.
CIVG198 (Modı®)	Gala x Liberty	S. Guiseppe di Comacchio (Ferrara), Italy	Deep, red skin color, high sugar content, very firm flesh apple, with scab resistance, long storage capacity and long shelf life.
CIVNIRED S.	limb mutation of CIVNI	Laces (South Tyrol), Italy	With greater extension of bright red overcolour.
CIV323 (Isaaq)	Galaxy x A3-7	Consorzio Italiano Vivaista, Ferrara, Italy	Precocious apple with good storability and resistance to apple scab.
CJ07	Mutation of Sciearly	Hawkes Bay, New Zealand	Early maturing with solid red colour.
Co-op 31 (WineCrisp).	Rock 41-12 x PRI 841-103	University of Illinois, Urbana, IL	Disease resistant, late-ripening apple, with full, fruity, sweet, mild-subacid flavor.
Dalinip	Whole-tree mutation of Pinova	Lezigne France	Intense striped, early maturing.
Daligris	Whole-tree mutation of Pinova	Lezigne, France	Differs in having smooth light brown to yellow russeting covers the stem cavity and spreads to cheeks, more prominent lenticels, and a red-orange rather than red-pink overcolour, consistently higher sugar content and pronounced anise flavour.
Dalirail	whole tree mutation of Imperial Gala	Lezigne France	Intense colour, early maturing
DAS-10	Whole tree mutation of Honeycrisp	Biglerville	Fruits that ripen 21 d earlier

Variety	*Parentage*	*Origin*	*Characters*
Delcored	Co-op 30 x Delblush	Malicorne, France	Late season, large, bright red-purple fruit, very firm, juicy, sweet,; 90 per cent redpurple over yellow ground color; flavor fruity, refreshing; very juicy; soluble solids moderately high, 14 °Brix; ripens 2 weeks after Golden Delicious. Tree: medium-to-large; semi-upright to spreading; resistant to apple scab (*Venturia inaequalis*) presumably due to *Vf* (*Rvi6*); moderately susceptibile to powdery mildew; moderate incidence of bitter pit.
Diva	McIntosh x PRI 674	Agriculture and Agri-Food Canada	Scab resistant cider apple for northern climates.
DS-41	Honeycrisp O.P.	Holman,WI	Fruit similar to Honeycrisp but larger, more stripy, higher in acidity, later ripening; stores up to 5 months in ambient atmosphere.
Dudek	Red sport of Caudle	East Wenatchee, WA	
El Nino	Red sport of Royal Gala	Royal T Farms, Royal City, WA	
Emmons	Chance seedling	found in orchard near Gleed, WA	Long storing apple with round fruit and dense, sweet flesh.
FUCIV51 (Fuji SAN-CIV(R))	Fuji sport (Fuji NAGAFU-12)	Consorzio Italiano Vivasti, Ferrara, Italy	Intense purple-red.
Foxtrot	Whole tree mutation of Tenroy Gala	Quincy,WA	Distinguished by having fruits that are large, more intensely red colored and earlier ripening than other forms of Gala.
Fugachee	Chance seedling	Bridgeport, WA,	Yellow-green apple with prominent russetted lenticels, conical shape, and a distinctive, low-acid, sweet flavor.
Fugachee Fuji	Earlier, redder sport of Fuji	Brewster, WA	
Fuji Fubrax (Kiku)	Red sport of Fuji	Eppan, South Tyrol, Italy	
Fujiko	Red sport of Nagafu 12 Fuji	Migliaro, Ferrara, Italy	
Gaia	Gala x A3-7	Consorzio Italiano Vivaisti, Comacchio, Italy	Similar to Gala but larger fruit more intensely flavoured and less subject to cracking, with resistance to apple scab.
Gala 2013	Solid purple-red Gala sport	GRIBA Baumschule Land-wirtschaftliche Gesellschaft, Terlan, Italy	Fruit: color 100 per cent intense purple-red, no stripes; young fruit fully red 40 d after full bloom, ripens 30 d before Golden Delicious, leaves have red petiole and central vein.

Variety	*Parentage*	*Origin*	*Characters*
Gala Schnico Red.	Intense red-skinned Gala sport.	Schniga, Bozen, Italy	Fruit: 100 per cent intense, dark red; 30 per cent more lenticels per unit area than Gala Schnico.Tree: blooms with Cox's Orange Pippin and Jonagold; pollinizers Red Delicious, Granny Smith, and Golden Delicious; tree ramified; growth habit spreading.
Galaval	Sport of Galaxy Gala	Corze, Loire Valley, France	Intense dark purple-brown skin over color having no defined stripes.
Gold Pink (Gold Chief)	Starkrimson x Golden Delicious	University of Bologna, Italy	Similar to Golden Delicious in flavor with no russeting.
GUN-300	Whole-tree mutation of Rogers Red McIntosh	Crown Point, NY	Early coloring, more upright.
Hwangok	Kogetsu x Yataka	The Rural Development Administration, Republic of Korea	Yellow-green, low-acid fruit, resistant to bitter rot (*Glomerella cingulata*) and leaf spot (*Botryosphaeria obtusa*).
Imara	Liberty x Granny	Better3Fruit, Rillaar, Belgium	Late season, bicolor fruit with aromatic, sweet flavor and multiple gene scab and powdery mildew resistance.
Jugala	Whole-tree mutation of Mitchgla Gala	Saint Paul D'Espis, France	Fruit maturing approximately 5 d earlier.
KAR 4	Paula Red x Novamac	Agriculture and Agri-Food Canada Kentville Research Station, Kentville, Nova Scotia	Early maturing red apple with apple scab resistance.
Lady in Red	Earlier, more colored sport of Cripps Pink	Hawkes Bay, New Zealand	
Lady Laura	Sport of Cripp's Pink	Borenore NSW, Australia	With earlier and more intense coloration.
Lentz Jonagold	Limb mutation of Rubinstar Jonagold	Honor, MI	With earlier and greater red fruit coloration as well as red coloration of the leaf petiole and mid-vein.
LJ-1000	Whole tree mutation of Honeycrisp	Quincy, Washington	With earlier and greater red coloring and higher sugar levels.
Lurechild	LubA793 x LubA264	Lubera, Buchs, Switzerland	Red-fleshed apple with high fruit quality. flesh red; flavor berry-like, sugar-acid balance good, eating quality better than LubA264; ripens 10-15 Sept. at Buchs, before Luresweet. Tree: blooms mid- to late April at Buchs; vigor low; hardy to -18°C; resistant to apple scab.

Variety	Parentage	Origin	Characters
Luregust	LubA793 x LubA264	Lubera, Buchs, Switzerland	Red-fleshed apple. Fruit: flesh red; eating quality good; aromatic, acidity high, similar to Luresweet, sweeter than LubA264, less sugar than Luresweet; hardy to -18 °C; resistant to apple scab
Luresweet	LubA688 x LubA275	Lubera, Buchs, Switzerland	Red-fleshed, high-quality apple. Fruit: sugar and acidity high; hardy to -18 °C; resistant to apple scab
MAIA1	Honeycrisp x Fuji	Wabash, IN	Late blooming and ripening, very crisp, sweet with long storage and shelf life potential, resistance to fire blight slight.
Majesty	Co-op 25 x CIVCP-142	Consorzio Italiano Vivaisti, Comacchio, Italy	Similar to Gala but larger, more intensely flavored, fruit less prone to cracking, with resistance to apple scab and stores up to 7 months in controlled atmosphere.
MinnB42	Deeply red-colored, crisp-textured Honeycrisp sport	University of Minnesota	Fruit: greater area of more intense red than Honeycrisp, texture crisp; skin thin, easily penetrated; blooms after Minnewasta; vigor moderate; cold hardy to; more resistant to apple scab than Minneiska.
MN55 (Rave™)	Honeycrisp x MonArk	University of Minnesota	Early ripening, crisp, juicy, red apple with long storage life.
Moonlight	Goldstar x Telemon	Inst. of Experimental Botany, Czech Republic	A late-maturing cultivar with columnar habit, mediumsized, yellow-green fruits having good storage quality and *Rvi6* (*Vf*) resistance to apple scab.
Mored	Chance seedling	Cultura Jean Moors, Bilzen, Belgium	Fully red-skinned, sweet apple with juicy, crisp texture. Blooms late April at Bilzen, with Golden Delicious; vigor moderate; spreading; hardy to -20 °C; tolerant to powdery mildew (*Podosphaera leucotricha*) and gloeosporium rot.
New York 1	Honeycrisp x NY752	Cornell University/ New York State Agric. Exp. Stat. Geneva, NY	Attractive red fruits that are sweet, juicy and have a crisp texture, ripen early mid-season and store 100–150 d.
New York 2	Braeburn x Autumn Crisp	Cornell University/ New York State Agric. Exp. Stat., Geneva, NY	Excellent fruit quality at harvest and after cold storage and precocious bearing.
Origin	Topaz x Fuji	Wadenswil, Switzerland at Agroscope Changins-Wadenswil	Red blush striped with weak stripes over light yellow ground color, no bloom or greasiness; flesh firm; texture fine; moderately sweet, very juicy. Tree: vigor medium; spreading; resistant to scab (*Vf/ Rvi6*)
Orion	Golden Delicious x Otava	Institute of Experimental Botany, \Czech Republic	Late, dessert-type, triploid apple, with good keeping quality and *Vf* scab resistance.

Variety	*Parentage*	*Origin*	*Characters*
Odysseus	Mutsu x Firiki	Naoussa, Greece	Produces high quality fruit with excellent fruit quality, particularly high nutritional value and a flavor more similar to the preferred Firiki.
PB-148	Gala x Jack Humm	Nevis Fruit Co. Ltd., Central Otago, New Zealand	Early season, highly colored apple having a good flavor and shelf life.
Pink Chief	Limb mutation of Cripps Pink	Fruit Varieties International, Grove, Tasmania, Australia	Pink-red Cripps Pink sport with columnar bearing habit.
Prem A280	Braeburn x Royal	Horticulturaeultural and Food Research Inst. of New Zealand	Dark-red fruited apple notable for its sweet flavor and crisp texture.
PV 1027	Earlier maturing sport of Arends	Rogers, Ohio	
Red Topaz	Redder, earlier-maturing sport of Topaz.	Pencin u Liberce, Czech Republic	
RS103-130.	Royal Gala x CPR7T90	Department of Primary Industry and Fisheries Research Station, Brisbane, Queensland, Australia	Compact, medium-sized apple tree, resistant to apple scab (Races 1 to 5), with variegated leaf colour and red fruit.
Ruby Pink	Later maturing, highly colored sport of Cripp's Pink	Main Ridge, Victoria, Australia	
Seohong	Tsugaru x Chukwang	National Institute of Horticulturaeultural and Herbal Science, Jangangu, Suwon, Korea	Early-maturing apple with high levels of phytochemicals.
Smaragd	Granny Smith x McIntosh Wijcik	Faculty of Agriculture, University of Novi Sad, Serbia	Columnar, late season, with green fruit
SPA493 (Salish™)	Splendour x Gala	Agriculture and Agri-Food Canada	Yellow with red blush, midseason, long storage apple
Stark Gugger	Limb mutation of Starking Delicious	Bressanone, Italy	Distinguished fruits with purple red coloration over 100 per cent of the skin, a long conic shape with pronounced crowning, and coloration 10 d earlier.
Thome Red	Limb mutation of Caudle McIntosh	Comstock Park, MI	Redder skin coloration and red coloration of the leaf petiole and veins.

Rootstock Breeding

The tools available to breeding programs for understanding how apple rootstocks can modify tree architecture, interface with the soil at the physical (root penetration/anchorage), chemical (pH, CEC, nutrient extraction) and biological level, resist infections and deal with climate stress are increasing. What is also increasing is the genetic base (source of new germplasm) that is available to discover more stable sources of a particular trait or new traits altogether. In recent years in the Geneva® breeding program, it has been learnt that rootstocks modify the scion at the most fundamental level: gene expression. A series of experiments performed by grafting the same scion onto several, genetically diverse apple rootstocks and then monitoring global gene expression by means of a gene chip revealed that rootstocks have a tremendous influence on activation/inactivation, up-regulation, and down-regulation of entire gene networks that involve basic tree physiological and cellular functions like sugar metabolism, lignin formation, photosynthesis, disease response, nutrient transport, cell cycle, *etc.* (Jensen *et al.*, 2003; 2010).

Among the technologies that will influence new rootstock development are: 1) Marker assisted breeding and quantitative trait locus mapping (Antanaviciute *et al.*, 2012; Gardiner *et al.*, 2012); 2) Gene transfer either from unrelated source or (more desirable) from the same species (Borejsza-Wysocka *et al.*, 2009; Malnoy *et al.*, 2010); 3) High throughput DNA/RNA sequencing (Chagne *et al.*, 2012); 4) Exploitation of knowledge about long range root to shoot (and vice versa) communication through plant hormones, small and micro-RNAs; 5) High throughput affordable phenotyping under different treatments, understood to be the ability to monitor the physiological status of tree tissues on hundreds of plants at the same time and the ability to monitor hormone, nutrient concentration, transport, protein activity in tissues of hundreds of plants (Luby *et al.*, 2010).

The most prevalent cultivar, JM7 (a dwarfing apple rootstocks in Japan), is highly susceptible to field vole (*Microtus montebelli* Milne-Edwards, 1872) gnawing in apple orchards because of poor in total phenolics and rich in nitrogen. Shinada and Moriya (2018) indicated that future breeding projects on vole-resistant dwarfing apple rootstock genotypes should aim to develop a cultivar having high total phenolics and low nitrogen, while keeping the advantageous traits of the JM7 rootstock, such as high productivity, and ease of propagation by hardwood cutting.

In the Geneva breeding program in collaboration with scientists at Penn State University, researchers have measured gene expression of 26,000 unique gene sequences on a full sib population to discover genes that were associated with important apple rootstock traits (Jensen *et al.*, 2012). They utilized the expression profiles for these genes to map expression QTLs (e-QTLs) (Holloway and Li, 2010) for all 26,000 unigenes and recorded the location in the apple genome of the DNA control regions that modulate the production of RNA for these genes. This process was followed by a compilation of genes/eQTLs that were in cis configuration with the traits of interest (correlated with the trait in the rootstock progeny), and contained DNA polymorphisms that also segregated with the trait. One of the traits of interest that has been studied with this method is the induction of flat branch angles (IFB) in the scion by the rootstock which was characterized in the Geneva germplasm (Fazio and Robinson, 2008a, 2008b). Using a genetic map and a breeding population segregating for this trait, they discovered a QTLs for IFB on the top of chromosome 3 of apple and proceeded to ascertain which eQTLs were associated with the phenotype. While several eQTLs were also discovered in the same region, only two had strong correlation with branch angle modification (APPLE-0F000019515 whose sequence is similar to a Mitogen Activated Protein Kinase and APPLE0F00019310 with similarities to a MYB Transcription Factor).

The list of apple varieties/rootstocks developed and registered in recent years are given in Table 29.5a and 29.b.

Table 29.5b: Recent Developed Rootstocks in Apple

Variety	*Parentage*	*Origin*	*Characters*
G.210 (Geneva® G.210)	Ottawa 3 (*Malus* x *domestica*) x Robusta 5 (*M.* x *robusta*);	New York State Agricultural Experiment Station and the USDA-ARS, Geneva, NY	Propagated by stool bed layering, unbudded stools have recurved leaves with wavy edges; clonally propagated mother plants in the stool bed produce rootstock liners that are 30 to 50 cm tall with few spines. Rootstock performance: grafted tree is semi-freestanding and intended for use with processing or weak scion cultivars; semidwarfing, 50–60 per cent of seedling; productive; produces scions similar to scions on M.7 and G.30; higher precociousness to scions than M.7; cumulative yield efficiency similar to M.9; good cold hardiness; resistant to fire blight, *Phytophthora cactorum*, and wooly apple aphid; tolerant to the replant disease complex
G.214 (Geneva® G.214)	Robusta 5 (*M.* x *robusta*) x Ottawa 3 (*M.* x *domestica*)	New York State Agricultural Experiment Station and the USDA-ARS, Geneva, NY	Propagated by stool bed layering, unbudded stools have recurved, pointed wavy leaves; stool beds have similar productivity to M.9.; stems have few spines; root systems express fine root trait. Rootstock performance: young grafted tree needs support; dwarfing, 30–35 per cent of seedling; precocious; productive; similar vigor and precocity to M.9, Nic 29, and M.26, but more productive; good cold hardiness; resistant to fire blight, Phytophthora root rot, and wooly apple aphid.
G.222 (Geneva® G.222)	Robusta 5 (Malus x robusta) x Malling 27 (M. x domestica);	New York State Agricultural Experiment Station and the USDA-ARS, Geneva, NY	Propagated by stool bed layering, unbudded stools have rounded, deep grey-green leaves; stems have some spines. Rootstock performance: young grafted tree needs support; semi-dwarfing, 45–55 per cent of seedling; productive; similar vigor to Malling 7 but more precocious and productive; good cold hardiness; resistant to fire blight (*Erwinia amylovora*), Phytophthora root rot (*Phytophthora parasitica*), and wooly apple aphid (*Eriosoma lanigerum*).

Variety	Parentage	Origin	Characters
G.890 (Geneva® G.890)	Robusta 5 (M. x robusta) x Ottawa 3 (M. x domestica);	New York State Agricultural Experiment Station and the USDA-ARS, Geneva, NY	Propagated by stool bed layering, unbudded stools have straight, long stems; few spines; recurved, green leaves. Rootstock performance: grafted tree is free-standing and intended for processing or weak scion cultivars; semi-dwarfing, 55–65 per cent of seedling; productive; similar vigor to M.7 and MM.106, but more precocious and productive; cumulative yield efficiency similar to M.9; good cold hardiness; resistant to fire blight, Phytophthora root rot and woolly apple aphid.
G.969 (Geneva® G.969)	Robusta 5 (M. x robusta) x Ottawa 3 (M. x domestica)	New York State Agricultural Experiment Station and USDA-ARS, Geneva	Propagated by stool bed layering, unbudded stools have straight long shoots, leaves are slightly recurved. Rootstock performance: grafted tree is free standing; semi dwarfing, 45–55 per cent of seedling; productive; similar vigor to M.7, but much more precocious and productive; good cold hardiness; resistant to fire blight, Phytophthora root rot and woolly apple aphid.

Source: Clark and Finn, 2010, 2012; Gasic *et al.*, 2016; 2018.

Walnut

Among nut crops, walnut is the most important crop, and has immense commercial importance even in the areas, situated in remotes without having good market and transport infrastructures. Because of long storability, growers have opportunity to hold the produce at room temperature for 2-3 months easily, and sell at intended time. Despite of huge genetic diversity, only few options are there with the growers to grow known walnuts in India. Poor supply of quality planting materials has further worsened the situation of walnut Industry in the country. The modern breeding objectives of walnut breeding are given in Table 29.6.

Table 29.6: Breeding Objectives of Walnut

Scion	Rootstock
Tree characters Late leafing, better overlapping of staminate and pistillate flowers, dwarf structure, lateral bearing habit, upright growth habit *etc.*	Easy to root, resistance to crown gall, Phytophthora rot, root knot nematode, CLRV *etc.*
Nut characters Nut yield, nut size (weight), grade, shell seal, shell thickness *etc.*	
Kernel characters Kernel recovery, kernel colour	
Resistance to biotic stress Bacterial blight, blackline (CLRV), deep bark canker, anthracnose, codling moth *etc.*	

Walnut Breeding Proramme in Different Countries

United States of America (California)

In USA, three major breeding goals are there *viz.*, increased yield and kernel quality, selection of appropriate leafing and fruiting date, and increased resistance to diseases. To do this, breeders have combined French cultivars for quality and late leafing with California cultivars, such as Payne, chosen for lateral bearing and precocity. Late leafing helps avoid walnut blight (*X. arboricola* pv. *juglandis*) spread by spring rains. Payne is in the background of all current University of California cultivars. After evaluation of about 6000 progeny, ten cultivars were released in 1968. In 1979, an additional three cultivars, Chandler, Howard, and Sunland were released. Tulare was released in 1993, and current Californian walnut production is largely derived from Chandler (74.2 per cent), Tulare (12.6 per cent), and Howard (8.1 per cent). The third phase of the breeding program has been led by Charles A. Leslie since 2009 with the continued goal of obtaining cultivars with early harvest dates, good yield, light kernel color, and resistance to walnut bacterial blight and cherry leafroll virus. Three new cultivars have been released recently: Ivanhoe (very early harvesting) in 2010, Solano in 2012, and Durham in 2016. All three of these cultivars have very light coloured kernels and harvest earlier than Chandler (Bernard *et al.*, 2018).

Rootstock improvement is concentrated on developing and releasing disease resistant Paradox (hybrids between *J. nigra* and *J. regia*) and improving the clonal propagation procedures needed for commercial production. In recent years, two new rootstocks have been released: RX1 (*J. microcarpa* × *J. regia*) with resistance to Phytophthora crown and root rots, and VX211 (*J. hindsii* × *J. regia*) which is very vigorous and shows tolerance to nematodes (McGranahan and Leslie, 2012).

France

In France, germplasm collections are maintained by the Institut National de la Recherche Agronomique (INRA) (in English: National Institute of Agricultural Research) of Bordeaux. Twenty-eight crosses were made mainly with French cultivars as female parents (Franquette, Grosjean, Marbot) because they have a late bud break and good fruit quality. Male parents were chosen for lateral bearing including California cultivars (Pedro, Chandler) (Germain 1997). Two cultivars were released in 1995: Fernette and Fernor, both lateral bearing cultivars, from a cross between Franquette and Lara. A third new cultivar, Ferjean, a lateral bearing cultivar with quite small kernels, became available in 1999 from a cross between Grosvert and Lara. In 2010, the last four cultivars from this program were released: Feradam, Ferbel, Ferouette, and Fertignac. Unfrtunately, the breeding program ended in 2007 due to a new orientation for INRA, following a decision to focus on a smaller number of model species, such as *Prunus* and *Malus*. A new research project began mid-2017 in cooperation between the INRA Bordeaux-Aquitaine Centre and the Centre Technique Interprofessionnel des Fruits et Légumes (Ctifl) (in English: Fruit and Vegetable Interprofessional Technical Centre), supported financially by the Nouvelle- Aquitaine region. The main objectives are evaluation of the genetic diversity of germplasm resources (about 270 worldwide accessions), identification of the genetic determinants of traits of interest, and establishment of the necessary tools for MAS achievement ((Bernard *et al.*, 2018).About 70 per cent of French walnut production is provided by Franquette, an ancestral cultivar with late leafing and terminal bearing habit, discovered in the early 1800s.

China

Chinese breeding goals include early fruiting, high yield with high quality, and tolerance to anthracnose (*C. gloeosporioides*) and other diseases. Each province has developed many local cultivars, but the main ones of national interest include, for instance, Bofeng, Bokexiang, Zha343, Luguang, Xiangling, Hanfeng, Wen185, the Jinboxiang series, the Zhongling series, the Jinlong series, and the Luguo series. Between 2006 and 2012, more than 20 bred cultivars were released, including Yangza No.1 (tolerant to walnut blight) and Yangza No.2 hao (for warm and humid areas) from the Forestry Bureau of Dali in Yunnan in 2006, Jinxiang from the Shanxi Academy of Forestry Sciences in 2007 (high yield), Zanmei (early bearing), and Zhenzhuxiang (small fruits) from the Agricultural University of Hebei in 2012 (Chen *et al.*, 2014). In addition, 17 newly bred cultivars were released by the Shandong Pomology Institute between 2003 and 2012, including Luguo 2 in 2012 from an intraspecific hybridization and Luwen 1 in 2009 from an interspecific hybridization (Zhang *et al.*, 2014). Rootstock improvement work also has proceeded in China, including the Jin RS-1 series released in 2011 by the Shanxi Academy of Forestry Sciences. Jin RS-2 and RS-3 have been evaluated for cold, disease, and pest resistance, and seem to be ideal for northern Chinese areas subject to frost (Wang *et al.*, 2014). The list of some newly bred Chinese cultivars are given in Tables 29.7 and 29.8.

Table 29.7: New Walnut Cultivars Developed in China

Cultivar	*Parents*	*Year*
Hybrids		
Luxiang	Shangsong 6 × Xinjiangzhipi	2001
Yubxin lines	Yunnanboke × Xinjiang	2001
90301, 90303	Yunnansantai × Xinzao 13	2002
Lufeng	Shangsong 6 × Akesu 9	2003
Daixiang	Liaohe 1 × Xiangling	2003
Yuanlin	Yuanfeng × Chandler	2007
Selections		
NY-83-05	Vina' walnut seedling	2001
Daihui	Xiangling' walnut seedling	2003
Jinboxiang 1~4	Xinjang walnut seedling	2004
Lvling	Xiangling' sprout	2005
Yihe 1	Seedlings of *J. hebeiensis* HU	2006
Qinquan 1	Walnut seedling	2006

Source: Wu *et al.*, 2010.

Table 29.8: Varieties Developed by Shandong Institute of Pomology, China

Cultivar	*Parents*	*Trait*
	Hybrid	
Daixiang	Liaohe 1× Xiangling	Lateral bud fruiting rate of 95 per cent; nut weight is 13.9 g, kernel rate is 58.9 per cent, and the whole kernel is obtained easily
Luguo 2	Shangsong 6 × Luxiang	Lateral bud fruitfulness of 73.6 per cent; kernel rate is 59.6 per cent
Luguo 7	Xiangling × superior plant of late fruiting walnut	Kernel ratio is 56.9 per cent; Lateral bud fruit fullness is 87 per cent
	Selection	
Daifeng	Fenghui	Kernel rate - 58.9 per cent.
Daihui	Xiangling	Kernel rate - 58.5 per cent
Luguo 6	Yuanfeng	Kernel ratio - 55.3 per cent,
Luguo 8	Daixiang	Kernel ratio - 55.1 per cent,

Source: Zhang *et al.*, 2014a.

Besides, Wang *et al.*(2014b) developed six dwarf walnut cvs. Xinwen 609, Xinwen 724, Xinwen 908, Xinwen 915, Xin 916 and Xinwen 917, and suggested that developing dwarf walnut cultivars by selecting dwarf individuals from a seedling population of second-bearing walnuts is not only feasible, but also an efficient method of developing new cultivars.

Turkey

In Turkey, walnut germplasm consists mainly of early leafing and terminal bearing cultivars. The production is made mostly in the Mid-eastern Anatolia and Western Black Sea regions. The first Turkish selection study dating to 1971 examined diverse areas of Turkey and 20 seedlings genotypes were selected, including Yalova 1,Yalova 3, Bilecik, and Sebin which has become a reference in many production areas of Turkey (Akca, 2014). The National Walnut Cultivar Breeding Program, initiated in 1998, introduced new cultivars with late leafing and lateral bearing. Then, in 2000, a new breeding program was started to develop cultivars with improved nut quality traits including greater kernel weight and shell thickness (Akca and Polat, 2007). In 2008, a variety breeding program directed at developing cultivars with higher yield, late leafing, improved nut quality, and resistance to blight made 13 crosses using Turkish cultivars ([a]ebin,Akçal, *etc.*) and foreign cultivars (Franquette and Hartley). From these, 1340 hybrids were obtained and are under evaluation (Akca *et al.*, 2016). From the previous mass selection program, three cultivars were released in 2009: Maras 18, Sutyemez 1, and Kaman 1 (Sutyemez, 2016).

Iran

In Iran, most walnut trees also originated from seed (Ghasemi *et al.*, 2012). Awalnut improvement program started there in 1983 with selection of superior genotypes from native germplasm with two major goals: to create new cultivars and to develop efficient vegetative propagation methods (Hassani *et al.*, 2014). Seven superior genotypes were used as parents, and progenies were evaluated along with eight foreign cultivars, including Serr, Hartley, *etc.* As a result, two cultivars were released in 2010: Jamal (protandrous) and Damavand (protogynous).

Haploids and di-haploids (DHs) are important for genetic and developmental studies as well as for plant breeding. They have potential use in mutation research, genetic analysis, transformation, and in the production of homozygous cultivar(s) that can be used to produce F_1 hybrids for exploitation of heterosis (Germana, 2009). Grouh *et al.* (2011) obtained haploid plants from 'Hartley', 'Pedro', Z63, and Z67 after pollination using pollen irradiated at 300 and 600 Gy. The haploid plantlets showed only one allele of their female parents. These profiles confirmed the parthenogenetic origin of the obtained haploid plants. The techniques used to induce haploid walnut plants by irradiated pollen were successful and could be used in breeding programs and accelerate genome analysis in this plant in which the genome size is approximately three times, the size of the human genome.

The list of newly released and registered walnut varieties are given in Table 29.9.

Table 29.9: Newly Released Varieties of Walnut

Variety	*Parentage*	*Origin*	*Characters*
Robert Livermore.	UC86-11 x Howard	University of California, Davis	A lateral-bearing walnut with red seedcoat. Nut: 14.1 g; 50 per cent kernel by weight; shell and seal strong; kernels plump, easily removed as halves; bears terminally and laterally; precocious; protandrous; ripens mid- to late-season 7-10 d before Chandler.
Fredrick.	Payne O.P	Stockton, CA	A walnut that harvests shortly after Payne with high yield and less blight. Nut: 11.4 g; smooth, ovate; light color; 54 per cent kernel, suitable for use in-shell; Kernel: 6.6 g; easily removed in halves; mostly extra light color. Tree: vigor high; moderately upright structure; bears on terminals; protandrous; yield high; harvests 1 week after Payne.
Ivanhoe.	UC67-13 x Chico	University of California-Davis	A very early harvesting walnut with extra-light kernel color. Nut: 13.5 g; smooth, oval; 57 per cent kernel; seals sometimes weak. Kernel: 7.7 g; easily removed in halves; high proportion of extra-light color. Tree: vigor moderate to low; bears on terminals and laterals; precocious; protogynous; yield high; harvest date very early, similar to Serr and 1 mo before Chandler; susceptible to blight.
Solano.	UC67-13 x Chico	University of California–Davis	Early-mid season harvesting walnut with extra-light kernel color. Nut: 14.7g; smooth, oval; seals strong; 55 per cent kernel; suitable for use in-shell; kernel: 8.0g; easily removed in halves; high proportion of extra-light color. Tree: vigor moderate; upright; bears on terminals and laterals; precocious; protandrous; yield high; harvest date mid-early, similar to Vina and 2 weeks before Chandler.

Variety	*Parentage*	*Origin*	*Characters*
Durham.	Chandler x PI159568	University of California, Davis,	Midseason harvesting walnut with extra-light kernel color, well-filled nuts, and smooth, light-colored shell. seals strong; 55 per cent kernel; suitable for use in-shell. Kernel: 8.4 g; easily removed in halves; high proportion of extra-light color. Tree: vigor similar to Chandler; growth habit upright; bears on terminals and laterals; protandrous; yield high; harvest similar to Tulare and 10 d before Chandler; occurrence of blight low.
Rootstock			
RX1.	*Juglans microcarpa* DJUG 29.11 x *J. regia* O.P	University of California, Davis and USDA-ARS	A phytophthora resistant hybrid clonal rootstock for Persian walnut. Plant: less vigorous than typically used *J. hindsii* x *J. regia* seedling hybrids; easily propagated by tissue culture. Rootstock performance: shows resistance to *Phytophthora citricola* and *P. cinnamomi* in greenhouse and field trials; performs well in replant situations and waterlogged conditions.
VX211.	*J. hindsii* PDS96-43 x *J. regia* O.P	Davis, CA, University of California, Davis and USDA-ARS	A vigorous, nematode-tolerant hybrid clonal rootstock for Persian walnut. Plant: notably more vigorous than typically used *J. hindsii* x *J. regia* seedling hybrids; readily propagated by tissue culture. Rootstock performance: very vigorous growth habit; grows well in the presence of nematodes (*Pratylenchus vulnus* and *Meloidogyne* sp.); moderate resistance to *P. citricola*.
Luwen 1	Wild walnut × Xiangling	Shandong Institute of Pomology, China	Strong resistance (fruit anthracnose, rot disease), and grafts well.
Luwen 8	Wild walnut × Xiangling	Shandong Institute of Pomology, China	Strong resistance, and walnut grafting affinity. Excellent walnut rootstock.

Source: Clark and Finn, 2010, 2012; Gasic and Preece, 2014; Gasic *et al.*, 2018; Zhang *et al.*, 2014a.

Table 29.10: Summary of Walnut Genomics and Transcriptomics Information on the NCBI Database

Species	*Total Nucleotide Sequences*	*Nucleo-tides*	*EST*	*GSS*	*Notes/References*
J. regia	223823	16918	5253	49405	40 ESTs unpublished/1-188ESTs involved in cellular regulation and global metabolism from leaf under irradiance or in darkness/2-5025 ESTs from seed coat tissues/3
J. hindsii x *J. regia*	13660	101	13559	0	2715 ESTs from leaves and 10844 from roots inoculated with *Pratylenchus vulnus* nematodes/4

Species	*Total Nucleotide Sequences*	*Nucleo-tides*	*EST*	*GSS*	*Notes/References*
J. nigra	211	129	78	4	78 ESTs involved in the regulation of heartwood production, related to programmed cell death, from stem/5
J. cathayensis	44	32	12	0	12 ESTs from fresh leaves, buds, female flowers and male flowers to study genetic diversity/6
All other taxa	601	601	0	0	
Total	238339	170043	18902	49409	

www.ncbi.nlm.nih.gov.

Genomic and Transcriptomic Studies

Molecular markers permitted the construction of the first genetic maps between 1994 and 2001 and genomics allowed to crack the walnut genome in 2016. The summary of walnut genomics and transcriptomics information is geven in Table 29.10.

REFERENCES

Akca, Y. 2014. Persian walnut. *Acta Horticulturae,* 1050: 139–142.

Akca, Y. and Polat, A.A. 2007. Present status and future of walnut production in Turkey. *Eur. J. Plant Sci. Biotechnol.,* 1: 57–64.

Akça, Y., Sütyemez, M., Yilmaz, S. and Karadag, H. 2016. The new walnut variety breeding program in Turkey. VIIth International Scientific Agricultural Symposium. Jahorina, Bosnia and Herzegovina, pp. 461–466.

Anon. 2017. Horticultural Statistics at a Glance 2017. Horticulture Statistics Division Department of Agriculture, Cooperation and Farmers Welfare Ministry of Agriculture and Farmers Welfare Government of India.

Antanaviciute, L., Fernandez-Fernandez, F., Jansen, J., Banchi, E., Evans, K.M., Viola, R., Velasco, R., Dunwell, J.M., Troggio, M. and Sargent, D. J. 2012. Development of a dense SNP-based linkage map of an apple rootstock progeny using the Malus Infinium whole genome genotyping array. *BMC Genomics,* 13.

Antofie, A., Lateur, M., Oger, R., Patocchi, A, Durel, C. E. and Van de Weg, W.E.2007. A new versatile database created for geneticists and breeders to link molecular and phenotypic data in perennial crops: The AppleBreed DAtaBase. *Bioinformatics,* 23: 882-891.

Bernard, A., Lheureux, F. and Dirlewanger E. 2018. Walnut: past and future of genetic improvement. *Tree Genetics and Genome,* 4: 106-133.

Borejsza-Wysocka, E., Norelli, J., Baldo, A., Malnoy, M., Farrell, R.E., Bassett, C.L. and Aldwinckle, H. 2009. Efficient generation of RNAi mutants of apple using multivector transformation. *Phytopathology,* 99: S191-S191.

Chagne, D., Crowhurst, R.N., Troggio, M., Davey, M.W., Gilmore, B., Lawley, C., Vanderzande, S., Hellens, R.P., Kumar, S., Cestaro, A., Velasco, R., Main, D., Rees, J.D., Iezzoni, A., Mockler, T., Wilhelm, L., Van de Weg, E., Gardiner, S.E., Bassil, N. and Peace, C. 2012. Genome-Wide SNP Detection, Validation, and Development of an 8K SNP Array for Apple. Plos One 7.

Chen, X., Xu, L., Zhang, S.L. and Liu, Z.Q. 2014. Walnut Genebank in China National Clonal Plant Germplasm Repository. *Acta Horticulturae*, 1050: 89–94.

Clark, John R. and Finn, Chad E. 2010. Register of New Fruit and Nut Cultivars List 45. *HortScience*, 45(5): 716-756.

Clark, John R. and Finn, Chad E. 2012. Register of New Fruit and Nut Cultivars List 46. *HortScience*, 47(5): 536-562.

Gardiner, S., Norelli, J., de Silva, N., Fazio, G., Peil, A., Malnoy, M., Horner, M., Bowatte, D., Carlisle, C., Wiedow, C., Wan, Y., Bassett, C., Baldo, A., Celton, J., Richter, K., Aldwinckle, H. and Bus, V. 2012. Putative resistance gene markers associated with quantitative trait loci for fire blight resistance in *Malus* 'Robusta 5' accessions. *BMC Genetics*, 13: 25.

Gasic, Ksenija and Preece, John E. 2014. Register of New Fruit and Nut Cultivars List 47. *HortScience*, 49 (4): 396-421.

Gasic, Ksenija, Preece, John E. and Karp, David. 2016. Register of New Fruit and Nut Cultivars List 48. *HortScience*, 51 (6): 620-652.

Jensen, P.J., Rytter, J., Detwiler, E.A., Travis, J.W. and McNellis, T.W. 2003. Rootstock effects on gene expression patterns in apple tree scions. *Plant Molecular Biology*, 53: 493-511.

Jensen, P.J., Makalowska, I., Altman, N., Fazio, G., Praul, C., Maximova, S.N., Crassweller, R.M., Travis, J.W. and McNellis, T.W. 2010. Rootstock-regulated gene expression patterns in apple tree scions. *Tree Genetics and Genomes*, 6: 57-72.

Jensen, P.J, Halbrendt, N., Fazio, G., Makalowska, I., Altman, N., Praul, C., Maximova, S.N., Ngugi, H.K., Crassweller, R.M., Travis, J.W. and McNellis, T.W. 2012. Rootstock-regulated gene expression patterns associated with fire blight resistance in apple. *BMC Genomics*, 13.

Germain, E. 1997. Genetic improvement of the Persian walnut (*Juglans regia* L.). *Acta Horticulturae*, 442: 21–32.

Ghasemi, M., Arzani, K., Hassani, D. 2012. Evaluation and identification of walnut (*Juglans regia* L.) genotypes in Markazi province of Iran. *Crop Breeding Journal*, 2: 119–124.

Grouh, M. S. H., Vahdati, K., Lotfi, M., Hassani, D. and Biranv and, N. P. 2011. Production of haploids in Persian walnut through parthenogenesis induced by gamma-irradiated pollen. *J. Amer. Soc. Hort. Sci.*, 136 (3): 198–204.

Hassani, D., Dastjerdi, R. and Haghjooyan, R. 2014. Genetic improvement of Persian walnut (*Juglans regia* L.) in Iran. *Acta Horticulturae*, 1050: 95–102.

Hias N., Leus L., Davey M.W., Vanderzande S., Van Huylenbroeck J. and Keulemans J. 2017. Effect of polyploidization on morphology in two apple (*Malus* × *domestica*) genotypes. *Hort. Sci.* (Prague), 44: 55–63.

Holloway, B. and Li, B.L. 2010. Expression QTLs: applications for crop improvement. *Molecular Breeding* 26: 381-391.

Huang, Z., Hu, H., Shen, F., Wu, B., Wang, X., Zhang, B., Wang, W., Liu, L., Liu, J., Chen, C., Zhang, R., Chen, R., Wang, Y., Wu, T., Xu, X., Han, Z. and Zhang, X. 2018. Relatively high acidity is an important breeding objective for fresh juice specific apple cultivars. *Scientia Horticulturae*, 233: 29-37.

Iglesias, I., Bonany, J., Batlle, I., Reig, G., Troggio, M., Volz, R., Kumar, S., Espley, R., Allan, A.C., Friend, A.P. and Chagne, D. 2016. Physiological and genetic control of red skin colouration in apples grown under warm and cool conditions. *Acta Horticulturae*, 1110: 27-34.

Kumar, S., Chagne, D., Bink, M. C. A. M., Volz, R. K., Whitworth, C. and Carlisle, C. 2012.Genomic Selection for fruit quality traits in apple (*Malus* x *domestica* Borkh.). *PLoS ONE*, 5: 1-10

Luby, J., Finn, C.E., Gasic, K., Iezzoni, A., Oraguzie, N., Brown, S.K., Byrne, D.H., Clark, J.R., Crisosto, C.H., Davis, T.M., Evans, K., Gradziel, T., Hancock, J.F., Bassil, N.V., Fazio, G., Main, D., Peace, C., Weebadde, C.K., Van de Weg, E. and Yue, C.Y. 2010. Germplasm Sets and Standardized Phenotyping Protocols for Fruit Quality Traits in RosBREED. *HortScience*, 45: S55-S56.

Malnoy, M., Boresjza-Wysocka, E.E., Norelli, J.L., Flaishman, M.A., Gidoni, D. and Aldwinckle, H.S. 2010. Genetic transformation of apple (*Malus* × *domestica*) without use of a selectable marker gene. *Tree Genetics and Genomes*, 6: 423-433.

McGranahan, G. and Leslie, C. 2012. Walnut. In: Badenes, M.L. and Byrne, D.H. (eds) *Fruit Breeding*. Springer, Boston, p. 827–846.

Mellidou, I., Keulemans J., Davey, M.W., Chagne, D., Gardiner, S.E. and Laing, W.2014. Nutritionally-enhanced apples: markers for molecular breeding for fruit vitamin c concentrations in apple. *Acta Horticulturae*, 1048: 163-167.

Peil, A., Kellerhals,M., Hofer, M. and Flachowsky, H. 2010. Apple breeding-from origin to Genetic engineering. *Fruit, Vegetable and Cereal Science and Biotechnology*, 8: 118-138.

Shimada, Takuya and Moriya, Shigeki. 2018. Preference of Japanese field voles (*Microtus montebelli*) for apple rootstock genotypes: correlations with bark characteristics. *Crop Protection*, 112: 239–245

Sedov E.N., Sedysheva G.A., Serova Z.M. and Gorbacheva, N.G. 2014. Autogamy of polyploid apple varieties and forms. *Russian Agricultural Sciences*, 40: 253–256.

Sedysheva, G.A. and Gorbacheva, N.G. 2013. Estimation of new tetraploid apple forms as donors of diploid gametes for selection on a polyploidy level. *Universal Journal of Plant Science*, 1: 49–54.

Sutyemez, M.2006. Comparison of AFLP polymorphism in progeny derived from dichogamous and homogamous walnut genotypes. *Pak J. Biol. Sci.*, 9: 2303–2307.

Wang, G., Wang, J.Y. and He, Q. 2014. Agronomic evaluation and heritability of Jin RS-2 and Jin RS-3 walnut rootstocks. *Acta Horticulturae*, 1050: 113–121.

Wang, N., Qu, C., Jiang, S., Chen, Z., Xu, H., Fang, H., Su, M., Zhang, J., Wang, Y., Liu, W., Zhang, Z., Lu, N. and Chen, X. 2018. The proanthocyanidin-specific transcription factor MdMYBPA1 initiates anthocyanin synthesis under low-temperature conditions in red-fleshed apples. *The Plant Journal*, 96: 39–55.

Wu, G.L., Meng, H.J., Hao, Y.Y., Liu, Q.L., Wang, D. and Tian, J.B. 2010. Thirty years of breeding walnut in China. *Acta Horticulturae*. 861: 109-118.

Zhang, M.Y., Xu. Y., Xiang, K. and Wang, X.L. 2014a. Review of walnut breeding research at the Shandong Institute of Pomology. *Acta Horticulturae*, 1050: 55–60.

Wang, G.A., Zhang, Q., Huang, M.M. and Yakup, A. 2014b. The breeding of six Xinjiang dwarf walnut cultivars. *Acta Horticulturae*, 1050: 151-159.

Xue, H., Zhang, F., Zhang, Zhi-Hong, Fu, Jun-Fan, Wang, F., Zhang, B. and Ma, Y.2015. Differences in salt tolerance between diploid and autotetraploid apple seedlings exposed to salt stress. *Scientia Horticulturae*, 190: 24–30.

Yamagishi, N., Kishigami, R. and Yoshikawa, N. 2014. Reduced generation time of apple seedlings to within a year by means of a plant virus vector: a new plant-breeding technique with no transmission of genetic modification to the next generation. *Plant Biotechnology Journal*, 12: 60–68.

Breeding of Perennial Horticultural Crops *Pages* **399–411**
Editors: S.K. Singh, V.B. Patel, A.K.Goswami, Jai Prakash and Chavlesh Kumar
Published by: **BIOTECH BOOKS, NEW DELHI**

Chapter **30**

Low Chill Temperature Fruits Improvement

M.K. Verma, V.B. Patel, Chavlesh Kumar and Tanushree Sahoo

Division of Fruits and Horticultural Technology,
ICAR-Indian Agricultural Research Institute, Pusa, New Delhi – 110 012

INTRODUCTION

India, situated north of the equator between 8° 4′ and 37° latitude and 68° 7′ and 97° 25′ longitude, is bounded on the southwest by the Arabian Sea and on the southeast by the Bay of Bengal. To the north and northeast lies the mighty Himalayan range. To the west lies Pakistan and to the east, Bangladesh and Myanmar. In the north, China (Tibet), Nepal and Bhutan share the international boundary with India. To the south Sri Lanka shares the maritime boundary and is separated from India by a narrow channel of the Bay of Bengal formed by the Palk Strait and the Gulf of Mannar. India is one of the largest countries of the world and covers an area of about 3,287,263 km^2. It measures 3,214 km from north to south and 2,933 km from east to west and has a land frontier of 15,200 km and a coastline of 7,516 km.

There are three main climatic seasons in India: monsoons (June-September in the southwest and, October-November in the northeast), summer (April-July), and winter (October-March). Winter prevails from October to March and is marked by cool day and night, cloudy as well as clear skies, and occasional hot days. In the northern Himalayan areas, the temperature falls by 0.6°C for every 100 m rise in altitude, which causes a variety of climates from nearly tropical in the foothills to tundra type above the snow line. There is sharp contrast between temperatures of the sunny and shady slopes, and a high diurnal range of temperature, temperature inversion and rainfall variability based on altitude. Most of the rainfall is in the form of snow during late winter and spring months. The area to the south of the great Himalayan range is protected from cold winds coming from the interior of Asia during winter. The leeward side of the mountains receives less rain while the well exposed slopes get heavy rainfall. The places situated between 1070 and 2290 m altitude receive the heaviest rainfall and the rainfall decreases rapidly above 2290 m. The Himalayan range has heavy snowfall during the winter months, particularly from December to February, at altitudes above 1500 m with a high diurnal range of temperature.

The temperate climatic zone area is found in the states of Jammu and Kashmir, Himachal Pradesh, Uttarakhand, and parts of Arunachal Pradesh, Meghalaya, Manipur, Sikkim and Nagaland. Because of an ideal climate this region offers tremendous opportunity for production of high quality temperate pome fruits (apple and pear), stone fruits (peach, plum, apricot, and cherry), and nuts (walnut, almond, chestnut, pecan nut, and hazelnut). About 80 per cent of the temperate region in India is rainfed. In most of the areas, the bulk of the precipitation takes place during three or four months (mostly in the monsoon period from June-September). However, the Kashmir valley is devoid of monsoon rains (June-September) and March-April is the rainy season, but precipitation is received during winter in the form of snow. The remaining periods are more or less completely dry and trees can suffer from serious moisture stress, a major factor for low productivity. Because rainfall occurs within a short period, and the terrain is hilly, there are special problems. Rain water travels down the hill slopes at high velocity, eroding the soil, making gullies, with little chance of being absorbed. The result is floods during the rainy season and drought or moisture stress in the dry season. In mountainous tracts, soils (cultivable area and depth) are less compared to the plain area. Relatively flat areas occur in small valleys, river banks, and ridge tops. In most areas, entire hillsides are dotted with villages set amidst terraced fields. The soil profile, depth, and characteristics change according to slope gradient; many soils in levelled areas have poor drainage.

Another potential area where low-chill temperate fruits can be successfully cultivated in many countries on commercial scale is lies in subtropical zone. The subtropics are the zones of the Earth immediately north and south of the tropic zone, which is bounded by the Tropic of Cancer and the Tropic of Capricorn, at latitude 23.5 ° north and south. The term 'subtropical' describes the climatic region found adjacent to the tropics, usually between 20 and 35 degrees latitude in both hemispheres, but occasionally found at slightly higher latitudes. In subtropical climates the winters are relatively warm, but not as hot as the summer season. These climates rarely - if ever - see frost or snow, and plants such as palm, citrus and many broadleaf evergreens flourish, in contrast to the hardier deciduous and coniferous trees which dominate mid-latitude climates. As one moves toward the tropical side the slight winter cool season disappears altogether, while at the polar threshold of the subtropics the winters become much cooler.

Productions of temperate-zone fruits in subtropics may seem fanciful to be initiated, but it is a practice that has existed in localized regions for generations. Seedling peaches are grown in Venezuela, northern Thailand, and southern Mexico. No one knows precisely when these species were introduced, but local selection has produced cultivars adapted to areas with little or no chilling temperatures. Rest either does not occur or is sufficiently shallow to be broken by stress induced by defoliation or drought. Commercial cultivations of temperate-zone fruits are well-established in subtropical areas, where winter temperatures allow partial breaking of the rest period. In areas of northern Mexico, southern Africa, and Israel, for example, chemical treatments are regularly applied to supplement the effects of limited chilling. Because temperature declines with increasing elevations, many microclimates exist in the subtropics where temperate-zone fruits are not currently grown, but could be. Subtropical regions where temperate fruits are grown include the southern third of California (Mediterranean type), the low deserts of the Southwest USA (hot arid type), the Gulf Coast and most of Florida (humid type), the southern Mediterranean and northern Sahara, northern India (monsoon), southeast China (humid), the middle part of South America (varied), and much of Australia (varied). Similar potential climatic conditions are also available in subtropics of India. These areas located in Afghanistan (Kabul), Australia (Brisbane and Sydney), Brazil (São Paulo, Curitiba, Florianópolis and Porto Alegre), China (Hong Kong, Macau and Shanghai), Europe (Lisbon, Barcelona, Rome, Dubrovnik, Athens and Sochi), India (Lower mountainous regions of J&K, Himachal Pradesh, Uttarakhand, Punjab, Haryana, western UP, Mizoram, Meghalaya, Arunachal Pradesh, Sikkim, and Nilgiri Hills), Iraq (Baghdad), Israel (Jerusalem), New Zealand (Auckland), North Africa (Cairo), Pakistan (Islamabad), Russia (Sochi), South Carolina, Orlando, Florida, Tucson, Arizona), Turkey (Mersin, Antalya, Rize), United States (Houston, Texas, Charleston, South Africa (Durban and East London) and Uruguay (Bella Unión).

The subtropical region in India lie between latitude 15° and 30° north of equator covering north Indian plains, the Gangetic basins, large parts of eastern states, eastern and western Ghats, Nilgiri region *etc.* In this zone spring is generally frost free and chilling availability ranges from 50 to 500 hours. This region is thus not considered suitable for temperate fruits but grows several low chilling varieties of peach, plum, apricot, almond, pear and nuts. The area and production of stone fruits in this region is estimated to be much higher than in the temperate zone in India. These areas being closer to the markets, offers better production potential than those in higher hills. The winters are relatively warm, but not as hot as the summer season. These climates rarely - if ever - see frost or snow, and plants such as palm, citrus and many broadleaf evergreens flourish, in contrast to the hardier deciduous and coniferous trees which dominate mid-latitude climates. As one moves toward the tropical side the slight winter cool season disappears altogether, while at the polar threshold of the subtropics the winters become much cooler. Rainfall patterns vary widely throughout the subtropics. India: North Indian plains, Gangetic basins, parts of eastern states, eastern and Western Ghats, Nilgiri region

Significance of Chilling

Dennis (1997), stated that several systems have been devised for growing temperate zone fruits in warm climates, where chilling is insufficient to break bud dormancy in "standard" cultivars. These are: using chemicals to break dormancy, grafting with low chill scions, moving the trees to cold storage or high elevations in containers, evaporative cooling applications, preventing the induction of dormancy by defoliation. Countries, such as Argentina, Australia, Brazil, Chile, Egypt, France Reunion Islands, India, Indonesia, Israel, Italy, New Zealand, Mexico, Morocco, Pakistan, Philippines, Portugal, Spain, South Africa, Sri Lanka, Thailand, Tunisia, Turkey, Zimbabwe, *etc.* have restricted conditions to grow temperate zone fruits. Since the temperatures are too high during winter periods, low chill cultivars are preferred and some preventations should be taken (Table 30.1).

Table 30.1: The Climatological Data of Various Places of Subtropical Climate (1951-1980)

Station	*Altitude (m)*	*Latitude*	*Longitude*	*Mean Temperature (°C)*	*Temperature (°F)*	*Chilling Accumulation (Chilling hours)*
Agra	169	27.1	78.02	14.65	58	415
Aligarh	187	27.53	78.04	14.2	58	415
Allahabad	98	25.77	81.44	16.15	61	253
Alwar	271	27.3	76.35	14.65	58	415
Amritsar	234	31.38	74.52	11.6	53	685
Bahraich	124	27.34	81.36	15.7	60	307
Bareilly	173	28.22	79.24	15.2	59	361
Barmer	194	25.45	71.23	18	64	121
Bhatinda	**	30.1	74.35	12.5	55	577
Bhopal	**	23.15	69.4	17.85	64	91
Chandigarh	347	30.44	76.53	13.25	56	523
Churu	291	28.15	74.55	13.25	56	523
Dehradun	682	30.19	78.02	12.6	55	577
Ferozepur		30.55	74.4	12	54	631
Jaipur	390	26.49	75.48	15.15	59	361
Kanpur (Up)	126	26.26	80.22	15.6	60	307
Lucknow	128	26.52	80.56	14.75	59	361
Meerut	222	29.01	77.58	14.2	58	415
Delhi (Safderjung)	216	28.35	77.12	14.2	58	415
Delhi (Palam)	233	28.34	77.07	13.5	56	523
Varanasi	85	25.27	82.52	15.85	61	253
Sriganganagar	177	29.55	73.53	13.2	56	523
Ludhiyana	247	30.56	75.52	12.95	55	577
Jalawar (Rajasthan)	321	24.32	76.1	17.1	63	145
Hisar (Haryana)	221	29.1	75.44	13.45	56	523

Source: IMD.

The temperate fruits, nuts and berry crops require an exposure to low, but above freezing, temperatures, for resumption of growth and fruiting, called the chilling requirement. In general, their commercial cultivation is limited by high latitudes, where too much cold kills plant tissues and by the low altitudes, where, the winters are not cold enough to fulfill the chilling requirement. Hutchins (1932) reported that the chilling requirement is met, only, if the plants are exposed to less than 7°C temperature. However, later on, it was found that low-chill cultivars/species of these fruits can partially meet the chilling requirements even at higher temperatures. In order, to find out the suitability of non-temperate regions for growing temperate fruits, various models/formulae were proposed to predict the number of available chilling hours (Chill unit) for a specific location. The model of Sharpe (1969) and Weinberger (1956), based on the mean temperature of the coldest month, is very simple and practical, as mean maximum and mean minimum temperatures are available for most of the locations. Sharpe (1969) has further suggested that Florida peaches can meet their chilling requirements at higher temperature, so his allotment of chill-units is higher than that of Weinberger (1956), who does not assign any chill unit to areas with mean monthly temperatures higher than 14.4°C. The states of Punjab, Haryana, northern Rajasthan, plains of U.P., Bihar and West Bengal falling in the subtropical zone. The areas being contiguous to Himalayas, the cool climate prevails during November to February (winter season), with January being the coldest month, when mercury may touch -1°C, at some places, for few hours for a few days. The ground frost/fog may occur from end December to early February. The mean monthly temperature of coldest month (January) at various locations in these plains varies from 11.6°C to 18°C at various places located in this zone. Therefore, using the model of Sharpe (1969) the chill units available at Agra, Aligarh, Alwar, Amritsar, Bhatinda, Churu, Dehradoon, Ferozpur, Meerut, Delhi, Sriganganagar, Ludhiana, and Hissar are ranging from 415 to 685 chilling units. So there exists the possibility of growing low-chill cultivars of temperate fruits. However, these area experience very hot summer, with 40-45°C air temperatures prevailing for 2-3 weeks during May-June, when sunscald occurs in some fruits. The highest temperature recorded is 50°C at Sriganganagar, which also receives about 523 chill units. Therefore, in addition to low-chilling requirement these cultivars should also be heat tolerant. The other abiotic stresses to be experienced in these areas include slightly alkaline/sodic saline soils, flooding during monsoons and high winds velocity. The region being near to major consumption centres and with well-developed physical connectivity, through network of rural roads, the returns from these fruits are comparable to that in higher hilly areas and other subtropical fruits. Many medium to large plantations of temperate fruits especially that of pear and peach exists in Punjab, Haryana and Western U.P. *etc.* and total area under stone fruits is more than in the hills

Chilling Zones

Much of the Southeast Asia region is within the tropics and thus receives no chilling temperatures. Beginning with about 20° latitude, there is a positive accumulation of chilling which increases with the latitude. Generally more than 1000 chilling units are accumulated above 26° latitude whereas in Australia low- to medium-chill areas are found at latitudes of up to 35° south (Table 30.2). The chilling accumulated is affected by the ocean currents and local geography, especially the altitude in tropical and subtropical zones. With cooler temperatures at the higher altitudes, it was estimated in Taiwan that with each increase in altitude of 100 m there is an increase in chilling accumulation of 27 chilling units. When using the expected temperature decrease with altitude of 1°F or 0.56°C per 100 m, the estimated chilling increase was 54–61 units per 100 m altitude.

Table 30.2: Estimated Mean Chilling Accumulation in Southeast Asia and Australia as a Function of Latitude

Latitude	*Chilling Range*	
	Southeast Asia	*Australia*
< 20	None	
20–23.9	0–630	0–700
24–27.9	100–1500	50–500
28–31.9	> 1000	300–1100
32–35.9	> 1000	450–1300
36–39.9	> 1000	> 850

Chilling accumulation estimated by using the following equation: CU = 3547 – 53 (mean coldest month temperature in degrees F.

Although the mean chilling received is useful, it is the year-to-year fluctuations that are critical to examine when choosing varieties that would be well adapted to the region. Thus, the most appropriate chilling range (and most consistent in production) would be varieties requiring between 450 and 600 CU. If cold weather during bloom was not a problem, lower chill varieties could also be used. Mild winter regions of the world are classified into three major adaptation zones: the transitional, medium, and low-chill zones by chill requirement of commercial cultivars.

1. ***The transitional-chill zone*** includes regions in which most commercial cultivars have a chill requirement of 650–950 chilling units (CU). This zone experiences years in which the higher chill varieties produce poorly due to mild winters. This is true of the southern San Joaquin valley in California on the West Coast and central Georgia and the coastal plain of South Carolina on the East Coast of the USA, some regions of southern Australia, and the southern and higher altitude regions of South Africa.
2. ***The medium-chill zone*** includes areas where cultivars with a chill requirement of about 400–650 CU can be grown consistently. This would include areas such as southern Georgia, northern Florida, southern Louisiana, south central Texas, southern Spain, the highlands in Mexico, Northern parts of subtropical India, and lower altitude areas of South Africa.
3. ***The low-chill zone*** is characterised by frequently receiving less than 400 CU. This includes almost all parts of the subtropical areas lies in India.

Historical Perspective of Low Chill Temperate Fruit Improvement

Although no temperate fruit is indigenous to this region, some apple (crab apple), peach, plum, pear and apricots have got established here. Some of these fruits may have come with invaders, traders, slaves and soldiers and later missionaries. For instance, pear and peaches were planted by Chinese hostages, during the period of King Kanishka, near Amritsar (Punjab) in the first century AD. While reviewing horticulture in ancient India, Firminger (1863) mentions that an apple, like White Joanneting variety, flowers and fruits at Ferozepur (Punjab). He also recorded that the apple and pears grow into big trees near Calcutta, flowers profusely but the entire fruits drop, but a dark kind of plum, eatable after cooking fruits near Calcutta. The systematic efforts to introduce temperate fruits in the northern plains were initiated in the later half of the 19th century with the establishment of Agro-Horticulture Society of India and later on the Agricultural College at Pune, Coimbatore, Faisalabad (Pak), Nagpur, Sabour and Kanpur. The introduced fruits varieties were planted at Govt. Botanical Gardens (Company Baghs) at Calcutta, Saharanpur and other places. In the first half of the 20th century large introductions were made at Palampur, Punjab and Kandghat (Patiala State).

There are a large number of temperate fruits which are grown in the temperate climate but all these fruits cannot be grown in the subtropical climate. Only those temperate fruits can be grown in tropical/subtropical climate which have comparatively low chilling requirement and have the ability to tolerate high summer temperature. Proper selection of temperate fruits should be done which can be successfully grown under subtropical climatic conditions. For example, crops such as pear (*Pyrus communis*), peach (*Prunus persica*) and strawberry (*Fragaria ananassa*) are ideal crops for subtropical climate followed by fruits such as plum (*Prunus salicina*), apricot (*Prunus armeniaca*), almonds (*Prunus dulcis*), olive and persimmon (*Diospyros kaki*) which can be grown in a wide range of subtropical and warm temperate climate. To some extent the cultivation apple (*Malus domestica*) can also be tried.

Low Chill Fruit Varietal Development

Selection of cultivars of the temperate fruits to be grown under subtropical climate is one of the key requirements for their successful cultivation in the subtropical regions. Only those cultivars which have low chilling requirement and ability to tolerate high summer temperature should be selected. Breeding programs have provided new cultivars selected for their short chilling requirements and tolerance to high temperature. One of the most successful cultivars is 'Anna' apple, developed by A. Stein in Israel. This is now extensively grown in areas with limited chilling. The introduction and the selection of superior types have remained major

Sunbest Nectarine

Sunmist Necatine

Suncoat Nectarine

Table 28.3: Temperate Fruits and their Cultivars Suitable for Cultivation under Subtropical Climate

Sl. No.	*Fruit Crop*	*Cultivars (*Chilling hours required; value given in bracket indicates the chilling requirement of the respective variety)*
1	Peach (*Prunus persica*)	Okinawa (50), Red Ceylon (50), UF Sun (100), FlordaGrande (100), Flordabelle (110), Flordared (110), Flordawon (110), FlordaPrince (150), FlordaGlo (150), Tropic Beauty (150), UF Beauty (200), Sunred (210), Flordabest (250), UFO (250), UF 2000 (300), UF Blaze (300), Flordadawn (300), Early Amber (310), Flordasun (310), UF Sharp (325), FlordaKing (350), Flordacrest (350), Gulfking (350), Desert Gold (350), Jewel (350), Gulfcrimson (400), Gulfprince (400), Sun Ripe (400), Florda Home (400), La Festival (400), La Pecker (400), Rio Grande (400), Tejan (420), Rechan (450), Tex Star (450), Whiterobin (500), Bonita (500), Sunfire (500), Ventura (500), May Gold Su (500), Gulfcrest (525), Flordaqueen (540), Sungold (540), Springtime (650), June Gold (650), Maygold (660), Junegold (660), Springtime (660), Armgold (660), Suwannee (660), Tropic Snow, Double Flowering, Parbhat, Pratap, Khurmani, Sharbati, Safeda Early Cream, Saharanpur Prabhat, Shan-i-Punjab, Shaharanpur No. 6, Ranjit Bagh Early, Safeda (LR Brothers), Shaharanpur Hybrid 3, China Flat, Babcock,
2	Nectarine (*P. persica var. nucipersica*)	Sunbest (225), Sunraycer (250), UF Royal (250), UF Queen (250), Sunmist (275), Sundollar, (350), Suncoat (375), Sunred
3	Apple (*Malus domestica*)	Dorsett Golden (250), Anna (300), Tropic Mac (300), Tropic Sweet (300), 88-20 (375), Ein Scheimer (400), 60-39 (400), Tamma, Neomi, Tropic Beauty, Gallia Beauty, Winter Banana, Tame, Vered
4	Pear (*Pyrus communis and P. pyrifolia*)	Patharnakh, Gola, Leconte, Keiffer, Smith, Baghugosha, China Pear, Pineapple, Baldwin, Tenn, Flordahome, Ayers Hood, Orient, Carnea, Tsu Li, Ya Li, *P. calleryana* (rootstock requires 400 chilling hours)
5	Strawberry (*Fragaria ananassa*)	Chandler, Tioga, Torrey, Selva, Belrubi, Fern, Pajaro
6	Plum (*Prunus salicina*)	Satluj Purple, Kala Amritsari, Jamuni Meeruti, Titron, Aloo Bokhara, Alucha Black, Titron Howe, Gulfruby, Gulfbeauty, Gulfblaze, Gulfrose
7	Apricot (*Prunus armeniaca*)	New Castle, Early Shipley, St. Ambroise, Benazir, NJ-13
8	Almond (*Prunus dulcis*)	California, Papershell, Hybrid 15, Pathick.s Wonder, JKS-55, H-98, Achak (266), Desmayo Largueta (309), Ramillete (326), Marcona (435), Marta (478), Antoneta (514), Ferragnes (558)
9	Sweet Chery (*Prunus avium*)	Stella (200-250), Cristobalina, Temprona del Sot, Precoce de Bernard, Sunburst, Lapin, (Rootstocks: Gisela-5, Mahaleb)
10	Persimmon (*Diospyros kaki*)	Hachiya, Fuiju, Jiro, Hyakuma
11	Olive (*Olea europaea*)	Arbequina, Barnea, Frantoio, Koroneiki, Leccino, Picual, Coratina, Picholine
12	Blue berry (*Vaccinium angustifolium*)	Sharpblue (150), Emerald (200), Jewel (200), Windsor (225), Springhigh (225), Chaucer (400), Woodard (400), Brightwell (400), Climax (450), Tifblue (550), Powderblue (550)

research activities at various institutes, as a result of which cultivars like Sharbati peach, Titron and Kala Amritsari plum, Pathernakh, LeConte and Baghugosha pears became commercially popular in some parts of the subtropical India.

Peach

The breeding for lower-chill peach and nectarine varieties was initiated in the transitional chill zone of California at the beginning of last century. These programs released 'Babcock'. The Armstrong peach ('Flamingo', 'June Gold', 'Robin', and 'Springtime') and the nectarine ('Armking', 'Panamint', and 'Palomar') varieties were later released for lower chill zones. From the late 1930s to the 1960s, peach breeding programs began in the USA (Florida, California, Georgia, Louisiana, and Texas), Brazil (Pelotas and Campinas), India (Saharanpur, Uttar Pradesh) and South Africa. A large number of promising hybrids of peach and plum were introduced at Ludhiana from Gainesville, Florida (USA) and out of these introductions peaches like, 'Flordasun' (obsolete now, because of shorter shelf life), '16-33', 'Partap', 'FlordaPrince', 'EarliGrande' *etc.* are gaining popularity. The other important cultivars are also found promising are:-

Parbhat, an early ripening, white-fleshed peach hybrid was released from Saharanpur. Its fruit size is smaller and shelf-life shorter, but absence of acidity in the flesh suites Indian palate.

Shan-e-Punjab is most popular cultivar of peach because of larger size, yellow flesh and skin colour and better keepability. It is resulted a potential variety to grow under Ludhiana conditions gave yield 8.5 t/ha of better quality in terms of fruit weight (86 g/fruit) and 12.23^{o} brix planted at spacing 6 x 1.5 m accommodating 1111 trees per hectare (Kanwar and Singh, 2004).

Early Grande: This is an introduced variety, which performs very well under Punjab conditions. Tree is productive (95.0 kg/tree). Fruit is large (91.0 g) with red blush on the surface. Pulp yellow, firm (firmness of 10.9 lbs/inch2) and some red colouration next to the pit and semi-free from stone when fully ripe. Fruit ripens in the 1st week of May and possesses excellent shipping quality.

Flordaprince: This is an introduced variety which suits well to the Punjab conditions. Fruit is medium to large (76.0 g), round with little or no tip, red blush on yellow ground, pulp melting, yellow with some red colouration, semi cling having good fruit firmness (10.0 lbs/inch^2). It is a high yielding (100 kg/tree) variety. Fruit ripens in the last week of April.

Florda Prince Peach

Earli Grande

Babcock Peach

Pant Peach-1: This is a chance seedling selection from the population of cv. Sharbati under Pantnagar conditions. It ripens about one week prior to Sharbati but fruit quality is similar to Sharbati.

Pratap: This variety was selected for subtropical conditions in Punjab. Tree is good yielder (70 kg/tree). Fruit is medium to large (66.0 g), yellow with red blush and pulp is firm (fruit firmness of 15.0 lbs/inch2) and yellow with red colouration. Fruit matures one week earlier (3rd week of April) than that of Flordasun.

Pear

In pear, 'Patharnakh', is predominant cultivar, because of longer tree life (more than 50 years) unless killed by ganoderma, heavy bearing capacity and shelf-life (about 3 weeks, at ambient conditions). However, it has little local demand as immediately after harvest the fruit is of unattractive colour and hard texture. It has been found that the ripening treatments like, exposing to 0°C for 3 days or dip in 1500 ppm Ethephon, followed by ripening at 20°C results in development of aroma, attractive colour, softer flesh and increase in soluble solids. The semi-soft pears, 'LeConte' and 'Baghugosha' are erratic and poor croppers. Under Pantnagar conditions 10-year old 'Baghugosa' plantings gave good yield 4.0 t/ha, 4.7 t/ha, 7.2 t/ha, 5.4 t/ha and 9.8 t/ha in square (5 m x 5m), paired (within the pair 2.5 m and between the pairs 5.2m), hedgerow (within the hedgerow 2.5 m and between the hedgerows 5 m), cluster (within the cluster 2.5 m and between the clusters 5 m) and double hedgerow (within the double hedgerow 2.5 m and between the double hedgerows 5 m) systems respectively (Bist and Yadav. 2004). However, the selection of regular bearing and superior quality pears, like, Punjab Beauty, is likely to revive pear industry of northern plains. The salient features of other important cultivars/ varieties are as under:-

Baghugosha: Tree is upright and vigorous. Fruit greenish yellow, stem-end tapering. Pulp white or cream coloured, sweet somewhat gritty. It ripens in August and stands transportation well. Average yield is 100 kg per tree.

Patharnakh: Tree spreading and vigorous. Fruit medium, round, green with prominent dots. Pulp gritty, crisp and juicy. Keeping quality good. It stands transportation well. It ripens in the last week of July. Average yield is 150 kg per tree.

Punjab Beauty: Tree is upright, medium in vigour, precocious, regular in bearing and good yielder (100 kg/tree). Fruit is medium to large (125.5 g) and firm (10.4 lbs/inch2 firmness). The ground colour of the fruit is yellow with red blush, pulp whitish, and good in taste (total soluble solids 14.0° B, acid content 0.30 per cent). Fruit becomes less gritty when mature and mellowed when ripe. It matures in the 3rd week of July.

Plum

Plum is also one of the potential fruit crop suitable to grow under subtropical conditions. In recent past a number of successful results revealed that under Amritsar (Punjab) conditions the selection named as 'Alu Bokhara Peshawari' showed its promise by producing large sized fruits (23.50 g) of attractive red colour having high TSS: acid ratio (11.08) coupled with some special merits like free stone, pink coloured flesh with uniform sweetness (TSS 13.3° brix) and thus is absolutely free from sourness near the pit, the serious most drawback of Indian plums. The maturity period is also quite earlier (2nd week of June) as compared to those of temperate plums (Bal, 2004). The other important varieties/cultivars are as under:-

Kala Amritsari: This is the most popular cultivar, particularly in Punjab. Fruit is medium (13 g), round and dark brown at maturity. Pulp is yellow, moderately juicy and excellent for jam making. This is self-fruitful but yield improves if pollinated with Titron. It ripens in middle of May and average yield is 40-45 kg per tree.

Pant Plum-1: This is a selection from the seedling population raised at the Department of Horticulture, GBPUA and T, Pantnagar (Uttaranchal). It is dwarf and bears yellow and sub-acidic fruits. It can be a dwarfing rootstock for plum cultivars.

Satluj Purple: Tree is self-incompatible and must be inter-planted with 'Kala Amritsari' as pollinizder. Average yield of fruits is 30 kg/tree. Fruit is quite big (28.0 g), crimson, pulp thick, excellent in shipping quality and ripens in the 2nd week of May.

Apple

Winter chilling requirements for most varieties are 500 to 1000 hr. below 45° F (7° C). There are relatively few varieties adapted to mild winter climates (chill hours below 500). To ensure successful apple production in mild winter zone, select from the following varieties that need less than 300 hr. of chilling: 'Beverly Hills', 'Gordon', 'Tropical Beauty', 'Anna', 'Dorsett Golden', and 'Ein Sheimer'. Under mid-hill conditions of Solan, 'Tropic Beauty', 'Parkin's Beauty', 'Red Baron', and 'Aziza' are found promising (Sharma, *et al.*, 2004)

Dorsett Golden Apple

Anna Apple

Tropic Sweet Apple

Almond

With the discovery of a peach x almonds hybrid in Nawanshahr (Punjab), and undue publicity to it, the almond improvement work was started at Ludhiana, under an ADHOC project of ICAR in early 70's. Some useful introductions – 'California Paper Shell', 'Pathich's Wonder'; 'Selections – 'J.K. 55' and Hybrids 'H-15', 'H-258' were identified, but with the termination of the project in 1978 the cultural practices for almond cultivation could not be standardized as result of which this promising nut could not establish in plains. The silent features of some other suitable cultivars are as under:-

California Papershell: It ripens in the last week of June. The thin-shelled nuts are bold, weighing about 2.1 g each and the average yield is 3.8 kg/tree.

Hybrid 15: The thin-shelled variety with bold nuts ripens in the first week of July. On an average, a nut weighs 1.7 g and the average yield is 3.7 kg/tree.

Pethick's Wonder: Thin shelled nuts of this variety are bigger than all other varieties. On an average, a nut weighs 2.5g, the average yield being 3.7 kg/tree.

JKS 55: This is an earliest ripening variety. Its nuts are harvested in the first week of June. The nuts are semi-hard and small and the average yield is 3.5 kg dry nuts/tree.

Future Scope

There are several success stories written that these could be cultivated under subtropical climate as in Venezuela, Israel, Vietnam, Australia and Mexico. Therefore, these can also be repeated in Indian conditions. A collaborative effort among researchers, extension workers and development department could lead to make a break through for their commercialization. This require a systematic approach includes, identifications of zones according to chilling accumulation and high summer temperature, selection of appropriate fruit crop, selection of appropriate cultivar/hybrid, awareness campaign by extension workers, and financial support by the financial institutions including public v/s private.

Recent attempts have led to the development of new cultivars of apple, peaches, and nectarines. These cultivars can be grown successfully under subtropical conditions. Locally adapted or developed cultivars/varieties of peach and some plum cultivars produce crops of high quality in many parts of Punjab, Uttarakhand and Uttar Pradesh.

REFERENCES

Bal, J.S. 2004. Genetic resources of plum under subtropical conditions of Punjab, India. *Acta Hort*.m 662: 147-150.

Bist, L.D. and Yadav, Ashish 2004. Long term studies on planting systems-cum-density of planting in low-chill soft pear under subtropical conditions. *Acta Hort.*, 662: 217-220.

Dennis F.G. Jr. 1997. Temperate zone fruits in out-of the way places- what does the future holds. *Acta Hort.*, 441: 23-27.

Dhatt, A.S. 2004. Prospectus of temperate fruit cultivation in Indi-Gangetic plains. *Acta Hort*. 662: 107-109.

Kanwar, J.S and Singh Harminder 2004. Scope of high density plantings of peach in the subtropics of India. *Acta Hort.*, 662: 221-224.

Mohammed Shafat 2004. Present status and future scope of growing temperate fruits in Rajasthan. *Acta Hort.*, 662: 101-105.

Sharma, G., Chua, G.D. and Sharma, O.C. 2004. Studies on evaluation and variability parameters in low chilling apple (*Malus x domestica* Borkh.). *Acta Hort.*, 662: 157-162.

Sharpe, 1969. Subtropical peaches and nectarines. *Proc. Fla. State Hort. Soc.* 82: 302-306.

Weinberger, J.H. 1956. Prolonged dormancy trouble in peaches in the southeast in relation to winter temperature. *Proc. Amer. Soc.*, 67: 107-120.

Breeding of Perennial Horticultural Crops *Pages* **413–428**
Editors: S.K. Singh, V.B. Patel, A.K.Goswami, Jai Prakash and Chavlesh Kumar
Published by: **BIOTECH BOOKS, NEW DELHI**

***Chapter* 31**

Rootstock Improvement

V.B. Patel, Chavlesh Kumar, M.K. Verma and S.K. Singh

Division of Fruits and Horticultural Technology,
ICAR-Indian Institute of Agricultural Research, Pusa, New Delhi – 110 012

INTRODUCTION

Rootstock is very important plant component of a composite grafted plant that give rise to root system. It is also referred to as understock as well as stock, a frequently used synonymous of rootstock. Rootstock bears roots that provide support to the whole plants as it anchor plants in the soil. It acquire water and nutrient from the soil and transport to all other organs of the plants for their growth and development. Roots vary markedly in its structure and functions. It also serves as storage organs and act as principal element of co-ordinating plant response to a range of biotic and abiotic stressors such as soil pests and diseases, water and nutrient deficiency or excess, temperature variations and other soil limiting conditions of acidity, alkalinity, heavy metals *etc.* Rootstocks are now indispensable in fruit industry and also for few vegetables because of their capability of providing opportunity to grow a cultivar to diverse set of production environment. Rootstocks support the scion for several or many traits for improving establishment, enhancing production, productivity and quality of produce by means of imparting tolerance to pest and diseases, adverse soil conditions, improving nutrient and water uptake, as well as other limiting growth conditions. The most striking advantage of use of rootstocks is its ability to modify vigour of the plants and thereby enabling the orchardist to plants the tree at closer spacing and reducing the cost of cultural operations.

Rootstock Diversity and Domestication

Although grafting has been a practice for propagation of elite fruit crop selections for at least 2000 years but during the initial years of grafting, the rootstocks were selected on the basis of its availability as well as compatibility with the scion. As a result, the orchardists had to manage undesirable traits such as excessive vigour, insects-pests and diseases *etc.* through pruning, fertilization, and pesticide application which enhanced the cost of production. The realisation of effect of selected root systems could be beneficial for increasing productivity of fruit crop necessities selection and domestication of rootstock. The certain rootstocks have been identified for their ability to dwarf trees and increase the fruit/wood ratio produced by the same trees. Increased understanding of dwarfing rootstocks in orchard production systems has become the foundation of a productive orchard, and that as the interface between the scion and the soil, and by providing vital elements such as anchorage, water, nutrients, and disease protection, they ultimately affect the sustainability of the orchard.

As the science progressed, rootstock had been chosen for the traits having enhanced productivity and disease resistance. Still today, the knowledge of rootstocks diversity for most of the crop, the geographic origins or current distribution of cultivated rootstocks, or their domestication is relatively scarce. Similarly phylogenetic relationship studies are also limited compared to the scion varieties used for commercial production. It has been evident from the literature domestication of rootstock follow the following trends as referred by Warschefsky *et al.* (2016):

1. Rootstock species are often closely related to but genetically distinct from the scion species they support;
2. Multiple species and their hybrid derivatives are often used to generate rootstocks for a crop, although
3. Despite these uses, relatively few rootstock genotypes are employed for a given crop
4. Rootstock selection is a function of both the scion genotype with which it is grafted and the environment in which the grafted plant will be grown
5. Rootstocks are selected are for their own traits and traits imparted to the scion.

The typical example of rootstock domestication linked with management of Phylloxera disease, which were introduced to Europe in mid 1800s devasted the grape industry dependent on *V. vinifera.* Use of north American grape species as rootstock namely *V. labrusca, V.riperia, V. rupestris* and *V. berlandieri* of which *V.riperia, V. rupestris* are easy to root while *V. berlandieri* provides tolerance to chalky soil. Similarly for apple, different rootstock are used to combat a varieties of biotic and abiotic stresses.

The study of mechanism shaping phenotypic variation gives the clue of domestication and improvement of perennial horticultural crops. It has been estimated that the perennial plants comprise 40 per cent of domesticated species, 60 per cent of seed plants, and 80 per cent of tropical plant species, and provide the key features that influence evolutionary processes in unique ways relative to annual systems. Latest development particularly high-throughput genomics and phenomics have shaped the understanding of perennial horticultural crops diversity and evolution. Rootstocks are largely derived from perennials, outcrossing wild relatives and exhibits some of the characteristic features of woody plant domestication that includes high level of heterozygosity and greater extent of clonal propagation

Ideal Rootstock

Rootstock for perennial horticultural crops are selected for its intended effect on the scion varieties. In general and ideal rootstock should possess following root and shoot characteristics:

1. Easy to propagate/multiply by clonal or by seeds
2. Easy to bud or graft and should be graft compatible
3. Should have wider adaptability
4. Should be precocious
5. Dwarfing/tree size control
6. Strong root system for anchorage to scion without any staking
7. Should be resistant to biotic and abiotic stresses
8. Induce higher yield and quality

Problems in Rootstock Breeding

The breeding of rootstocks is considered more distinct and difficult compared to annual and perennial horticultural crops due to typical floral biology, highly heterozygosity, sterility, parthenocarpy, apomixes, polyembryony, poor fruit set, excessive fruit drop, out crossing, *etc.* It also has long juvenile phase and require large space for evaluation of progenies. The genetics and inheritance pattern of the quantitative traits are not fully understood though the significant progress has been made. Further rootstock breeding takes more time to breed and select the desirable rootstock compared to the scion breeding.

Improvement of Rootstocks: Latest Trends

Much like in breeding of perennial horticultural crops, rootstock breeding often target wild and semidomesticated species, feral individuals, and landraces that are well adapted to specific environments and known to be biotic and abiotic stress resistant. The breeding of new productive, diseaseresistant, dwarfing rootstocks using modern selection techniques is very important objective in most of the breeding programmes with aim of increasing orchard productivity and profitability. Breeding of rootstocks is a lengthy, resourceintensive activity and require multistage long drawn breeding process which starts with a large progenies from planned crosses that reduced down to a handful of "elite selections" by multiple rounds of inoculation with diseases, the application of molecular markers for important orchard traits as well as multiple replicated evaluations for orchard traits such as yield, yield efficiency, tree vigor, suckering, nutrient uptake efficiency, and response to any other unique stress events. Now a days breeding program further requires genomic and bioinformatic tools for markerassisted breeding of rootstocks while leveraging discoveries in whole tree phenomics and metabolomics. Elite selections are further subjected to extreme soil temperatures, replant soils, multiple strains of diseases of concerned crops, viruses, and graft union strength with multiple scion varieties.

Different methods adopted for improvement of rootstocks are Introduction, Selection, Hybridization, Mutation, Polyploidy and use of precise breeding and biotechnology tools like marker assisted breeding, Genetic transformation, Gene Editing *etc.* Rootstock breeding clearly targets specific traits. Some of the success and attempts made by the breeders are dealt hereunder;

Modulation of Scion Behaviour: Tree Architecture

Rootstocks generally affect the canopy vigour of the tree by shortening internodal length, altering the angle of the branches, time and rates of active growth. Dwarfing rootstocks are preferred characters having potential of suppressing scion growth. Rootstocks can induce or modify a scion variety for their growth particularly trunk cross-sectional area (TCSA), plant height, canopy volume, structure of branching, suckers, fruit size, and production efficiency (Layne,1994). It also affects distribution of the dry matter and starch contents in trees (Caruso *et al.*, 1997). The reduction in canopy vigour and tree size enable tree to plant at closer spacing for high density planting system leading to greater productivity per unit area. A lot of efforts have been made to breed and select the genotypes those are either dwarf or impart dwarfness to the scion. Extensive work has been conducted in temperate fruit through selection and other breeding methods. The list of dwarfing rootstocks is given hereunder in few selected crops (Table 31.1)

Table 31.1: Dwarfing Rootstocks Developed in Selected Fruit Crops

Name of Fruit Crops	*Important Rootstocks (Dwarfing)*
Apple	B.9 (M.8 X 'Red Standard' (Krasniz Standard), B.10 (Malling 27 X Robusta 5), G.16(Ottawa 3 X *Malus floribunda),* G.41 (Malling 27 X Robusta 5), G.65 (Malling 27 X 'Beauty' crab apple), G.214, G.935 (Ottawa 3 X Robusta 5), JM -7, M.9(Sel from Juane De Metz), M.27(M.13 x M.9), Mark (M9 x unknown), Ottawa 3(Robin crab x M9), P.2(M.9 X Common Antonovka), P.22(M-9 x Common Antonovka), Supporter 1(M9 X *Malus baccata*), Supporter 4(M9 X M4), V.3 (Kerr' crabapples x M.9)
Apricot	Hybrid P -2308, Prunus beyesiyi
Ber	*Zizyphus nummularia*
Cherry	Gisela 10, colt, Charger, Rubira
Citrus	Flying dragon, Trifoliate orange and its hybrids 059 hybrid(Sunki mandarin (*C. sunki*) x [Rangpur lime (*C. limonia*) x Troyer (*P. trifoliata*)]; F and A 418, #23, #24 (Troyer citrange X Cleopatra mandarin)
Guava	Pusa Srijan, *P. pumilum, P. friedrischathalianum*
Mango	Kurukkan, Olour, Vellaikolumban (for Alphonso only), Rumani (for Dashehari) Anupam (As interstock) Khom(Taiwan), Creeping Latra,Totapuri Red small, 13-1, Saigon 119 (Indonesia), Piva(South Africa)

Name of Fruit Crops	Important Rootstocks (Dwarfing)
Peach	K146-43(Controller 5, *Prunus salicina* Lindl. × *Prunus persica* L.), Reuter's Red Leaf, Prunus beyesiyi, Krymsk 1(Chinese Bush Cherry x Myrobalan Plum), Siberian C
Pear	Oregon-211, 249, Quince C
Plum	Pixie or Pixy, St. Julien2,3, Plumina, Jaspi

Modulation of Biotic Stress Tolerance

Development of rootstocks resistant to various biotic stresses such as insect-pests, diseases are most important breeding objective globally. A large number of pests and diseases are affecting the horticultural crops. The list of major insect pests of fruit crops targetted for rootstock development has been given in the Table 31.2. Management of these biotic stresses through chemical control are considered to cause environmental hazards, pesticides residue in the produce as well as development of resistance of these chemicals among causal organisms. Therefore, use of breeding tools to develop biotic stress resistance rootstocks becomes very important. At the same time, identification and utilisation of sources of resistance are basic requirement for breeding disease resistance. Several sources of resistance to major diseases have been identified in different fruit crops and are given in Table 31.3. Considerable progress has been made in the identification of sources of resistance to various biotic stresses including pests and diseases affecting perennial horticultural crops. The various breeding strategies for creating vertical and or horizontal resistance in perennial horticultural crops have been used. Several rootstocks have been identified and released for commercial use.

Table 31.2: Economically Important Pests Targeted for Rootstock Development and their Resistance Source

Crop	Pests	Resistance Source
Almond	Nematode	Almen 1; Almen 88, Almen 20, Hensen 2168, Hensen 536, Hyb. GF 557
Apple	Wooly apple aphids	Northern Spy(Er1, LG7), *M. baccata, M. sieboldii(Er3), Robusta5(Er2, LG17)*
Apricot	Nematode	Nemaguard, Rancho, Resistant,S-37, Okinawa, Nemared, Marianna GF 8-1,Marianna 2624
Cherry	Nematode	M×M clones, *P. cerasus*
Citrus	Nematode	*P. trifoliata*
Grape	Phylloxera	*V. rupestris, V. riparia, V. cinerea, M. rotundifolia, V. champinii, V. berlandieri*
	Nematode *Xiphinema index* *Meloidogyne* spp.	*V. rufosomentosa, M. rotundifolia, V. bchampinii, V. longii*
Peach	Nematode	Nemared, Nemaguard, Flordaguard, Halford, Hyb.GF 557, Okinawa, Shalil, Guardian, MP-29
Plum	Nematode	Marianna 2624, Marianna GF 8-1, Nemaguard, Hyb. GF. 557

Table 31.3: Economically Important Diseases Targeted for Rootstock Development and their Common Resistance Source

Crop	Diseases	Common Resistance Source
Apple	Fire blight	*M. robusta* (Robusta 5; QTLs on LG3 and LG7; peoxidase gene: MxdPrx8))
	Apple Scab	*M. floribunda* (vf resistant gene)
	powdery mildew	*M. robusta*
Citrus	Citrus Tristeza Virus	*P. trifoliata*, partial resistance to CTV of some pummelos and kumquats and tolerance of mandarins, lemons, and several rootstocks such as Cleopatra mandarin, *C. amblycarpa*, Rangpur lime, Rough lemon, and Volkamer Lemon
	Citrus Canker	*C. junos* Siebold ex Tanaka and some mandarins (Satsuma, Dancy)
	Phytophthora	*P. trifoliata* and its hybrids, *swingle citrumello, Bento cintrange* (*hybrid*), Microcitrus, and *C. amblycarpa,* Certain pummelos and mandarins, sour orange, Volkamer lemon
Grape	Downey mildews	*V. Muscadinia, V. rotendifobia, V. amorensis*
	Powdery mildews	*V. labrusca, V. rupestris, V. riparia, V. cinerea*
	Bacterial diseases	*V. amurensis, V. Labrusca, V. caribea, V. simpsonii, M. rotundifolia, V. coriacea*
Guava	Wilt	*Psidium friedricsthalianum*
Pear	Fire blight	*Pyrus communis*

Woolly apple aphid (*Eriosoma lanigerum,* Hausmann) is one of the most important pests of the apple throughout the world. Both root and shoot are affected by this pest, however, root is damaged more compared to shoot portion. Systematic rootstock breeding for wooly apple aphid resistance was initiated jointly by East Malling Research Station, UK and John Inn Research Institute at Merton using Northern Spy as resistant source. The collaboration yield a series of rootstocks MM 101 to MM115 of which MM 104 (M2 x Northern Spy), MM 106 (M1 x Northern Spy), and MM 111 (Merton 793 x Northern Spy) are still popular. Further immunity to WAA was identified in *M. robusta* (Robusta 5) which has *Er*2 resistant gene. In 1968, Cornell University, Geneva campus initiated apple Rootstock Breeding program later merged with USDA in 1998, which are mainly focused on biotic stresses such as woolly apple aphid, fire blight and other diseases as well as tree size control. They used different Merton series, *M. domestica* and *M. robusta* and released rootstocks having multiple level of tolerance and varying degree of tree size control (Table 31.4).

In grape, Phylloxera, *Daktulosphaira vitifoliae* (Fitch) has been a most problematic pest in grape growing in the Europian and American countries. Introduction of the American grape species namely, *V. riparia, V. rupestris and V. berliendieri* as rootstocks could solve the problem to some extent. Thereafter the breeding programme focussed to develop Phylloxera resistant rootstock and as a result a number of resistant rootstocks have been developed such as SO4, 5BB and 5C using *Vitis berlandieri* Planch. X *V. riparia* Michx. resistant to phylloxera. Some of the prominent rootstock developed through interspecific hybridization have been given having the Phylloxera resistance as well as other important rootstock characters and are used

Table 31.4: Apple Rootstock having Multiple Biotic Stress Tolerance Developed by USDA and Cornell University, Geneva

Rootstocks	*Percent Tree Size*	*Equivalent Malling Size*	*Fire Blight Resistant*	*Replant Tolerance*	*Woolly Resistant*
G.11 (M26 X Robusta 5	< 30	M.9	Yes	Partial*	NO
G.41 (Malling 27' X 'Robusta 5)	30	M.9	Yes	Yes	Yes
G 214 (Robusta 5 x Ottawa 3)	35	M.9 EMLA	Yes	Yes	Yes
G.16 (Ottawa 3 X Malus floribunda),	35	M.9 EMLA	Yes	Partial	No
G.222 (Robusta 5 x Malling 27)	40	M.9 EMLA	Yes	Partial	Yes
G.935 (Ottawa 3x Robusta 5)	40	M.9 EMLA	Yes	Yes	No
G.969 (Robusta 5 x Ottawa 3)	40	M.9 EMLA	Yes	Yes	Yes
G.210 (Ottawa 3 x Robusta 5)	40	M.9 EMLA	Yes	Yes	Yes
G.202 (Malling 27' X 'Robusta 5)	45	M.26	Yes	No	Yes
G.890 (Robusta 5 x Ottawa 3)	45	M.26	Yes	Yes	Yes
G.30 (Robusta 5 x M9)	45	M.26	Yes	Yes	No

Source: Tom Avhill (2016),http://treefruit.wsu.edu/article/geneva-rootstock-performance-2016-rootstock-trial-update/

worldwide for grape growing (Table 31.5). Nematode is also a problem in grape growing. A number of rootstocks tolerant to resistant against different nematode have been developed *e.g.* O39-16, Schwarzmann, Freedom, Boerner, 1616C, Ramsey, 101-14 Mgt, Fercal *etc.*

Table 31.5: Development of Rootstocks through Interspecific Hybridization

Sl.No.	*Rootstock*	*Parentage*	*Salient features*
1.	SO4	*V. berlandieri* x *V. riparia*	Vigorous, good for neutral or mild alkaline soils. Good nematode and phylloxera tolerance.
2.	Kober 5 BB	*V. berlandieri* x *V. riparia*	Vigorous, highly resistant to phylloxera and tolerance to nematode.
3.	99 Richter	*V. berlandieri x V. rupestris*	Drought tolerant, can be used in in acid soils and does not tolerate salt, High resistance to phylloxera and root knot nematodes and moderate resistance to dagger and lesion nematodes.
4.	140 Ruggeri	*V. berlandieri x V. rupestris*	drought tolerant, well adapted to acid soils, and resistant to salinity. Highly resistant to phylloxera. Moderate resistance to root knot nematodes.
5.	110 Richter	*V. berlandieri* x *V. rupestris*	Vigorous, tends to delay maturity, drought tolerant and tolerant of up to 17 per cent lime in the soils.
6.	1103 Paulsen	*V. berlandieri* x *V* *V. rupestris*	Vigorous (similar to 110 R), adaptable to claylime soils and reported to be somewhat salt tolerant.
7.	41 B	*V. berlandieri* x *V. vinifera*	Moderate vigorous, early fruit maturity. Possesses exceptional resistance to highlime soils.

Sl.No.	Rootstock	Parentage	Salient features
8.	Dogridge	*V. berlandieri* x *V. vinifera*	Extremely vigorous rootstock with good resistance to nematodes, moderate tolerance to phylloxera and high-lime. High level of suckering is a commercial drawback.
9.	Fercal	*V. berlandieri* and *V. vinifera*	Suitable for high lime containing European soils.
10.	Riparia Gloire de Montpelier	*V. berlandieri* and *V. vinifera*	Not appropriate for calcareous soils and dry sites. Very high resistance to phylloxera. Advances scion maturity.
11.	Rupestris St. George	*V. berlandieri* and *V. vinifera*	Drought resistance, High resistance to phylloxera. Very sensitive to rootknot nematode, sensitive to dagger, and moderately resistant to root lesion nematodes. Very susceptible to the root-rot fungi and Fan leaf degeneration.

Breeding for Abiotic Stress Tolerance

Perennial horticultural crops are grown in a wide diversity of soil and climatic conditions; therefore, trees are subjected to important abiotic stresses that limit the production such as acidic, saline and alkaline soils, flooding and drought conditions; extremely low and high temperatures. Use of certain rootstocks avoid the problem of abiotic stresses to a great extent. Further, breeding of new rootstocks that are tolerant to abiotic stress conditions so as to improve the existing rootstock. Some of the common resistant wild species are given below in Table 31.6.

Table 31.6: Major Traits and Resistance Sources for Abiotic Stress Tolerance

Crop	Traits	Resistance Source
Grape	Winter hardiness	V. amurensis, V. riparia
	Lime chlorosis	V. vinifera, V. berlandieri
	Salinity	V. berlandieri
	Drought/Hot climate	V. arizonica, V. monticola, V. aestivalis, V. lincecumii
Citrus	Iron Chlorosis	Rough lemon (*C. jambhiri* Lush), *C. macrophylla* Wester, Volkamer lemon (*Citrus limonia* Osbeck), and *C. amblycarpa* (Hassk.) Ochse
	Salinity	Rangpur lime (*C. limonia* Osb.) and Cleopatra mandarin both are Cl^- excluder and *Poncirus trifoliata* is Na^+ excluder, Attani-1 (*C. rugulosa*), RLC-6 and sour orange
	Cold	Satsuma mandarins, the Kumquats and *P. trifoliata*

Drought Tolerance

Limiting water condition or drought stress is one of the ominous abiotic factors limiting the productivity of perennial horticultural crops. Three mechanism are operative in the plant system for drought resistance namely drought escape, drought avoidance, and drought tolerance. Plant respond to drought tolerance through enhancing osmotic adjustments, cell membrane stability, epicuticular wax, partitioning and stem reserve mobilization, manipulation and stability of flowering processes *etc.* Rootstock selection and improvement has been done in different fruit crops. A number of drought tolerance species/hybrids have been identified in different crops. These include Rangpur lime (C. *limonia*), Volkamer lemon (C. *volkameriana* V. Ten. and Pasq.), and rough lemon (C. *jambhiri* Lush.), *C. reshni, C. aurantium, C. limonia, and P. trifoliata* of Citrus are drought resistant and increase the fruit yield of the grafted scion.

Some highly, moderately drought tolerant species/hybrids are listed in Table 31.7.

Table 31.7: Drought Resistant Rootstock in Fruit Crops

Crop	*Rootstock*
Annona	*A. reticulata* and *A. montana*
Apple	M 9, M 26, M 111
Cherry	Mazzard, Mahaleb and M×M series
Peach	Almond, Peach × Almond and Peach × *P.davidiana*hybrids, GF 677
Citrus	Flying Sunki, Carrizo Citrange, Swingle Citrumelo Rangpur lime (C. *limonia*), Volkamer lemon (*C. volkameriana* V. Ten. and Pasq.), and rough lemon (C. *jambhiri* Lush.), *C. reshni, C. aurantium, C. limonia, and P. trifoliata*
Grape	V. berlandieri (110R, 140Ru (420A, 5BB), V. champinii (Dog Ridge, Ramsey (Salt Creek), Freedom, Harmony) V. rupestris (St. George, 1103 Paulsen)
Mango	Kurukkan (India) Kaew and Ta-Lub-Nak (Thailand), BAU 6, 7 and 8 (Bangladesh), Tsar-Swan(Taiwan), Criollo de Cholucanas (Peru)

Salinity Tolerance

Salinity is one of the important edaphic factors mostly prevalent in arid and semi-arid tracts of the world. About 6 per cent of cultivated land has salinity problems worldwide, where NaCl is considered to be the most abundant and soluble salt in the soil solution. However, ions causing problems in such conditions are mostly chloride, sulphate, Bicarbonate, sodium, calcium, magnesium, nitrate, potassium and to limited extent boron. Different fruit crops tend to manifest a tolerant to moderate to high sensitivity to salinity, however majority of fruits like citrus, grape, apple, pear, peach, cherry, apricot, strawberry, plum, *etc.* are susceptible to moderate susceptible to soil salinity. Examples of selected tolerant/resistant rootstocks and their sources are given in Table 31.8.

Table 31.8: Salinity Tolerant Rootstock and Species in Fruit Crops

Crop	*Tolerant/Resistant Rootstock*	*Species*
Citrus	Attani1 and Attani 2, Swingle Citrumelo, Troyer Citrange, Rusk Citrange,US-897, X 639, Carrizo Citrange	Rangpur lime (*C. limonia* Osb.) and Cleopatra mandarin (C resini) both are Cl^- excluder; *Poncirus trifoliata* is Na^+ excluder, Attani-1 (*C. rugulosa*), RLC-6 and sour orange
Grape	Salt Creek, Freedom, 140Ru 1103P, SO4, 41B	*V. rupestris, V. cinerea, V. champini* and *V. berlandieri*
Mango	Gomera-1, 13-1, 13-3, Sabre, Olour, Bappakai, Kurukan, Kohuamba (Srilanka)	*Mangifera Kasturi,* M. *zeylanica*
Prunus	HS-314 and GF-677	
Avocado	Borchard, Duke-7 and G755C	

The rootstocks are now selected and suggested for different situations in many fruit crops. For examples, grape rootstock has been suggested for different abiotic stress has been given in the Table 31.9.

Table 31.9: Suggested Grape Rootstock for different Abiotic Stress Situations

Situation/Problem	*Rootstock*
Soil EC more than 2 m mhos/cm and water EC more than 1 m mhos/cm	Ramsey, Dog ridge, 140 RU, 99 R, 110 R
Soil ESP more than 15 per cent and/or water SAR more than 8	140 RU, 1613, Ramsey, Dogridge
Free calcium content of soil is more than 12 per cent	140 RU, SO4, 420 A
Chloride content of water is more than 4 meq/l	Ramsey, Dog ridge B, 140 RU. Teleki 5-C
Poor vigor of the variety without any soil/water problem	Dog ridge, St. George, SO4, 140 RU
For increased nitrogen and potassium uptake	Dog ridge, St. George, 34 EM, Ramsey

Source: National Research Centre for Grapes, Pune.

Mutation Breeding

Mutations seem to have been an important means of variation in fruit crops and have played an important role in the evolutionary history as spontaneous mutations are frequently observed in cultivar and rootstocks of perennial horticultural crops. The mutations reported to provide important characteristics for rootstock development such as hardiness, resistance to pest and disease *etc.* Some of the achievements are development of mutants of the grapevine variety Gravesac (5306, 5327, 5338 and 5339), issued from a 30 Gy dosis of Cs^{137} irradiation (Silva *et al.*, 2010).

Somatic Hybridization

Somatic hybridization has been exploited in perennial horticultural crops by fusion of somatic protoplast derived from two different cultivar, species or genera. The technique include isolation of protoplasts from desired genotypes by mechanical or enzymatic methods, their fusion followed by culturing the hybrid cells, and finally regeneration of hybrid plant. In this process, cell wall is degraded to obtain protoplasts. Enzymes, such as pectinase, macerozyme, cellulase, *etc.* having ability to hydrolyze the plant cell wall are used and mostly leaf protoplast are used for hybridization. Electrofusion has been found better compared to Polyethylene Glycol fusion(PEG) fusion as it is not toxic to the cells being fused. somatic hybridization has been very promising for few rootstock breeding (citrus) for combining genotypes having different tolerance to abiotic stress. For development of hybrids allotetraploid somatic hybrids are produced that are combined with diploid rootstocks having complementary traits and the process is called cybridization. It induces phenotypical variations in highly heterozygous species or genotypes, without breaking their complex multilocus structure. It helps in manipulating nucleo-cytoplasmic male sterility, to transfer tolerance traits with cytoplasmic determinisms. The somatic hybridization at tetraploid level enables mixing of the genes from three to four proven rootstocks such as sour orange + Carrizo, Cleopatra + trifoliate orange and sour orange + Palestine. Tetrazygs name was given by Grosser to identify tetraploid progeny from such crosses.

Genetic Transformation

Direct gene transfer is very promising tools for improvement of perennial crops scion and rootstock. The incorporating of the desirable genes into plant cell on the basis of molecular, biological and genetic engineering techniques offers great potential for the genetic improvement. Much progress has been made in the last few years in the genetic manipulation of perennial horticultural crops. Latest work related to rootstock development has been given below:

Transgraft

In the woody fruit tree species, grafting is generally practiced to take the advantages of both the scion as well as rootstock. In the grafting, scion and rootstock unites and formed a composite plant. In the grafting, rootstock influences the various phenotypes of the scion genotypes *i.e* imparting dwarfing vigour, influencing the fruit set *etc.* but the scion and rootstock do not mix genetically and retain their genetic integrity. To explore theses phenomenon, the concept of transgraft techniques came into existence. In the transgraft, the rootstock is genetically modified for desirable trait and the scion the genotype is non genetically modified. The non-genetically modified scion genotypes acquire all the benefits from their genetically modified rootstock while the product of the scion do not contain any transgenes (Schaart and Visser, 2009; Haroldsen *et al.*, 2012; Lemgo *et al.*, 2013; Limera *et al.*, 2017). Thus, transgraft can be successful used in the modern orchard cultivation for overcoming the several horticultural problems. Few of the success of the transgraft is enlisted in the table below in the Table 31.10.

Table 31.10: Successful Transgrafting in Perennial Horticultural Crops

Plant Species	*Name of Gene*	*Source*	*Trait*	*References*
Apple (*Malus domestica*)	rolB	Apple (*Malus domestica*)	Control of scion vigor and reduce plant height	Smolka *et al.*, 2010
Grapevine (*Vitis vinifera* L.)	Shiva-1 lytic peptide	Grapevine (*Vitis vinifera*)	To control Pierce's disease (PD) (*Xylella fastidiosa*)	Dutt *et al.*, 2007
Sweet cherry (*Prunus avium*)	PNRSV	*Prunus* necrotic ringspot virus (PNRSV)	Resistance to PNRSV in non-transgenic scions	Zhao and Song, 2014

Cisgenesis and Intragenesis in Rootstock Developments

Several bio-safety issues have been arises regarding the transgenic crops thus an alternative, cisgenesis and intragenesis developed as new tools for the genetic modification of the crop species (Espinoza *et al.*, 2013). The term cisgenesis defined by the Schouten *et al.*, 2006b that "a crop genotype that is genetically modified with one or more genes which is derived from their crossable crop individuals". The cisgenes to be incorporated should contain the promoter, introns and terminators in the normal sense orientation.

In case of intragenesis, the genetically modified plant species contain intragenes derived from the same species or from the crossable individuals. Intragenes are of hybrid origin that constituted by genetic elements from different genes and loci (Rommens *et al.*, 2007). The cisgenesis and intragenesis tools have potential to overcome the limitation of traditional breeding in crop species and avoid the genetic drag which generally happened in the traditional breeding system. Further, both tools have offer the way for the faster gene transfer and reduces the number of generation in crop breeding (Holme *et al.*, 2013). Cisgenesis and intragenesis explored for the improvement of perennial fruit crops, however it needs to be explore in the rootstock improvements. The list of cisgenesis and intragenesis in fruit crops are presented in the Table 31.11.

Table 31.11: Successful Examples of Cisgenesis and Trangenesis in Horticultural Crops

Fruit Crop	*Trait*	*Source*	*Name of Gene*	*References*
Apple (*Malus domestica*)	Resistance to Apple scab (*Venturia inaequalis*)	Apple (*Malus domestica*)	HcrVf2	Vanblaere *et al.*, 2014
	Resistance to Apple scab (*Venturia inaequalis*)	Apple (Malus floribunda 821)	Rvi6	Krens *et al.*, 2015
	Resistance to fire blight (*Erwinia amylovora*)	Apple cv Gala Galaxy	FB_MR5	Kost *et al.*, 2015
Grapevine (*Vitis vinifera* L.)	Resistance to Powdery mildew (*Erysiphe necator*)	Grapevine (*Vitis vinifera*)	VVTL-1	Dhekney *et al.*, 2011

RNA Interference

In the plant system the RNA Interference (RNAi) is recently discovered natural mechanism to regulate the gene expression and offer to the breeders for development of varieties/rootsicks with great precision and accuracy (Jagtap *et al.*, 2011). RNAi is an endogenous cellular process by which the undesirable or unwanted specific nucleic sequences are switch off naturally or can be regulate the gene expression before the process of translation (Baum *et al.*, 2007; De Alba *et al.*, 2013). The presence of double-stranded RNA molecules activate the suppression or inhibition of genes expression (Voinnet, 2008; Parent and Vaucheret, 2012). Thus, RNAi eventually explored for creating the possibility of knock down of gene expression or function (Limera *et al.*, 2017). RNAi employed in woody perennial fruit tree species to obtain the pathogen resistance. The success of RNAi in the fruit crop species are listed in the Table 31.12. The process of RNAi can be explored in the rootstock improvement to combat the various soil borne pathogen.

Table 31.12: Use of RNAi Techniques in Fruit Crops

Fruit Crop	*Trait required*	*Responsible gene*	*Source*	*References*
Grape fruit	Resistance to CTV	CTV	Citrus tristeza virus (CTV)	Febres *et al.*, 2008
Apple	Dwarfing	MdGA20-ox	Apple (*Malus domestica*)	Zhao *et al.*, 2016
	Powdery mildew resistance	MdMLO19	Apple (*Malus domestica*)	Pessina *et al.*, 2016
Pear	Early Flowering	MdTFL1	Apple	Freiman *et al.*, 2012

Table 31.13: Application of Clustered Regularly Interspaced Short Palindromic Repeats (CRISPR) and CRISPR-Associated Protein (Cas9) in Perennial Horticultural Crops

Plant Species	*Name of Gene*	*Source*	*Trait*	*Achievements*	*References*
Citrus sinensis Osbeck	EBE_{PthA4} of the of the CsLOB1 promoter	Wanjincheng orange (Citrus sinensis Osbeck)	Mutation in the EBE_{PthA4} effector to induce citrus canker resistance	High rate of resistance to citrus canker by mutate t EBE_{PthA4} effector	Peng *et al.*, 2017
Grapevine (*Vitis vinifera* L.)	L-idonate dehydro-genase gene (IdnDH)	Grapevine (Vitis vinifera L.)	Genome editing and targeted gene mutation	100 per cent mutation frequency in the transgenic cell mass (CM) as well as corresponding regenerated plants expressing sgRNA1/Cas9	Wang *et al.*, 2018 Ren *et al.*, 2016
Grapevine (Vitis vinifera L.) and apple (Malus domestica) protoplasts	Grape gene MLO-7 and the apple genes DIPM-1, 2, and 4	Grapevine (Vitis vinifera L.) and apple (Malus domestica)	Resistance to powdery mildew in grape and resistance to fire blight disease in apple	Efficient targeted mutagenesis in the protoplasts of both grape MLO-7 and the apple DIPM-1, 2, and 4	Malnoy *et al.*, 2016

Genome Editing

The genome editing techniques enables the researchers to integrates, delete or mutate the targeted genes in to the organism. The genome editing techniques not only create the novel genotypes while precise manipulation of the genetic information but also make the arena for the gene function studies. The various genome editing tools like zinc finger nuclease (ZFN) and tale nucleases (TALENs), clustered regularly interspaced short palindromic repeats (CRISPR) and CRISPR-Associated Protein (Cas9), and oligonucleotide-directed mutagenesis (ODM) used for the precise genome editing of the organism (Limera *et al.*, 2017). The genome editing tools are explored in the woody tree species. The genome editing technique have potential to develop the novel rootstock while precisely edit the gene of interest. However, the improvement in woody fruit tree species through the genome edited techniques is in nascent stage. The few of the examples of genome editing the woody perennial are listed in the Table 31.13.

Rootstock improvement of perennial fruit crop is challenging because so many desirable traits are required to be combined into a hybrid progeny rootstock. Therefore primary strategy of rootstock breeders has been to identify the donor parents having desired traits. A second strategy has been to combine these traits using either traditional breeding or biotechnological approaches.

REFERENCES

Baum, J. A., Bogaert, T., Clinton, W., Heck, G. R., Feldmann, P., Ilagan, O., *et al.* (2007). Control of coleopteran insect pests through RNA interference. *Nat. Biotechnol.,* 25: 1322–1326. doi: 10.1038/nbt1359

De Alba, A. E. M., Elvira-Matelot, E. and Vaucheret, H. (2013). Gene silencing in plants: a diversity of pathways. *Biochim. Biophys. Acta Gene Regul. Mech.,* 1829: 1300–1308. doi: 10.1016/j.bbagrm.2013.10.005

Dhekney, S. A., Li, Z. T. and Gray, D. J. (2011). Grapevines engineered to express cisgenic *Vitis vinifera* thaumatin-like protein exhibit fungal disease resistance *in vitro* cell. *Dev. Biol. Plant,* 47: 458–466. doi: 10.1007/s11627-011-9358-3

Dutt, M., Li, Z. T., Kelley, K. T., Dhekney, S. A., Van Aman, M., Tattersall, J., *et al.* (2007). Transgenic rootstock protein transmission in grapevines. *Acta Hortic.,* 738: 749–754. doi: 10.17660/ActaHortic.2007.738.99

Espinoza, C., R. Schlechter, D. Herrera, E. Torres, A. Serrano, C. Medina, and P. Arce-Johnson (2013). "Cisgenesis and intragenesis: new tools for improving crops." *Biological Research,* 46(4): 323-331.

Febres, V. J., Lee, R. F. and Moore, G. A. (2008). Transgenic resistance to Citrus tristeza virus in grapefruit. *Plant Cell Rep.,* 27: 93–104. doi: 10.1007/s00299-007-0445-1

Freiman, A., Shlizerman, L., Golobovitch, S., Yablovitz, Z., Korchinsky, R., Cohen, Y., *et al.* (2012). Development of a transgenic early flowering pear (Pyrus communis L.) genotype by RNAi silencing of PcTFL1-1 and PcTFL1-2. *Planta,* 235L 1239–1251. doi: 10.1007/s00425-011-1571-0

Haroldsen, V. M., Chi-Ham, C. L. and Bennett, A. B. (2012). Transgene mobilization and regulatory uncertainty for non-GE fruit products of transgenic rootstocks. *J. Biotechnol.*, 161: 349–353. doi: 10.1016/j.jbiotec. 2012.06.017

Holme, I.B., Wendt, T. and Holm, P.B. (2013). Intragenesis and cisgenesis as alternatives to transgenic crop development. *Plant Biotechnology Journal*, 11(4): 395-407.

Jagtap, U.B., Gurav, R.G. and Bapat, V.A. (2011). Role of RNA interference in plant improvement. *Naturwissenschaften*, 98(6): 473-492.

Jia, H. and Wang, N. (2014). Targeted genome editing of sweet orange using Cas9/sgRNA. *PLoS ONE*, 9: e93806. doi: 10.1371/journal.pone.00 93806

Kost, T. D., Gessler, C., Jänsch, M., Flachowsky, H., Patocchi, A. and Broggini, G. A. L. (2015). Development of the first cisgenic apple with increased resistance to fire blight. *PLoS ONE*, 10: e0143980. doi: 10.1371/journal.pone.01 43980

Krens, F. A., Schaart, J. G., Van der Burgh, A. M., Tinnenbroek-Capel, I. E., Groenwold, M. R., Kodde, L. P., *et al.* (2015). Cisgenic apple trees; development, characterization, and performance. *Front. Plant Sci.*, 6: 286. doi: 10.3389/fpls.2015.00286

Lemgo, G., Sabbadini, S., Pandolni, T. and Mezzetti, B. (2013). Biosafety considerations of RNAi-mediated virus resistance in fruit-tree cultivars and in rootstock. *Transgenic Res.*, 22: 1073–1088. doi: 10.1007/s11248-0139728-1.

Limera, C., Sabbadini, S., Sweet, J.B. and Mezzetti, B., 2017. New biotechnological tools for the genetic improvement of major woody fruit species. *Frontiers in Plant Science*, 8: 1418.

Malnoy, M., Viola, R., Jung, M.-H., Koo, O.-J., Kim, S., Kim, J.-S., *et al.* (2016). DNA-free genetically edited grapevine and apple protoplast using CRISPR/Cas9 ribonucleoproteins. *Front. Plant Sci.*, 7: 1904. doi: 10.3389/fpls.2016.01904

Nishitani, C., Hirai, S. K., Masato, W., Kazuma, O., Keishi, O., Toshiya, Y., *et al.* (2016). Ecient genome editing in apple using a CRISPR/Cas9 system. *Sci. Rep.*, 6: 31481. doi: 10.1038/srep31481

Parent, J.S. and Vaucheret, H. (2012). The origin and eect of small RNA signaling in plants. *Front. Plant Sci.*, 3: 179. doi: 10.3389/fpls.2012.00179.

Peng, A., Chen, S., Lei, T., Xu, L., He, Y., Wu, L., *et al.* (2017). Engineering canker resistant plants through CRISPR/Cas9-targeted editing of the susceptibility gene CsLOB1 promoter in citrus. *Plant Biotech. J.*, doi: 10.1111/pbi.12733 [Epub ahead of print].

Pessina, S., Angeli, D., Martens, S., Visser, R., Bai, Y., Salamini, F., *et al.* (2016). The knock-down of the expression of *MdMLO19* reduces susceptibility to powdery mildew (*Podosphaera leucotricha*) in apple (*Malus domestica*). *Plant Biotechnol. J.*, 14L 2033–2044. doi: 10.1111/pbi.12562

Ren, C., Liu, X.J., Zhang, Z., Wang, Y., Duan, W., Li, S.H. and Liang, Z.C. (2016) CRISPR/Cas9 mediated efficient targeted mutagenesis in Chardonnay (*Vitis vinifera* L.). *Sci. Rep.*, 6: 32289.

Rommens CM, Haring MA, Swords K, Davies HV and Belknap WR (2007). The intragenic approach as a new extension to traditional plant breeding. *Trends Plant Sci.*, 12: 397-403.

Schaart, J. G. and Visser, R. G. F. (2009). Novel Plant Breeding Techniques Consequences of New Genetic Modication-Based Techniques in Comparison to Conventional Plant Breeding. The Netherlands Commission on Genetic Modication. COGEM Research Report number 2009–02.

Schouten, H.J., Krens, F.A. and Jacobsen, E. (2006b). Do cisgenic plants warrant less stringent oversight? *Nat. Biotechnol.*, 24: 753.

Smolka, A., Li, X. Y., Heikelt, C., Welander, M. and Zhu, L. H. (2010). Effects of transgenic rootstocks on growth and development of non-transgenic scion cultivars in apple. *Transgenic Res.*, 19: 933. doi: 10.1007/s11248-010-9370-0.

Vanblaere, T., Flachowsky, H., Gessler, C. and Broggini, G. A. (2014). Molecular characterization of cisgenic lines of apple 'Gala' carrying the *Rvi6* scab resistance gene. *Plant Biotechnol. J.*, 12: 2–9. doi: 10.1111/pbi.12110.

Voinnet, O. (2008). Post-transcriptional RNA silencing in plant-microbe interactions: a touch of robustness and versatility. *Curr. Opin. Plant Biol.*, 11: 464–470. doi: 10.1016/j.pbi.2008.04.006

Warschefsky, E.J., Klein, L.L., Frank, M.H., Chitwood, D.H., Londo, J.P., von Wettberg, E.J. and Miller, A.J., 2016. Rootstocks: diversity, domestication, and impacts on shoot phenotypes. *Trends in Plant Science*, 21(5): 418-437.

Xianhang Wang, Mingxing Tu, Zhi Li, Yuejin Wang and Xiping Wang (2018) Current Progress and Future Prospects for the Clustered Regularly Interspaced Short Palindromic Repeats (CRISPR) Genome Editing Technology in Fruit Tree Breeding. *Critical Reviews in Plant Sciences*, 10: 1080/07352689.2018.1517457, (1-26).

Zhao, D. and Song, G. Q. (2014). Rootstock-to-scion transfer of transgenederived small interfering RNAs and their effect on virus resistance in nontransgenic sweet cherry. *Plant Biotechnol. J.*, 12: 1319–1328. doi: 10.1111/pbi. 12243

Zhao, K., Zhang, F., Yang, Y., Ma, Y., Liu, Y., Li, H. *et al.* (2016). Modification of plant height via RNAi suppression of MdGA20-ox gene expression in apple. *J. Am. Soc. Hortic. Sci.*, 141: 242–248.

Breeding of Perennial Horticultural Crops *Pages* **429–439**
Editors: S.K. Singh, V.B. Patel, A.K.Goswami, Jai Prakash and Chavlesh Kumar
Published by: **BIOTECH BOOKS, NEW DELHI**

Chapter **32**

Breeding Vegetables for Protected Cultivation

Harshawardhan Choudhary, T.K. Behera and A.D. Munshi

Division of Vegetable Science,
ICAR-Indian Institute of Agricultural Research, Pusa, New Delhi – 110 012

INTRODUCTION

Greenhouse vegetable production is an intensive agricultural production system which requires more labour and other inputs than open-field condition. The success of protected cultivation of vegetables relies on the production of higher yields with better fruit quality and ensuring the fresh product availability during off season which fetch higher prices in the market. The investment and production costs are substantially higher which necessitates use of high quality input to maximize the output as well as the profit. It enables cultivation of crops that can give maximum yield on specific days and exotic crops that can't be grown in normal open field cultivation (*e.g.* coloured capsicum, parthenocarpic cucumber, speciality melons *etc.*). Therefore, cultivation with engineered vegetable varieties especially for protected environment has to be utilized to enhance the yield and quality of different crops. The introduction of protected cultivation technology in India has a time lag in comparision to many countries. China started protected cultivation technology almost at the same time when India made its beginning. But now China, Japan and USA are the leading countries having large area under polyhouse for horticultural crops in the world and have also developed a number of tomato and parthenocarpic cucumber varieties especially suitable for polyhouse. Despite of all these efforts, India is still lagging behind. There appears ample potential and scope for increasing area under low cost polyhouse in periurban areas for production of high value vegetables during off season for taking advantage of high price of the available market in nearby cities.

Breeding Approaches for Polyhouse

A great range of temperature variation *i.e.,* 10-45°C used to exist in Indian greenhouses depending upon the winter to summer seasonal fluctuations. In each location, conditions vary enormously during the growing season and changes can occur very rapidly. Therefore, it is necessary to breed varieties having a wider adaptation to varying climatic conditions and wide range of temperature for cultivation under polyhouse conditions round the year. Breeders are considering 3 main aspects; poor light conditions, low temperatures in the early stage of the growing season, high temperatures in the later stage of the production. Genotypes are usually chosen after several evaluation trials in different types of greenhouses or other protected structures like polyhouse or net house. Efficiency of selection would be increased if the breeders could use simple methods for measuring varying responses of the plant to the different climatic conditions. Evaluation criteria must be easy to practice rapidly on a large number of genotypes, without destruction of the plant, if possible at an early stage. Vegetable improvement shows many characters which require consideration by the breeders in order to get varieties which in turn can give the highest returns for all the investments made in the greenhouse cultivation. The genes required for crop improvement are present in different lines, varieties, strains or populations of the crop species and their relatives which could be wild forms or wild relatives. Some useful genes, often critical genes, are contributed by the wild relatives of the crops like the genes for insect and disease resistance, tolerance to abiotic stresses like drought, salinity, cold *etc.*, and even for quality traits and yield. Various lines, strains, populations of a crop species and its related wild species constitute the germplasm of the crop. The genes useful for the improvement of existing varieties are needed to be explored and incorporated for their further improvement. Backcross method, whose objective is to improve one or two specific defects of a high yielding variety, which is well adapted to the area and has other desirable characters can be utilized for the purpose. The characters lacking in the existing variety are transferred into it from a donor parent without changing the genotype of the existing variety, which is well adapted to the area and has other desirable characters. In the backcross method, the hybrid and the progenies in the subsequent generations are repeatedly crossed to one of the parents of the $F_{1.}$ As a result, the genotype of the backcross progeny becomes increasingly similar to that of the parent to which the backcrosses are made. At the end of 6-8 backcrosses, the progeny would be almost identical with the parent used for backcrossing. Thus the end result of the backcross programme is a well adopted variety with one or two improved characters.

Populations of cross pollinated crops are highly heterozygous as well as heterogeneous. Their genetic makeup is such that they show variable inbreeding depression, which in some cases may be severe. Therefore, population improvement becomes a vital step to overcome the problem. In population improvement, mass selection or its modification are used to increase the frequency of desirable alleles, thus improving the characteristics of populations.

The recurrent selection schemes were devised in relation to heterosis breeding. This idea is to ensure the isolation of superior inbreds from the populations subjected to recurrent selection for their ultimate utilization in the production of hybrid and synthetic varieties. This is a population improvement method followed in cross pollinated vegetable crops with objective to concentrate favourable genes distributed among individuals in a best population. This method is effective in improving yield and other quantitative characters by the accumulation of additive genetic variance, and is highly efficient for creating superior germplasm. When a population is attained off its peak potential, recurrent selection may be useful in breaking out new types. The source population for recurrent selection may be an open pollinated variety, F_1 hybrid, synthetic variety, composite variety or inter crossed population of selected inbred lines.

Breeding or Parthenocarpic Cucumber

Cucumber is one of the most popular vegetable of cucurbitacae family and grown worldwide for its freshly harvested fruits. Fruit yield in cucumber is suppressed due to its fruiting habit. Fruit developing from the first pollinated flower inhibit the development of subsequent fruits. This phenomenon is broadly known as crown fruit dominance or first fruit inhibition. Therefore, in mechanical once over harvest of seeded cucumber, only one to two fruits per plant are developed. Fruit set inhibition is less in seedless cucumber when compared to its seeded counter part. Cucumber is highly thermo sensitive crop and more successfully grown under protected cover as compared to open-field condition. Indeterminate type continues to grow until the plant dies, with the internode length relatively constant throughout the length of the vine. The growth habit are important in breeding programmes as these help in increasing yield and availability period to a greater extent. Ideotype breeding along with incorporation of useful genes for parthenocarpic character can be utilised on a large scale in poly-houses. Parthenocarpic gynoecious cucumber varieties are suitable for poly-house cultivation as these varieties develop fruits automatically without any pollination. Breeding effort should be concentrate on several fruit characteristics such as shape, colour, spine type (coarse or fine), spine colour (white or black), skin thickness and surface warts, high TSS, crispness, enhanced shelf life, resistance to biotic and abiotic stress, highly responsive to fertilizer, photo and thermo insensitiveness. The emphasis however is to develop parthenocarpic gynoecious hybrids. Development of suitable Ideotype novel genes for biotic and abiotic stress resistance is the need of the hour for breeding cucumber varieties/F_1 hybrids suitable for protected cultivation. The extension of protected cultivation and breeders' work has been linked by a close relationship. The increase in return in protected cultivation has made a greater effort in plant breeding possible, which has encouraged the expansion of crops. Further, plant breeding efforts have permitted to solve specific problems of varieties suitable for protected cultivation as well as general problems which also benefit the open field cultivation.Ideotype breeding along with incorporation of useful genes in crops like parthenocarpic can be utilised on a large scale in poly-houses. Gene *'pc'* responsible for parthenocarpy, *'F'* responsible for short inter-nodal length can be utilised through pure line and back cross breeding methods to establish them in plant population in parthenocarpic cucumber.Parthenocarpic gynoecious cucumber varieties are suitable for polyhouse cultivation as these varieties develop fruits automatically without any pollination.

Now a days many of the private seed companies are selling varieties/F_1 hybrid of parthenocarpic cucumber at a very high price as their seeds are being sold on per seed basis.

The crop shows several flowering habits. Most of the cultivars are monoecious with separate male and female flowers. Gynoecious cultivars produce only female flowers. The biological processes leading to sex expression in plants are of tremendous practical significance for fruit production of many agricultural and horticultural crops. Sex-expression studies in cucumber showed that the different sex types are determined by three major genes: *M/m*, *F/f* and *A/a*. The *M/m* gene in the dominant condition suppresses stamina development and thus leads to female flowers. The *F/f* gene in the dominant condition shifts the monoecious sex pattern downwards and promotes femaleness by causing a higher level of ethylene in the plant. The cucumber, with its diverse sex types, represents a perfect model plant for studying the molecular and physiological basis of sex expression. All floral buds contain staminate and pistillate primordial at early stages of their development; however, abortion of either staminate or pistillate parts results in female or male flower, respectively, while in hermaphrodite flowers the staminate and pistillate primordial gets matured. The distribution of these three different flower types (staminate, pistillate and hermaphrodite) on the plant results in the different sex types in cucumber. These three sex type are mainly determined by three major genes *F/f*, *M/m* and *A/a*. The gene *F* promotes femaleness on the whole plant level while the gene *M* determines whether flowers are unisexual (*M_*) or bisexual (*mm*). Thus the genotype *M_FF* is gynoecious; the genotype *M_Ff* sub gynoecious due to incompatible dominance of *F*; and the genotype *mmF_* is hermaphrodite. If *F/f* is homozygous recessive, the sex expression is influenced by the gene *A/a; MMffA_* genotypes are monoecious; *MMffaa* genotypes are androecious. Although the three major genes determine the basic regulation of sex types, the sex expression is influenced by additional modifying genes and environmental factors such as photoperiod, temperature and irradiation. Parthenocarpy is the development of fruit without fertilization. This phenomenon was first studied in cucumber by Noll (1902) who introduced the term parthenocarpy. In parthenocarpic lines, a larger proportion of photosynthates could be diverted to fruit tissue instead of producing seeds. Partheonocarpic habit has become indispensable to greenhouse cucumber production because it alleviates the need for bees. Individual plants in the greenhouse produce many fruits simultaneously, with many fold higher yield in comparison to field grown crop. Greenhouse or parthenocarpic cucumber production is very popular in many areas of the world. The fruits of parthenocarpic cucumber are mild in flavour, seedless and have thin edible skin that requires no peeling. Fruits are generally 10-14 inches in length, and 200 g in weight. The parthenocarpic varieties require no pollination for fruit setting. If pollination occurs, the fruit will form seeds, shape of fruit will be destroyed and develop bitter taste. It is therefore essential to prevent bees and other pollinators entering the green houses. Cucumber cultivars are selected for commercial use based on quality and yield characteristics. An important quality factor is the shape of the cucumber referred to as the length to diameter (L/D) ratio. This becomes important to meet specifications of the number of whole cucumber units per container. The cool marine environment causes the

L/D ratio of many cultivars to be undesirably long and therefore many of the useful cultivars for other production regions have been eliminated.The presently available parthenocarpic cucumber cultivars have been bred and developed in Europe under similar climatic conditions and produce fruit with desirable L/D ratios. The parthenocarpic cultivars have potential for improvement of both the L/D ratios and fruit interior quality. Almost all slicing cucumber varieties cultivated in greenhouse in Western Europe are parthenocarpic, and parthenocarpic pickling cucumber varieties are also available in the market. In Germany the use of parthenocarpic varieties in production for the processing industry is a major part of the total cultivation area of pickling cucumber. Parthenocarpic fruit setting results in an earlier and much more regular production rhythm (Zwinkels, 1987). Parthenocarpy circumvents the inhibitory effect of seed formation on subsequent fruit development. Parthenocarpy must be combined with constant femaleness, because the fruits formed after fertilization in genetic parthenocarpic varieties become misshapen, are of no economical value and lead to loss of production. It is possible to induce artificially parthenocarpic fruit set by treatment with morphactin (Chlorfluorenol) as reported by Robinson and Cantliffe (1971). Parthenocarpic varieties do not require pollination to set the fruit and the fruits are seedless. So maintenance is a major problem in these lines. The parthenocarpic lines are usually maintained by the use of plant growth regulators and chemicals *e.g.* Silver nitrate and GA_3. Kasrawi studied the effect of silver nitrate on sex expression and pollen viability in parthenocarpic cucumber. He investigated the effect of $AgNO_3$ concentration, number of sprays and time of application on sex expression and pollen variability of parthenocarpic cucumber. He found that greatest numbers of staminate flowers are produced on plants when $AgNO_3$ was sprayed twice with 300 ppm concentration at one week interval with initial spray at first true leaf stage. He also found that 2-3 sprays of 200-500 ppm $AgNO_3$ starting at first true leaf stage produce the maximum number of staminate nodes. When the plants were sprayed at first true leaf stage production of staminate flowers began with the 1st node and at the time of onset of flowering. Experiments reported the effect of silver nitrate on induction of staminate flowers of gynoecious parthenocarpic flowers and found that $AgNO_3$ concentration and numbers of sprays affect the induction of staminate flowers. All the treatments using one spray were infective for producing staminate flowers. The treatment with 100 ppm $AgNO_3$ failed to induce staminate flowers The greatest numbers of staminate nodes were produced on plant by 2-3 sprays of 400-500 ppm $AgNO_3$. The initial spray was applied at first true to leaf stage and subsequent treatment was applied at one week interval. The plants showed injury a few days after spray of 400-500 ppm $AgNO_3$ and recovered within 7-10 days.

Genetics of Parthenocarpy

Parhtenocarpy can be regarded as a qualitative character *i.e.* the ability to develop fruits without pollination (apomixes excluded). For breeding purposes parthenocarpy should preferably be regarded as a quantitative character *i.e.* the ability to produce a certain yield of parthenocarbic fruits. There has been conflicting reports regarding the inheritance of parthenocarpy in cucumber. Wellington and Hawthorn (1928) reported that there was incomplete dominance of parthenocarpy which they later concluded as recessiveness (Hawthorne and Wellington, 1930). Pike and Peterson (1969) studied parthenocarpy as a qualitative character and claimed to be governed by one incompletely dominant gene in the homozygous condition PP produces parthenocarpy fruit early, with the first fruit developing generally by 5th node. Heterozygous Pp plant produces parthenocarpic fruit later than homozygous plant and fewer in number. The homozygous recessive pp produces no parthenocarpic fruits. Trapping of pistillate flowers was effective in identifying homozygous pp plant but failed to identify heterozygous plants. Screen cages and field isolation of gynoecious lines are effective means for accurately identifying all parthenocarpic plants.

Juldasheva (1973) communicated that one recessive gene was responsible for parthenocarpy. The comprehensive experiments of Kvasnikov *et al.* (1970) led to the conclusion that parthenocarpy is governed by many incompletely recessive genes. De Ponti and Garretsen (1976) explained the inheritance of parthenocarpy by three independent, isomeric major genes with additive action, together with a non-allelic interaction of the homozygote- heterozygote type. This interaction must be explained by the phenomenon that in genotypes with one + allele for parthenocarpy, this allele dose not cause any phenotypical expression. In view of the rather larger variance in parthenocarpy expression of segregating population, there is a possibility that minor genes also contribute to the degree of parthenocarpy. Indications have been found for linkages between genes that govern parthenocarpy, femaleness and the spined/hairy fruit-character. They reported the probability that one of the genes for parthenocarpy and one of those for femaleness (Kubicki, 1969) were located on the same chromosome and that a second gene for parthenocarpy and the gene for the spine/hairy fruit character are both located on another chromosome.

Breeding of Parthenocarpic Pickling Cucumbers

A programme for development of parthenocarpic hybrid pickling cucumber for field production was started at Department of Horticulture, University of Wisconsin. The parthenocarpic germplasm was derived from a gynoecious population referred previously as gynoecious synthetic (GS). This population was developed in 1977-78 from a series of gynoecious lines collected from Europe, US and America through numerous breeding programmes. In early 1980, a half sib recurrent selection programme was combined with a pedigree selection programme to develop parthenocarpic population from gynoecious synthetic material. The parthenocarpic population was screened for parthenocarpy under three different photoperiods. Initial selection was made for parthenocarpy, yield, fruit quality with little emphasis on disease resistance and other characters. After several generations of breeding, methodology was redirected to inbred line, F_1 hybrid development. The inbreds were selected for parthenocarpy, gynoecious stability, fruit quality, yield, disease resistance and other horticultural traits.

Varieties

There are many cultivars available for greenhouse cucumber production from private sector. Cultivar selection is one of the most important decisions made during crop production process. Several factors should be considered when selecting cultivars including disease resistance, plant vigour, earliness, yield, fruit size, quality and fruit colour. The available parthenocarpic cultivars are generally produced by European seed companies, however, a few number of companies have sales and technical support representative in other parts of the world and Asia. Keeping in view the above facts, concept of polyhouse vegetable breeding programme was developed at Pantnagar during 2002 in cucumber and tomato. Some of good genotypes were isolated and pure line and hybrid breeding programme were adopted for improvement of these genotypes. After three years of multilocation testing at tarai, midhills and higher hills under polyhouse condition, four genotypes were identified and released, which are in parthenocarpic cucumber, *i.e.* Pant Parthenocarpic Cucumber-2 (1755 q/ha) and Pant Pant Parthenocarpic Cucumber-2 (1605 q/ha) were released by Uttarkhand State Variety Release Committee during May, 2011 for commercial cultivation at farmers' field. The programme on development of parthenocarpic gynoecious varieties for protected cultivation was undertaken during 2010 at Division of Vegetable Science IARI, New Delhi. The line DPaC-6 was observed as most promising as it expressed 25.0 per cent and 16.7 per cent higher yield than check Pant Parthenocarpic Cucumber–2 and Asma, respectively. The yield obtained by DPaC–6 (126.0 t/ha) can be considered as quite high since it was obtained during off-season (winter season) under low cost polyhouse without using any energy. On the basis of it consistent superior performance DpaC-6 was identified as Pusa Seedless Cucumber-6 by IARI Variety Identification Committee in 2016. The inheritance studies of fruit skin colour and parthenocarpy was also conducted by crossing DPaC-6 and monoecious cucumber variety Pusa Uday. The F_1 progeny showed intermediate colour between dark green DPaC-6 and light green Pusa Uday. The F_1 progeny showed true gynoecious parthenocarpic behaviour as its fruits were seedless and developed without pollination which suggested that gynoecious parthenocarpic trait is governed by single dominant gene. The F_1 progenies were advanced to F_2 and also simultaneously backcrossed with parthenocarpic line DPaC-6 to confirm the monogenic dominant nature of parthenocarpy.

Genomic Assisted Breeding for Introgression of Gynoecious Trait in Indian Monoecious Cucumber (*Cucumis sativus*) cv. Pusa Uday

Indian cucumber cultivars exhibit monoecious sex expression coupled with crown fruit inhibition which is supposed to be the main reason for lower productivity. Gynoecism has been extensively exploited in heterosis breeding for enhancing productivity of cucumber worldwide but exotic gynoecious lines are not stable at higher temperature and they possess very few lateral branches which are not suitable for open field condition. The introgression of gynoecious trait with marker assisted breeding can improve the productivity of Indian cucumber. The introgression of gynoecious trait into Indian cucumber varieties requires molecular mapping of gynoecious trait in crosss between Indian monoecious cucumber and exotic gynoecious line. The F_2 plants of the cross G 421(exotic gynoecious) x Pusa Uday (Indian monoecious) were used to study of the inheritance and molecular mapping of gynoecious trait (*F*). Out of 233 SSR markers screened for parental polymorphism among the parents Pusa Uday and G-421, five primers pairs (SSR-

13251, SSR-02021, SSR-15818, SSR-15516, and SSR- 18718) were found polymorphic. Another functional dominant SCAR marker CsACS1G specific to the *F* locus was also used for mapping. The developed SCAR marker amplifying the promoter sequence of the gene CsACS1G showed amplification of 430 bp fragments in *FF* and *Ff* plants but no amplification in monoecious (*ff*) plants in all F_2 population. These six makers (5 SSR and 1 SCAR) were used for mapping of F_2 population with 209 individuals. The linkage studies using single marker analysis was performed using JoinMap 4.0 software. Two co-dominant markers, namely SSR-13251 and SSR-15516 were located on the same side of the gene of interest. The markers SSR-13251 and SSR- 15516 were tightly linked to the *F* locus at 1.5 and 4.5 cM, respectively and *F* locus positioned at the top of the chromosome 6. The gene based SCAR marker (amplifying the promoter sequence of the gene CsACS1G) clearly distinguished gynoecious and monoecious plants in all F_2 population. These three molecular markers can be effectively utilized for marker assisted introgression of gynoecious trait from exotic line to Indian monoecious cucumber. These 3 markers were used for foreground selection from 192 $BC1F_1$ individuals and 18 plants heterozygous for *F* locus and similar to Pusa Uday with gynoecious trait were selected. These plants were selfed to produce the $BC1F_2$ and backcrossed with Pusa Uday to develop $BC2F_1$ population and genotyping of two populations with these 3 markers were carried out in next cycle of growing and phenotypic selection was made for plant growth and fruit traits of Pusa Uday. Genotypes homozygous for *F* locus, stable phenotypic expression of gynoecious trait and growth and fruit characters matching to Pusa Uday were selected which will be further evaluated in different environment for varietal development.

Tomato

Tomato (*Solanum* lycopersicon L. 2n=2x=24) is being extensively grown as warm season fruit vegetable annual crop all over the world. In India, tomato has wider coverage in comparison to other vegetables. The leading tomato growing states are Uttar Pradesh, Karnataka, Maharashtra, Haryana, Punjab and Bihar. Tomato is a very good source of income to small and marginal farmers and contributes to the nutrition of the consumers. It is a rich source of vitamins and organic acids. Tomato is a warm season crop and thrives in temperatures between 1-30°C and is neither tolerant to frost, nor to waterlogged conditions. It is well suited in different cropping systems which include cereals, grains, pulses and oilseeds. The differences are observed in production and productivity in polyhouse bred tomato (350-450 tonnes/ha) and field bred tomato (150-200 tonnes) under polyhouse condition. The average productivity of open field bred variety in field condition is 250 q/ha whereas under polyhouse condition is about 800q/ha (AICVIP Annual Report).

Breeding Objectives

- Large fruit size (150-200g)
- Thick pericarp (0.8-1cm)
- 4-5 fruits per cluster
- Long fruiting period (8-9 months)

- ☆ Long plant height (20-25 feets)
- ☆ Potato leaf type
- ☆ Ability to set fruits in wide temperature range (10-45°C)
- ☆ Resistant to fruit and shoot borer
- ☆ Long storage life under room temperature
- ☆ Indeterminate, single stemmed plants
- ☆ Medium sized bright colored fruits
- ☆ Short internode
- ☆ More number of fruits per plant
- ☆ High TSS and acidity
- ☆ Thermo-insensitivity
- ☆ Resistance to white fly and other important insects and pests.
- ☆ Exploration of male sterility having specific marker trait. (green cotyledon)

The genus *Solanum* includes both self-incompatible and self-compatible species, with the later varying in their degree of outcrossing. Tomato is self-pollinating, but can easily hybridize within the species or cross with wild relatives under appropriate conditions, thus permitting gene introgression from wild relatives. Throughout the last century, tomato breeding has been based on various standard methods, including the pedigree method of hybridization followed by selection and backcrossing of desired traits from one parent into another, which has resulted in the generation of improved tomato varieties and hybrids with high quality and yield. Tomato improvement has increased by the exploitation of exotic resources and the introgression of the new valuable genes into the tomato genepool. Classical breeding has not only developed cultivars with monogenic and dominant resistance for controlling certain plant pathogens or by the combination of F_1 hybrids, but has also enabled the acquisition of good, value-added agronomic traits, such as high fertility and fruit setting, earliness, uniformity, adaptation and firmness and long shelf life appropriate for shipping to distant markets. The tomato flowers are normally perfect. There are four to eight flowers in each compound inflorescence. There is a light protective anther cone surrounding the stigma leading to self pollination. Emasculation is usually done one day prior to anthesis. At this stage the sepals start getting separated and the anthers and corral begins to change from light to dark yellow colour. The stigma is fully receptive at this stage allowing for pollination even immediately after emasculation. Tomato is a typical day-neutral plant. It is mainly self pollinated but a certain percentage of cross pollination also occurs. If the pollen is shed as the style grows up through anther tube, self fertilization is mandatory. Self pollination also takes place when style is short and stigma is not extroverted beyond connivent anthers. A certain degree of cross pollination happens when stigma protrudes outside the level of extrovert anther. The progeny of a single homozygous self fertilized plant forms the pureline. Such lines remain homozygous as long as self pollination takes place. The success of this breeding method depends on the existence of variability in the original population. This method is very much potential for creation of parental lines to produce F_1 hybrids.

Hybrid breeding is another important tool in improvement of tomato under protected condition. Hybrids offer opportunities for improvement in productivity, earliness, uniformity, quality and for rapid deployment of dominant genes for resistance to disease and pest. For large scale production of F_1 hybrids there should be some naturally occurring mechanism for ensuring cross pollination between parental lines with minimum of self fertilization as genetic male sterility prevails in tomato. Shelf life of tomato fruit can be greatly improved by breeding for increased firmness, by utilizing "ripening inhibitor" genes or by a combination of both. Incorporation of improved firmness, a partially dominant character, resulted in cultivars with almost doubled shelf-life. As such fruits can be picked at later stages of ripening their flavour and colour is either normal or improved in comparison with normal cultivars. The partially dominant gene 'rin' in rin/+ combinations was found to add about 20 per cent to 30 per cent to length of shelf-life. While this increase is of little value in normal soft genotypes, fruit of rin/+ in genetically firm backgrounds exhibit a shelf-life increased by 2-4 days. Flavour of these genotypes was found normal and colour intensity only slightly reduced. The partially dominant mutant gene 'nor' has a stronger effect on shelf-life than 'rin'. Fruits of nor/+ genotypes picked at the early pink stage can be kept for about 3 weeks without cooling. The disadvantages of nor/+ types usually are the inferior flavour and decreased final colour ratings. It was found, however, that the general genetic background and certain mutant genes as well as growing conditions greatly affect these quality characters.

The combined effect of the ripening mutant genes in rin/+ nor/+ genotypes is a pronounced slowdown in fruit ripening. At 20° to 24°C such fruits may have a shelf-life of 4-5 weeks but their flavour is sometimes inferior and their colour orange to light red.

Varieties

The successful production of greenhouse vegetables is extremely dependent on the use of proper varieties. Although seed costs may be somewhat higher for these varieties, this represents a relatively small investment when considering the total cost of production. The important genes favouring polyhouse cultivation were also exploited. Gene 'sp' responsible for indeterminate character was utilized for the development of these varieties. By adopting appropriate selection methods and hybrid breeding procedure two varieties namely, Pant Polyhouse Tomato-2 and Pant Polyhouse Hybrid Tomato -1 were released first time in India among the public sector. After three years of multilocation testing at tarai, midhills and higher hills under polyhouse condition, Pant Polyhouse hybrid tomato-1 (1616 q/ha) with average fruit weight of 130-140 g and Pant Polyhouse hybrid tomato-2 (1291 q/ha) with average fruit weight of 100-105 g. were released by Uttarkhand State Variety Release Committee during May, 2011 for commercial cultivation at farmers' polyhouse. Recently, One cherry tomato variety was developed from Division of Vegetable science, ICAR- IARI, New Delhi. It is first indigenous cherry tomato variety suitable for green house environment. It attains average vine length varying from 9-12 m. Its fruit berries are round shaped with uniform ripening pattern and deep red colour. It has 18.5 average flower truss per plant. The average fruit weight is about 13 g with approximately average fruit yield 22 kg/plant with yield potential of 4-5 tonnes/1000 m^2.

Designing Novel Genotypes of Melon for Protected Cultivation

Protected agricultural systems could benefit vegetable growers in northern India if fresh harvested produce will be of high value and grown in efficient production systems to ensure that high returns are gained in relation to investment costs. New potential crops such as speciality melon could open new markets for future and existing greenhouse growers in India. Different types of high quality melons like Sun melon, Galia melon or locally called Sarda melon are becoming more preferred among consumers because of its attractive fruit appearance, better fruit quality and longer shelf life in comparison to indigenous muskmelon varieties. These fruits can remain on the shelf of retail outlets for 3-4 weeks without any deterioration in its quality depending on their storage temperature. These melon fruits fetch higher prices in the market mainly due to its off season availability in the market along with superior fruit quality and longer storage life. Most of the seeds of these varieties are being imported and their production is also limited to the areas of moderate climatic conditions. Few years ago, trait specific breeding programme was initiated at IARI, New Delhi for designing novel genotypes/varieties of melon which can be grown in protected cultivation. Large number of germplasm of *Cucumis melo* from different horticultural groups were collected from India and abroad and several genotypes could be developed which were evaluated under different environmental conditions during different seasons under protected structures for specific traits of economic importance. Recently, Pusa Sarda variety of speciality melon (*C.melo* var. *inodorous*) have been released by Delhi variety release committee for cultivation in protected environment. Its yield potential is 5.44t/1000 m^2 with TSS 13.6°brix in net house during main season. Other promising genotypes of speciality melon from inodorous group with high fruit yield, longer shelf life and better fruit quality which are suitable for protected cultivation are DHM-162 (4.90 t/1000 m^2 with TSS 13.5°brix), DHM 159 (4.60 t/1000 m^2 with TSS 13.8 °brix).The fruit rind colour of DHM-159 is white/creamish which makes it a novel genotype with very less seed cavity size. The hybrids between two horticultural groups were developed to exploit the longer shelf life, thick flesh, higher carotenoids traits of *inodorous* and musky flavour of *cantaloupensis* and 3 most promising hybrids for protected cultivation were identified as DMH-12 (5.5tonne/1000m^2), DMH-16 (5.1tonne/1000m^2), and DMH-5 (4.5tonne/1000m^2), with better fruit quality traits and post harvest storability. Hybrid DMH-5 can be cultivated in open field condition also during summer season. Suitable technology for production of specilaity melon in protected structures have been developed so that it can be successfully grown 3 times a year in north Indian plains.

Breeding of Perennial Horticultural Crops *Pages* **441–447**
Editors: S.K. Singh, V.B. Patel, A.K.Goswami, Jai Prakash and Chavlesh Kumar
Published by: **BIOTECH BOOKS, NEW DELHI**

Chapter **33**

Fruit and Vegetable Varieties for Processing

Ram Asrey

Food Science and Postharvest Technology Division,
ICAR-Indian Institute of Agricultural Research, Pusa, New Delhi – 110 012

INTRODUCTION

A number of improved horticultural crops varieties have been developed during past few decades, which ranged from the high yielding to various biotic and a-biotic stress resistances. In fruits and plantation crops more than 280 high yielding varieties suitable for various purposes have been developed. In vegetables, approximately 500 varieties have been developed from various institutes and agricultural universities. While developing these varieties, processing and export quality traits were either ignored or got the least priority.

Indian processing sector is still at low ebb which process about 3 per cent (2 per cent in organized sector and 1 per cent in unorganized sectors) of total fruits and vegetables production. Economic reasons and habit of the people make such differences, besides many other processing related factors. Processing and value addition for a product can be in terms of economic gain, time and money saving in preparation, quantity and quality improvement or modification of raw ingredients for specific desirable characteristics. With increased incomes, urbanization and changing eating habits, the demand for value added processed foods has increased manifold. In efforts to diversify products and to meet the market requirements, there is an increased need to add value to our products in order to fetch better prices. India is the largest producer of mango, banana, papaya, okra and ginger and ranks second in over all fruit and vegetable production in the world. Despite this impressive standing in production, India's share in the global market is still miniscule and touching around 1 per cent only. Varieties of a particular fruit and vegetable make the substantial difference in the fresh produce acceptance and processed product quality, which have been realizing by processors and scientists. The difference may be physical, biochemical or sensory characteristics affecting various operations during fruit processing. Besides, several factors, varietal differences also influence the shelf life of the finished processed products. Selection of a wrong variety for processing may yield poor processed products in terms of color, texture, flavour and overall quality. Hence, it is imperative to know various processing quality attributes of fruits and fruit cultivars for processing. Debate is going on doubling farmers income by 2020 through various agri-innovations so with more through horticulture sector. And if it is realized, value addition and marketing will be key changer in this endeavor.

Variety Selection Criterion

Colour

Carotenoids play important role in product quality of canned mango, citrus and pineapple. Colour of mature fruit is mainly due to xanthophylls ester and carotene. In orange, market demand is more for juice with a deep orange colour especially in the concentrate. Although some varieties are excellent in taste and aroma, yield a pale coloured juice and also pigmentation varied with season. These require blending of pale coloured juice with richly coloured juices. Similarly in mango, pulp from pale coloured varieties may be blanded with those from highly coloured varieties such as Amrapali and Dashehari to produce nectar of acceptable quality.

At lower pH colour is intense red but with rise in pH this colour change to blue or purple. Product colour is affected by raw material and processing due to presence of lycopene and anthocyanin. Pear variety, Bartlett does not show pink discolouration due to low level of anthocyanin. This discolouration may be avoided by varying the processing conditions. Two types of browning occur in fruits are given under.

Enzymatic browning is common during peeling and slicing operations. Banana, grapes, apples, persimmon *etc.*, turn brown due to the action of polyphenols oxidase enzyme. This enzyme acts on a suitable phenolic substance in the presence of oxygen. Hence, control is based on the inactivation of at least one compound. Some varieties of fruit crop having lesser tendencies to discolour during preparation for processing due to low polyphenolic substrate like Sunbeam peach.

Size and Shape

Shape of fruits is also important for processing industry, which reduces loss during mechanical trimming and handling. Most of the fruits *e.g.* Mango, papaya, guava, *etc.* are often of inconvenient shape or size this create problems of handling and while processing. Therefore, selection of proper varieties in a fruit for proper shape is important.

Texture

Varieties of fruits for processing should be firm so that they can withstand the necessary heat treatment. Viscosity is important in fruit varieties with consistency with point of view of products such as juices, catsup, jams, jellies and preserves. Insoluble cellular material in fruits cause lack of flavour and colour, those fruits such as mango, papaya can directly converted into pulpy products by disintegrating the whole fruit to make nectars.

Flavour

Usually aroma is lost during processing. Some varieties of crop retain flavour even after processing like Alphonso in mango. A range of aromatic principles are involved fro fruit flavour some of which are given in Table 33.1

Table 33.1: Aromatic Principles of Fruits

Fruits	*Compound*
Apple, ripe green	Ethyl-2-methylbutyrate hexenal, 2-hexenal
Banana green Ripe Overripe	2-hexenal Eugenol Isopentanol
Grape fruit	Nootakatone
Lemon	Citral, Neral
Orange	Valencene
Raspberry	1-(p-Hydroxyphenyl)-3-butanone
Strawberry and pineapple	Ethyl butyrate
Grape (For flavour)	Methyl Anthranilate
Grape (Muscat flavour)	Methyl Salicylate

Defects

Defects in fruits used for processing lowers the quality of products, which are otherwise of very high quality. Both hereditary and environmental factors are associated with defects.

Nutritive Value

Consumers now days pay more attention to the nutritive value of products due to health consciousness. Fruits high in total solids generally more nutritious and have more stored-up total food values. Deep orange or orange pulp generally contains more β-carotene than pale coloured or whitish pulp.

Canning

The fruits should be firm ripe and evenly matured as in case of plums, grapes, kiwi fruits, whole olives, peaches and cherries. Apple, pear and banana are harvested at mature stage. In general fruits for canning are harvested at mellow-ripe stage, which retain maximum nutrients and typical flavour to that fruits.

Fermented Beverages

The beverages produced from fruits after fermentation is termed as wines and their distillates are termed as brandies. Such fermented beverages have a characteristic flavour of the particular fruit. Almost all fruits are used for production of fermented beverages. Proper maturity is important for fine quality wine production with a proper sugar-acid balance. For wine making from grapes, varieties should be chosen that posses sufficient sugars, high Total soluble solids (>24 per cent) and acidity ranged between 0.50-0.70 per cent. some of the fruit cultivars used for wine production.

Dehydration

Some of the fruits are very successfully preserved by dehydration. Fruits for drying are harvested at proper maturity with care to minimize damage to fruits. The most well known fruits preserved by drying are grapes. *Aonla,* apple, apricot, banana, date, figs, mango, papaya, pear, peach, plums, pomegranate *etc.* are also used to develop dehydrated products. Fruits for drying require relatively good flavour, firm, texture, regularity in shape and a high sugar water ratio. In general, high total acidity does not have much importance for dried products because low to moderate acidity are concentrated during drying (Laslo, 1992). French prunes cultivars as well as imperial sugar varieties are the most commonly selected for dehydration. For raising preparation cultivars with thin skin, firm flesh, high sugar content and moderate to low acidity is preferred. Berries loosely arranged in cluster accelerate drying process and saves the energy.

Status of Processing Industry

In India, less than 3 per cent of fruits and vegetables produced are processed as against 65 per cent in the US, 70 per cent in Brazil, 78 per cent in the Philippines, 80 per cent in South Africa and 83 per cent in Malaysia. The value addition is only 7 per cent in India as against 23 per cent in China and 88 per cent in UK. A number of processed products from fruits and vegetables are given in Tables 33.2 and 33.3.

Table 33.2: Processed Products from Vegetables

Sl.No.	*Vegetables*	*Existing Products*	*Newer Products*
1	Bitter gourd	Dehydrated	Juice
2	Carrot	Juices, dehydrated	Colour extraction, osmotically dried shreds
3	Cauliflower	Dehydrated	Minimally processed
4	Chili	Chutneys, pickle, dehydrated	Puree, essential oil
5	Cucumber,	Slices, whole in brine solution	Pickled
6	Mushroom	Canned, dehydrated, ketchup, pickle	Phytochemicals
7	Okra	Minimally processed, dehydrated	Coated and battered
8	Onion, garlic	Dehydrated, paste	Ready-to-use gravy base
9	Shelled beans and peas	In brine, frozen, canned	Protein isolates
10	Tomato	Juice, puree/concentrate	Ready-to-use gravy base, lycopene extraction
11	Potato, cassava	Chips, starch	Alcohol

Table 33.3: Existing and Newer Products from Fruits

Sl.No.	*Fruits*	*Existing Products*	*Newer Products*
1	Aonla	Preserve, pickle	Juice and concentrate, osmo-dried segments, powder, salted segment preserve
2	Bael	Pulp	Preserve, juice, canned
3	Citrus	Canned slices, pickle, squash, extraction of citric acid, essential oil from peels	Limonene extraction, flavours
4	Grape	Raisins, juice, wine	Canned grapes, concentrate
5	Guava	Jelly, juice, nectar	Pulp, concentrate, bars, powder
6	Kinnow mandarin	Blended juice	Debittered juice, concentrates
7	Litchi	Juice, squash, canned	Wine, juice concentrate, IQF
8	Mango (Green)	Pickle, chutney, dried slices, powder	Drink, juice concentrate
9	Mango (Ripe)	Canned slices, pulp, juices, nectar jam, bar	Frozen slice, mango powder, concentrate, wine, vinegar,IQF
10	Papaya (raw)	Tutti-fruitty	Papain/pectin, IQF

It is recommended to use suitable processing variety for particular type of processed fruit and vegetable product to maintain uniformity in quality. The Table 33.4 shows a few recommended varieties of fruits and vegetables for use in specific product type.

Table 33.4: Processable Varieties of Fruits and Vegetables

Crop	*Processable Variety*	*Processed product*
	Fruit	
Mango	Alphonso, Arka Puneet, Arka Anmol, Amrapali	Canned juice
	Totapari, Arka Aruna	Osmotic dehydration
	Alphonso, Amrapali	Canned pulp
	Ramkela	Pickle
Guava	Red fleshed exotic guava-7-12-EC-147036, Allahabad safeda	RTS beverage
	Sardar, Apple colour, Chittidar	Jelly
Citrus	Nagpur santra, Khasi mandarin	Canning
	Kinnow, Coorg mandarin	Juice
	Kagzi lime, Eureka, Kagzi kalan, Nepali round	Pickle
Litchi	Shahi, Rose scented	Canning
Pomegranate	Amlidana, Basein seedless	Anardana
	Papershel, Kandhari	Squash
Grape	Beauty seedless, Black champa	Wine, juice
	Arkavati	Raisin

Crop	*Processable Variety*	*Processed product*
Aonla	Banarsi aonla	Murabba/preserve
	Krishna	Candy
	Kanchan	Pickle
Vegetable		
Tomato	Pusa Gaurav, Pusa Uphar, Pusa Hybrid 2	Sauce, ketchup, chutney
Methi	Pusa Kasuri	Dehydrated
Paprika	KT-PL-19	Oleoresin, powder
Amaranth	Pusa Lal Chaulai	Pigments for natural dyes, food additives
Beet root	Detroit Dark Red	Pigments for natural dyes and food additives
Paprika	KT-PL-19	Oleoresin, powder
Ash gourd	Pusa Shakti, Pusa Ujwal	Candy sweet
Bitter gourd	Pusa Vishesh, Pusa Hybrid 1	Dehydrated, pickle
Pumpkin	Pusa Vikas	Sauce
Pea	Arkel, Pusa Pragati	Frozen, Canned
Cauliflower	Pusa Snowball Kt-1, Pusa Snowball Kt-25	Dehydrated, pickle, frozen curds

Export Destination

The major destinations for Indian fruits and vegetables are UAE, Bangladesh, Malaysia, Netherland, Sri Lanka, Nepal, UK, Saudi Arabia, Pakistan and Qatar. Major crops and varieties for exports are given in Table 33.5 and also the handling conditions in Table 33.6.

Table 33.5: Varieties for Export

Crop	*Variety*	*Characteristics*
Fruit		
Mango	Alphanso, Kesar, Banganpalli, Dashehari, Langra, Chausa	Colored peel, pulp taste having desired blend of acid and sugars along with flavour.
Pomegranate	Bhagwa, Arakta, Arka Mridula	Soft seed, Stable anthocyanin, Higher Juice recovery (> 48 per cent), free from sunburn, bacterial spot.
Grape	Thompson Seedless, Baglore Blue, Perlette, Sharad Seedless, Dilkhus	Higher TSS, firmness, globose or elongated berry, green rachis with firm pedicel anchorage.
Banana	Dwarf Cavendish and Rashthali	Firm texture, uniform fruit with 2.7 cm diameter and length of 21cm.
Vegetable		
Okra	Parbhani Kranti, Arka Anamika, Varsa Hybrid, Varsa Uphar	Green tender, 6-7 cm long

Crop	*Variety*	*Characteristics*
Chili	G-4, Pusa Jwala	Green, long, 6-7 cm long
Cluster been	Pusa Navbahar, Astrelk Gonel, Contender	Tender green, slender, 7-10 cm
Bitter Guard	MVTH-1	Dark green, small neck, 25-0-25 cm long
Onion	Agrifound Dark Red, Agrifound Light Red, Arka Niketan, N2-4-1, N53, Granex-33, Agrifound Rose, Arka Bindu	For gulf countries – 4-6 cm oval in shape For Asian counties – light red, 3-4 cm long For European countries – Oval shapes and 7-8cms long, light yellow in colour
Watermelon	Sugar Baby, Great lex	Oval in shape having individual fruit weight of 2-2.5 kg
Bottle guard	Pusa Naveen, Hybrid Co 1, Arka Bahar	Light green, straight, cylindrical, 25-30 cm long

Table 33.6: Handling Conditions for different Vegetables during Transit and Export

Crop	*Handling Conditions*	*Packaging Material*	*Weight of Individual Packing (kg)*
Brinjal	Pre-cool after harvesting at 7-10°C as soon as possible. Maintain RH around 65 per cent.	Wrap the individual fruit with tissue paper and place in CFB box.	5-6
Capsicum (Shimla Mirch)	Pre-cool after harvesting at 7-10°C within 24hrs. Maintain RH around 90-95 per cent.	Wrap the individual fruit with poly pack and place in CFB box.	5-6
Musk melon	Pre-cool after harvesting at 20°C as soon as possible. Maintain RH around 85-90 per cent during storage.	Pack in CFB boxes with cushioning and lining material.	5-10
Okra	Pre-cool harvested produce as soon as possible and maintain 8-10°C temp., 90-95 per cent RH during transportation and storage	Pack in ventilated boxes and store at 2°C temp., and 65-70 per cent RH.	2-5
Onion	Bulbs should be properly cured and ensure that the net of the bulbs are tightly closed having length of 0.75-1.0cm.	Pack in ventilated sacks of nylon fiber and store at 2°C temp., and 65-70 per cent RH.	5-25

Problems Associated with Export of Fruits and Vegetables

- Lake of suitable fruit and vegetable varieties accepted globally.
- Lake of air cargo and higher freight charges.
- Uncertainty in continuity of trade ties.
- Infrastructure shortage at airport and seaport.
- Average high price of fruit and vegetables due to frequent fluctuation in production.

Breeding of Perennial Horticultural Crops *Pages* **449–455**
Editors: S.K. Singh, V.B. Patel, A.K.Goswami, Jai Prakash and Chavlesh Kumar
Published by: **BIOTECH BOOKS, NEW DELHI**

Chapter **34**

Variety/Hybrid Maintenance

B.S. Tomar

Division of Vegetable Science,
ICAR-Indian Institute of Agricultural Research, Pusa, New Delhi – 110 012

INTRODUCTION

Seed is considered as a basic input in agriculture. Hence, the application of other inputs is depends upon the quality of seed used. The use of quality seed helps in ensuring the higher production and quality of the produce used as per the crop. The traits of quality seed comprise of genetic purity, physical purity, germination and seed health and are greatly influenced by the production practices, environmental condition and handling during harvesting, threshing, processing, packaging and storage of seed. Therefore, it is importance to ensure the supply of high quality seed or plantings materials to the growers for commercial cultivation.

Seed is a living material and defined by various ways; it is a mature ovules or a mature ovule encapsulated with seed coat and supported by food supporting tissue (endosperm or cotyledons) and capable to give rise a normal seedlings upon sowing/ planting. It is not only a sexually produced but may be tuber, bulbs, rhizomes, roots, cuttings, all type of grafts and other vegetatively propagated material.

The characteristic of seed consists of genetic purity, physical purity, germination percentage, seed health, seed vigor and uniformity (shape, size and colour *etc.*). These traits are mentioned in commercially available class of seed *i.e.*, certified or truthfully label seed. During the maintenance of variety following factors need to consider whether it is variety, hybrid, synthetic or composite or e lone, thus understanding of the above is an essential.

Variety Maintenance

Maintenance of the variety is considered life line of variety or hybrid and its life is depending upon the maintenance during production. Through maintenance, the yield and quality traits advantage could be harnessed by the farmer during commercial production getting higher productivity thereby more economic return and consumer can also take full advantage of improved quality traits in the commercial product. It has been realized in field crops that the variety could run in seed production chain even more than 2 to 3 decades. While, maintenance breeding in vegetable crops is not very largely practiced because of the limited involvement of the seed scientist as well as more challenging than field crops are being having variable breeding behavior. During the maintenance the type of material to be handled required different approaches to maintained the genetic constitution by the maintenance personnel.

Variety

The term variety, equivalent to cultivar is defined in the international code of nomenclature of cultivated plants (1969) as "an assemblages of cultivated plants which is clearly distinguished by any characters (morphological, physiological, cytological, chemical or others) and which when reproduced (sexually or asexually), retains its distinguishing characters. It is homozygous and homogenous in self pollinated crop and helonagyanozus and homogenous in cross pollinated.

Hybrid

An off spring between two un-identical parents which may be superior over the parents or better than the better parent. Genetically, it is heterozygous and homogenous in nature.

Synthetic Variety

Sprague, 1955 defined synthetic variety as the advance generation of multiple hybrids that is increased by open pollination, synthetic variety is heterozygous and homogenous in nature.

Composite Variety

It is generally include various breeding materials that have been put together on the basis of yield potential, maturity, disease resistance or other known characteristics.

For these reasons, seed production practices have been developed for maintaining the genetic constitution of crop varieties.

General Considerations in Seed Production

The seed of self fertilized crop, such as garden pea, lettuce, methi, lablab bean *etc.* may theoretically be re-sown for several generations without genetic deterioration. In practices, however progressive deterioration of the original stocks occurs rather rapidly through dilution by mixture with other varieties and species. Therefore, every four to six years new seed produced under strict control is required to maintain the yield potential of an improved variety.

Seed of cross fertilized crops has to be replace more frequently, since mixing as a results of natural cross pollination with other varieties from other field. Seed of synthetic and composite varieties (cauliflower and cabbage) may have to be replaced after three to five years, whereas hybrids must be replaced each year if their original yield potential is to be maintained.

Procedure of Maintenance of Variety

To maintained the variety/inbred line in its original form, the size of population and method of evaluation based upon the traits of interest *viz.*, yield, quality and among the quality traits the quantification of vitamins, minerals and nutraceutical present and compared with initial value as described by breeder at the time of release for the variety or hybrid.

Size of Population and Evaluation Approach in Maintenance of Varieties/Lines CMS/SI Lines

Population Size

In self pollinated crop *viz.*, garden pea 350-500 plants, tomato/brinjal/chili/capsicum (30-40 plants), often cross pollinated crops like okra and broad bean (150-200 plants) and cross pollinated crops i.e, cauliflower, cabbage, turnip, radish, onion (40-50 plants/bulbs) are selected on the basis of morphological, flowering and economical part (root, head, knob and leaf traits *etc.*) and grown in isolation. However, in CMS/SI lines in cauliflower/cabbage 10-12 plants are enough to maintain the genetic constitution of the variety.

Evaluation of Selected Plants

In garden pea; plant to row progeny, tomato/brinjal/chilli/capsicum; plant to plot progenies and okra; plant to row or pod to row may be adopted. However, in case of highly cross pollinated crops *viz.*, cauliflower, cabbage, onion and root crops in which inbreeding depression is pronounced in selected plants random mating are allowed in isolation. In case of cucurbits; 10-20 true to type plants are selected before flowering and pollination is done by hand. Finally, the best fruits confirming the varietal traits are selected and their seeds are mixed.

In wind pollinated crop, palak, amaranth and chenopod, the selection of plants are based on certain traits *i.e.*, leaf shape, intensity of colour in leaves and bolting traits and covered with sattan cloth based net or from centrally located plants are tagged to maintained the variety as suggested by Dark (1969).

Criteria of Selection in different Crops

The selection criteria vary crops to crop and the criteria followed in different crops have been described below:

- ☆ **Tomato, brinjal, chill and capsicum:** Plant types, foliage colour, flower colour, fruit shape, colour of fruit and tolerance/resistance to diseases if any.
- ☆ **Garden pea, broad bean, lablab bean, cluster bean:** Plant types (pole/bush), flower colour, pod shape and colour.
- ☆ **Leafy vegetable (palak, methi, amaranth, veg. mustard):** Leaf shape, size, colour, presence/absence of pigments on leaf/stem, bolting behavior (early or late).
- ☆ **Cole crops (cauliflower, cabbage, broccoli and knol-khol *etc.*):** Plant type (spreading, semi spreading, erect, very erect), stalk height, curd colour, shape, presence or absence of bract leaf, compactness and bolting behavior *etc.*).

Root Crops

- ☆ **Carrot**: Root colour, shape, foliage size, core colour, core size, pigment on shoulder (purple/green), smoothness of root and forking.
- ☆ **Radish**: Root shape, colour of shoulder, foliage size, sponginess, texture and pungency.
- ☆ **Turnip**: Root shape, colour, foliage size, pithiness, texture and pungency.
- ☆ **Beet root**: Root shape, smoothness, foliage size, colour intensity and zoning.

Bulb Crops

- ☆ **Onion:** Bulb shape, No. of glumes, colour, TSS and storability behavior (bolting, smut or rot).
- ☆ **vGarlic:** Colour of bulb and clove, shape of clove, number of cloves and TSS.
- ☆ **Cucurbits (Gourds and melons):** Fruit shape, colour, ridge (continuous/broken), presence of hairs on fruit, flesh colour and TSS (muskmelon, watermelon)
- ☆ **Okra**: Foliage colour, inter nodal length, pod shape, pod colour, resistance to YVMV and ELCV.

Besides, the above it is also important to consider the level of vitamins, minerals and phyto-nutrients during the maintenance of the variety. Since, vegetables are considered as protective foods. The important vitamins and phyto-nutrients present in different vegetable crops are given in Table 34.1

Maintenance of Special Traits in Vegetable Crops

Gynoecious Lines

The gynoecious trait in cucumber is determined by a single dominant gene "F", since these plants have only female flowers, hybrid seeds may be produced using gynoecious maternal lines without the requirement for male flower emasculation.

Table 34.1: important Vitamins and Phyto-nutrients Present in different Vegetable Crops

Crop	*Vitamins*	*Minerals*	*Phyto-nutrients*
Brinjal	-	Potassium, Sodium	Chlorogenic acid, Nasunin
Bitter gourd	Vit.C	-	Momordicin, Charintin
Beet root			Betalines
Carrot (Red)	-		Lycopene
Carrot (Yellow)	Vit.A (Beta carotene)	-	-
Carrot (black)	-	-	Anthocyanin
Chilli	Vit. C		Olereosin (Capsanthin – red)
Capsicum (Red/Yellow)	Vit. C	-	Lycopene/Caretonoids
Garlic	-	-	Allicin
Onion	-	Copper, Seleinum	Querecetin, Fructanse (pre-biotic compounds)
Tomato	Vit.C	-	Lycoepne
Water melon		Potassium	Lycopene, Citrulin
Musk melon	Vit.A	-	-
Radish/turnip (Red)		Iron, Calcium	Anthocyanidins
Leafy vegetables (Palak, methi, veg. mustard, chinopods)	Vit. K	Potassium, Calcium, Iron	-
Kale	Vit.A	Potassium	Leutin
Cauliflower and Cabbage	-	-	Glucosinolates
Broccoli	Vit.C	-	Leutin
Lettuce	Vit.C, Folic acid	-	Leutin
Pumpkin	Vit. A	-	-
Knol khol	Vit.K	-	-
Okra	-	Iodine	-
Beans	Foliate	Potassium, Calcium, Iron, Magnesium, Zinc	

When the temperature exceeds beyond 30°C the stability of gynoecious sex expression is affected. Unfortunately, the temperate gynoecious lines are unstable for gynoecy under high temperature and long photoperiodic conditions. Tropical gynoecious lines (DC-102, DC-103) having stable sex even at temperatures around 40°C have been developed at IARI and utilized in the hybrid breeding programme.

Gynoecious line are maintained by application of GA_3 @ 1500-2000ppm, Silver Nitrate @ 250-300ppm, Silver Thiosulphate @400ppm at 2-4 true leaf stage. A problem observed with gibberellic acid application is that different gynoecious lines vary in response to GA application and, in some cases, the number of induced male flowers was not sufficient for hybrid seed production. Additionally, GA applications typically cause excessive stem elongation or malformed male flowers. Because of the erratic male flower induction by use of gibberellic acid, application of silver compound such as silver nitrate is done to induce male flowers. Silver ions inhibit ethylene action and thus promote male flower formation in gynoecious cucumber plants. However, due to phytotoxic effects of silver nitrate such as burning of plants, silver thiosulphate (400 ppm) is now widely used by seed producers for the maintenance of gynoecious cucumber lines. It induces male flowering of cucumber plants over a longer period and is less phytotoxic compared to silver nitrate.

SI Lines

The SI lines could be maintained by bud pollination at experimental level or when small quantity of seed is to be produced. However, at commercial scale the SI lines are maintained and multiplied with enhanced level of CO_2 (400 – 600 ppm, for 6-8 days) under closed structures *viz.*, glass house or poly house to overcome the self incompatibility at flowering temporarily and pollination is done by the pollinators.

Variety Developed for Protected Conditions

The varieties developed for protected conditions in tomato, capsicum, cherry tomato, cucumber and muskmelon should be maintained and multiplied under protected structures to obtain the full expression of traits of interest.

Management of Bitterness in Cucumber and Luffa

Most cucurbit species have bitter compounds due to presence of oxygenated tetra cyclic triterpene compounds namely cucurbitacins in their foliage. Cucurbitacins are thought to be toxins produced by these plants as a defense against insects and herbivores. The cucurbit cultivars have been selected by plant breeders to have low cucurbitacin content in the fruits except for ornamental gourds where higher level of cucurbitacin is desirable. Bitter fruits in cucumber are occasionally produced when plants are exposed to drought during fruiting. Another way of producing bitter fruit is when bitter pollen grains fertilized the non-bitter ovule, the resulting fruit will be bitter in taste because the gene Bt (bitter fruit) is dominant over bt (bitter free) gene and the phenomenon is known as metaxenia. Generally, during development, varieties are tested to verify that the leaves and fruits are not bitter-tasting, and if maintained by self pollination (or sib-pollination in an isolation block), there will be no problem of bitterness.

In order to examine the bitterness in selected plant two fruits in each plant are developed through controlled pollination (full-sib) and at attaining the marketable stage, the second fruit is harvested and tasted for the presence of bitter compound. The seed obtained from the bitter free fruits are pooled for further multiplication.

Conclusion

Breeding new varieties and hybrids are a continuous process to cater the need of the society and consumers. The development of new varieties required 8 to 10 years of time with good amount of financial involvement and manpower where as hybrid development takes 3 to 5 years. However, the life of the variety and hybrids depends on the maintenance of varieties and parental lines of hybrids in its original potential as well as genetic makeup, thus variety maintenance is considered as life line of variety. The life span of the variety range from 5-6 years in self pollinated crops, 3-4 years in case of composite and synthetic varieties whereas, hybrid required replacement every year. The maintenance of the varieties largely determined by breeding behaviours of the crop concerned and the number of plants to be selected also important for the maintenance of total gene pool of the variety. Although, variety maintenance has been considered a routine job but the maintenance of the variety is vital for the longevity of the variety in commercial production and it has been practically utilized in maintenance of the many cereal, pulses and oil seed crop. The maintenance of the variety in vegetable crops is much more tedious and energy requiring activity than field crop being the economical part is not the seed. In vegetable it may be leaf, stem, modified stem, root and compact group of buds, inflorescence and immature/mature fruits accordingly the emphasis need to be laid down to maintain the original traits of the variety keeping in view of the importance economical part of the crop. In recent time the vegetable have been bred for higher nutrition and nutraceuticals values, hence require the laboratory support to assess the minimum level of nutrition and nutraceuticals including vitamins and minerals.

In bulbs crop, bulb shape, size, glume, sprouting behaviours during storage and removal of dry rot, basal rot and black smut affected bulbs have to be considered. In root crop *viz.*, carrot (root size, shape, root colour, core size, core colour, foliage size and duration), radish and turnip (root size, shape, pigmentation on shoulder, spongy tissue, foliage size and duration) and beet root (root shape, smoothness of root skin and zoning). Similarly, in cole crop plant type, heads colour, shape and bolting pattern need to be considered in maintenance of the variety and parental lines. Among the leafy vegetables leaf shape, pigmentation and bolting behaviours of plants should be looked upon. The evaluation of the lycopine content, anthocynin content, beta carotene, betalines, antioxidants compounds and level of minerals specially potassium, iron and magnesium etc. depending upon the crops has to be examined for their level in the selected plants/bulb/roots.

Breeding of Perennial Horticultural Crops *Pages* **457–470**
Editors: S.K. Singh, V.B. Patel, A.K.Goswami, Jai Prakash and Chavlesh Kumar
Published by: **BIOTECH BOOKS, NEW DELHI**

Chapter **35**

Protection of Perennial Horticultural Crops through Intellectual Property Rights

Chandan Roy

Department of Plant Breeding and Genetics,
Bihar Agricultural University, Sabour, Bhagalpur – 813 210, Bihar

INTRODUCTION

Intellectual property rights (IPR), an exclusive right given to the owner of an invention so that the inventor may exclude others from unauthorized use of his or her invention. Intellectual property "intangible" in nature means it cannot be defined with its physical parameters. It is expressed in some discernible way to be protectable. Patent is the oldest form of intellectual property right exist. In the year 1447, the first patent statute was established in Venice. The word patent originates from the Latin word *Patere* means "to lay open". Currently, there are several forms of intellectual property rights exist that protects different kind of properties created by the human being. Paris convention in the year 1883 has led to the industrial property rights (priority rights and Unfair competition) that followed by 1886 Berne convention that took care of literary and artistic rights. Gradually, several IP laws were established for the protection of almost all kind of invention.

India is the one of the rich nation for plant genetic resources. It carries 7.6 per cent of all mammalian, 12.6 per cent of all avian, 6.2 per cent of all reptilian, 4.4 per cent of all amphibian, 11.7 per cent of all fish, and 6.0 per cent of all flowering plant species in the world (Lal, 1989). India carries a distinct identity for its pattern and distribution of biodiversity region. It contains a great diversity of its natural ecosystem from evergreen tropical rain forest of Western Ghat; North Eastern region; Andaman and Nicobar island to dry alpine scrub high in the Himalayan region. Greater amount of species diversity has been observed over the longitudinal decrease or increase in these regions. Nearly 21.05 per cent areas of total geographical land of India under forest cover. Deforestation has led to the greater loss of forest biodiversity. According to the World Resources Institute around 450 million hectares forest areas declined in the last 30 years. Worldwide, nearly 15 million hectares tropical forest degrades that also creates 4.6 gigatons of carbon dioxide. Increase rate of urbanization and industrialization is a threat to the climate. Presently, climate change is a major concern both for developed and developing nations. Conservation and sustainable use of plant genetic resources could be a boon for handling the climate change and also to minimize the deforestation issue (Table 35.1).

Table 35.1: Establishment of International Agreements and Major IP Regulations

International Agreement	*Year of Establishment*	*Related Intellectual Property*
Paris Convention	1883	Industrial property rights: Priority Rights and Unfair competition
Berne Convention	1886	Literary and Artistic Rights
Madrid Agreement	1891	International Registration of Mark
Lisbon Agreement	1958	Appellation of Origin
World Trade Organization (WTO)	1994	TRIPS agreement, PVP, Integrated circuits, Trade secrets

Harmonizing the patent laws was first attempted in the Paris convention (1886) that covers industrial property including patent and trademark. However, copyright issue was not considered in the Paris convention which was later on taken care in the Berne convention (1886). With the establishment of World Intellectual Property Rights (WIPO) in the year 1976 the international IPRs regime has strengthen up and well coordinated among the countries. One of the strongest multilateral instruments for IPRs established as TRIPS (Trade Related Aspects for Intellectual Property Rights) agreement among the member state of WTO (World Trade Organization). PGR are considered the product of nature and excluded from patentability. Earlier plant genetic resources were shared freely among the researcher over the region and continents. Presently, it is considered as the property of the territory where it evolved and adapted over the time.

Why to Protect Plant Varieties

- ✰ To increase production by making availability of new, superior varieties with novel characters like stress/disease resistance or nutritional value
- ✰ To increase breeding activity and encouragement of new breeders in the public and private sectors or even farmers.
- ✰ Improving farmers, growers, consumers and industry with overall economic benefit.

- ✰ Improving competitiveness in foreign markets and the rural economic status
- ✰ Protection of useful genetic resources which are in the process of extinction

Several IP regulations are existing for the protection of plant genetic resources at national and international level. Some of the important agreements are discussed briefly:

International Convention for the Protection of New Varieties of Plants (UPOV)

Establishment of UPOV (International Convention for the Protection of New Varieties of Plants) in the year 1961 was the first of such kind where new plant varieties and seed materials were considered for protection. Main objective of UOPV was to provide an effective system to protect new plant varieties. This encourages plant breeders for development new varieties; quality seed distribution to the farmers and provide benefit to the society. Breeder's right and farmer's right were also considered later in different amendment of UPOV (1978 and 1991) (Table 35.2).

Table 35.2: IP Regulations Related PGR Management and their Utilization

Year	*IP Regulations*
1930	1st attempt to recognize IPR by the enactment of Plant Patent Act by USA
1961, 1978, 1991	UPOV Convention on Plant Variety Protection, Breeder's Rights
1983-1993	FAO International Undertaking on PGR (IUPGR)- Breeder's Rights, Farmer's Rights
1986-1994	Uruguay Round of Negotiations on GATT/attention on agriculture; long term reform of agricultural trade and domestic policies aimed at increased market orientation
1992	Rio Conventions on Biological Diversity (CBD)
1993	Revision of IUPGR set in – to harmonize with CBD
1995	WTO came into existence/Agreement on TRIPs requiring protection of plant variety by countries either by patents or effective *sui generis* system or suitable combination thereof
2001	International treaty on PGR for Food and Agriculture Multilateral System for access and benefit sharing, PPV and FR Act, 2001 in India
2010	Nagoya Protocol of CBD on Access to Genetic Resources and the Fair and Equitable Sharing of Benefits Arising from their Utilization

Convention of Biological Resources (CBD)

Conservation of forest genetic resources their utilization benefit sharing was given more concern with the establishment of convention of biological diversity (CBD) in the year 1992. The excess rate of diminishing biodiversity is alarming food security of the present and future generation. Identification of biodiversity rich sites and their protection was the focus of the CBD. Besides, sustainable use of genetic resources was one of the main agenda of the convention. The convention tried to make a common understanding among the countries for exchange of genetic resources. However, one concern is the genetic resources exchanged between the countries before the convention came into force. Identification of those territories and the genetic resources exchanged are needs to be found out. The convention reaffirmed the sovereign rights of countries over their natural resources, access to genetic resources with national governments and is subject to national laws.

International Treaty on Plant Genetic Resources for Food and Agriculture (ITPGRFA)

Food and Agricultural Organization adopted International Treaty on Plant Genetic Resources for Food and Agriculture in the year 2001. The main aim of the treaty is to conserve the all kinds of plant genetic resources, their sustainable use and legal exchange of genetic resources. Presently, growing demand in the timber industries, the volume of income through good quality wood has increased largely. ITPGRFA provides a multilateral system for free access of genetic resources. The treaty also recognizes the contribution of farmer in the creation, protection of PGR and also the traditional knowledge associated with them. It keeps the interest of the farmers in participating in the national decision making process and ensure their share from the benefit arising from their contribution.

Nagoya Protocol

The agreement was undertaken in the year 2010 in Nagoya, Japan that deals with access of genetic resources and benefit sharing out of their utilization. This protocol fulfils one of the objectives of CBD through providing a transparent legal framework. Genetic resources by the researchers or for commercial purpose must be lawfully acquired. Besides, it also covers the traditional knowledge associated with it. Benefits generate through commercialization of genetic resources or its traditional knowledge must be equally shared among the conservators or the inventors. Many indigenous tree species of India and Southeast Asia are of high value crop. Teak, Neem and Sandalwood trees are indigenous to India and high value crop. Nagoya protocol as n undertaking of CBD is providing a legal instrument to regulate the exchange of such high value crops.

Protection of Plant Variety and Farmer's Rights Act (PPV and FRA)

The Protection of Plant Variety and Farmer's Rights Act (PPV and FRA) was enacted in the year 2001. It was adopted by the Government of India as a *sui generis* system of IPR after the Trade Related Agreement of Intellectual Property (TRIPs) agreement of World Trade Organization (WTO). Article 27.3 (b) of the TRIPs agreement mentioned that all the member countries have to protect their plant genetic resources by the patent law or any *sui generis* system. The Indian Patent Act established in the year 1970 has excluded the method of agricultural and horticultural production from patentability. As a result PPV and FR act is first act enacted in India to protect the plant genetic resources. This is one of the unique act enacted keeping in the mind of different provisions mentioned under UPOV, CBD and TRIPS agreement. Breeder's rights, Researcher's rights, Farmer's rights and benefit sharing has been included into this act (Pratibha *et al.*, 2004; Roy, 2016).

Objectives

- ☆ To provide an effective system for protection of Plant varieties and rights of farmers and plant breeders
- ☆ To recognize the farmers in respect of their contribution made at conserving, improving and making available plant genetic resources for development of new plant varieties
- ☆ To protect plant breeders' rights to stimulate investment for R and D and develop new varieties
- ☆ To facilitate the growth of seed industry to ensure production and availability of high quality seed/planting material.

Important provision of the act has been discussed below:

Breeder's Rights

Breeder of a variety upon registration of the variety may have an exclusive rights to sell, produce, distribute, marketing, export import and transfer the right of the variety.

Researcher's Rights

Section 30 of the act provides the use of any variety registered under this Act by any person using such variety for conducting experiment or research. The use of a variety as an initial source of variety for the purpose of creating of other varieties: provided that the authorization of the breeder of a registered variety is required where the repeated use of such variety as a parental line is necessary for commercial production of such other newly developed variety.

Farmer's Rights

An exclusive right given to farmers in this act as to keep the interest of the farmers for continuing conservation and development of germplasm in the farmers filed. The role of farmers in the current agricultural growth and development is enormous. Most the varieties released in the latter half of last century was through direct selection from the farmers' materials or improved through mass selection of available high yielding landraces. Following rights have been given to a farmer in this act:

1. A farmer has the right to save, use, reuse, exchange and sell of its farm produce including seed of a registered varieties.
2. Farmers can claim compensation from the breeder of a registered variety if it fails to do expected performance as mentioned at the time of registration.
3. Farmers are exempted from any fee for registration and in any proceedings under the act before registered or Authority or Tribunal or High Court.
4. If a variety is evolved, cultivated and conserved by a group of farmers or by a community, can be registrable for the group or for the community.

5. Farmers are entitled for benefit sharing if their genetic materials are used for development of new registered variety; he or she can apply for benefit sharing.
6. If a registered variety is unable to meet the requirement of the farmers with reasonable price than he or she can apply for compulsory license, where the right for seed production and marketing of that variety will be given to other agencies.
7. Farmers are exempted from the compensation if the registered variety is fails to do expected performance under suitable condition.
8. Farmers are protected from the innocent infringement if proven that at the time of infringement he or she was unaware about the act.
9. Farmers are entitled for recognition and rearward as Plant Genome Savoir Farmer Reward and Recognition if their conserved materials are used as donor for genes in the varieties registrable under this act. When a farming community is engaged in conservation, improvement and preservation of genetic resources of economic plant or their wild relatives; the community would be eligible for the Plant Genome Savior Community Award.

Benefit Sharing

Benefit sharing upon commercialization of registered plant variety is an important provision of this act. As per the section 26 of the act the claims under benefit sharing can be submitted by Citizens of India or Firms/NGOs formed or established in India. Such claims after due examination and after receiving counter-claims will be disposed by the PPV and FR Authority. The amount thus determined shall be deposited by the breeder in the manner referred to in clause (a) of sub-section (1) of section 45 in the National Gene Fund.

Varieties Eligible for Registration

1. **New varieties**: Newly developed variety which is not available in the public domain or not been distributed to the farmers.
2. **Extant Varieties**: A variety available in India which is either
   - ☆ Notified under section 5 of Seeds Act, 1966, or
   - ☆ A variety about which there is common knowledge, or
   - ☆ Any other variety which is in public domain.
3. **Essentially derived varieties**: A variety which is predominantly derived from other initial variety keeping all the essential quality of the initial variety except the traits for which it has been newly developed. In example a transgenic cotton variety contained additional one resistant gene for lepidopteron insect into the initial variety may be registered as essentially derived variety.
4. **Farmer's Variety**: A variety which has been traditionally cultivated and developed by the farmers in their field or a landrace or wild relative of a crop about which farmer has common knowledge.

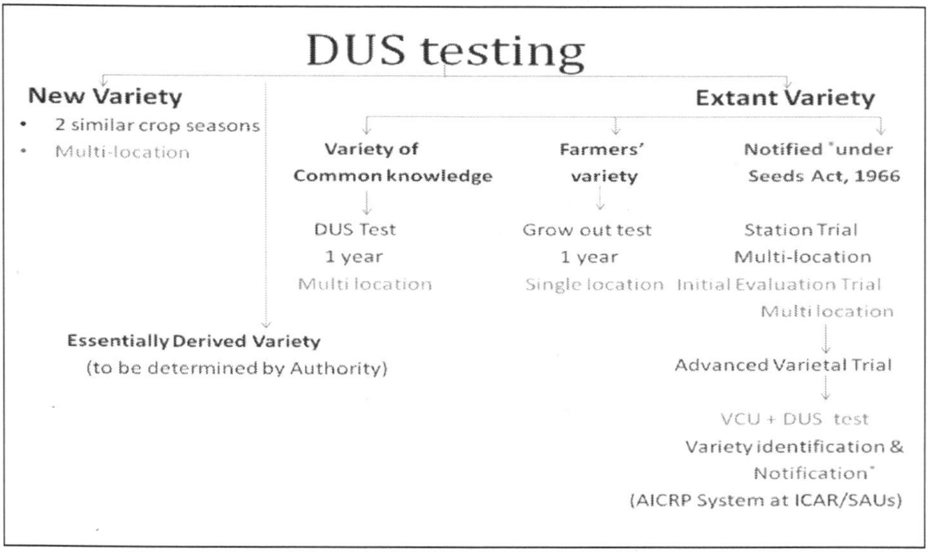

Figure 35.1: DUS Test Carried Out for different Kinds of Varieties under PPV and FR Act.

Essential Criteria for Registration

There are mainly three essential criteria under the act to test a variety before registration. These are distinctiveness, uniformity and stability (DUS). DUS testing guidelines are determined by the authority and till now DUS guidelines of 102 crop species (including agricultural, horticultural, trees and vines crops) have been developed where varietal registration process is continuing. For more details visit www.ppvfra.gov.in

1. **Distinctiveness:** Distinguishable by at least one essential characteristic from a variety which is a matter of common knowledge or available in the public domain. In the picture rice varieties can be easily distinguish from one another by their appearance of color.
2. **Uniformity:** Sufficiently uniform in its essential characteristics. In the above picture black husked rice variety in the middle is more uniform than the other two varieties in the form of color and grain size.
3. **Stability:** When the essential characteristics of the variety is remain unchanged generation after generations after repeated propagation.

Duration of Protection

Depending upon the crop species the duration of protection of a registered variety is varied. In general annual crops are protected for duration of 15 years while the duration of varieties of trees and vines is 18 years (Table 35.3).

Application and Future Scope of the Act in Horticultural Crop

Registration of crop species starts from the date of notification in the official gazette. Registration of crop varieties started since 2006 with 12 crop species (Roy,

Application for registration

↓

Receive application and allotment of PVP Number

↓

Acknowledgment letter send to the applicant

↓

Application transferred to the concerned Registrar

↓

Examination of application by the Registrar

↓

Acceptance or rejection of application

↓

Allotment of registration number (If accepted)

↓

Dispatch seed to the DUS test center

↓

DUS testing of seed for specific time period

↓

Data analysis and submission of report

↓

Publish in PVJ for pre grant opposition

↓

Issuing of registration certificate

Figure 35.2: Varietal Registration Process under PPV and FR Act-2001 (*Source* plantauthority.gov.in).

Table 35.3: Duration of Protection of Crop Varieties in India

	Period of Protection (in Years)		
	Total duration	*Initial*	*Extended*
Trees and Vines	18 years from date of registration	9 years	9 years
Other Crops	15 years from date of registration	6 years	9 years
Extant Notified varieties	15 years from date of notification of variety under the Seeds Act, 1966 by the Central Govt.		

2018). Currently, 150 plant species are allowed for registration. DUS guidelines for each of the crop species have been developed for testing of their respective crop varieties. Till now, DUS guidelines for fifty six (56) crop species related to perennial horticultural crop including fruit, vegetable, flowers and spices crops have been developed and open for registration (Table 35.4).

Table 35.4: List of Perennial Fruit, Vegetable, Flower and Spices Crops Open for Registration in PPV and FR Act (up to 2018)

Sl.No.	*Crop*	*Scientific Name*	*Number of Varieties Registered*
		Fruit Crops	
1	Mango	*Mangifera indica* L.	2
2	Coconut	*Cocos nucifera* L.	-
3	Pomegranate	*Punica granatum* L	6
4	Almond	*Prunus dulcis* (Mill.) D.A. Webb	-
5	Apple	*Malus domestica* Borkh	-
6	Pear	*Pyrus communis* L.	-
7	Apricot	*Prunus armeniaca* L.	-
8	Cherry	*Prunus avium* L.	122
9	Walnut	*Juglans regia* L.	-
10	Grapes	*Vitis* spp.	-
11	Indian jujube (Ber)	*Ziziphus mauritiana* Lamk.	-
12	Acid Lime	*Citrus aurantifolia* Swingle	-
13	Mandarin	*Citrus reticulata* Blanco	-
14	Sweet Orange	*Citrus sinensis* (L.) Osbeck	-
15	Banana	*Musa* spp.	-
16	Papaya	*Carica papaya* L.	-
17	Peach	*Prunus persica* L Batsch.	-
18	Japanese Plum	*Prunus salicina* L.	-
19	Strawberry	*Fragaria x ananasan* Duch	8
20	Bael	*Aegle marmelos* (L.) Correa	2
21	Jamun/Black plum	*Syzygium cumini* (L.) Skeels.	-
22	Nutmeg	*Myristica fragrans* Houtt.	-
23	Custard apple/Sugar apple	*Annona squamosa* L.	-
24	Indian Gooseberry	*Emblica officinalis Gaertn.*	2
25	Guava	*Psidium guajava L.*	-
26	Litchi	*Litchi chinensis* Sonn.	-
27	Cashew	*Anacardium occidental L.*	-
28	Arecanut	*Areca catechu L.*	-
29	Tamarind	*Tamarindusindica* L.	-
30	Cashew	*Anacardium occidental L.*	-
31	Arecanut	*Areca catechu L.*	-
32	Tamarind	*Tamarindusindica* L.	-
33	Date Palm	*Phoenix dactylifera* L.	-

Sl.No.	Crop	Scientific Name	Number of Varieties Registered
		Flower Crops	
1	Gladioulus	*Gladioulus* L.	-
2	Rose	*Rosa* spp.(other than *R.damascena*)	-
3	Jasmine	*Jasminum auriculatum.* L.	-
4	Carnation	*Dianthus caryophyllus* L.	1
5	Orchid	*Paphiopedilum Pfitz.*	-
6	Canna	*Canna* L.	-
7	Jasmine/Mogra	*Jasminum sambac* L.	-
8	Marigold	*Tagetesspp.* L.	-
9	Jasmine	*Jasminum multiflorum* L.	-
10	Bougainvillea	*Bougainvillea Comm. Ex* Juss.	-
11	Orchids	*Vanda*	1
12	Orchids	*Dandrobium*	-
13	Orchids	*Cymbidium*	-
14	Orchids	*Cattleya* Lindl.	-
15	Orchids	*Phalaenopsis* Blume	-
		Vegetable Crops	
1	Elephant Foot Yam	*Amorphophallus paeoniifolius*	1
2	Taro	*Colocasia esculenta*	-
3	Giant SwampTaro	*Cyrtosperma chamissionis/C.merkusii*	-
4	Sweet potato	*Ipomoea batatas* (L.) Lam	-
5	Cassava	*Manihotesculenta* Crantz.	-
		Spice crops	
1	Turmeric	*Curumma longa* L.	3
2	Ginger	*Zingiber officinale* Rosc.	8
3	Small cardamom	*Elettaria cardamomom* Maton	53

Thirty three (33) fruit crops, 15 flower crops, 5 vegetable crops and 3 spices crop are included. Recently, in the 29th authority meeting, date palm (*Phoenix dactylifera* L.) has been included for registration from 11th July, 2018. Including all crop varieties 15,647 applications were submitted for registration. Most of the applications were submitted as farmer's varieties. As per the report of PPV and FR Authority 3441 plant varieties were registered. Year wise details of registered varieties have been depicted in the Figure 35.3. Maximum numbers of 1796 varieties were registered in rice (1796) crop followed by maize (238), sugarcane (228) *etc.* among them 1531 varieties were registered as farmer's varieties. Contribution of cereals and pulses crop are high. Nearly 6 per cent of the total variety registered is perennial horticultural crops. Altogether, 217 varieties were registered as perennial horticultural crops including fruits, vegetables, flower and spices. Maximum numbers of 122 varieties were registered for cherry followed by small cardamom (53). India is rich in horticultural biodiversity carries huge genetic diversity in

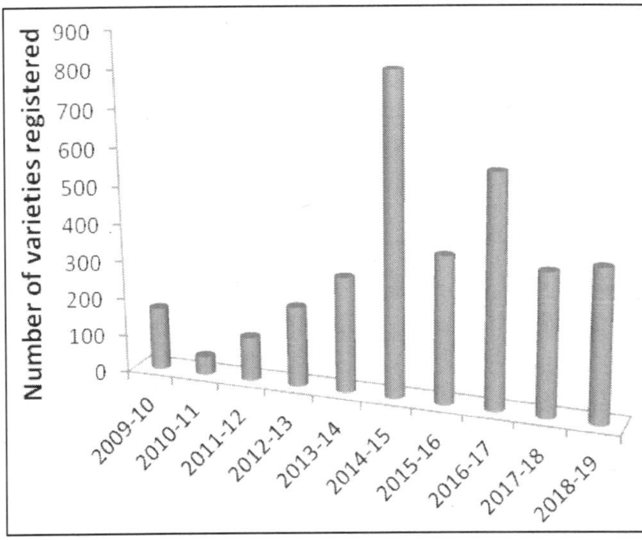

Figure 35.3: Year-wise Details of Number Varieties Registered under PPV and FRA.

almost all the horticultural crops. Nearly, 417 species of horticultural crops are cultivable in India (Dhillon, *et al.*, 2005). Around 27 species in fruits and nuts, 23 in vegetables, 15 in plantation and tuber crops and 16 in spices and condiments are cultivated on regular basis. Extend of wild relatives in these crop species are also large. Almost 331 in fruits and nuts, 215 in vegetables, 161 in spices and condiments and 154 in plantation and tuber crops (Singh 2010). Awareness generation among the researchers, breeders and farmers regarding the varietal protection, access and benefit sharing of genetic resources would be effective.

Geographical Indication Act, 1999

Geographical indication (GI) of goods, as the name implies, is an indication in the form of a name or a sign for goods that have a specific geographical origin and possess qualities or a reputation that are due to the place of origin. As per the article 22.1 of the TRIPS agreement the goods or resources having relationship with a geographical location must be protected as Geographical Indication (GI) of goods. According to TRIPS agreement- "Geographical indications are, for the purpose of this Agreement, indications which identify a good originating in the territory of a member, or a region or locality in that territory, where a given quality, reputation or other characteristic of the goods is essentially attributable to its geographic origin".

A sign or tag are used which indicates the product originates from the place of its repute. The GI right holder can exclude others from using this sign or tag for their product which have not been produced from the farm of right holder or not up to the standard as mentioned at the time of GI filling. The quality of the produce is essentially derived from the place of its origin. GI helps the buyer to ensure about the product is coming from authenticate source.

Paris convention (1883) mentioned about the source of indication and appellation of origin. Later on the term geographical indication was used. Source of indication means the any sign or indication that the product originates from a country or region or a specific place. For example "Made in India" indicates the produced originates in India. Whereas appellation of origin indicates the essential qualities of the product has relation with its place of origin. For example "Darjeeling tea" means the name indicates the tea grown at specific farms of Darjeeling, West Bengal and maintains the quality as applicable.

Impact of Geographical Indication Act

In India, the GI Act was passed in the year 1999 and Darjeeling tea is the first GI goods registered under GI act in the year 2004. As of year 2018, 577 products have been registered as GI goods. Out of which 90 products belongs to Agricultural goods. Darjeeling tea was first product that was registered as GI goods in India from West Bengal. Recently, Intellectual Property Rights Cell, Bihar Agricultural University, Sabour has facilitate four farming communities to get registered of their respective products as *Mahgahi Pan* of Betel Vine; *Bhaglapuri Jardalu* of mango, *Katarni Dhan* of rice and Shahi litchi popularly grown in the state of Bihar (Roy *et al.*, 2018). Till now 52 produces related to perennial horticultural crops have been registered under GI (Table 35.5). Maximum number of products was registered from the state of Maharastra. It provides strong support to the rural and tribal comminutes for strengthening the economic status through identification of important biological resources available in a particular geographical locations and promotion of the quality food produce and generating income.

Table 35.5: List of Perennial Horticultural Crops Registered as GI Goods in India

Sl.No.	*Registration Number*	*Geographical Indications*	*State of Origin*
1.	33	Coorg Orange	Karnataka
2.	35	Nanjanagud Banana	Karnataka
3.	69	Mysore Jasmine	Karnataka
4.	70	Udupi Jasmine	Karnataka
5.	71	Hadagali Jasmine	Karnataka
6.	72	Spices – Alleppey Green Cardamom	Kerala
7.	78	Coorg Green Cardamom	Karnataka
8.	110	Eathomozhy Tall Coconut	Tamil Nadu
9.	111	Laxman Bhog Mango	West Bengal
10.	112	Khirsapati (Himsagar) Mango	West Bengal
11.	113	Fazli Mango grown in the distric of Malda	West Bengal
12.	109	Naga Mircha	Nagaland
13.	124	Virupakshi Hill Banana	Tamil Nadu
14.	126	Sirumalai Hill Banana	Tamil Nadu
15.	125	Mango Malihabadi Dusseheri	Uttar Pradesh
16.	130 and 141	Vazhakulam Pineapple	Kerala

17.	131	Devanahalli Pomello	Karnataka
18.	132	Appemidi Mango	Karnataka
19.	133	Kamalapur Red Banana	Karnataka
20.	143	Guntur Sannam Chilli	Andhra Pradesh
21.	154	Mahabaleshwar Strawberry	Maharashtra
22.	165	Nashik Grapes	Maharashtra
23.	129	Byadagi Chilli	Karnataka
24.	185	Gir Kesar Mango	Gujarat
25.	229	Ganjam Kewda Flower	Odisha
26.	238	Madurai Malli	Tamil Nadu
27.	211	Bangalore Blue Grapes	Karnataka
28.	385	Nagpur Orange	Maharashtra
29.	375	Arunachal Orange	Arunachal Pradesh
30.	376	Sikkim Large Cardamom	Sikkim
31.	377	Mizo Chilli	Mizoram
32.	435	Assam Karbi Anglong Ginger	Assam
33.	436	Tripura Queen Pineapple	Tripura
34.	479	Chengalikodan Nendran Banana	Kerala
35.	438	Tezpur Litchi	Assam
36.	465	Khasi Mandarin	Meghalaya
37.	466	Kachai Lemon	Manipur
38.	437	Memong Narang	Meghalaya
39.	489	Vengurla Cashew	Maharashtra
40.	494	Beed Custard Apple	Maharashtra
41.	495	Jalna Sweet Orange	Maharashtra
42.	520	Uttarakhand Tejpat	Maharashtra
43.	471	Waigaon Turmeric	Maharashtra
44.	500	Purandar Fig	Maharashtra
45.	502	Solapur Pomegranate	Maharashtra
46.	473	Bhiwapur Chilli	Maharashtra
47.	493	Dahanu Gholvad Chikoo	Maharashtra
48.	498	Jalgaon Banana	Maharashtra
49.	499	Marathwada Kesar Mango	Maharashtra
50.	241	Banaganapalle Mangoes	Andhra Pradesh
51.	551	Bhagalpuri Zardalu	Bihar
52.	552	Shahi Litchi	Bihar

Conclusions

India is one of the twelve mega biodiversity country in the world, it carries two biodiversity hotspots at Himalayan Region and Western Ghat. Besides, North-

Eastern hilly areas; Andaman and Nicobar Islands are also rich in huge biological diversity. Phyto-biodiversity and species richness have been observed with the longitudinal decrease or increase in these regions. Plant genetic resources (PGR) are considered as heritage of mankind. However, urbanization, industrial growth, deforestation and climate change are threatening the plant genetic resources (PGR) which led to many plant species to be extinct or endangered. Protection of PGR, their sustainable utilization is crucial to mitigate the adverse effect of climatic change, their access and benefit sharing. Protection of plant biodiversity particularly new varieties was first time considered with the establishment of Union for Protection of Plant Varieties in the year 1961. Protection of Plant Varieties and Farmer's Rights act (PPVFRA) was a *sui generis* system adopted in India for protection of plant species. Presently, 58 crop species of perennial horticultural crops including fruits, vegetable spices and flowers are accessible for registration under this act. However, most of the genetic resources available in the public domain are untouched that needs intensive work plan to provide protection to those valuable germplasms.

REFERENCES

Dhillon, B.S; Rana, J.C. and Singh, A.K. 2005. Collection, conservation and utilization of indigenous horticultural crops germplasm. In: Crop Improvement and Production Technologies of Horticultural Crops Vol. 1, (Eds.) K.L. Chadha, B.S. Ahluwalia, K.V. Prasad and S.K. Singh. Malhotra Publishing House, New Delhi, 10–35.

Pratibha Brahmi; Saxena Sanjeev and Dhillon B. S. 2004. The Protection of Plant Rights Act of India. *Current Science*, 86(3): 392-398.

Roy Chandan 2016. Regulatory measures for the protection of Plant varieties in India, In. Sarkar *et al.,* Technical Handbook on Intellectual property Rights in Agricultural Biotechnology, ICAR-IIAB, Ranchi.

Roy Chandan 2018. The Role of Intellectual Property Rights in Agriculture and Allied Sciences. Apple Academic Press, USA.

Roy Chandan; Kumari Prerna and Ranjan Rakesh Deo 2018. The Role of Intellectual Property Rights in Conservation, Utilization and Benefit Sharing of Plant Genetic Resources. *Current Journal of Applied Science and Technology*, 31(3): 1-6.

Singh, Anurudh K. 2010. Biodiversity in horticulture and future needs. In: Savonier 4th Indian Horticulture Congress-2010.

Index

A

Abiotic stress tolerance 11, 420

Acacia senegal 55

Advanced backcross-QTL (AB-QTL) strategy 125

Aegle marmelos 35, 41, 49, 51, 88

Agathi 23

Agricultural production 149, 159, 429

Allele mining 140

Allopolyploids 198, 199

Amla 35

Annonas 50, 59

Apple 3, 5, 6, 11-15, 17, 19, 73, 76, 77, 81, 115, 152, 185, 202, 210, 221, 236, 239, 321-327, 377-382, 388, 396, 397, 406, 409, 416-425, 443, 465, 470

Apple scab 3, 71, 76, 79, 152, 156, 237, 382-386

Apricot 6, 13, 20, 93, 185, 406, 416, 417, 465

Arid and semi-arid fruits 49

Arid fruits 2, 41, 43, 49, 59

Arid zone fruits 4, 52, 58

Artificial crossing 343

Association mapping 28, 33, 101, 103, 119, 126, 129

B

Backcross method 26

Backcross population 133

Bael 3, 21, 35, 43, 51

Banana 2, 3, 14, 16, 17, 20, 21, 38, 73, 185, 190, 207, 210, 241-249, 406, 442, 443, 446, 465, 468, 469

Banana improvement 241

Berry colour formation 300

Biochemical markers 205

Biotechnological approaches 25, 426

Biotic stress resistance 192

Biotic stress tolerance 14, 417, 419

Bitterness 271, 273, 454

Blood red guava 18

Boradi 49

Bottleneck 367

Breeder's rights 459, 461

Breeding cycle 227, 235

Breeding objective 256, 297, 323

Breeding strategies 212, 342
Bulk segregant analysis (BSA) 28, 135
Burrowing nematode 16, 247

C

Cactus pear 49
Capparis deciduas 49, 54
Capsicum annuum 75
Carissa carandus 54
Cassava 23, 24, 27, 444, 466
Chakurmani 23
Chemical mutagenesis 186
Chemical suppression 281
Chilling zones 403
Chironji 35
Chlorophyll fluorescence imaging 164
Chochlearia damica 86
Chow-chow 24, 26
Chromoaphis juglandicola 74
Chromosome number – genome mutations 188
Chromosome rearrangements 188
Cisgenesis 76, 77, 149-153, 156, 423, 424, 426
Cisgenics 58, 152
Citrus 2, 3, 5, 11, 13, 14, 19, 21, 35, 59, 63, 73, 88, 103, 153, 154, 185, 188, 199, 200, 207, 210, 215, 218-224, 251-267, 350, 416-421, 425, 426, 445, 465
Citrus butterfly 14
Citrus genomics 264
Citrus improvement 200, 201, 251, 264
Citrus nematode 16
Citrus reticulata 39, 40, 264
Citrus scion varieties 253
Clonal plants 203, 204
Clonal selection 4, 25, 213, 257, 296, 343
Cold tolerance 12, 373
Commercial micropropagation 203
Composite interval mapping 125
Conservation of horticulture genetic resources 37
Conservation strategies 62
Conventional breeding 23
Conventional methods 2
Convention of biological resources (CBD) 459
Cordia spp. 55
Corm borer 14
CRISPR 30-32, 78, 80, 82, 84, 108, 110, 147, 176, 217-222, 225, 425-428
CRISPR/Cas9 technology 30
Crop improvement 2, 71, 128, 149, 184, 326, 470
Crop management 175
Crop mutagnesis 194
Crossing technique 242
Crowned rot 17
Cryopreservation 46, 61, 66, 67, 69, 70
Cucurbits improvement 269
Culture procedures 91
Curry leaf 24, 26
Custard apple 5, 6, 210, 469
Cytological markers 205
Cytoplasmic genome 197

D

Dagger nematode 16
Daktulosphaira vitifoliae 15, 288, 299, 303, 418
Date palm 49, 52, 53
Disease resistance 16, 74, 77, 78
Distant hybridization 344
DNA based marker 113, 131
DNA conservation 68
DNA fingerprinting 19, 225
DNA molecule 61, 119
Dormancy 89

Doubled haploid 27, 124, 133, 167, 222
Downy mildew 17, 82, 154, 223, 275, 288, 290, 298, 301, 303, 309, 312
Drought tolerance 11, 420
Drumstick 23, 28, 31
Dry domains 49
Dwarfing rootstocks 71, 236, 328, 414-416

E

Economically important pests 14, 417
Elite clones 4, 5
Elite cultivars 2, 237
Emblica officinalis 35, 41, 50
Embryo rescue 58, 85, 259
Encoding phytoene synthase 29
Endosperm culture 89
Environmental sustainability 159
European food safety authority (EFSA) 150
Exotic crops 429
Exploit genetic resources 57
Express sequence tags 100

F

F_1 families 231
Farmer's rights 459-461, 470
Fermented beverages 443
Floral biology 270, 316, 329, 368
Fruit and vegetable varieties 441
Fruit breeding 2, 5, 18, 80, 81, 101, 194, 216, 221, 314, 397
Fruit crop improvement 1, 4, 8, 71, 149, 216
Fruit fly 14
Fruit species 3, 21, 35, 37, 41-46, 81, 152, 156, 427
Functional genomics 140, 143, 145

G

Garciania species 40
Gene editing techniques 30
Gene function analysis 307
Gene modification 30
Gene tagging 131
Genetically modified (GM) 149
Genetic engineering 83, 306
Genetic maps 114
Genetic resource conservation vii, 61
Genetic resources 35, 41, 45-49, 52, 57, 61, 62, 64, 68, 70, 82, 113, 208, 317, 362, 459, 470
Genetic transformation 423
Genetic variations 292
Gene transfer 228
Genome editing 78, 107, 222, 426, 428
Genome sequencing 101, 306
Genome wide association mapping 104
Genome wide selection 28
Genomics assisted breeding 99
Genomic selection 99, 176, 379
Genomics in fruit crops 20
Genomics-led breeding 248
Geographical indication act, 1999 467
Germplasm collection 49, 53, 54, 68, 128, 209
Germplasm collection 212, 337
Germplasm evaluation 343
Globe artichoke 24, 26
Grape 2-6, 10-17, 20, 94, 102, 153, 210, 283, 288, 293, 298, 300, 308, 314, 417-422, 425, 443-446
Grape improvement 283
Grape varieties 310
Grewia subinaequalis 49, 55

Guava improvement 315, 326
Gynoecious sex form 279

H

Hand pollination 278
Haploids and double haploids 18
Herbicides tolerance 78
Heterosis breeding 276
High throughput plant growth analysis 173
Hippophae rhamnoides 56
Horticultural genetic resources 61
Hybrid breeding 197
Hybridization 6, 8, 25, 57, 145, 213, 228, 232, 234, 242, 252, 258, 274, 295, 324, 330, 343, 355, 369, 375, 416, 419, 422
Hybrid maintenance 449
Hybrid seedling recovery 259
Hybrid seedlings 261

I

Ideal rootstock 415
Important cultivars 319, 320, 323
Inbred selection 274
Indian bean 24, 27
Individual plant selection 273
Inheritance pattern 341
Inheritance studies 327
Insect pest resistance 72, 73
In silico phenotyping 172
Integrated analysis platform 174, 179
Intellectual property rights 457, 458, 468, 470
Intergeneric hybridization 57
International convention for the protection of new varieties of plants (UPOV) 459
International treaty on plant genetic resources for food and agriculture (ITPGRFA) 460
Inter-simple sequence repeats 28
Interspecific hybridization 57, 419
Inter-varietal hybridization 324
In vitro conservation 65
In vitro culture 259, 306
In vitro mutagenesis 189, 190
IR thermal imaging 164
Isozyme analysis 346
Ivy gourd 24, 26

J

Jackfruit 5, 24, 26, 40
Jamun 3, 21, 35, 43, 47, 53
Jerusalem artichoke 24, 26

K

Karonda 3, 35, 43
Khirni 35, 41, 46
Kumut 49

L

Lasora 35, 49
Lesion nematodes 16, 419, 420
Linkage disequilibrium 28, 103, 104, 119, 124, 126, 139, 302
Linkage mapping 104, 121, 265
Litchi improvement 333, 341
Log of odds (LOD) ratio 135
Low chill fruit varietal development 405

M

Malabar tamarind 35
Mango (*Mangifera indica*) 3, 4, 5, 7, 14, 17, 18, 20, 21, 37, 38, 45, 102, 185, 207, 210, 222, 357-363, 416, 421, 442, 465, 469
Mango biotechnology 360
Mango improvement 357
Mapping populations 121, 122, 131, 132, 137
Marker-assisted backcrossing 99, 305

Marker assisted breeding 18, 28
Marker-assisted selection (MAS) 19, 28, 69, 76, 99, 101, 128, 139, 217, 291, 296, 303, 312, 348, 361, 362, 391
Medicinal plant conservation area 63
Microbudding 236
Mildew resistances 301
Modern breeding tools 301
Molecular mapping 113
Morphological markers 205
Multiple interval mapping 125
Mutation breeding 27, 183, 185, 189, 192, 245, 261, 347, 422
Mycoplasma 16

N

Nagoya protocol 459, 460
National research centres 1
Navel orange worm 74
Near infrared (NIR) imaging 164
Near-isogenic lines (NILs) 134
Net assimilation rate (NAR) 170
Next generation phenotyping (NGP) 159, 160
Next generation sequencing 20, 100
NGS 100, 101, 114, 115, 160, 217
Non-conventional breeding 27
Nucellus culture 88
Nucleo-cytoplasmic interactions 197

O

Oligonucleotide-directed mutagenesis 426
Open pollinated selections 213
Opuntia ficus indica 49, 56
Osmotic stress 13
Ovary and ovule culture 87

P

Papaya 4, 7, 14, 17, 20, 73, 74, 115, 185, 206, 210, 365, 367-375, 445, 465
Papaya improvement 365, 367, 371
Parthenocarpic cucumber 280, 431, 435
Parthenocarpic pickling cucumbers 434
Parthenocarpy 107, 108, 367, 432, 433
Passion fruit 8
Peach 8, 11-13, 19, 21, 93, 185, 237, 406-408, 417, 421, 465
Perennial crop breeding 209
Perennial horticultural crops improvement 1
Perennial vegetable crops 23, 24, 26
Pest resistance 14, 72, 73
Phalsa 3, 35, 43, 56
Phenome data bank 173
Phenome-wide association studies (PheWAS) 166
Phenomics 159-162, 166, 170-179
Phoenix dactylifera 49, 52, 53, 465, 466
Physical map construction 117
Physical mutagenesis 186
Phytoenedesaturase 29
Pilu 35, 49, 56
Pineapple 4, 10, 185, 190, 206-210, 406, 468, 469
Pink pulped papaya 18
Plant genetic resources 2, 61, 470
Plant mutagenesis 184
Plant tissue culture 204, 221
Ploidy breeding 8
Ploidy manipulation 215
Podosphaera leucotricha 71, 82, 379, 385, 427
Pointed gourd 24, 26
Pollination 24, 88, 243, 270, 278, 281, 294, 343, 350, 354
Pollination behaviour 270
Polyembryonic species 259
Polygalactouridinase (PG) gene inhibiting proteins (PGIPs) 155

Polyploidy 8, 27, 32, 202, 257, 263, 325, 416
Polyploidy breeding 27
Pomegranate 2, 3, 5, 8, 14, 185, 210, 221, 445, 465, 469
Population structure 127
Powdery mildew 17, 71, 76, 82, 153, 275, 288, 298, 301, 303, 309, 312, 378, 383, 385, 418, 425, 427
Pre-selection indices 212
Processing industry 444
Project directorates 1, 2
Protected cultivation 280, 429, 439
Protection of plant variety and farmer's rights act (PPV and FRA) 460
Pseudostem weevil 14
Psidium guajava 315-318, 328-331

Q

QTL mapping 115, 117, 119, 122, 139, 167
Quantitative trait locus (QTL) 103, 118, 122-125, 134, 138, 146, 167-172, 217, 265, 303, 304, 361, 380, 387, 397, 418
Quantitative trait locus (QTL) Mapping 138

R

Random amplified polymorphic DNA (RAPD) 19, 28, 32, 43-46, 103, 113, 116, 136, 190, 206, 208, 216, 246, 248, 261, 265, 302, 346-350, 352, 355, 360, 362
Recombinant inbred lines (RILs) 124, 134
Relative growth rate (RGR) 170
Researcher's rights 461
Rhubarb 26
RNA interference (RNAi) 29, 31-33, 75, 77, 146, 395, 424-428
RNA-sequencing 147
Root-knot nematodes 16
Rootstock breeding 8, 211, 298, 300, 386, 415, 418
Rootstock improvement 323, 413
Rosaseous crops 105

S

Salinity tolerance 421
Salt tolerance 12
Salvadora oleoides 35, 56
Scion breeding 380
Scion improvement 323
Scion varieties 9, 253
Seabuckthorn 56
Seed conservation 64
Seedless mango 18
Seedlessness 9, 10, 20, 96, 108, 258, 297, 304, 330, 339, 370
Seedling selection 5
Seed sterility 89
Sequence-characterized amplified regions (SCARs) 217
Sex-determination 28
Sex modification 271
Short wave Infrared (SWIR) 164
Simple interval mapping 125
Simple sequence repeats or microsatellites (SSR) 216
Single nucleotide polymorphism (SNPs) 28, 104, 217, 347
Site-specific nucleases 77
SlAGL6 108-111
Somaclonal variations 203-206
Somatic embryogenesis 191
Somatic hybridization 58, 234, 422
Stem borer 14
Stone weevil 14
Strawberry 10, 17, 20, 21, 26, 73, 153, 155, 206, 208, 220, 318, 406, 443, 465, 469
Sub-tropical fruits 2

Sweet pepper 75, 84
Sword bean 24, 27
Syzygium cuminii 35, 47, 53

T

Targeted genome editing 77
Taro 23
Temperate fruits 2-4, 69, 314, 406
Tissue culture techniques 27
Tissue tolerance 13
Tolerant hybrids 11, 17
Tomato 111, 115, 137, 224, 436-438, 444, 446, 452, 453
Traditional breeding 149
Trait dissection 166
Trait pyramiding 166
Transgenic approach 29, 73, 348
Tree architecture 10, 416
Triploid seedless hybrids 200
Tropical fruits 2, 35, 41-47, 59, 353

U

Underutilized fruits 41, 45, 46, 59

V

Varietal wealth 3
Vegetable banana 23
Visual-near infrared (VNIR) 164
Vitis cinerea 211, 284, 292, 303
Vitis vinifera 13, 21, 77, 78, 80, 88, 95, 104, 153, 156, 173, 207, 221, 225, 283, 285, 288, 296, 303, 313, 423-427

W

Walnut improvement 377
Water-use efficiency (WUE) 170
Wild jujube 49
Wild relatives of plants 2
Wild species 9, 17, 326
Winged bean 24, 27
Woolly apple aphid 15, 389, 418

Z

Zinc finger nuclease 30, 426
Ziziphus mauritiana 49, 52
Ziziphus nummularia 147, 148